Autodesk®
Architectural Desktop:
An Advanced
Implementation Guide

Autodesk®
Architectural Desktop:
An Advanced
Implementation Guide

PAUL F. AUBIN

autodesk® press

THOMSON

DELMAR LEARNING

Australia • Canada • Mexico • Singapore • Spain • United Kingdom • United States

autodesk®
press

Autodesk® Architectural Desktop: An Advanced Implementation Guide
Paul F. Aubin

Autodesk Press Staff

Business Unit Director:
Alar Elken

Executive Editor:
Sandy Clark

Acquisitions Editor:
James DeVoe

Developmental Editor:
John Fisher

Editorial Assistant:
Mary Ellen Martino

Executive Marketing Manager:
Maura Theriault

Channel Manager:
Fair Huntoon

Marketing Coordinator:
Sarena Douglass

Executive Production Manager:
Mary Ellen Black

Production Manager:
Andrew Crouth

Production Editor:
Stacy Masucci

Art and Design Coordinator:
Mary Beth Vought

Library of Congress Cataloging-in-Publication Data
Aubin, Paul F.
  Autodesk Architectural Desktop: An Advanced Implementation Guide/ Paul F. Aubin.
      p. cm.
  ISBN 1-4018-8876-3
  1. Architectural drawing--Computer-aided design. 2. Autodesk Architectural desktop. 3. AutoCAD I. Title.

  NA2728 .A84 2002
  720'.28'402855369—dc21
                      2002071961

Notice To The Reader

Trademarks

CONTENTS

FEATURES OF THIS EDITION

Autodesk Architectural Desktop: Advanced Implementation Guide is a concise manual focused squarely on making Autodesk® Architectural Desktop (ADT) a success in your organization. Like its predecessor, *Mastering Autodesk Architectural Desktop*, this book emphasizes the process of creating projects in ADT. It treats the ADT tool set as an interconnected series of objects and procedures, rather than independent commands and routines. This manual delves deep into techniques that support proven strategies to deploy ADT throughout your entire organization and make it successful. The book follows a logical progression through issues of installation, setup, configuration, training, customization and standardization. Beginning with a thorough look at your firm's work style and project types, you are guided through how to mesh existing practices with the ADT Building Model concept. This manual gives you the tools required to assess your current CAD standards and to build a checklist of action items useful throughout the deployment process. Several topics critical to successful implementation are dealt with both philosophically and in numerous hands-on exercises. Layering standards, the Display System, building a Content Library, creating Details and sharing files with consultants are just some of the topics covered. At the conclusion of this manual, you will be armed with all the necessary tools to build a successful implementation plan uniquely suited to your practice.

UNITS

Many of the tutorials in this book reference both Imperial and Metric units. Symbol names, scales, references and measurements are given first in Imperial units, followed by the Metric equivalent in square brackets [ ]. For example, when there are two versions of the same symbol or file, they will appear like this within the text: **Aec3_Room_Tag_P** [**M_Aec3_Room_Tag_P**], or this "Open the file named *Small Room Imperial.dwg* [*Small Room Metric.dwg*]." When the scale varies, a note like this will appear: **1/8"=1'-0"** [**1:100**]. If a measurement must be input, the values will appear like this: **3/32"** [**2.5 mm**]. Please note that in many cases, the closest logical corresponding metric value has been chosen, rather than a "direct" mathematical translation. For instance, text is often drawn at 3/32" in height in Imperial drawings. If this were translated directly to millimeters, the result would be **2.3812499999999996**. Naturally, this is not a correct value for text height in Metric drawings. Therefore, the more commonly accepted 2.5 mm height has been chosen instead. (*Every attempt has been made to make these decisions in an informed manner. However, it is hoped that readers in countries where Metric is the standard will forgive the American author for any poor choices or translations made in this area.*)

Unfortunately, time did not permit the inclusion of every tutorial in both Imperial and Metric units in the book's printed form. However, all tutorials that were produced in Imperial units only in the book will be available in Metric for download from *paulaubin.com* and the *OnLine Companion* (See below) in PDF format. Please feel free to download and print them at your convenience.

STYLE CONVENTIONS

Style Conventions used in this text are as follows:

Step-by-Step tutorials:	I. **Perform these steps.**
Commands and Menu Picks:	**Design>Walls>Add Wall**
Dialog Box elements:	Change the delimiter of the **Major** field to **None**.
Keyboard input:	Press ENTER.
User Input:	Type **this**
File and Directory Names:	*L:\Content*
Multi-unit Tutorials/Text References	1/8"=1'-0" [1:100]

WHO SHOULD READ THIS BOOK?

The primary audience of this book is advanced users of ADT with AutoCAD® experience. The book is written using ADT R3.3, but it is equally applicable to R3.0. If you use ADT to produce Architectural Construction Documentation or Design drawings, Facilities layouts or Interior Design studies and documentation, this book is for you. CAD managers, job captains, architects, interior designers, design build professionals, facilities planners and building industry CAD professionals all stand to benefit from the information contained within. If your copy of ADT is still sitting on the shelf or if you have only just begun using Walls and Doors, this book is *especially* for you. This book assumes that you are already familiar with and have been using ADT at least at the basic level. A good thorough knowledge of AutoCAD basics and tools is also assumed. If you lack this prerequisite experience or are migrating from another non-AutoCAD-based CAD system, it is recommended that you first acquire and read *Mastering Autodesk Architectural Desktop*, which assumes only a working knowledge of AutoCAD and no prior knowledge of ADT. It will provide an excellent starting point for you before you embark on the challenges set forth in this manual.

HOW TO USE THIS BOOK

Rather than offer a single comprehensive solution to each problem, the book takes a prescriptive approach to ADT implementation. This means that the book will systematically walk through the steps required to begin reaping benefits from ADT. You will first learn how to objectively assess your current CAD system, and then you will learn tools to help you analyze what needs to be done to begin maximizing your firm's ADT productivity. Several suggestions are made throughout that will build upon previous lessons and topics. For this reason, it is recommended that you read the chapters in order. However, a particular chapter may be focused on a particular type of user and/or job function. This will be indicated at the beginning of the chapter in a special note immediately following the introduction. In those instances, if you do not personally fill the indicated role within the firm, you may feel free to skip that chapter or topic. However, even if you do not envision personally performing the functions outlined in a particular chapter, go ahead and skim through it anyway, time permitting, as you will gain valuable insight into the potential and limitations your colleagues might face.

The following types of users have been identified based on their typical roles and responsibilities within a firm:

- **IT Support Person** – This individual(s) is usually responsible for the procurement, installation, configuration and maintenance of all of the firm's computer hardware, software and network components. In larger firms, CAD software and other specialized software and equipment often have additional personnel dedicated to just those items, such as a CAD Manager for CAD software.

- **CAD Manager** – Due to the highly specialized nature and needs of CAD software, there is often a dedicated individual (or more) whose responsibility includes the installation, configuration and support of all of the firm's CAD software needs. This person will often devise a set of global guidelines and standards to be used in all projects firm wide. The CAD Manager usually provides day-to-day support to CAD Software Users as well. In some cases, this individual may be the same person as the IT Support Person (typical in smaller firms).

- **Project Manager** – This individual (often a licensed architect, though not required) is the person in charge of all aspects of a particular project's successful completion. This includes devising schedules, procuring resources and personnel, and devising production strategies, including (but certainly not limited to) how to best utilize CAD software tools in design and production. The success or failure of a given project ultimately rests with this person; therefore, for CAD software to be successfully implemented in a project flow, this individual must understand the software package's strengths, weaknesses and impact on the project. This individual may or may not have any personal CAD experience but does need to access and manage CAD data and CAD personnel.

- **Project Data Coordinator** – This individual (not necessarily a licensed architect) is responsible for mapping out the specific use of all tools, CAD and traditional, to be used on a given project. This is not to imply that the members of a team should not be engaged in decisions about how and what tools to use on a project. It is simply to say that someone must ultimately hold the responsibility for making a definitive and binding decision regarding workflow and procedure on any given process. Regardless of technique and toolset chosen, this individual will build the plan and establish the guidelines used to complete a project. Depending on a variety of factors such as firm size, corporate culture and the level of CAD expertise in the firm's management, this individual may also serve one of the other roles mentioned here, as well as manage the communication with consultants and contractors.

- **CAD Software User** – This person is proficient in the use of the CAD software used in the firm. A CAD software user must be proficient in all tools used in a typical office project. Specialists may evolve naturally (or even be groomed deliberately) based on firm specialties, needs and corporate culture. However, in general, it is desirable that all individuals expected to use CAD software to perform their daily tasks be proficient in 90 percent of those tasks that are considered typical for your firm's standard body of work.

- **CAD Data Consumer** – This individual does not create CAD data, but rather must regularly access that data. These individuals may be employees of the firm, sub-contractors, vendors or clients that do not use CAD software but rely on CAD data to perform their job function (such as spec writers, estimators, or firm principals). This book is not really targeted to the CAD Data Consumer, but it is an important group to consider in building a successful implementation plan.

Note: The titles above are meant to be indicative of an individual's role and/or level of responsibility in the firm. Every firm inevitably devises its own titles for it personnel which may or may not indicate that individual's job function. Please use this list as a guide to help you assess which roles you and your colleagues typically fill within your organization. It is also quite common for a particular individual to fill several of the roles above within the same or different projects. Again, the list is meant to be a guide. However, recognizing that each of these roles does exist in some form in your organization and identifying those individuals who currently fill or would best fill them is critical to achieving a successful implementation.

As a final note on usage of this book, keep checklists as you work through the chapters of action items required. Use these lists to help you complete the tasks that are required in your own office's standards and procedures.

HOW TO USE THE CD

Files used in the tutorials throughout this book are located on the CD ROM included with the book. Insert the CD into your computer and a menu of available resources will display.

Should the menu fail to load double click on the MY COMPUTER icon, locate the CD ROM drive icon, and double click on it. The menu will then display.

Once the menu is displayed, simply click the appropriate button for the resource you wish to view and the resource will be presented to you.

 Note: If changes are made to any of the drawing files or other resources you must save the modified file to a new location on your computer, otherwise the changes you made will be lost.

ONLINE COMPANION

Additional resources related to the content in this book are available on line. Log on to our Web site for complete information at:

http://www.autodeskpress.com/onlinecompanion.html

The author maintains his own Web site where you can find resources related to this and his other books. Log on at:

http://www.paulaubin.com

There are several additional online resources listed in Appendix A. Shortcuts to these URLs are also contained in the *Resources.exe* archive on the CD.

INTERNATIONAL EXTENSIONS

Some of the functionality added to Architectural Desktop starting in Release 3 is included in a collection of International Extensions. The International Extensions do not install automatically; they must be chosen during the initial installation of the software. In R3.3, this is the default choice when the install is run. AEC Dimensions, Scale-dependent Display Reps, Live Sections and Area objects all rely on the International Extensions' having been installed. It is highly recommended that you do install the International Extensions, and it is assumed that the International Extensions have been installed throughout this text.

AEC Dimensions are live dimension objects, which remain linked to the ADT objects that they are dimensioning. Changes to the model are instantly reflected in the AEC Dimension strings without need for a separate edit. Additionally, AEC Dimensions may be displayed and printed at different architectural scales. With scale-dependent Display Reps, symbols can be made to adjust size when displayed at different scales. Also included in the International Extensions are Area objects, Area Groups and Live Section Display. Area objects and Groups allow for complex area take-offs and rule-based calculations in construction documents. Reports generated from Areas may be exported to Excel. Live Section Displays can be used to work on ADT models from a section vantage point with materials properly rendered at the cut line.

Unfortunately, if you have already installed ADT using the default install, you will need to uninstall and then reinstall the software to be able to install the International Extensions. Weighing the inconvenience of having to uninstall with the additional functionality outlined above is a tough decision, but once you begin reaping the benefits of these tools, the minor disruption caused by this one-time situation will be long forgotten. If you find yourself in this situation, please proceed to Chapter 1. There you will find a detailed tutorial on installation that includes the International Extensions.

SERVICE PACKS

It is important to keep your software current. Be sure to check on line at *www.pointa.autodesk.com* on a regular basis for the latest updates and service packs to the Architectural Desktop software. Having the latest service packs installed will help you ensure that your software will continue to run trouble free. If you use the Autodesk Architectural Desktop R3 Today window as your startup screen (this is the default for the software), Point A is accessible directly from this window. This is an even faster way to check and see if your software is current.

 Caution: Please make sure you have at least ADT R3 with Service Pack 2 installed before attempting to perform the tutorials in this book. ADT R3.3 with Service Pack 1 is preferred.

ABOUT THE AUTHOR

Paul F. Aubin is the author of *Mastering Autodesk Architectural Desktop* and this book, both published by Autodesk Press. Paul has a background in the architectural profession spanning over fifteen years. Paul's long-standing affiliation with Consulting for Architects, Inc. (CFA), an Autodesk Systems Center (ASC) and Premier Training Center (ATC), has earned him the title of Associate. He currently serves as the moderator for CADence magazine's online CAD questions forum and is a frequent lurker on the Autodesk ADT Discussion Forum. He is a regular speaker at Autodesk University and was the recipient of the Autodesk Central Region 2001 Architectural Award of Excellence. Paul has been a guest speaker at various events for The American Institute of Architects, ADT Users' Groups and The Association of Licensed Architects. Throughout his six years with CFA, Paul has successfully trained over 1100 professionals in Architectural Desktop, AutoCAD and Autodesk VIZ. Prior to joining CFA, Paul served as CAD Manager with an interior design and architecture firm in downtown Chicago and amassed many years of hands-on project architect–level experience. The combination of his experiences in architectural practice, as a CAD Manager and as an instructor give his writing and his classroom instruction a fresh and credible focus. Paul is an Autodesk Certified Instructor (ACI) and an Associate member of the American Institute of Architects. He received his Bachelor of Science in Architecture and his Bachelor of Architecture from The Catholic University of America. Paul lives in Chicago with his wife Martha, their sons Marcus and Justin and daughter Sarah Gemma.

DEDICATION

This book is dedicated to my wife Martha. Your love and support make it possible for me to chase my dreams. I love you with all my heart.

ACKNOWLEDGEMENTS

I would like to thank several people for their assistance and support throughout the writing of this book.

Thanks to Jim Devoe, John Fisher and all of the Delmar team. As always, you are a super bunch of people to work with! I appreciate your continued faith in me.

Several firms and individuals provided an invaluable source of research and commentary for this project. Thanks to the staffs of Niles Bolton Associates, Gensler, HOK, Eppstein Uhen Architects, VOA Associates TPS Consulting and Laing. Thanks also to Jeff Agla, Chris Janoch, Mario Guttman, Christopher Holm, Bret Tushaus, Eric McKinney, Steve Jolly, Mervin Richards and James Smell who provided input and reviewed advance chapters.

There are far too many folks in Autodesk Building Industry Division to mention. Thanks to all of them, but in particular, Bob (Batch) Batcheler, Julian Gonzalez, Paul McArdle, Chad Ames, Charles Spink, Chris Yanchar, Bill Glennie, Matt Dillon, John Janzen, Tatjana Dzambazova, Simon Jones, Michael Nachtsheim And all of the folks at Autodesk Tech Support.

The author would also like to acknowledge the following people:

Technical Editor: Eric Stenstrom, VOA Associates, Inc.

Technical Editor: Jerry Jackson, Autodesk, Inc.

Copy Editor: Gail Taylor

Composition: John Shanley, Phoenix Creative Graphics

A special thanks to David McFadden and my colleagues at CFA (especially "Man Garvin")

Thanks to the staff of the Oak Lawn, Illinois Starbucks for the use of their facility and the bottomless coffee cup while writing.

I am grateful for the boundless love and support of my mom, Maryann Minichiello, and my dad, Del Aubin, and my brothers Marc and Tom. Thanks to all my family and friends for putting up with me. And a special thanks to Marcus Angelo, Justin Angel and Sarah Gemma. You three are the greatest kids any dad could hope for.

Finally, this book would not have been possible without the preponderance of "The Clay Freeman."

Thank you for purchasing *Autodesk Architectural Desktop: An Advanced Implementation Guide*. You are about to embark on a process designed to enable you and your firm to begin reaping the benefits that Autodesk Architectural Desktop has to offer. Regardless of your firm's size and composition, this manual will assist you in deploying, installing, configuring, customizing and utilizing ADT successfully in design and production.

> ## WHO SHOULD READ THIS SECTION?
>
> This short introduction provides a "bird's eye" view of the process and results advocated by this book. This section is intended for everyone.

THE BENEFITS OF MODEL-BASED DESIGN

In recent years, technological advances have continued to press forward at an ever-increasing rate, while total cost of technology-ownership continues to decrease. It is now possible to create very realistic, dimensionally accurate models of a building and its systems. Software technology enabling the creation of these types of models has actually been available for several years; it is the combination of software technology, common software platform (AutoCAD) and the ever-increasing speed and associated price drop in today's PC hardware that have allowed the Building Model concept to enter the mainstream.

A *virtual building model* is quite simply a digital replica of a complete building including, at minimum, all walls and major enclosures and, ideally, all major mechanical systems and structural components. The purpose of such a model is to serve the construction of the actual building and the ongoing maintenance of the facility during and after project completion. Architects have long recognized the value of creating study models during the design process. Models have been used for centuries as a way to "mock up" a building design concept and make it easier for clients and architects alike to visualize. These traditional models were built from material such as wood, cardboard or foam core. However, they typically served only the design or marketing phases of a project. With the advent of computer-generated 3D models, the potential inherent in a 3D virtual building model reaches much further than the design and production phases. With a virtual building model, several solid benefits to all players in the building process now become available.

BENEFITS TO THE ARCHITECT AND ENGINEER

- Design decisions can be made with greater confidence and conviction as you can truly "see" the impact of form and intent.
- Design intent can be communicated more clearly early in the process, facilitating better feedback from senior architects and other key team members.

- Clients can be brought into the design process earlier and "shown" the design because they understand 3D. Clients have a much easier time visualizing the building if they are shown a mock-up than they will from a collection of two-dimensional drawings.
- Much better coordination between trades can be achieved when compared to traditional methods. With the virtual building model, wall, structural and mechanical systems can all be included. With all of these items contained in a single composite building model, interference or clashes can be detected in the design phase rather than in the field as expensive change orders.
- The Construction Documentation process is easier, more reliable and often faster. This is true because the information is extracted from a single source. Changes are easier to execute from parametrically based building model components and once changed, all views and references to this single source respond to the change.
- The building model can be "queried" for critical information. As more data that is "real" is imbedded in the building model, more useful design analysis, schedules and reporting can be extracted directly from the model. This has the potential to greatly reduce the time and errors spent calculating this information by traditional means.

BENEFITS TO THE BUILDING OWNERS

- A virtual building model reduces the time between the commitment of funds and the opening of the facility. With a better coordinated and less error-prone design process, building construction cost and schedule predictions will become much more accurate and reliable. This gives an immediate cost-savings benefit to the owner, while giving a competitive advantage to the architect.
- The more potential errors captured during design, the fewer will occur during construction. This raises the potential to reduce construction costs and again provides benefit to both owner and architect.
- With the leveraging of the building model, the accuracy of time/cost predictions will be improved. This is an important consideration for banks and lending institutions that commit funding to building projects.

BENEFITS TO THE CONTRACTOR

- With a true virtual building model, contractors can create much better cost estimates. This is because all of the information needed is contained in the model.
- Clash conflict errors can be caught early in the construction process so that coordination of personnel and resources can be better achieved.
- The virtual building model can help facilitate detailed construction scheduling and coordination. With this "4D" scheduling optimization, the construction process can be "simulated" over time to catch scheduling problems and physical site constraints such as limits of staging areas, and crane and truck delivery space requirements.
- The design is better communicated of the design to subs and crews. Again, most people can visualize 3D more easily than two-dimensional drawings. Many sub-contractors may even prefer a 3D model to traditional 2D drawings.

WHERE ARE WE NOW?

There remain many cultural and procedural obstacles in the way of our achieving all of these benefits within the construction industry. Fear of litigation and questions about ownership of building design data are just some of those obstacles. In addition, some of these "high-level" benefits require buy-in from parties outside of our sphere of control. Some, such as "4D Scheduling" and "Clash Detection," may required third-party software applications in order for the goal to be fully realized. Many innovative companies in the United Kingdom and Europe have developed products or are currently developing products that interface with ADT to produce some of these benefits, when used in conjunction with today's ADT.

Despite these apparent limitations, all of these ideas are presented here to help you understand the full potential and promise of the building model. None of the benefits is possible with a set of "flat" CDs. They become possible only with an object-oriented, parametric, hybrid (2D/3D) virtual building model. The "hybrid" nature of ADT objects is a critical piece of the equation. A hybrid object works simultaneously as both a 2D drafted entity and a 3D model component. (Details on hybrid qualities of ADT can be found in Chapter 2.) Autodesk Architectural Desktop is capable of delivering the first generation of this virtual building model today. The immediate benefits to architects and designers are faster design cycles, better design visualization, better coordinated CDs, less re-work and fewer errors. To achieve these benefits, we must embrace the ADT Building Model philosophy and workflow. To do that well requires some "re-tooling" of existing standards and procedures. Assisting you with that process is the goal of this *Advanced Implementation Guide*.

WHY SHOULD WE SPEND TIME TO "IMPLEMENT" ADT?

Every attempt has been made by the developers of ADT to create useful and usable content and configurations "out-of-the-box." However, despite their best efforts, you will not always find the provided resources satisfactory. This situation will naturally lead to the desire to customize some or all of the ADT environment. There are three factors most crucial to successful ADT implementation:

- Commitment to the Building Model philosophy
- Training in the tools and new methodologies
- Comprehensive and integrated setup

It is critical to attain "buy-in" from yourself and your team members. Even more important is "buy-in" from upper management within your organization. Without the commitment of the decision-makers of your firm, your ADT implementation is unlikely to succeed. If management understands and supports the building model concept, you are much more likely to receive a commitment of time and resources required to complete an ADT deployment. Be certain that those resources include a budget for training. ADT is a powerful and complex software package. However, "complex" need not mean "complicated" or "too hard." A little bit of focused training can go a long way in overcoming obstacles posed by the ADT "paradigm" shift.

Finally, although usable out-of-the-box, ADT will serve you much better if you standardize and tailor your configuration to suit your firm's specific needs. To make this case, let's take a high-level look at the benefits that can result from a carefully planned and deployed ADT implementation. The specific goals of your firm will vary depending on a variety of factors, but some of the typical benefits of a well-planned ADT deployment are as follows:

- **Use a single standard software package** – ADT is able to serve as both the 3D design and 2D drafting platform.

- **The software will manage your standards** – With tools like the Display Manager, Style Manager and Layer Manager, many of the common opportunities for error and non-compliance with standards can be eliminated.

- **Reduce/eliminate custom programming** – Many of the areas previously requiring custom programming and add-on routines are replaced by native functionality that is part of ADT. The burden of maintenance of custom routines is no longer the firm's responsibility. This can represent significant cost savings in private software development and tech support requirements.

- **Extract benefits from 3D design** – Many of the benefits of the virtual building model mentioned above begin to become possible with the introduction of ADT. With ADT's powerful "hybrid" approach to 3D design, 2D production can occur seamlessly while a true 3D model is worked on.

- **Bring building engineers onto same platform** – Products such as Autodesk Building Mechanical and Building Electrical are built on an ADT foundation. If your consultants are using these packages, this streamlines coordination efforts and eliminates time-consuming data translation.

- **Capture more design intent** – Again, the 3D virtual building model allows very accurate visualization of actual design conditions. This enables better design decisions to be made and coordinated, with real-time feedback.

Throughout the course of this book, our goal will be to deploy and configure Architectural Desktop to enable you to reap many of these benefits. Careful configuration of templates and other critical tools will help set you and your firm up for "repeatable" success. Achieving good results from a pilot project staffed with all of the "star" CAD gurus from your firm is beneficial, but not necessarily "repeatable" by future projects. Thorough and easy-to-understand standards and procedures are the key to repeatable success. The following tutorial will give you a "bird's eye" view of the benefits that we hope to achieve.

Install the CD Files and Open a Sample File

1. Install the files for the Introduction located on the Autodesk Architectural Desktop: Advanced Implementation Guide CD ROM.

The archive for the Introduction is named *Chapter00.exe*. A folder named *Chapter00* will be created on your hard drive. Refer to the "How to Use the CD" section in the Preface for instructions on installing the sample files included on the CD.

2. In the *Introduction* folder, **Move** the *Office Standard.stb* file to your *Plot Styles* folder.

This will make this sample Plot Style Table available to ADT. The default location of the Plot Styles folder is *C:\Program Files\Autodesk Architectural Desktop 3\Plot Styles*.

3. In the *Introduction* folder, **Open** the file named *A-101.dwg*.

4. **Zoom** and **Pan** around the drawing a bit.

Load Display Configurations

The currently active Display Configuration is "Work." A Display Configuration establishes the display parameters and "look" of all ADT objects in a drawing.

1. From the **Desktop** menu, choose **Select Display**.

2. Choose **Plot Plan 48** and then click **OK** (see Figure I–1).

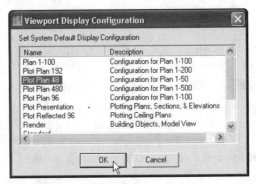

Figure I–1 *Load a Display Configuration*

    3. From the **View** menu, choose **Regenall**.

Note the change to the drawing. Pay particular attention to the change in the Doors and the level of detail displayed within the Walls.

    4. From the **Desktop** menu, choose **Select Display**.

    5. Choose **Plot Plan 96** and then click **OK**.

    6. From the **View** menu, choose **Regenall**.

Again, note the change in detail displayed for both Doors and Walls. Also note the appearance of Tags for Doors and the change in size of the Room Tags. The Plot Plan 48 Configuration is intended for plotting drawings at a large scale, while Plot Plan 96 is used for a smaller scale. (This topic is covered in Chapters 5 and 9.)

    7. From the **Desktop** menu, choose **Select Display**.

    8. Choose **Plot Reflected 96** and then click **OK**.

    9. From the **View** menu, choose **Regenall**.

Layout Tabs

Notice that the Space objects disappeared, the Ceiling Grids and Lights appeared and the Doors turned gray and dashed.

    1. Click the Normal Plot layout tab.

There are three viewports in this layout tab. One is set to **Plot Plan 48** at 1/4"=1'-0" [1:50], another to **Plot Plan 96** at 1/8"=1'-0" [1:100] and the third to **Work**. Notice that the Room Tags appear the same size regardless of the respective viewport scales. These are special Room Tags designed to display correctly at two or more scales. This is called "Scale-dependent" and is covered in Chapter 7. Notice also that the Door Tags are visible in only the Plot Plan 96 viewport.

2. At the bottom of the screen in the AutoCAD status bar, click the LWT (Lineweight) toggle button (see Figure I–2).

Figure I–2 *Toggle on the lineweight display*

This toggles the lineweight display on in the current drawing. Since we are in a paper space layout, lineweights display properly on screen relative to the current sheet size. **Zoom** and **Pan** around the drawing to get a closer look.

3. From the **Desktop** menu, choose **Layer Management>Layer Manager**.
4. On the left side, click the **Xref** node in the tree.
5. On the right side, select the **A-GR03** XREF (number 1 in Figure I–3), and then click the word ***Varies** in the **Plot Style** column (number 2 in Figure I–3).
6. Choose **Halftone** and then click **OK** (number 3 in Figure I–3).

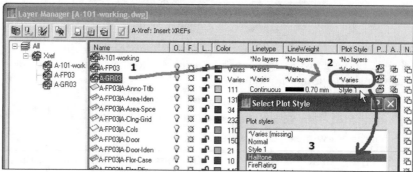

Figure I–3 *Screen the display of the entire XREF*

7. Click **OK** and then **Regenall**.

Notice that the Column Grid is now displayed as a halftone gray. We are using a "Named" Plot Style Table to achieve this effect. (Similar results can be attained with Color-dependent Plot Style Tables.) The topic is covered in detail in Chapter 3.

Entity Display Props

1. Open the file named *A-FP03.dwg*.
2. From the **View** menu, choose **3D Views>SW Isometric.**

Notice how all of the objects in the model have changed color.

3. Zoom in and select one of the masonry Walls in the restrooms, right-click and choose **Entity Display.**

On the Display Props tab, be sure that the "Model*" is showing in the list at the top (as shown in number 1 in Figure I–4).

4. On the Display Props tab, click the **Edit Display Props** button (number 2 in Figure I–4).

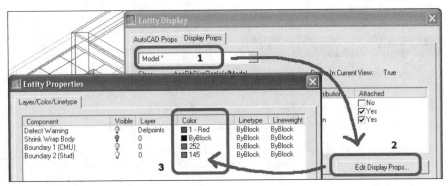

Figure I–4 *Explicit color assignment in the Model Display Rep*

Note that each material has been assigned its own "explicit" color mapping. (Number 3 in Figure I–4)

5. **Cancel** all dialog boxes to return to the drawing.

6. From the **View** menu, choose **3D Views>Top.**

Note that all colors return to their previous settings.

7. Zoom in and select the same masonry Wall in the restroom, right-click and choose **Entity Display**.

On the Display Props tab, be sure that the "Plan*" is showing in the list at the top (as shown in number 1 in Figure I–5).

8. On the Display Props tab, click the **Edit Display Props** button (number 2 in Figure I–5).

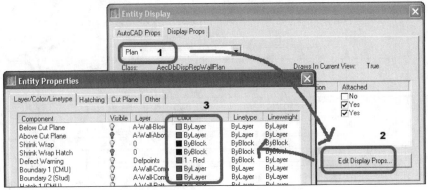

Figure I–5 *ByLayer color assignment in the Plan Display Rep*

Notice that all components are now assigned to layers, which have their colors assigned to ByLayer. (Number 3 in Figure I–5) With the ADT Display System, we can achieve traditional

CAD management goals that advocate ByLayer color assignment in Plans while simultaneously assigning explicit colors to the 3D Model for VIZ material mapping.

9. **Cancel** all dialog boxes to return to the drawing.

Modify a Wall with Property Data

1. Select the back Wall of Closet 3013 (click directly on the hatching in the Wall to select), right-click and choose **Wall Modify**.
2. From the **Style** list, choose **A1** and then click **OK**.
3. At the command line, type **objrelupdate** and then press ENTER twice.

Notice the change to the Wall and the Wall Tag. (OBJRELUPDATE was necessary to update this tag in this drawing session. Had we closed and reopened the drawing, the tag would also have updated.)

Stair Display Configuration

1. In the *Introduction* folder, **Open** the file named *A-CM01.dwg*.

The file contains a Stair core and restrooms. The file is made of a few XREFs and represents a three-story building core. Using customized Display Configurations already contained within the file, we can correctly display the Stair on the ground floor (going up only), the middle floor (going both up and down) and on the top floor (going down only).

2. From the **Desktop** menu, choose **Select Display**.
3. Choose **Plot Plan First Floor** and then click **OK**.
4. From the **View** menu, choose **Regenall** (see Figure I–6).

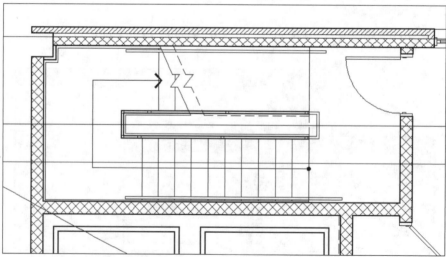

Figure I–6 *Stairs showing the up direction only*

Note that the Stairs now display going up only.

5. Repeat the above steps, choosing **Plot Plan Top Floor** this time.

Note that the Stairs display going down only this time.

6. Repeat the above steps returning to **Work** this time.

7. **Pan** over to the right of the drawing to reveal a Section object running through the Stairs.

Section Styles and Design Rules

This Section was generated from the staggered Section Line running through the plan. The Section object is currently assigned to the "Standard" Section Style. The Standard Style does not take advantage of lineweights. In addition, some of the line work has been manually erased, however, the erased lines are still visible and show up red in the Standard Style. Applying a new Style to this Section can remedy both issues. Section Styles are one of the great "untapped" powerhouses of ADT.

1. Select the Section, right-click and choose **Update**.

2. In the **Result Type** area, next to **Style to Generate** choose **Construction Documents 96** from the list (see Figure I–7).

3. Place a checkmark in the "Save Copy of User Edits" checkbox, and choose "Standard" from the "Style for Copy" list (see Figure I–7).

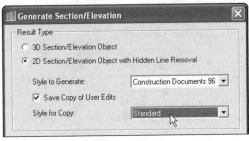

Figure I–7 *Swap a more detailed Section Style*

Pay close attention to the railing balusters and the swing direction indicators in the toilet stall doors.

4. Click **OK** to re-generate the Section.

The Door swing direction lines have become dashed and the balusters have changed color to indicate that they are now using a lighter lineweight designated by the Construction Documents 96 Style. Notice however, that the red "erased" lines still appear to have no effect. Actually, there are now two Sections, one generated from the model and the other containing only the "User Edits" (in this case the red erased lines.)

5. Click directly on the Stairs to select the Section, right click and choose "Merge Linework."

6. When prompted to "Select entities to merge," click on the red vertical line on either side of the Section and then press ENTER.

Notice that wherever there was a red line in the original Section, the line work has now been erased. (In reality, the "Erased" component of this Style is simply turned "Off." The other significant feature of Construction Documents 96 Style is that it has assigned appropriate lineweights to the line work of this Section. You can visualize this effect much better from a paper space layout.

7. Click the Normal Plot layout tab.

8. At the command line, type **lts**, press ENTER, and then type **1** and press ENTER again.

This will reset the LTScale for proper viewing in Layouts.

9. Click the small LWT (Lineweight) toggle button on the AutoCAD Status bar at the base of the screen.

This will turn on the Lineweight display in this drawing.

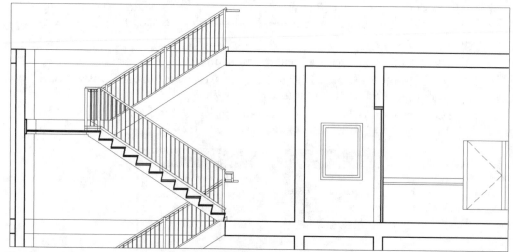

Figure I–8 *A Section object with Design Rules and standard lineweights*

10. **Zoom** and **Pan** around the drawing (see Figure I–8).

Previewing the drawing from a layout this way, with the lineweights toggled on, gives a very accurate representation of how the drawing will actually look when plotted.

11. Return to model space and choose **Front** from the **View>3D Views** menu.

Note the colors of the various objects. The color of these objects causes a reaction within the Section Style. For instance, the Railing Balusters are Color 123. The Section Style uses a "Rule" that tells it to "lighten" the lineweight of any object in the Model whose Color is 123.

LET'S GET STARTED...

This short tour was designed to show you some of the many benefits to a well-planned and executed ADT implementation. Many of the fine details of execution are saved for the coming chapters. It is important to note that none of the steps we have just completed requires any custom programming to implement. All of these benefits and the many more contained in the coming pages require only a bit of time and pre-planning.

Now that your appetite has been sufficiently whetted, it's time to get to work on your ADT implementation. In Chapter 1, we will discuss recommended installation procedures and configuration.

Planning

This section is designed to help you build a plan for your firm's implementation of Architectural Desktop. It will help you assess what actions are required to move forward with ADT implementation. The core concepts and methodology of ADT R3.3 are also reviewed.

Section I is organized as follows:

WHO SHOULD READ THIS SECTION?

This section is intended for the **CAD Manager**, the **Project Data Coordinator** and the **Project Manager**. The **CAD Software User** should also be engaged in the discussions that these chapters would prompt. Therefore, this section is important to that group as well.

CHAPTER 1

Installation and Configuration

INTRODUCTION

What better place to begin than with the installation and configuration of Autodesk Architectural Desktop R3.3. Installing ADT is not difficult, and the default installation is quite usable right out of the box. However, if our goal is to get the most out of ADT, we want to take the time to carefully plan every aspect of our implementation. That begins with a well-thought-out installation strategy. There are many ways that we can customize our Architectural Desktop software. In this chapter, we will explore installation and configuration of ADT. Even if you have already installed ADT, you may want to reinstall it or tweak your installation after reading this chapter. If you skipped the Preface, please go back and at least skim the "Style Conventions," "How to Use This Book," "How to Use the CD" and "Service Packs" topics before proceeding.

OBJECTIVES

The main objective of this chapter is to build an ADT installation plan and a set of reusable resources, to make the initial installation and any future *reinstallations* proceed as smoothly as possible. In this chapter, we will explore the following topics:

- Understand the options in the setup wizard
- Build a remote Client Deployment
- Set up network-based installation resources
- Build an Office Standard configuration and user profile
- Create an Office Standard ADT program icon

WHO SHOULD READ THIS CHAPTER?

This chapter is intended for the **IT Support Personnel** and the **CAD Manager**. However, if you have questions about ADT installation and configuration that you always wanted to know but were afraid to ask, then this is the place to start.

ADT SOFTWARE INSTALLATION

You will need to perform several tasks to fully install ADT and all of your office standard resources. They are summarized as follows:

1. Set up network shares
2. Decide on installation type (network or stand-alone)
3. Build a Client Deployment (User Install)
4. Build a stand-alone full installation (CAD Manager Install)
5. Set up network resources
6. Build the Office Standard profile
7. Complete client setup

SET UP NETWORK SHARES

Network topology is beyond the scope of this book, but it is recommended that a dedicated server be used to store project files. For the purposes of this book, it will be assumed that all client workstations have access a single dedicated server machine. It will further be assumed that there are mapped network drives for use in storing project files, library resources, office standards and installation files. For the purposes of this book, the following drive letter designations will be used (as shown in Figure 1–1):

- Drive letter **P** for project files
- Drive letter **L** for library files
- Drive letter **I** for installation files

Figure 1–1 *Recommended network drive mappings*

The projects share should contain a root folder for every active project in the firm. Sub-folders within each project folder are permissible. The details of these sub-folders are discussed in the CAD Standards section in the next chapter. The library share should contain all of the office-standard content and resources that are *not* project specific and *not* related to installation. This includes blocks, symbols, styles, details and layer standards. The install share is the location for all files required to install ADT and any menus, profiles, add-ons, patches and extensions. Each of these locations will be discussed in detail in coming topics, chapters and sections. If possible, these drive mappings could be located on three different physical servers or, at the very least, be three separate shares on the same server. The specifics are not important to our discussion. Feel free to use different letters and designations as is appropriate for your actual hardware, software and network topology.

If you are not sure how to set up Network Shares, please consult your network support personnel. The I and L Drive Network Shares ought to be made Read Only to all but yourself and any other IT– or CAD Manager–level employees. This will prevent users from accidentally modifying the contents of the office standard library.

If you use ADT in a stand-alone configuration without a server, you may want to consider setting up a simple peer-to-peer network with a dedicated PC acting as a server. However, if you run a single machine configuration with a single license of ADT and have no plans to set up a server, it would still be wise to set up each of the above-recommended folders at the root of your C Drive. Simply name them *Projects*, *Library* and *Installation*. Use them in place of the mapped drives recommended throughout this book. In this way, you can still reap the benefits of easy customization and quick reinstallation of ADT. In addition, your customizations and project files will be easy to back up since they are contained within three discrete folders.

OFFLINE FILE ACCESS

If your workstations use Windows 2000 Professional or Windows XP Professional, you can take advantage of the Offline Files feature. With this feature, users can make any network resource "Available Offline." This feature will guarantee that network resources are always available by making a mirror copy to the local hard drive of all the files within a specified folder or folders. This feature can be used on desktop machines to guard against lost network connections. It is especially useful for those individuals who use laptops. The benefit to the IT and CAD Manager is that a laptop can be set up the same as all other machines without worry that, when users disconnect from the network, they will be unable to use ADT and all its networked resources. More information on the Offline Files feature can be found in the Windows Help Files or at *www.microsoft.com.*

INSTALLATION OPTIONS (NETWORK OR STAND-ALONE)

Architectural Desktop can be installed one of three ways: It can be installed on the individual workstation as a stand-alone application, it can be configured to be run and licensed remotely from the server, or it can be configured to run locally on the desktop, with access to the server for authorization only. The most common ways to install ADT are completely stand-alone installed and authorized locally, or installed locally with authorization acquired from the license server. Due to performance considerations, it is not recommended that the remote installation be used.

Deciding whether to configure ADT as a stand-alone install or to use the network license server can be an important issue. If you have enough ADT licenses for the number of users in your firm, and you have fewer than twenty-five users, you ought to seriously consider the stand-alone installation. The extra time taken authorizing each machine separately can be weighed against the increased cost of maintaining a license server required in the network installation. However, the greater the number of users and the more users whose need for the software is less than full time, the more cost savings benefit you will reap from the network installation. In addition, if you wish to track license usage, consider the network deployment.

The primary benefit of network installation comes when there are more users than available licenses. In a firm of say fifty users that owns only forty licenses, use of the network license server allows the forty licenses to be distributed as needed to the users on a first-come, first-served basis. All fifty users would have ADT installed on their local machine. All machines would access the license server when ADT was launched on their desktop. Any user could use ADT, so long as forty or fewer persons were logged into ADT at any given time. If the forty-first person were to log in, that person would be refused a license and ADT would not run until some one else quit the application and "freed" a license. If your environment grows, then additional licenses can be purchased and added to the license server.

Note: The ADT license manager allows for many features, including auto-save and "time out" features if an individual leaves their desk. Refer to the license manager documentation for more information on these features.

In firms where many users do not need to be in ADT all day long, the license server can provide an attractive alternative to purchasing dedicated stand–alone licenses. There are also additional benefits for the larger offices. Use of the License Manager gives the CAD Manager or IT Support Person a means to measure license usage for all users and Autodesk products. With this information, informed recommendations can be made to management regarding purchasing and upgrading of future licenses.

 Note: Some of the advanced reporting functions require the CAD Manager Extension. If you do not automatically receive extensions, contact your Autodesk Reseller regarding the subscription program.

For additional information on the specifics of the network installation options, consult the *Network Administrator's Guide*, the online help files and your local Autodesk authorized reseller. The *Network Administrator's Guide* is an HTML document called *ACAD_netguide.htm*, and it can be found in the *Netsetup\support\AdLM\Docs* folder on the ADT 3.3 installation CD ROM.

No matter which method of licensing you choose, it is highly recommended that common resources be placed in a series of shared folders on your network server (as suggested in the "Set Up Network Shares" section earlier). This gives your CAD Manager or network support person a single point of access to update, back up and maintain all company resources. Details on building these network resources are covered later in the "Set Up Network Resources" section.

It is important to note that the specific mode of licensing employed by your firm has no effect on your ability to access ADT shared resources from a common location on your company network. Menu files, blocks, LISP and VBA routines, styles and content of all kinds can be located on the network server and accessed by all users regardless of method of installation. The Network License Manager affects *only* how ADT receives its authorization code.

BUILD A CLIENT DEPLOYMENT

When a network installation of ADT is built, the Network Setup Wizard (*netsetup.exe*) will be used for your installation of ADT rather than the standard *Setup.exe* installer. Using the Network Setup Wizard, you will build a "Client Deployment Image" that will contain all of your preferred settings and installation options. When this deployment is installed on the client workstation, it requires no user input. This can help streamline the task of installing several workstations throughout the office and guarantee consistency.

 Note: The installation and configuration of the Network License Manager is not covered in this book. Please refer to the appropriate documentation provided with ADT.

Using the Network Setup Wizard

1. Insert your Autodesk Architectural Desktop CD into your CD ROM drive and browse to the Netsetup directory (such as *D:\Netsetup*) folder.
2. Double-click **netsetup.exe** (Network Setup Wizard) as shown in Figure 1–2.

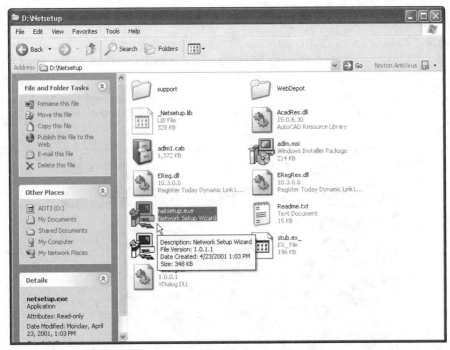

Figure 1–2 *Locate the Network Setup Wizard and double-click*

3. On the Network Installation Selection screen, choose **Client and Autodesk License Manager Installation** and then click **Next** (see Figure 1–3).

 Note: If you wish to view the Network Administrator's guide, click the **View the Network Administrator Guide** button.

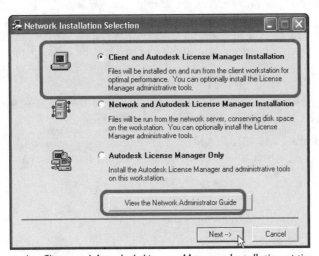

Figure 1–3 *Choose the Client and Autodesk License Manager Installation option*

18

This option will create the Client Deployment Image. You will be required you to install the License Manager later, if you have not already done so. If the License Manager is already installed and running, it is not necessary to reinstall it. However, if ADT is installed, you may decide to re-build your Client Deployment Image and reinstall after reading the recommendations in this chapter.

4. On the Autodesk License Manager Installation screen, choose the **Install Autodesk License Manager files later** option and then click **Next** (see Figure 1–4).

5. On the Installation Folders screen, type in the location of the network server (including sub-folders if appropriate) where you want the image files to be stored.

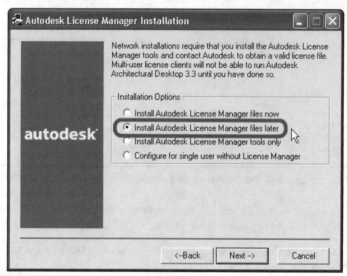

Figure 1–4 *Choose the option that installs the License Manager later*

The format follows the Universal Naming Convention (UNC) **Server\Share\Folder**. Users will install ADT from this location. Although you must use the UNC path to create the image, users can use the drive letters you established. For example, *MyServer\ADTNet\ ADT-Deployment* would appear as *I:\ADT-Deployment* from users' workstations.

6. Place a check mark in the **Client installations will be run in Silent mode** check box (see Figure 1–5) and then click **Next**.

The Silent mode setting will install ADT on the user's workstation without any installation prompts. This guarantees that the installation is consistent across every machine in the office.

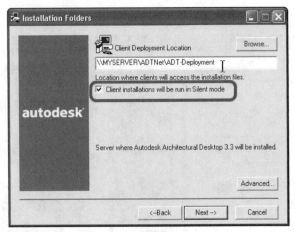

Figure 1–5 *Choosing deployment location and Silent mode*

The Server Configuration screen requires the name and ID of the machine that will distribute licenses to your ADT clients. According to the notes on this screen, "The Server Host Name is the name of the computer running the AdLM (Autodesk License Manager)." Information on how to choose and configure your license server is located in the Network Administrator's Guide. This server does not have to be running in order to build the Client Deployment. However, it will need to be installed and running before the client installation will function.

7. On the Server Configuration screen, type the name of your license Server.

The notes on the server configuration also describe how to locate the Server Host ID.

8. In the *NetSetup\support\AdLM* folder of the Autodesk Architectural Desktop CD, double-click **lmtools.exe** (see Figure 1–6).

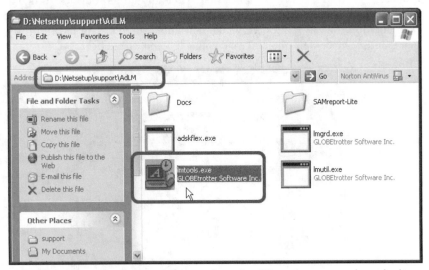

Figure 1–6 *Launch the lmtools.exe application from the CD on the system where the License Manager is (or will be) installed*

9. In the LMTOOLS window, click the System Settings tab (see Figure 1–7).

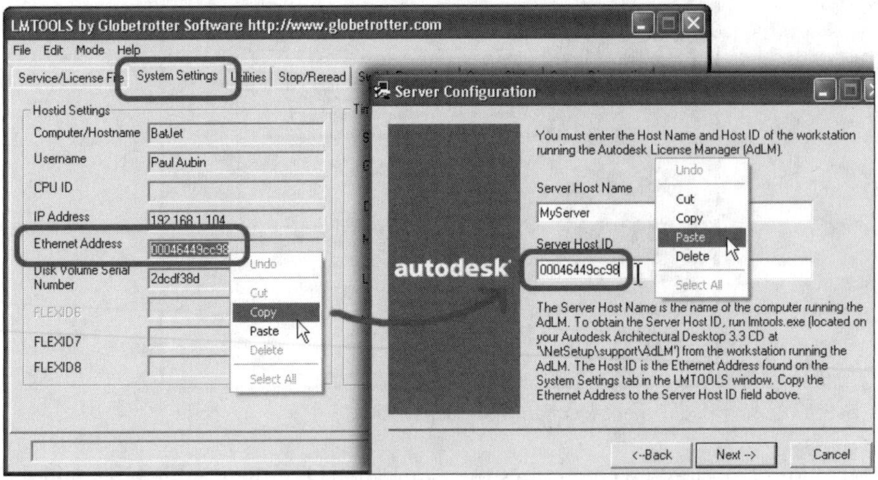

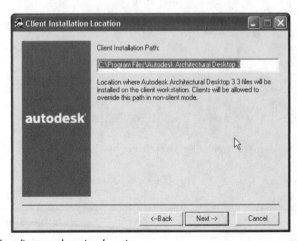

Figure 1–7 *Locate the Ethernet Address and copy it to the Server Host ID field*

The Host ID is the Ethernet Address found on the System Settings tab in the LMTOOLS window.

10. Copy the **Ethernet Address** to the **Server Host ID** field on the Server Configuration screen (see Figure 1–7) and click **Next**.

11. On the Client Installation Location screen, type the path on the *local user machine* where ADT should be installed and click **Next** (see Figure 1–8).

The default location is *C:\Program Files\Autodesk Architectural Desktop 3*. It is recommended that you accept this default location. It is possible to install ADT in virtually any location on the client workstation. However, it is a certainty that ADT and all its routines and content were thoroughly tested from the default path. There is a very long list of items to attend to in your ADT implementation; unless you are very adamant about an alternate location, it is recommended that you use the default one and save yourself some extra effort.

Figure 1–8 *Input the client workstation location*

Note: If you chose Silent mode, users will not be able to change this location or any other options during the install.

12. On the Client Setup Type screen, choose **Custom Installation** and then click **Next** (see Figure 1–9).

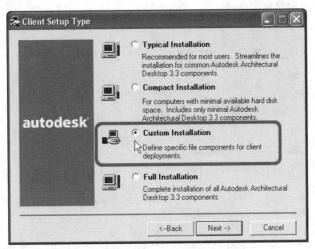

Figure 1–9 *Choose Custom Installation to designate the files to install*

Custom installation allows you to choose from a list of components you wish to have installed on the client workstation. For instance, it is not necessary to install the sample files on every user's workstation. You could have a copy of the sample files on the server in some common location that users could access instead. Another example is AEC Content. In Chapter 8, we will discuss ADT Content in detail. It is desirable to have all of the content for the entire office installed in a single central server location. For this reason, it is recommended that you install the default ADT Content on the server only and choose *not* to install any of the content locally through the network deployment. The same is true of the texture maps. Refer to the topic below on setup of the network resources for more information.

13. Clear the check boxes next to **Texture Maps**, **Architectural Content** and **Samples** and all three of the Content choices grouped below each of them, and then click **Next** (see Figure 1–10).

14. On the Serial Number screen, input your Serial Number and CD Key and then click **Next**.

You will find these among the documentation or packaging shipped with your ADT.

15. On the Personal Information screen, input the required fields and then click **Next**.

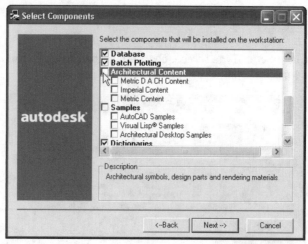

Figure 1–10 *Choose not to install content and sample files on the local machines*

The Autodesk Architectural Desktop 3.3 Today window launches when the program is first run. This window includes file open and creation functions, the CAD Manager's Bulletin Board and optional access to the Autodesk Point A Web Portal. In the Configure Today screen, you can predefine the behavior of the Today window at the client workstation. You *must* choose a file for the Today CAD Manager's Bulletin Board. It can be almost any file type. Click the **Browse** button and locate an HTML file to use for this feature, or any file such as a text file or a graphic file. The file can be located in any folder on your network server, such as *I:\Bulletin Board*.

 Note: Be careful to choose an actual file and not just a folder. If the file you designate here cannot be found by ADT when launched by the user, a series of error messages will appear.

You can create an HTML page (or complete web site) for your CAD Manager's bulletin board feature. There are dozens of software tools designed to assist you in this endeavor. If you have no HTML acumen, there is a very simple HTML CAD Manager's Bulletin Board page provided on the *Autodesk Architectural Desktop: Advanced Implementation Guide* CD ROM. The file is named *MasterADT.htm*. Follow the steps outlined in the "How to Use the CD" section of the Preface and install the files for Chapter 1. A folder named *Chapter01* will be created in the *ADT_AIG* folder on your C Drive. Included with this file is a file named *Logo.jpg*. Be sure to copy both files, *MasterADT.htm and Logo.jpg*, to your bulletin board folder on the server (*I:\Bulletin Board*). This simple text file can be edited using Windows Notepad, Front Page or any other text or HTML editor. Simply open it, add your company name and replace the provided image reference with your company logo. See the sidebar for details on this process.

If you wish to control what users see when the Today window appears, check **Apply CAD Manager Control to Today**. You may optionally clear the check box next to **Allow access to Autodesk Point A services**. This will remove direct access to the Autodesk Point A web portal through the ADT Today window. Alternatively, you can allow Point A to appear, but control how and if users see messages informing them of product updates and patches to ADT (see Figure 1–11).

EDITING A BULLETIN BOARD HTML FILE

To create a simple HTML file to use as a CAD Manager Bulletin Board, you can create your own, or open the one included in the *Chapter01* folder from the CD. Create a folder on your I Drive named *I:\Bulletin Board* and copy both *MasterADT.html* and *Logo.jpg* into it. You can use Windows Notepad or Front Page to edit the file. Locate the following line:

<p><font size="6"><strong>**Company Name**</strong></font></p>

Replace the text **Company Name** with the name of your company. If you want to change the web page link, replace the **web address** with your own company's web site.

<p><a href="**http://www.paulaubin.com/**">

The current image named *Logo.jpg* is used as a stand-in for your company's logo. If you wish to use your company's logo, use an image editor to save a copy of your company logo in JPG format. Make sure to change the pixel dimensions in the HTML file to match the logo JPG you create. (They are currently set to 150x186 for the sample image.)

<img border="0" src="**Logo.jpg**" width="**150**" height="**186**"></a></p>

Save the changes to the file and test it by double-clicking it to view it in your web browser. If something does not work correctly, reopen the file in your editor and make any corrections. Once you are certain that the file is working properly, you may assign it to your Bulletin Board.

16. Make your choices and click **Next**.

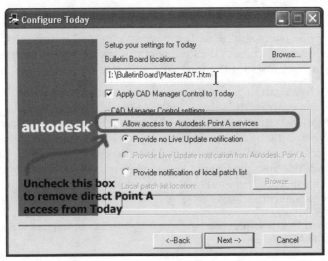

Figure 1–11 *Configuring the Today window to have no Point A access*

 Note: Removing the direct access to Point A in Today does not prevent users from accessing this most useful design information portal through their normal web browser.

17. On the Architectural Program Options screen, choose your preferred units format (Imperial or Metric) the organization of DesignCenter content and whether to install the International Extensions, and then click **Next** (see Figure 1–12).

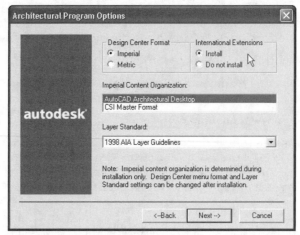

Figure 1–12 *Choose your Architectural Program Options*

Design Center Format – Your choice will determine which Content tree will be called from the commands in the drop-down menus. Choose whichever unit system you use most frequently. This setting only affects the *default* behavior of the DesignCenter menu commands. You will be able to access all nodes of the DesignCenter (Imperial, Metric and D A CH) regardless of the choice made here.

International Extensions – It is highly recommended that you *do* install the International Extensions. They provide many additional features and potential to the software. Refer to the "International Extensions" section in the Preface of this book for more information. This book will assume that you have installed the International Extensions.

Imperial Content Organization – If you choose **Imperial** for **Design Center Format**, you will be able to choose between two variations of folder organization for the DesignCenter folders and accompanying menu structure. The two choices are the default **Autodesk Architectural Desktop** organization and **CSI Master Format**. Be certain to choose the same format here as you plan to use when you install your content folders to the server. (Remember that we have excluded the content files from the Client Deployment, and will install them to the server later. More information on this can be found below.) The default setting, **Architectural Desktop**, is the setting used throughout this book.

Caution: This choice can only be made during installation. You cannot change it later without reinstalling ADT.

Layer Standard – A Layer Standard and Layer Key file will determine the default layering of all ADT objects. You may choose between AIA and BS 1192 standards during installation. Your choice here is not critical as many more options will be available later, such as "DIN 276" and "ISYBAU." For now, choose the option closest to the standard you plan to use. Layer Standards will be covered in much more detail in Chapter 4.

MANUALLY EDITING THE INSTALLATION OPTIONS

All of the choices made in the Network Setup Wizard are recorded to a text file named *appsetup.inf*. This file will be created in the *IMAGE* folder of your Client Deployment location (*I:\ADT-Deployment\IMAGE* in the above example). The *Readme.txt* file located in the *D:\Netsetup* folder of the CD includes detailed instructions on all of the possible variables allowed in this file. Open *appsetup.inf* in Windows Notepad and edit any values as appropriate. For example, if you want the default Layer Standard to be one of the standards mentioned above as not being included in the list, you can manually edit it here.

On some occasions, the Netsetup.exe application fails to write this file completely. This is another instance when manually editing this file will prove advantageous. You can also maintain more than one version of this file with alternate options for those clients who need slightly different settings. For example, you may have a project team that is using Metric units, while the rest of the office is based on Imperial.

You should receive a Setup Confirmation screen next.

18. Verify all of the settings and then click **Next**.

At this point, the Network Setup Wizard will build your Deployment Image (see Figure 1–13). Go grab a cup of coffee while it proceeds.

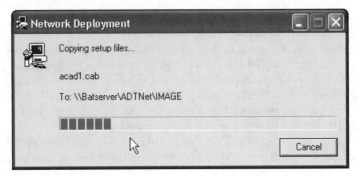

Figure 1–13 *File copy progress*

19. Click **Finish** on the last screen to finish installation and view the *readme* file.

Once the creation of the install image is complete, users simply need to log in to the network share where the Image is located, and double-click the **setup.exe** (see Figure 1–14). No further intervention by the user is required. ADT will install to their local machine with all of the settings pre-established. However, before proceeding to the client workstations, some other items require configuration on the server side. The next topic will cover the specifics. Even though this task could be relegated to the user, complete installation will still require a few additional steps. For this reason, you may wish to perform the complete installation yourself.

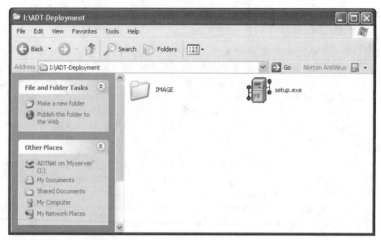

Figure 1–14 *The newly created setup.exe ready to install a client*

 Note: The primary purpose of the preceding tutorial was to suggest the recommended settings for a client workstation installation. To produce a fully functioning installation, you will need to install the Network License Manager and License Manager Tools on an appropriate server machine. You will also need to follow the instructions included in the Network Administrator's Guide to acquire and install your authorization codes.

BUILD A STAND-ALONE FULL INSTALLATION

In the last sequence, we built a network deployment for client installation (with network licensing) of Autodesk Architectural Desktop. This will be the primary installation used for user workstations. The CAD Manager will still need to install ADT locally from the standard ADT *setup.exe* installation program. (This is located at the root of the ADT Installation CD ROM.) This stand-alone *full install* will be used to access the content libraries provided with ADT. In the Client Deployment installation, we chose not to install the content and sample files. You may have also opted not to include other items as well. In this installation, we will perform the "Full" installation. Once complete, we can manipulate and reconfigure the items installed as necessary.

Create a Stand-alone Full Install

1. Run **setup.exe** from the root of the ADT 3.3 Installation CD ROM.

 Note: Do **not** use the *setup.exe* in the Netsetup folder this time.

Follow the prompts on each screen of the wizard.

2. When prompted for the type of installation to perform: Typical, Compact, Custom or Full, choose **Full**.

As stated above, use the default installation folder (*C:\Program Files\Autodesk Architectural Desktop 3*) unless you really need an alternate location. Complete the installation and read the *readme* file.

3. Double-click the new **Autodesk Architectural Desktop R3.3** icon created on your desktop to test your installation.

4. Once you are certain the installation is running properly, exit ADT.

SET UP NETWORK RESOURCES

You should now have a single full installation of ADT. (We will refer to this as the "Management Install.") In addition, you should have a Client Deployment ready to install client workstations from your network share (we will refer to this as the "User Install"). You will perform all of your standardization and implementation tasks from the Management Install.

Some of the folders within the ADT installation folder of the Management Install need to be made accessible from the server by all users.

1. **Copy** (not move) the *Content, Template, Textures, Plotters* and *Plot Styles* folders to the root of your library share (L Drive.)

Be sure to copy the entire folder so that each of the folders and all of their sub-folders and contents will be copied. The reason for copying rather than moving is that you will maintain a copy of all resources on your Management Install that you can use as working copies for ongoing maintenance until you are ready to publish them up to the server.

We will also make an adjustment to the default *Content* folder structure. By default, the *Layers* folder is included within the *Content* folder. When you browse the *Content* folder from within ADT in the DesignCenter, this *Layers* folder will appear empty. This often causes confusion among users. If we move the *Layers* folder to the root of the L Drive, it will no longer appear in the DesignCenter. (Later, we will be certain that we correctly path our Layer Standards settings to the Layer Standards file in its new location.)

2. **Move** (not copy) the *Layers* folder from the *L:\Content* folder to the root of L Drive (*L:\Layers*).

 Note: These are "Best Practice" recommendations that will provide the greatest benefit to users and managers alike. However, these recommendations assume a network infrastructure is in place that can handle the increase in traffic. You may wish to test your specific access speeds from a user standpoint and make or plan any necessary upgrades to your network infrastructure.

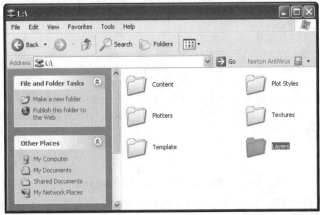

Figure 1–15 *Copy the Content, Template, Textures and Plotting Resources folders to the L Drive; move the Layers folder to the Root*

Potentially there are also items within the *Support* folder that you may wish to place on the server as well depending on the level (if any) of customization you would like to do to those items. Support files that are frequently customized include the following:

- **Font Mapping File (FMP)** – This file instructs AutoCAD to substitute fonts included in this file for other font files on your system. This is useful when the fonts in the drawing are unavailable or undesirable. Use a text editor such as Notepad to edit.

- **Linetype Definition File (LIN)** – This file defines the patterns of any linetype available in AutoCAD. Use a text editor such as Notepad to edit.

- **Menu Files (MNU, MNS, MNL, MNC)** – These files define the contents of all menus, drop-down, cursor and right-click, available in AutoCAD. Several menu files may be loaded at once. Use a text editor such as Notepad to edit.

- **Hatch Pattern Definition File (PAT)** – This file defines the patterns of any hatch available in AutoCAD. Use a text editor such as Notepad to edit.

- **Program Parameters File (PGP)** – This file defines any keyboard shortcut (command alias) available in AutoCAD. Use a text editor such as Notepad to edit.

- **Custom Dictionary File (CUS)** – This file defines a custom list of words used by the spell checker. (This is a great place to include your entire list of commonly used abbreviations.) Use a text editor such as Notepad to edit.

- **Custom Applications (LSP, ARX, DVB)** – These files are application programs and plugins to AutoCAD. They require programming knowledge and software to edit and create.

For some of these items like the Font Mapping File, Program Parameters File and the Custom Dictionary File there can be only a single file loaded by ADT when it launches. There may be several available to choose from, saved in various locations on the local drive and the network, but only one will ultimately be loaded into any particular AutoCAD/ADT session. In contrast, several different Menu Files and Custom Applications may load at the same time. Finally, most of these items could be set up either way, and often are. Hatch Pattern Definition Files and Linetype Definition Files are good examples of files that often have a default file that is used most often and then additional files that users can load when needed. All of these items can be installed on the server and accessed remotely by ADT users. If you intend to customize these resources, you may want to consider this option, as it will make management and update of these resources simpler.

Once you have decided which items to copy to the server, you next have to decide in which network share they are placed. There is an argument to be made for these items being regarded as library resources (and being placed on the L Drive) and another argument to be made for their being considered installation items, (and being placed on the I Drive). Either argument could be considered equally compelling. Since we are still on the topic of installation, we will treat these items like installation resources for the purpose of our discussion here. Therefore, we will create a *Support* folder on the I Drive. We will refer to this folder as the "Common Support" folder.

3. Create a new Folder on the I Drive named **Support**.

4. **Copy** only those items that you intend to customize from the above list to this Common Support folder.

For instance, if you plan to build a new Linetype file, you would want to place a copy of the *acad.lin* file in the *I:\Support* folder. Later you can open and customize that file. If you would like more information on customizing menus, linetypes, hatch patterns and much more, consult *Customizing AutoCAD 2002* by Sham Tickoo, published by Autodesk Press.

Next, we will create a folder named *C Drive Folders*. This folder will ultimately contain those items that we wish to store locally on each user's workstation. For now, it will remain empty.

5. Create a folder named **C Drive Folders** at the root of the I Drive.
6. On the root of the C Drive of your Management Install machine, create a folder named **Local Content** and another called **Temp**.

We will address the usage of these folders a little later.

 Note: A discussion of folder structure in the project files share (P Drive) appears in the next chapter.

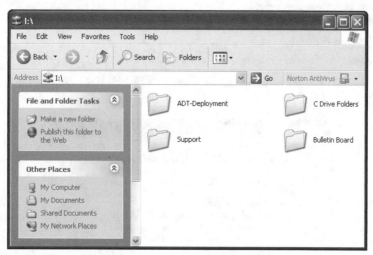

Figure 1–16 *Folders on the Install drive*

CUSTOMIZE THE ADT INSTALLATION

Once you have created your various resource folders on each of your network shares and populated them with content and resources, it is time to build an AutoCAD profile that "points" to each of these resource paths. An AutoCAD Profile is a tool used to manage a collection of system-wide AutoCAD settings. Profiles are managed from the **Options** command available on the **Tools** menu.

Build an Office Standard Profile

1. Launch Autodesk Architectural Desktop from your Management Install.
2. From the **Tools** menu, choose **Options** and click the Profiles tab.
3. Click the **Add to List** button.

4. In the Add Profile dialog box type **OfficeStandard** for the Profile name. For the description type **Office Standard Profile – Do not Edit.**

5. Click **Apply & Close**.

The new entry will appear in the list.

6. Double-click **OfficeStandard** to make it current.

 Tip: You can also select it and click the **Set Current** button.

At the top of the dialog box next to Current Profile, *OfficeStandard* should now appear.

 Note: It is important to create the new profile before configuring any settings. In this way, you preserve the original profile should you need to reset and start over.

Support File Search Paths

1. Click the Files tab.
2. Expand the Support File Search Path by clicking the plus sign (+) next to its entry.

There is a minimum of four entries in the Support File Search Path: Support, Fonts, Help and Layers (see Figure 1–17).

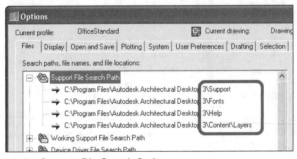

Figure 1–17 *Entries in the Support File Search Path*

- The **Support** path contains the Menu, Linetype, Hatch, PGP, Font Mapping and Custom Application files.
- The **Fonts** path contains all of the fonts unique to AutoCAD/ADT, (Compiled Shape Fonts). ADT also makes use of any fonts available to standard Windows programs.
- The **Help** path contains the AutoCAD and ADT help files.
- The **Layers** path contains the default ADT Layer Key files.

There is no need to change the location of the Help file and the Font file support paths. Both of these types of resources can be saved in any of the paths listed and they will all be available. For instance, if you have acquired a custom font that you wish to make available to the firm, it can be placed in a support path other than the Fonts path and it will become available along with the fonts in the Fonts path. However, some resource types cannot have more than one. One such type is the PGP file. AutoCAD/ADT will use the first PGP file it encounters as it loads the files from each support path in the list. It is therefore very important that these types of resources and their associated paths be located strategically. If you intend to modify the default menu file (*acad.mnu*), the default hatch pattern file (*acad.pat*), the default linetype file, (*acad.lin*), the default program parameters file (*acad.pgp*), the default Custom Dictionary File (*user.cus*) or the default font mapping file (*acad.fmp*), then place your customized versions of these files in your Common Support folder. (This is the folder you created above: *I:\Support*.) Once you have copied the resources there, you may add a support path to the *OfficeStandard* profile.

> **Note:** It is not necessary to customize the support files now. If you intend to create custom versions of these files, simply copy the default ones to the Common Support folder now, and customize them later when you are ready. For purposes of creating and testing the *OfficeStandard* profile, it is necessary only to have the files located in the Common Support folder.

3. Switch to Windows; copy the support files you wish to customize to the Common Support folder (*I:\Support*). See Figure 1–18.

Leave ADT running with the Options dialog box open as you do this.

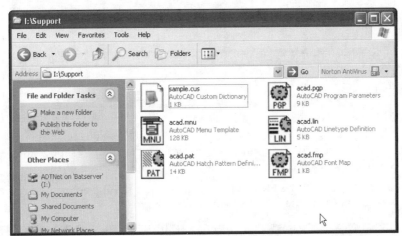

Figure 1–18 *Copy common resources to the Common Support folder*

4. Return to the ADT Options dialog box and click the **Add** button on the Files tab (see Figure 1–19).

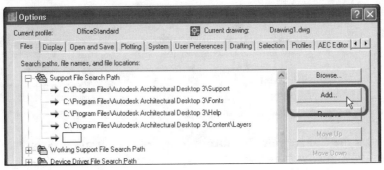

Figure 1–19 *Add a support path*

5. Click the **Browse** button, navigate to **I:\Support** and then click **OK** (see Figure 1–20).

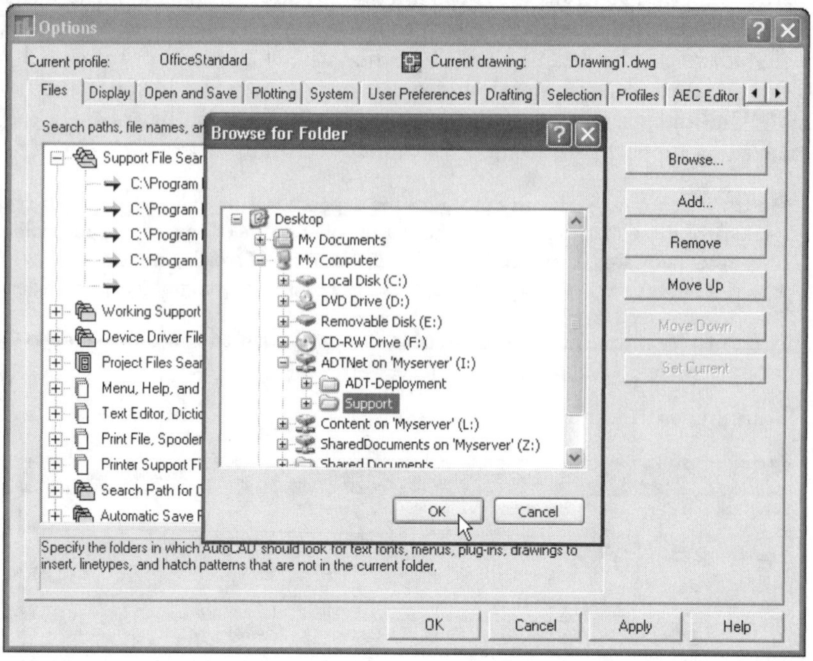

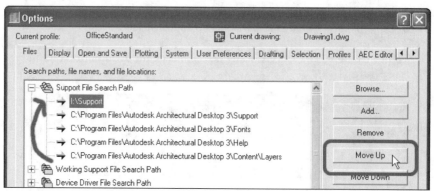

Figure 1–20 *Browse to I:\Support*

You have now added the new path, but in order for your customized support files to be read first by ADT when it launches, you need to move this path to the top of the list.

6. On the right, click the **Move Up** button until the **I:\Support** path moves all the way to the top of the list (see Figure 1–21).

Figure 1–21 *Move the new support path to the top*

7. Select the **C:\Program Files\Autodesk Architectural Desktop 3\Content\Layers** support path and then click **Browse**.

8. Browse to **L:\Layers** and then click **OK** (see Figure 1–22).

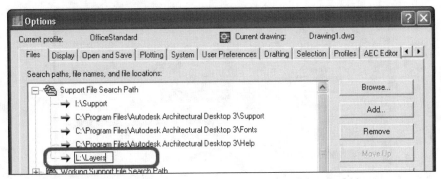

Figure 1–22 *Set the path for Layer Standards to the library server Layers folder*

 Note: This will use your network version of the Layer Standard file. We will discuss this file, its function and how to customize it in Chapter 4.

We are almost finished with the support search path settings; we still need to add one additional path. At the end of the "Set Up Network Resources" topic above, we created some new folders on the local hard drive. It will be very important to create the same folder on each user's hard drive as well. This is why we made the "C Drive Folders" resources on the I Drive. By copying its contents to the local C Drive of each user's machine, we can create the *Local Content* and *Temp* folders quickly and easily. The *Local Content* folder will be used by each user to save content and resources that users create for themselves. This local content will include resources that, for whatever reason, will not be made part of the office standard library. This can include the user's own personal PGP file, content that they are developing that has not been "approved" for the Office Standard library yet or shortcuts to the project folders of their projects. This is also the ideal place to store a user's custom toolbars.

9. Switch to Windows, and open the Local Content folder.

Leave ADT running with the Options dialog box open.

10. In this folder, right-click and choose **New>Text Document** from the context menu.

This will create a new Notepad document ready to be renamed.

11. Type **User.mnu** for the name and then press ENTER.
12. In the Windows warning message that appears about changing the file extension, simply click **Yes** (see Figure 1–23).

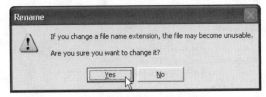

Figure 1–23 *Rename the file to an AutoCAD menu template*

13. Right-click the file (*User.mnu*) and choose **Open With>Notepad** (see Figure 1–24).

Figure 1–24 *Open the new file with Notepad*

14. Type ******Menugroup=UserToolbar*** and press ENTER (see Figure 1–25).

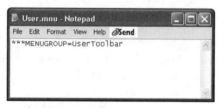

Figure 1–25 *Add a menu group header to the file*

15. **Save** and **Close** the file.
16. Return to the ADT Options dialog box and click the **Add** button.
17. Click the **Browse** button, browse to **C:\Local Content**, and click **OK**.

If you wish to allow your users to build their own custom PGP files, move the Local Content path statement to the top. This will allow them to create their own version of the PGP file and save it to the *C:\Local Content* folder. In this way, every user in the firm could have their own unique shortcuts without affecting other users.

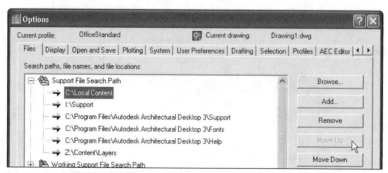

Figure 1–26 *Move the Local Content path to the top*

"LOCAL CONTENT" WON'T WORK ON OUR NETWORK...

In many network environments, the procedure above would be strictly prohibited. This is because users would become responsible for their own backup of personal resources, a user's local development would be consistently wiped out during a rebuild of a station for disaster recovery purposes and ghosting techniques used by many firms would make this process counterproductive. Reassignment of users to other seats would also potentially create problems on what users "leave behind" for the new person. However, you do not have to completely abandon the concept of Local Content if any of these concerns are issues for your firm. In many network environments, users are assigned their own "home" drive (usually something like "U Drive") on the network, to which only they have access. You can substitute U Drive for C Drive in the previous steps and achieve the same functionality. Just be certain that each person's "U Drive" is unique to his or her personal login, and it will work just fine. In fact, this will be far superior to the C Drive solution, as it will eliminate all issues related to backup, ghosting and disaster recovery.

 Tip: You can test what we have done so far by comparing the items listed under the Support File Search Path with those listed in the Working Support File Search Path. The Working Support File Search Path shows you the currently active paths. If you expand that item, you will likely notice that the two lists do not yet match. However, if you click **OK** to apply the changes and exit the Options dialog box, and then return to the dialog box, the two lists should now match. If they don't, you may have chosen an invalid path somewhere. Go back and review the steps above to try to rectify the situation.

Set Up the Remaining File Paths

There are several other path statements on the Files tab. Let's continue to configure these settings to strike an optimal balance between the resources we wish to manage centrally from the server and those that we will leave pointing to the local hard drive. As a general rule of thumb, do not change any paths that you don't intend to customize in any way. This will help keep your daily management tasks to a minimum.

1. If you have left the Options dialog box, return there now and click the Files tab.

The following summarizes the functions of the remaining paths on this tab:

- **Device Driver File Search Path** – This location contains the drivers for plotters and other hardware devices. If you intend to install and manage plotters globally for the entire office, you can safely ignore this path. (This is recommended.)

- **Project Files Search Path** – Here you can add a path for each project to help manage relative paths of External Reference files (XREFs.) More information on this topic can be found in Chapter 2. This setting will likely be different for every user. Do not edit it now.

- **Menu, Help and Miscellaneous File Names**
   a. **Menu File** – If you intend to build a custom *acad.mnu* file, then this is the place to reference it. Change this path to point to *I:\Support*. Do *not* change this path if you do not intend to customize *acad.mnu*.

b. **Help File** – This path points to the AutoCAD and ADT help files. There is no need to change this path.

c. **Default Internet Location** – This path is used by the built-in AutoCAD browser. Users will likely never use the built-in browser, so this path is *not* critical.

d. **Configuration File** – This file contains some configuration information for AutoCAD. It is recommended that you not change this path.

e. **License Server Path** – If you installed the network license manager, you may need to point to it here.

- **Text Editor, Dictionary and Font File Names**

  a. **Text Editor Application** – The default is the internal editor. It is possible to use another text editor such as Notepad or Microsoft Word. It will be assumed that the internal editor is being used for purposes of this book.

  b. **Main Dictionary** – This is the dictionary used by the spell checker. This path can be left alone.

  c. **Custom Dictionary File** – This is one of the file types mentioned in the previous topics. The file *sample.cus* is included as a sample. Copy this file to the *I:\Support* folder and edit it with Notepad. (You may change the name of the file if you wish as well.) The format of the file is self explanatory. Add all of the common abbreviations used in your firm (such as AFF, TYP, VIF). Change this path to point to this modified file. This will prevent those terms from being flagged as misspelled when a user performs a spell check.

  d. **Alternate Font File** – This entry contains a single font file listing. This font is automatically substituted for any font that is unavailable on a user's system when a drawing is loaded. The default is *simplex.shx*. You may change this default if you wish.

  e. **Font Mapping File** – This is one of the file types mentioned in the previous topics. The file *acad.fmp* is included as a sample. This file looks for instances of particular fonts when drawings are opened. If it locates one of those fonts, it will substitute it with the one listed in this file. If you wish to customize this file, follow the recommendations above to copy it to the server and edit it. Change this path to point to the edited file.

- **Print File, Spooler and Prolog Section Names** – There are three listing here that have to do with legacy plotting devices. All three values are empty by default. Change these values only if it is necessary for your particular hardware devices.

- **Printer Support File Path**

  a. **Print Spooler File Location** – This defaults the user's *Temp* folder (in Documents and Settings). We will set this to the *C:\Temp* folder created above in the "Set Up Network Resources" topic. (Do not use a network path for temporary files. This will cause a noticeable performance drain.)

  b. **Printer Configuration Search Path** – This is the location of all of the plotters and PC3 files. Change this path to point to the location where you intend to store office standard plotting resources, such as *L:\Plotters*. If you would like to give your users the ability to add their own custom plotters, you can place a shortcut within this folder to the "*C:\Local Content*" folder or a network project folder.

c. **Printer Description File Search Path** – This location contains the descriptions for plotters. If you intend to install and manage plotters globally for the entire office, you can safely ignore this path. (This is recommended.) If you changed the driver path above, change this path to the same location.

d. **Plot Style Table Search Path** – this is the location of all of the Plot Style Tables. Change this path to point to the location where you intend to store office standard Plot Style Table resources, such as *L:\Plot Styles*. If you would like to give your users the ability to add their own custom Plot Style Tables, you can place a shortcut within this folder to the "*C:\Local Content*" folder or a network project folder.

- **Search Path for ObjectARX Applications** – This path can be used to point to custom applications. It is empty by default.

- **Automatic Save File Location** – This is the location where AutoCAD will save the temporary SV$ (Autosave Backup Files). As we did above, set this to the *C:\Temp* folder. (Do not use a network path for temporary files. This will cause a noticeable performance drain.)

- **Data Sources Location** – Used for the DB Connect functionality. This is beyond the scope of this book.

- **Drawing Template File Location** – We will spend a great deal of time discussing drawing template files throughout this book. Template files will be saved on the server. In the section above, we copied the template folder to the root of the L Drive. Point this path to that folder: *L:\Template*.

Note: It is not important to have any template files in this folder at this time. We just need to have our OfficeStandard ADT profile mapped to this location. We will create the drawing template files later.

- **Log File Location** – If you wish to create a log file of the command line activity, place the path here. *C:\Temp* is a good choice if you choose to do this. The log file is a text file with an LOG extension.

- **Temporary Drawing File Location** – This is the place for AutoCAD to create its temporary files. As with the others, use the *C:\Temp*. (Do not use a network path for temporary files. This will cause a noticeable performance drain.)

- **Temporary External Reference File Location** – This location is used for temporary files created when XREF demand loading is enabled. More information on this topic can be found in Chapter 2. Use *C:\Temp* again. (Do not use a network path for temporary files. This will cause a noticeable performance drain.)

- **Texture Maps Search Path** – Used to store the texture maps for AutoCAD's built-in rendering. In the section above, we copied the *Textures* folder from your Management Install to the L Drive. Set this path to that location: *L:\Textures*.

You have now completed your customization of the Files tab.

2. Click the **Apply** button at the bottom of the dialog box to save these changes to the *OfficeStandard* profile.

Your Files tab should now look like the one pictured in Figure 1–27.

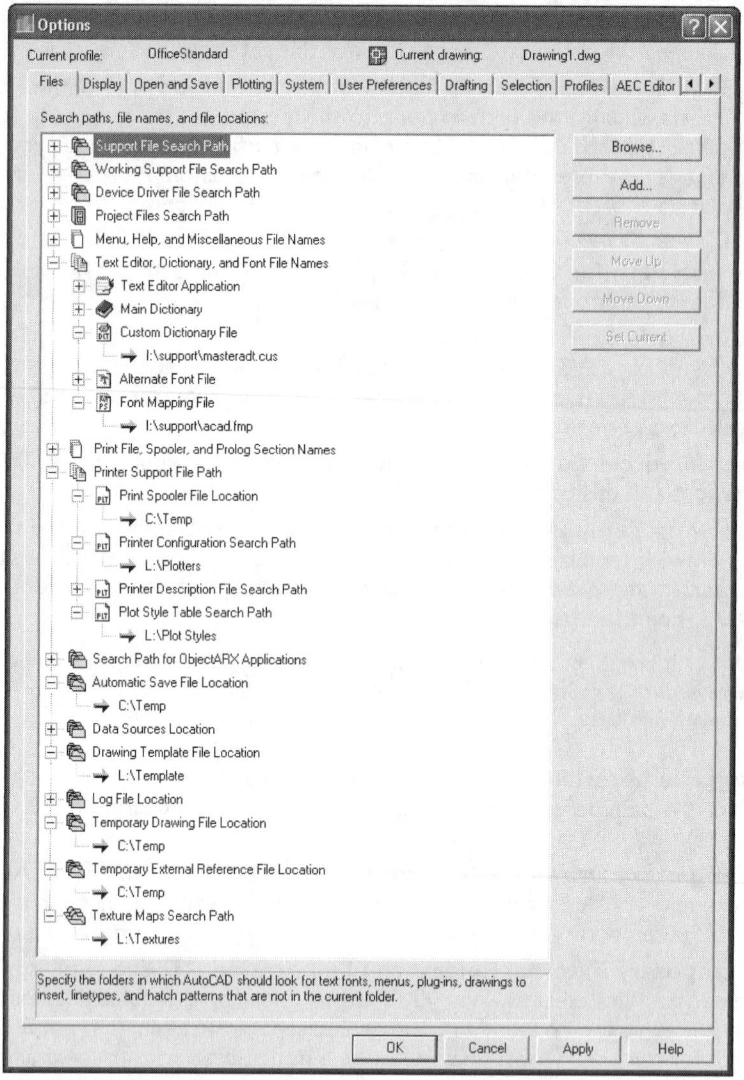

Figure 1–27 *Completing customization of file paths*

Set Up the AEC Content Path

The final path statement that we need to configure is located on the AEC Content tab. This single path controls where ADT will look for the AEC enabled content.

    1. Locate and click the AEC Content tab.

There can be only one path listed here; however, we can place as many sub-folders in this path as we wish. Therefore, we have complete control over the way we store all of the AEC Content.

Customization of AEC Content (sometimes also referred to as "ADT Content,") is a lengthy topic, which is covered in detail in Chapter 8. Therefore, for this exercise, we will remain focused on the configuration of our *OfficeStandard* profile and save specifics of customization for Chapter 8. In the "Set Up Network Resources" topic earlier, you copied the *Content* folder from your Management Install to the L Drive. We will now map a path to this network resource.

 2. Click the **Browse** button, navigate to **L:\Content** and then click **OK**.

 3. Choose either **Imperial** or **Metric** units for your Content menu default.

Regardless of your choice, users will be able to access both Imperial and Metric content. This setting simply establishes the default for the drop-down menu commands.

 4. Clear the **Display Edit Schedule Data Dialog During Tag Insertion** check box (see Figure 1–28).

When users use the ADT Schedule Tag routines, a dialog box will appear after each tag insertion with this box checked. This can become tedious and unproductive. It is recommended that this be turned off in your *OfficeStandard* profile. Users can always change this setting later if they wish. For more issues related to use of Schedules, refer to Chapter 6.

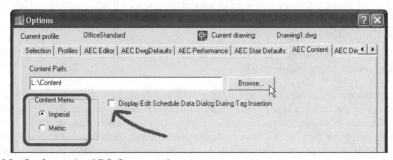

Figure 1–28 *Configure the AEC Content tab*

 5. Click **Apply** to save the AEC Content choices to your *OfficeStandard* profile.

Set Up System Variables

Before we finish our work in the Options dialog box, we need to configure some additional settings on the remaining tabs. Because we are working on our *OfficeStandard* profile, we are concerned only with global settings. Drawing-specific settings *cannot* be controlled with a profile. For this reason, do not concern yourself with any setting flagged by the blue AutoCAD drawing icon (see Figure 1–29).

Figure 1–29 *Drawing-level settings cannot be controlled by a profile*

1. Click the Display tab.
2. Clear the **Display scroll bars in drawing window** check box (see Figure 1–30).

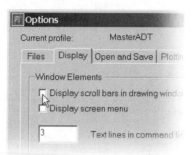

Figure 1–30 *Turn off the display of scroll bars in the drawing window*

 Note: The rationale behind this suggestion is that tools such as the use of a wheel mouse work much better than the scroll bars, and turning them off allows you to increase the size of the drawing area a bit. If you prefer to use scroll bars, ignore this suggestion and leave them on.

3. Click the **Open & Save** tab.
4. In the File Safety Precautions area, under **Automatic save**, change the **Minutes between saves** value to **20**.

 Caution: Automatic Save actually functions as an automatic backup. If the file had not been saved by the time entered in the **Minutes between saves** field, a backup file will be created in the temporary folder on your system, (configured in the Files tab, and which we set to *C:\Temp*). It is better to think of the Automatic Save feature as a Save Reminder. Should the AutoSave message appear at the command line, then perform an actual **Save** command immediately to avoid losing work. Also, note that by performing an actual save, ADT creates an additional backup file with a ".BAK" extension. This file will be saved to the same directory as the DWG file. If a DWG file should become corrupt, rename the BAK to a DWG extension and open this file. You should be able to recover anything up until your last save. Just remember to save often!

5. In the **File Open** area, change the **Number of recently used files to list** value to **9**.

This list appears on the File menu and the Today Window and provides quick access to recently used files. Nine is the maximum.

6. In the **Demand load XREFs** area, choose **Enabled with copy** (see Figure 1–31).

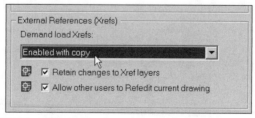

Figure 1–31 *Changing XREFs to Enabled with copy*

Note: This setting is very important to team environments. If **Disabled** is chosen, the Demand Load feature for XREFs will be turned off, and performance may suffer. If the **Enabled** setting is chosen, the user will enjoy better performance, but they will be locking every XREF file they load, thereby preventing any one else on the project team from opening those files. **Enabled with Copy** is a logical "compromise" choice, especially in team environments. Be sure to also consider the effect of Layer and Spatial indexing covered in Chapter 2.

SOMETIMES, "DISABLED" IS BETTER...

When **Enabled with Copy** is chosen, a copy of the file being XREFed is copied to the local machine's "Temp" directory and then opened. This simulates the effect that **Enabled** would have by opening the file, but it prevents the file locking associated with **Enabled**, by opening a copy of the file. (The copied file is always kept in sync with the original should a new version be saved.) When **Disabled** is chosen, the entire XREF file is loaded across the network into the RAM on the user's workstation. This will be true even if only a small portion of the XREF objects or Layers is needed.

Essentially, you are balancing network bandwidth against workstation RAM. If you find that your workstations have sufficient RAM and you do not intend to use XREF Clipping (XCLIP) very often, then consider **Disabled** instead. However, you will lose the potential speed gains afforded by using Layer and Spatial Indexes in conjunction with XREF demand loading and XCLIPing. In addition, all stations *must* to be set the same way, or you may get inconsistent results between users. No matter which setting you choose, you should avoid the AutoCAD default setting of **Enable**.

If you are uncertain which setting to use, it is recommended that you choose **Enable with Copy**. If you experience persistent cleanup problems in XREFs and Sheet files, include an OBJRELUPDATE/all statement in your *Acad.lsp* file.

The Plotting Tab

1. Click the Plotting Tab.

   Here you can establish a bit of control over the settings used when plotting. Bear in mind that all we are configuring here are overall defaults. Users will be able to change any or all of these settings as they see fit for a particular plot job.

2. In the **Default plot settings for new drawings** area, choose the device that you envision will be used most often by your users from the **Use as default output device** list.

Note: if you have not configured any plotters yet, you may wish to **Apply** your settings up to this point, exit the Options dialog box and add at least the plot device that you wish to use for a default choice. More information on plotting is available in Chapter 3.

The most important decision we need to make at this point is to decide whether you will use Color Dependent or Named plot styles. It is likely that there is some standard for this in place within your firm already. It is even more likely that this standard dictates the use of Color Dependent plot style tables. This is because Color Dependent plotting is most compatible with legacy plotting systems and outside consultants. However, since you are embarking upon the implementation of ADT and its new processes, this may be the ideal time to consider the switch to Named plot styles.

Important: There is no difference in functionality between the two types of plot style table. The difference lies solely in how they are applied to the objects within the drawing. ADT can be successfully implemented using either type.

If your firm is ready to break the non-intuitive link between color and lineweight from previous releases, you are ready for Named. However, if you envision more disruption than benefit from this switch, then stick with Color Dependent plotting. This topic is covered in much greater detail in Chapter 3.

If you are having a difficult time making this decision, do not despair—this particular setting in the Options dialog box, although a part of the profile, applies only to new drawings created with the Start From Scratch option. Therefore, even if you decide later to choose Named over Color Dependent or vice versa, you will be able to apply that new choice to the office standard template file. The setting in the template files is far more critical. Therefore, when the time comes to build your template, be sure you have read Chapter 3 and made your final decision.

3. Make your choice for **Default plot style behavior for new drawings**.
4. Click the **Add or Edit Plot Style Tables** button.

 Note: The Plot Style Manager should open to the folder *L:\Plot Styles*. If it does not, return to your Files tab and recheck your settings.

5. If you chose **use color dependent plot styles** in the previous step, right-click and **Copy** the file named **acad.ctb**. If you chose **use named plot styles** in the previous step, right-click and **Copy** the file named **acad.stb**.
6. Right-click again and choose **Paste**.
7. Rename the file just pasted to **Office Standard**.

 Caution: Be sure to leave the file extension (either CTB or STB) intact as you rename.

8. Close the Plot Style Manager window, and back in the Options dialog box, choose the new **Office Standard** file from the **Default plot style table** list.

Figure 1–32 shows *Office Standard.stb* being chosen as the default.

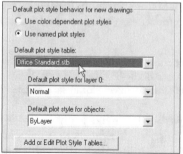

Figure 1–32 *Assign the Office Standard Plot Style table as default*

9. Click the **Apply** button.

The System Tab

1. Click the System Tab.

Most of the settings on the System tab are set optimally by default. Most of today's ADT users do not use a digitizer tablet; therefore, the default setting of **Current System Pointing Device** for the **Current Pointing Device** is appropriate. In the Layout Regen Options area, the default option of **Cache model tab and all layouts** will yield the greatest boost in performance when switching between layout tabs. It is recommended that this setting be maintained as is.

In the General Options area, next to the **Startup** option, you may choose to disable the appearance of the Autodesk Today on startup. The choice for this item is largely a matter of personal preference. However, do consider keeping the default setting of **Show TODAY startup dialog.** The Today window offers more functionality than the Traditional Startup Window, and certainly more than disabling it altogether. If you disabled Point A when building your Client Deployment, this is not an issue.

 Note: Regardless of your choice for startup option, the Today window is always available from the Tools menu.

2. In the Check Autodesk Point A for Live Enablers area, choose **Never**.

You have the option of having Live Enablers for product updates and Object Enablers to alert the user for installation. If you intend to administer patches and plugins globally, you will likely want to set this option to **Never**. This will prevent the alerts from coming directly to the user. However, you may want to leave it set to **When Autodesk Point A is available in Today** or **Always** on your Management Install. This way you will receive the alert on your station, and you can then download the update and install it on all machines in an appropriate time frame.

 Tip: If you are installing ADT for a small or one-person office, you may want to let Point A alert you for updates.

The final settings to configure on this tab will tweak the 3D performance of ADT.

3. Click the **Properties** button in the Current 3D Graphics Display area (see Figure 1–33).

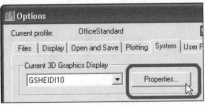

Figure 1–33 *Click the 3D Graphics Display Properties Button*

4. Place a check mark in the **Adaptive degradation** check box, and then place check marks in both **Flat shaded** and **Wireframe**.

These settings will allow the graphic display to degrade first to Flat shaded graphics and then to Wire frame as necessary to maintain 3D orbit performance. This is useful if you wish to preserve the interactive feedback and smooth continuous motion of the **3D Orbit** command over the shading mode. The shading will always return to the correct mode when the orbit is finished. If you use the materials and rendering functions built into AutoCAD, then perform the following steps:

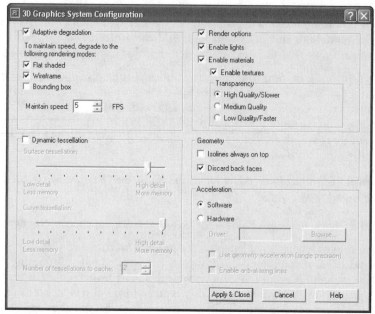

Figure 1–34 *Suggested settings for the 3D Graphics Display Properties*

5. Place a check mark in the **Render options** check box.
6. Place a check mark in any or all of the **Enable lights**, **Enable materials** and **Enable textures** check boxes.
7. Choose the level of **Transparency** quality you wish.

With the settings configured as shown in Figure 1–34, you will see lights, materials and textures in the highest quality directly in the AutoCAD viewport when using a shaded mode. This can be very helpful in assessing material and lighting conditions without the need for time-consuming trips to the Render dialog box.

 Tip: If you wish to use the **Mapping** command on ADT objects, you must have the Render display configuration active in your current viewport. For more information on the Display System, refer to Chapter 5.

8. Click the **Apply** button

The User Preferences Tab

1. Click the User Preferences tab.

Some veteran AutoCAD users new to ADT R3.3 (and AutoCAD 2002) are tempted to disable the **Shortcut menus in drawing area** option, or to revert the function of the right mouse button to the **Repeat Last Command** option. Neither of these choices is recommended. Autodesk Architectural Desktop relies heavily on the functionality of the right mouse button. It is highly recommended that you leave the default settings for shortcut menus and right-click customization intact.

NOTE TO VETERAN AUTOCAD USERS

You can still use SPACE to "Repeat Last Command" or "enter." Using SPACE with the ADT default mouse setting is *highly* recommended. This allows users to retain right-click button functionality and speed. If you are new to this use of the right click, you will only need to endure a day or two of discomfort before you begin to wonder how you ever lived without this critical functionality.

---

2. In the AutoCAD Design Center area, choose your units preferences.
3. At the bottom right portion of the window, click the **Lineweight Settings** button.
4. Choose your preferred **Unit for Listing** for Lineweights: **Inches** or **Millimeters**.

The setting you choose here will determine which measurement system is used to display the list of lineweights within the AutoCAD interface.

5. Set your choice for **Default** Lineweight.

This setting will be used to display and print all objects that do not have another lineweight assigned to them. When lineweights are displayed in model space, the scale is relative to other objects and set in screen pixels. You can adjust the overall scale of this display with the **Adjust Display Scale** slider. However, to get a true preview of lineweight scale, view them from paper space layouts.

6. Make the change, if you wish, to **Adjust Display Scale**, and then click **Apply & Close** (see Figure 1–35).

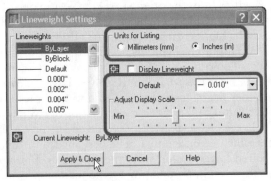

Figure 1–35 *The Lineweight Settings dialog box*

The **Adjust Display Scale** slider adjusts the relative thickness of the complete spectrum of lineweights when displayed on screen in model space. This setting has no effect on display in paper space or in plotting.

Drafting and Selection Tabs

1. Click the Drafting tab.
2. Choose your preferred settings.

If you do not have any personal preferences regarding these settings, accept the default choices.

3. Click the Selection tab.
4. Choose your preferred settings.

If you do not have any personal preferences regarding these settings, accept the default choices.

The AEC Editor Tab

1. Click the **AEC Editor** tab

All of the settings on this and any tab labeled "AEC" are unique to ADT. The AEC Editor and the AEC Content tabs are the only AEC tabs that contain profile settings. Earlier we already covered the AEC Content tab. All of the other AEC tabs contain drawing level settings and therefore have no impact on the *OfficeStandard* profile. In the **Diagnostic Messages** area, it is recommended that you not enable any of these messages for the *OfficeStandard* profile. For curiosity's sake, you may want to experiment with their being on in your Management Install to understand their function. When any of these messages is enabled, the command line will report a steady stream of messages indicating every action that is executed with each command. It tends to be much more information than the average user needs or wants to see. In the **Similar - Purpose Dialogs** area, both **Use Common Position** and **Use Common Viewer Position and Size** are turned on by default. It is recommended that you accept these defaults. This will have the effect of displaying all Add and Modify dialogs boxes in the same physical location on screen regardless of the command executed.

When a user explodes an ADT object, an anonymous AutoCAD block is created in its place. This block derives its properties from the original ADT object in one of two possible ways. By default, the **Block Properties of Exploded Objects** is set to **Maintain Resolved Layer, Color, Linetype**. This means that the block created from the ADT object will have the layer, color and linetype of the original ADT object. Typically, CAD standards dictate against this type of scenario in favor of having all objects set to ByLayer for color and linetype. If this is the case in your firm, you may wish to clear the **Maintain Resolved Layer, Color, Linetype** check box. There is, however, one other factor to consider. If the user further explodes the AutoCAD block into simple primitive AutoCAD entities, the resolution of layer, color and linetype propagates to this level as well. Therefore, several entities could end up on Layer 0 and set to ByBlock. This may not be the desirable outcome either. The best solution may be to leave the default setting of **Maintain Resolved Layer, Color, Linetype** turned on. This will correctly resolve the layers, which tends to be the most important property to maintain. You may then simply select all of the exploded entities and change their color, linetype and lineweight to ByLayer globally.

This is a very important topic when deciding how to share files with outside consultants. It is important to note that rules governing exploded ADT objects should be covered in the office CAD Standards or Guidelines documentation. Users may explode objects not understanding the repercussions of their actions. A clear statement and guideline regarding ADT objects should be communicated to the staff. Be sure to clearly state the firm's position regarding whether ADT objects should be exploded and, if so, the preferred procedure. For more information on the topic of exploding ADT objects, refer to Chapter 12.

2. In the **Block Properties of Exploded Objects** area, choose your preferred setting.

When the graphics used to display ADT objects becomes complex, object snap performance can suffer. To help alleviate some of this performance drain, you can turn on the **Optimize for Speed** check box in the **Object Snap** area. When this setting is turned on, only the Node and Insertion point object snaps will work on certain ADT objects like Stairs and Multi-View Blocks. Unless you anticipate frequent problems with performance in your firm related to this issue, it is recommended that you not activate this feature until and unless it becomes a problem for users (see Figure 1–36).

 Note: As models grow in size and complexity, multiple object snaps used in conjunction with the grid and snap functions can cause performance to suffer. If users begin to report "slow" network or drawing performance, it may be time to review some of these settings.

The final setting, **Optimize for Speed** in the **Layer Manager** area, refers to slowness that can occur in the ADT Layer Manager when a drawing layer list and the complexity of layer filters increase, or if the **Apply Changes Immediately** check box in the ADT Layer Manager is chosen. If you anticipate using the Layer Manager frequently, you may want to enable this setting.

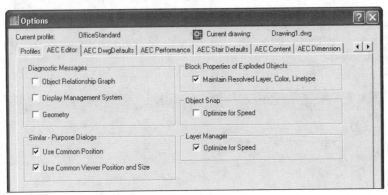

Figure 1–36 *Suggested settings for the AEC Editor tab*

3. Click **OK** to apply the settings and close the Options dialog box.

DRAWING SETUP

The Drawing Setup dialog box is used to configure some basic Architectural Desktop settings. There are four tabs in this dialog: Units, Scale, Layering and Display. Units, Scale and Display can be used by users to configure settings in the current drawing. Settings may also be pre-set here, which will be applied to drawings created with the Start From Scratch option. These topics, particularly Display, require more discussion than space in this chapter permits. Since most of these settings will be configured through drawing template files, we will have little concern for how these defaults are set. However, the setting on the Layering tab is important to our aim here.

1. From the Desktop menu, choose **Drawing Setup**.

2. Click the Layering tab.

It is important to note two options on this screen: **Layer Standards/Key File to Auto-Import** and **Save As Default**. A Layer Standards file contains a collection of layer preferences. In Chapter 4, we will discuss layer standards in detail. The ability to "Auto-Import" from one of these files means that layer standards can be imported into a drawing automatically. This can be very helpful in maintaining office standards. There are some issues related to achieving this "Auto-Import" successfully and reliably. We will save those for the discussion in Chapter 4. For now, we want to make sure that this path is properly pointing to our preferred network share and file.

CUSTOMIZING THE LAYERING STANDARD

ADT ships with a default Layer Standards file named *AecLayerStd.dwg*. This file is located within the *Layers* folder on the L Drive. It contains some industry-standard layering schemes such as AIA and BS1192. If you intend to edit this Layer Standards file (see Chapter 4), you may wish to open it now and add a Layer Key Style for your firm. You do not need to edit the style now; simply creating it will be enough. To do this, open the *AecLayerStd.dwg* file in ADT, and choose **Desktop>Layer Management>Layer Key Styles**. Create a new style and give it a name. **Save** and **Close** the file.

Click **OK** to close the Style Manager and apply the change (see Figure 1–37). If you wish, you may also rename the *AecLayerStd.dwg* file. The easiest way to do this is to simply **SaveAs**, and assign a new name. Choosing a name containing your firm's initials is a common strategy. (For example, ABC Architects, Inc. might use the name: *ABC_LayerStd.dwg*.) This method will help to identify files that you have customized. You do not need to perform these steps if you are planning to adopt an industry-standard layering scheme such as AIA.

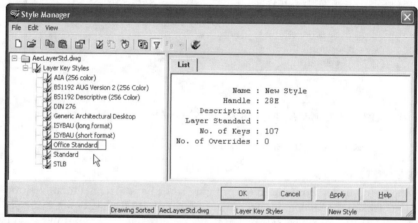

Figure 1–37 *Add an Office Standard layer key style*

3. Optional – If you are planning to edit the Layer Standards file, perform the steps in the "Customizing the Layering Standard" sidebar.

4. Click the small browse button (...) next to the **Layer Standards/Key File to Auto-Import** field.

5. Browse to **L:\Layers** and select **AecLayerStd.dwg** (or your customized version if you created one).

6. From the **Layer Key Style** list, make your choice.

Note: This will be your Office Standard layer key style. Choose either an industry standard such as AIA and BS1192, or a custom one if you created one.

7. At the bottom of the window, place a check mark in the **Save As Default** check box and then click **OK** (see Figure 1–38).

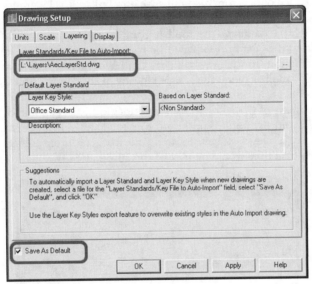

Figure 1–38 *The Drawing Setup Layers tab*

MENUS AND TOOLBARS

Our work in building the *OfficeStandard* profile is nearly complete. A profile contains all global system variables (that we already set in the Options dialog box), the list of drop-down menus and all of the active toolbars and their onscreen positions. Therefore, before we finalize work on our profile, let's make certain that all the preferred drop-down menus and toolbars are correctly installed and positioned.

First, if you intend to install any custom menus or plugin tools, this may be a good time to get them installed and the preferred paths added. Installation procedures vary considerably between developers. Therefore, consult the documentation provided with any third party software or plugin before installing it in your *OfficeStandard* configuration. For instance, if you are going to install the AutoCAD Express Tools, you will want to install them now, so that they become part of the *OfficeStandard* profile. This will make the rollout of these tools much easier. For information on acquiring the AutoCAD Express Tools or any of the Autodesk Extensions, contact your local Autodesk Reseller.

If you wish to activate or deactivate any of the toolbars loaded by default, do so now. In addition, in the "Support File Search Paths" topic above, we built a custom menu file named *User.mnu*. We will now load this menu file and build a custom toolbar for users to collect all of their favorite commands.

Build the MyFavorites Toolbar

1. From the Tools menu, choose **Customize>Menus**.
2. In the Menu Customization dialog box, type **User** and then click **Load** (see Figure 1–39).
3. Click **Close** to dismiss the Menu Customization dialog box.

Now that we have loaded our custom menu file, we now will add a custom toolbar to it. Toolbars in AutoCAD are saved within menu files. It is very important that we add our new toolbar to this custom menu file. This keeps the toolbar completely separate from the standard ADT toolbars, which allows the custom toolbar to be backed up and restored independently of the standard ADT installation. In this way, all users can have their own custom toolbar of favorite commands. Since this toolbar menu file is saved locally in the user's Local Content folder, each user will be responsible for backing up his or her own menu file. In fact, the user is responsible for the complete contents of their Local Content folder. This fact should be made very clear to all users, in order to avoid any unpleasant future consequences a misunderstanding on this topic could inevitably cause. (Unless you have located your Local Content folders in the user's home directories, in which case they would be part of the automatic office backup procedures.)

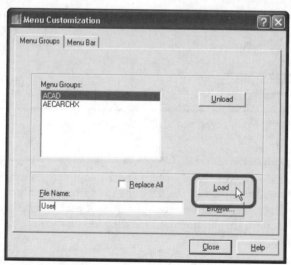

Figure 1–39 *Load the User menu file*

4. Right-click in the blank gray space next to any docked toolbar.

Figure 1–40 *Customize toolbars*

5. Choose **Customize** (see Figure 1–40).

The next step is the most critical part of this exercise:

6. From the **Menu Group** list, choose **User** and then click **New** (see Figure 1–41).

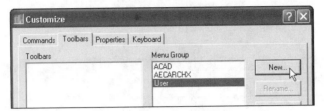

Figure 1–41 *Create a new toolbar in the User Menu Group*

7. In the **Toolbar Name** field, type **MyFavorites** and then click **OK** (see Figure 1–42).

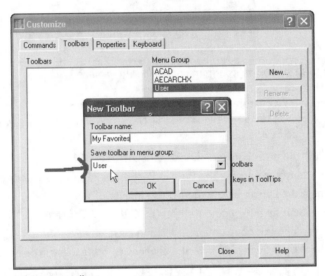

Figure 1–42 *Name the new toolbar*

Tip: Be certain that the **Menu Group** chosen in this step reads **User**; otherwise you will be adding your new toolbar to the wrong menu and users will not be able to customize it.

This toolbar can be used to store frequently used commands, or custom commands. You may want to add a few buttons to get the user started. For instance, the **Toggle Wall Graph Display** command is very helpful when working with Wall Cleanup. This command is located on the Wall Tools toolbar. Many of the other commands on this toolbar are not used very frequently. Therefore, it might be nice to copy just the **Toggle Wall Graph Display** command to our custom toolbar.

8. In the **Menu Group** list, click the **AECARCHX** Menu.
9. Scroll the toolbar list to the bottom and place a check mark in the **Wall Tools** entry.

10. With CTRL held down, drag the **Toggle Wall Graph Display** icon to the custom toolbar (See Figure 1–43).

11. Close the Wall Tools toolbar.

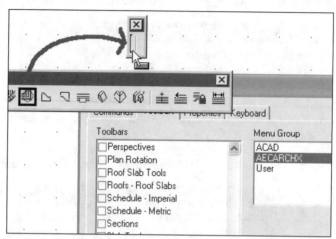

Figure 1–43 *Use* CTRL *to copy an icon from one toolbar to another*

When you build your own custom toolbars, the icons will turn to "smiley faces" if you do not save a bitmap of the button image in the same folder as the menu file. This will happen even to the button that we copied.

12. Right-click the new icon and choose **Customize**.

13. Click the button to access the Button Properties tab of the Customize dialog box.

14. Click the **Edit** button below the Button Image.

15. Click the **Saveas** button and navigate to *C:\Local Content*.

16. Input a name for the button image file and click **Save**.

17. Click the **Apply** button to save the changes to the toolbar icon.

Another very useful ADT command is **Object Relationship Update** (OBJRELUPDATE.) This command is like REGEN for ADT objects. It does not have an existing ADT icon. However, we can build own custom icon for this command.

18. Click the Commands tab in the Customize dialog box.

19. Scroll to the bottom of the list and select **User Defined**.

20. Drag the **User Defined Button** item in the **Commands** list to the custom toolbar (see Figure 1–44).

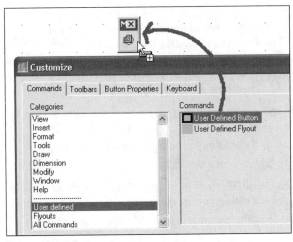

Figure 1-44 *Add a User Defined Button*

A blank button will be added to the MyFavorites toolbar.

    21. Right-Click the blank button and choose **Properties**.

    22. In the **Name** field, type **ADT Object Regen**.

The user sees this text in the tool tip for this button.

    23. In the Description field, type **ObjRelUpdate, all Objects**.

The user sees this text in the Status bar at the bottom of the ADT window.

    24. In the Macro field, type **^C^C_ObjRelUpdate;;** (see Figure 1-45).

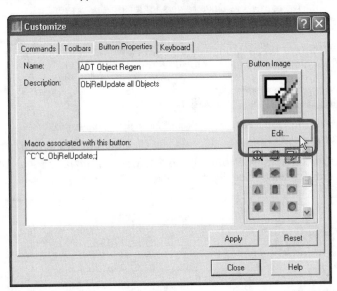

Figure 1-45 *Set up a custom button command string*

This is the actual command string. The ^C^C is used to cancel any active commands. The underscore ensures that the English language version of the command is used. If you don't work internationally, this is not required, but it won't hurt either. A semicolon is used to press ENTER. There are two in this case. When you type OBJRELUPDATE at the command line, the prompt requests a selection, or press ENTER to select all objects. The first semicolon selects all objects in response to this question. The second one executes the command.

25. If you wish you can choose an icon, or click the **Edit** button to paint your own (see Figure 1–45).

26. Don't forget to **Saveas** as in the previous example.

27. Click **Apply** to complete this icon.

28. Add any additional "starter" buttons you wish to the toolbar and then click **Close** in the Customize dialog box to return to the drawing window.

29. If you wish, dock the new toolbar to an edge of the screen.

Make the MyFavorites Toolbar Part of the Profile

1. Return to the Menu Customization dialog box.

2. Select **User** and then click **Unload**.

3. Click in the **File Name** field, type **User** and then click **Load**.

4. Click the **Close** button in the Menu Customization dialog box.

The reason for these steps is to ensure that the compiled version of the menu file (MNC) is loaded the next time we launch ADT and not the template menu (MNU). The compiled version actually contains the toolbar information; the template version does not.

5. In Windows, **Copy** the *Local Content* folder (and all of its contents) from the root of the C Drive to the *I:\C Drive Folders* folder on the server.

6. Manually create a new folder within *I:\C Drive Folders* called *Temp*.

By copying the contents of Local Content to the server, we can now easily copy it (and all of the MyFavorites toolbar menu code) to each local workstation.

Export the OfficeStandard Profile

1. Return to the Options dialog box and click the Profiles tab.

2. Be sure **OfficeStandard** is selected and click the **Export** button.

3. Type **OfficeStandardProfile** in the **File name** field, navigate to *I:\ADT-Deployment* and click **Save** (see Figure 1–46).

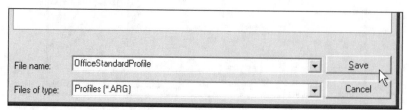

Figure 1–46 *Export a copy of the OfficeStandard profile*

When you export AutoCAD profiles, an ARG file is created. ARG stands for Autodesk Registry file. All of the information in the profile is actually saved in the Windows Registry. When we install a user workstation, we will use the Client Deployment that we created at the beginning of this chapter, and then apply the *OfficeStandard* profile to establish all of the preferred registry settings to the new installation. A few issues need to be exposed to make this procedure work seamlessly.

EXTENSIONS, PATCHES AND PLUGINS

If you are on the Architectural Desktop Subscription program, you will be periodically receiving new extensions to ADT. In addition, there are service packs and other patches released from time to time to address various issues. These items will need to be installed after you have installed the base software from the Client Deployment. The recommended order for a clean install is to install the Client Deployment first, patches and service packs next, followed by extensions and third party plugins, and finally any office customization, including the *OfficeStandard* profile.

 Note: Some extensions or third party plugins may perform better if installed prior to any service packs and patches. Be sure to read any documentation or readme files included with the software before installing it.

CREATE AN ADT DESKTOP ICON

The final step in achieving a complete "office standard" installation is to create a Desktop (and/or Start Menu) icon that launches ADT and the *OfficeStandard* profile. We will also make this icon load an office standard template file. However, since we have not yet built any template files yet, we will first create a "dummy template." We will point the icon to this template, and like all of the other resources, we can customize it later.

Throughout the course of this book, we work on the development of an Office Standard template file, each part with its own unique purpose. The only decision we need to make here is to choose a name for our standard template. For most users, it is likely plans, sections and elevations are the most common drawing types produced. Therefore, a standard model file template optimized for these drawing types is likely the most appropriate. Please keep in mind that users will always be able to choose any template they wish to begin drawings. Having ADT load a preferred template upon launch simply saves a step in many cases. It also serves to force ADT-specific functions to demand load automatically upon launch. This is often more desirable than having to wait for ADT functions to demand load the first time an ADT-specific command is executed. If you do not see a benefit to having ADT automatically load an office standard template file, you may omit this step. To create a custom ADT icon, do the following:

Build a Dummy Template

1. In Windows, navigate to the template folder that we copied earlier to the server. It should be in *L:\Template*.
2. From the *AutoCAD Templates* sub-folder, copy the file named *ACAD -Named Plot Styles.dwt* (*acad.dwt* if you decided to use Color Dependent plotting above) to the root of the template folder (*L:\Template*).

3. Rename the *ACAD -Named Plot Styles.dwt* (or *acad.dwt*) file you just copied to **Plan-Sect-Elev.dwt**.

 Note: Include the name or initials of your firm if you wish. You can also use a different name if you wish.

4. Close the template folder window.

Create a Desktop Shortcut

1. Minimize all running applications and right-click the Autodesk Architectural Desktop 3.3 icon.
2. From the right-click menu, choose **Copy** (see Figure 1–47).

Figure 1–47 *Copy the default ADT shortcut*

3. Elsewhere on the Desktop, right-click and choose **Paste**.

This will create a copy of the icon named "Copy of Autodesk Architectural Desktop 3.3."

4. Right-click this new icon and choose **Properties**.
5. Click the General tab and type the name you would like to appear for this short-cut, such as ABC Architects ADT (see Figure 1–48).

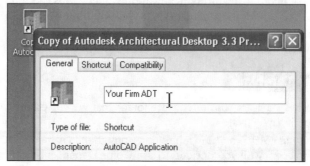

Figure 1–48 *Change the name of the icon on the General tab*

6. Click the Shortcut tab.

The **Target** field contains the instructions needed to launch ADT. There is the path to the ACAD executable and two command line switches.

 Note: The executable for ADT is actually *acad.exe*, (AutoCAD). All of the ADT functionality loads after the base AutoCAD package loads.

By default, the following text occupies the **Target** field:

```
"C:\Program Files\Autodesk Architectural Desktop 3\acad.exe" /t "Aec Arch
    (Imperial - Intl).dwt" /p "C:\Program Files\Autodesk Architectural Desktop
    3\adt.arg"
```

 Note: If you chose a different default location for ADT installing, the **Target** field will show that location instead.

We need to edit the information contained in the two command line switches: /t and /p.

7. In the **Target** field, immediately following the **/t** switch, replace the existing text, "Aec Arch (Imperial - Intl).dwt" with the following: **"Plan-Sect-Elev.dwt"**

It is important not to disturb the any of the other text already in the **Target** field. By adding the text above, we are instructing ADT to load the template file we just created upon launch. It is very important to type in this instruction exactly as it appears. *Do* include the forward slash (/), the SPACE and the quotes (""). See Figure 1–49.

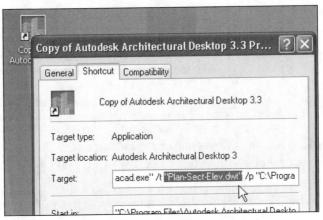

Figure 1–49 *Reference the office standard template file*

The remaining switch (/p) loads a profile. We will edit this to load the *OfficeStandard* profile that we just created.

8. Immediately following the switch above, replace all text following **/p** with the following: **"I:\ADT-Deployment\OfficeStandardProfile.arg"**

As before, it is important to preserve the /p, the **space** and the **quotes**. See Figure 1–50.

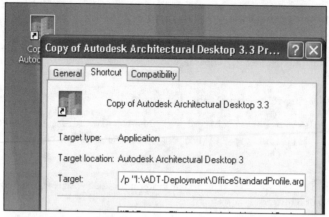

Figure 1–50 *Input the path to the OfficeStandard profile*

 Note: This is the path and profile saved in the "Export the *OfficeStandard* Profile" sequence above. If you used a different name and/or location, then substitute the actual path and file name you chose.

 Tip: There are additional command line switches that can be used in command string of your ADT icon. For more information, search for "command line switches" in the online help and then view the topic named "Customize Startup" for a complete list.

The final command string in the **Target** field should look like this:

```
"C:\Program Files\Autodesk Architectural Desktop 3\acad.exe" /t "Plan-Sect-
    Elev.dwt" /p "I:\ADT-Deployment\OfficeStandardProfile.arg"
```

 Caution: Be sure that your Management Install and your Client Deployment are installed to the same drive letters and directories, or the preceding exercise will not work.

9. Click **OK** to save the changes and exit the dialog box.
10. Double-click the new icon to test it out (see Figure 1–51).

Your screen should look very similar to Figure 1–51. Check the My Favorites Toolbar, the Today Window and the Template loaded to be sure everything is working properly.

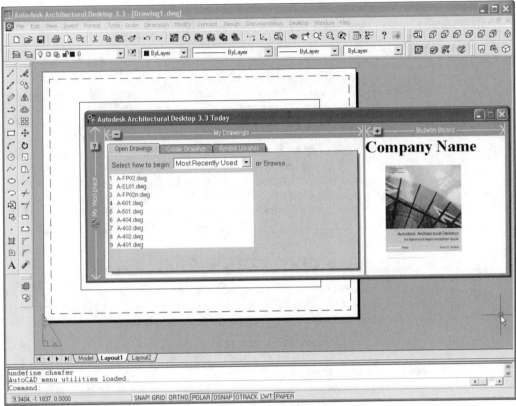

Figure 1–51 *The ADT screen using the new custom configuration.*

11. **Copy** this shortcut icon to the root of the I Drive (see Figure 1–52).

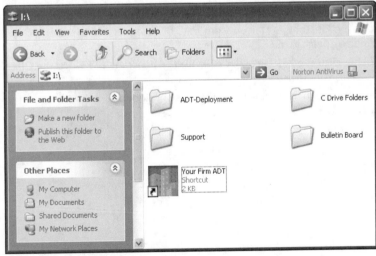

Figure 1–52 *Copy the custom icon to the I Drive*

ARGGGG... WHY WON'T THE ARG WORK?

The letters "ARG" stand for "Autodesk Registry" file. This is nothing more than a text file containing instructions for ADT, which are saved to the Windows Registry. The very first time a user double-clicks the icon we have created here, the /p switch will automatically import the *OfficeStandard* profile that is saved in the *OfficeStandardProfile.arg* file. However, this is a "one-time" operation. You cannot "re-import" an ARG file if a profile with the same name already exists on the system. Therefore, you must either delete the existing one or overwrite it. You can delete a profile in the Options dialog box, but you cannot delete the current profile, so you will need to make another one active first. An alternative is to overwrite the existing profile from outside AutoCAD/ADT. To do this, rename the ARG file to a REG file. When Windows warns you about changing the file extension, click **OK**. Simply double-click the newly renamed REG file to add the information within to the Registry and overwrite the existing profile, if it already exists. If it does not exist, it will be created.

ROLL OUT THE OFFICE STANDARD INSTALLATION

All of the network and installation resources that are necessary to roll out your customized installation of Autodesk Architectural Desktop R3.3 are now in place. You will still need to visit each user's workstation to complete the installation, but the steps you must perform are minimal.

1. From the client (User) workstation, map each of the network drives: I, L and P. (A simple batch file could be used to accomplish this; discuss the specifics with your IT or Network support personnel.)

2. Navigate to the root of the I Drive and copy the items within the *C Drive Folders* folder to the root of the local user's C Drive.

3. Navigate to the *I:\ADT-Deployment* folder and run *Setup.exe* to install the base ADT install on the user's workstation. (This is the "silent" mode executable created at the start of this chapter.)

4. Install any Patches, Service Packs, Plugins, and Extensions.

5. Copy the Custom ADT icon from *I:\ADT-Deployment* to the user's Desktop and/or Windows Start menu.

6. Double-click the ADT icon to execute the program and write the *OfficeStandard* profile to the Registry.

7. Test out the path listings and the custom icons to be sure everything is working properly.

SUMMARY

Create mapped network drives for installation resources, content library and project files.

ADT can be installed completely Stand-alone, Stand-alone with Network Licensing, or Run from the Network.

To simplify installation on multiple computers, or to back up a customized installation, use the NetSetup application and build a Client Deployment.

Save all your global AutoCAD/ADT preferences in an OfficeStandard profile. This profile can be loaded onto other machines to simplify installation and setup.

Create a custom toolbar to store favorite commands. Save the code in a local menu file in the Local Content folder.

Place all customized resources on the server and map paths to them in the Options dialog box.

Make a custom desktop icon to launch ADT with your customized office standards.

Systematically install all client workstations.

CHAPTER 2

Project and Building Model Structure

INTRODUCTION

This is a "big picture" chapter. Our goal here is to assess the type of work most prevalent in your firm's practice and identify any special setup needs that those project types will have. For instance, a single-family residence will likely be set up differently from a large new hospital building. Although there will be many similarities in all project types, there will be some issues that will pertain only to a particular class of building or project size. The size of the project team also plays into the overall setup and procedure of the project. We will address those issues as well.

This chapter addresses many issues critical to the ADT Building Model concept. Although it is long, all of the information contained herein is important to your firm's implementation process. References will be made throughout to compliance with industry standards such as the U.S. National CAD Standards. Depending on which country you practice in, you may prefer to reference an ISO, BS or DIN standard, such as BS 1192 or DIN ISO 13567, instead. This manual is not intended to supplant any of these industry standard publications, and in fact, you are encouraged to procure and consult the particular standard having jurisdiction in your area.

OBJECTIVES

In this chapter, we will address the ADT Building Model concept and explore some common project types and their specific needs and requirements. In this chapter the following will be covered:

- Become familiar with the ADT Building Model philosophy
- Evaluate the Building Model tools and strategies
- Work with Blocks and XREFs
- Assess your firm's most prevalent project types
- Address issues specific to each of several common project types

WHO SHOULD READ THIS CHAPTER?

This chapter is intended for the CAD Manager, the Project Data Coordinator and the Project Manager. The CAD Software User should also be engaged in the discussions that this chapter would prompt. Therefore, this chapter is important to that group as well.

PARAMETRIC DESIGN

Creating models in Architectural Desktop is an exercise in practicing parametric design. *Parametric design* refers to the direct manipulation of various object parameters available within each of the objects as design decisions are being formulated. Objects in ADT are programmed to represent the real-life objects for which they are named. All real-life objects have a series of defining characteristics that determine their shape, size and behavior.

To get a better sense of what these parameters might be, think of the characteristic to which you would refer if describing the object verbally to a colleague without the benefit of a drawing. Consider, for instance, a door. If discussing a particular door needed in a project with the contractor over the telephone, we would rely on descriptive adjectives and verbal dimensions such as "the door is a particular width, and a particular height, it is solid core and has a hollow metal frame." Once we had settled on the door required and hung up the phone, we would then need to convey graphically in our drawing documents the decisions we had just made regarding that door (and any other like it). In traditional CAD (AutoCAD), this would mean translating dimensions and materials into corresponding lines, arcs and circles that represent the required dimensions and materials in the drawing. In object-oriented CAD (ADT), dimensions and other specifications are simply input into a series of fields on forms that store the data for the object. This data remains accessible throughout the life of the object and the drawing. Therefore, the next time we phone the contractor and realize that circumstances on the site have forced us to spec a different door and size, rather than redraw the door and manually adjust the wall in which it sits, as we would in traditional CAD, we now simply re-open the data form for that door and input the new values. Not only does the door itself update because of this change, but the wall in which it is inserted updates as well. This is just one example of parametric design. Object parameters are always available for editing; data never needs to be recreated, only manipulated. Some principles of parametric design are as follows:

- **Draw Once** – In traditional CAD, each object needs to be drawn for each view; therefore the same door or wall may need to be drawn two, three or more times. With ADT, objects need only be drawn once. They are then "represented" in each of the required views of Plan, Section and Elevation.

- **Progressive Refinement** – Complete final design information is rarely known at the early stages of a design project. Changes occur frequently and often several times. In traditional CAD, it is easy to add new information. However, when major design changes occur, drawings must often undergo time-consuming re-drafting. With ADT, designs can be progressively refined over the life of the project. As new data is learned or design changes occur, ADT object parameters may be adjusted appropriately without the need to erase and re-create the drawing. The objects are drawn once, and then modified as required.

- **Style Based versus Object Based** – Most ADT objects make use of styles. A style is a collection of object parameters saved in a named group. When styles are assigned to objects, all properties of the style are transferred to the object in one step. If the style parameters change later, all of the objects using the style will change as well. This is the same behavior exhibited by Text and Dimension styles in traditional AutoCAD. ADT simply utilizes many more styles. Some object parameters however, are assigned directly to the individual objects and not controlled by the style. Consider again the Door object as an example. The Door style would be used to designate the type of door, such as a hinged double door. However, double-hinged doors can come in a variety of sizes; the door type double hinged is a style-based parameter, while the size is an object-based parameter.

- **Live versus Linked** – Some drawing types in ADT are edited directly on the "live" model data. This is the case with floor plans or live sections. The Display System (see below) controls what displays on the screen as we are working in ADT. Plans are live. If changes are made to the objects within the plan, those changes will be seen simultaneously on the live model in other views. However, some drawing types, namely sections, elevations, and schedules, are "linked." Rather than being a live view of the model, these separate drawings function as "graphical reports" of the model that remain linked and need to be periodically updated.

UNDERSTANDING ADT OBJECTS

The creation of a floor plan in traditional AutoCAD involves a process of drafting a series of lines and curves parallel to one another to represent walls and other elements in the architectural plan. This process is often time consuming and labor intensive. When design changes occur, the lines must be edited individually to accommodate the change. Furthermore, a plan created this way is two-dimensional only. When elevations and sections are needed, they must be created from scratch with additional lines and circles, which maintain no link to the lines and circles that comprise the original plan.

In contrast, Architectural Desktop includes true architectural objects designed to behave as the specific "real world" objects for which they are named. These objects coexist in AutoCAD alongside traditional lines, arcs and circles. Rather than draft lines as in the example above, ADT includes a true Wall object. This object has all of the parameters of a real wall built into it. Therefore, one need only assign the values to these parameters to add or modify the wall within the drawing. In addition, the Wall object can be represented two-dimensionally or three-dimensionally, in plan or in section, all with a single drawing element. This means that unlike traditional drafting, which requires the wall to be drawn several times, once for plan, once for section, and again for elevation, an ADT Wall need only be drawn once, and then "represented" differently to achieve each type of drawing: plan, section and elevation. An even greater advantage of the ADT object is that if it is edited, it changes in all views. This is the advantage of its being a single object, and it provides a great productivity boon. With lines, each view remains a separate drawing; therefore, edits need to be repeated for each drawing type—a definite productivity drain. The intelligence of the object extends even further. ADT objects may have data attached to them, which can be linked directly to schedules and reports. Objects also adhere to rules built into them that control their behavior under various circumstances. Doors know they should cut holes in Walls. When the Wall is moved, the Doors associated with that Wall move with it as well. Spaces (rooms) know to grow and shrink when their controlling edges are reshaped. Columns know to move when the column grid line they are attached to moves. Stairs remain constrained to restrictions placed on them by building codes, and their Railings link automatically to these constraints as well. These and many other relationships are programmed into the software and are directly controlled by the user.

ADT OBJECT COMPOSITION

Let's begin a detailed exploration of ADT objects with a look at some terminology:

▶ **ADT Object** – An individual drawing entity, designed to represent a particular object or fullfill a specifc function, created and modified based on unique built-in parameters. These parameters can be physical, graphical or data properties.

▶ **Parametric** – Based on parameters. Parameters are simply the set of physical properties and defining characteristics of objects. A parametric or physical property will *also* be expressed graphically and in property data.

▶ **Display (Graphic) Property** – Affects only the display of an object, not its physical characteristics. It is a property that is visible *only* from a specific view of the object, such as Plan or Elevation.

▶ **Data Property** – A piece of information added to an object for reporting, computational or analytical purposes. Data properties are not required to relate to an object's physical or graphical characteristics.

▶ **Display Representation** – A set of Display (graphic) properties comprising a graphic representation of a particular object as it relates to a specific view or a particular type of drawing, such as Plan or 3D Model.

▶ **Property Set** – A group of related Data properties gathered together for a common purpose. Property Sets are the foundation of Schedule Tables.

In addition to the preceding definitions, which correspond to all ADT objects, the complete collection of ADT objects can be broken down into four classifications: **Architectural, Helper, Annotation** and **Multi-Purpose**. Figure 2–1 shows how the complete list of hard-coded ADT objects can be classified in order to understand their respective use, function and inter-relationships. These classifications will be used throughout this book.

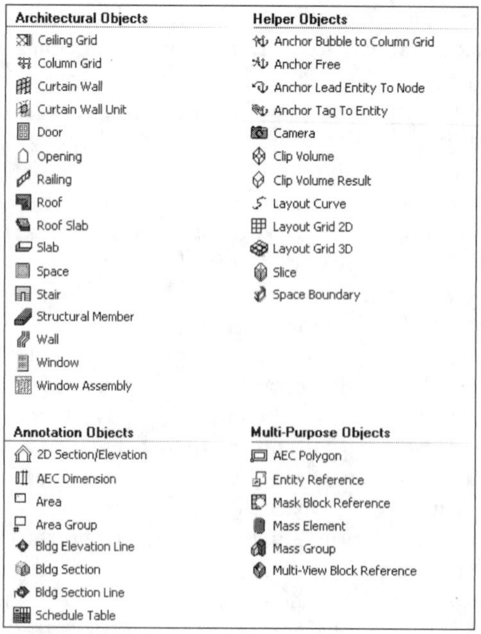

Architectural Objects	Helper Objects
Ceiling Grid	Anchor Bubble to Column Grid
Column Grid	Anchor Free
Curtain Wall	Anchor Lead Entity To Node
Curtain Wall Unit	Anchor Tag To Entity
Door	Camera
Opening	Clip Volume
Railing	Clip Volume Result
Roof	Layout Curve
Roof Slab	Layout Grid 2D
Slab	Layout Grid 3D
Space	Slice
Stair	Space Boundary
Structural Member	
Wall	
Window	
Window Assembly	
Annotation Objects	**Multi-Purpose Objects**
2D Section/Elevation	AEC Polygon
AEC Dimension	Entity Reference
Area	Mask Block Reference
Area Group	Mass Element
Bldg Elevation Line	Mass Group
Bldg Section	Multi-View Block Reference
Bldg Section Line	
Schedule Table	

Figure 2–1 *ADT object classifications*

As indicated above, ADT objects are intelligent entities designed to behave as the specific "real world" objects for which they are named. This is actually much better characterized as the description of the Architectural objects as shown in Figure 2–1. Therefore, you would expect the Architectural objects shown on that list, such as a Wall, to behave like and consequently to be used to represent a real life object, in this case a wall.

The Helper objects are often generated as an intermediary step in some larger building model process. Such is the case with Slice and Space Boundary objects. Helper objects also provide useful diagnostic information ("Anchor Tag to Entity" and "Anchor Free") on other objects or provide a way to maintain a particular relationship between two or more objects. This is the purpose of the Anchor and Layout objects.

Annotation objects provide a variety of notation functions within ADT. Most of these object types have obvious functions. Schedules and AEC Dimensions are examples of this.

Multi-Purpose objects do not fit neatly into any of the other categories. This is because objects of this type can be used for a variety of purposes. For instance, a Mass Element can be used as an Architectural object if it is used to model some item that is not easily represented with the other Architectural objects. It can also be a Helper object when used to modify some other object, as is the case when being used as an interference condition.

Each of the objects, regardless of classification, has physical, graphical and data parameters. The most obvious parameters are the physical dimensions Length, Width and Height. Depending on the specific type of object, there are dozens of additional parameters controlling all manner of object behavior. For instance, Swing Direction is a parameter of Door objects that would be meaningless for Wall objects. However, both Doors and Windows have the frame dimension parameters of frame length and depth. Parametric properties affect the object in all views. If you change the width of the Wall when viewing it in plan, you would expect that the change would also be evident in elevation and 3D model. This is in fact the case. Parametric properties are edited from either the Modify or Properties dialog box for an object (see Figure 2–2).

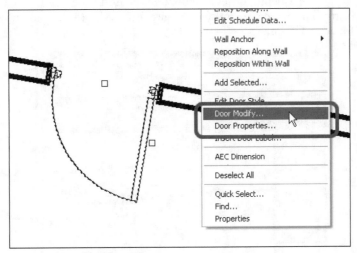

Figure 2–2 *Accessing parametric properties*

TO BE OR NOT TO BE PARAMETRIC...

As you begin to learn which features are parametric and which are not, you may be left wondering, "Why did the programmers do it like that?" Two factors will dictate the level of parametric control available to us when using ADT. The first consideration is clearly the requirements of the object. A Door requires a frame, and therefore one would expect to be able to control the basic dimensions of the frame parametrically. However, Doors may also have louvers, hardware and lites. Each of these items can easily be represented in ADT; however, the procedure involves a few more steps than those required to change the parametric properties.

These items raise a second issue: object maturity and/or robustness. A number of factors will influence the software developers when deciding the level of features to develop for a particular object. These will include the average frequency with which a particular property needs to be changed, system performance, the impact that change may potentially have on other objects and the overall priority of the feature relative to any additional functionality gain within the big picture scenario.

All parametric properties also affect object display, but rarely do Display properties influence the object's physical parameters. Window muntins can be added to a Window object; however, they cannot be added parametrically. Muntins can only be added through Display properties; therefore, they must be added separately to each view in which we wish them to display. Furthermore, if we desired to count these muntins or report their size on a schedule, we would also need to add a separate manually input data property as well.

In another example, hatching added to a wythe of brick in plan view of a Wall is only visible in plan. The elevation and model will not have the hatch pattern nor will the schedule report it as brick. Custom hardware can be added to a Door object, but this is applied as a Display property directly to the view where it would be seen, such as elevation and model. These are all examples of Display properties. Display properties are edited in the Display Props tab of an object, available from the Entity Display command (see Figure 2–3). More information on the Display System can be found in Chapter 5.

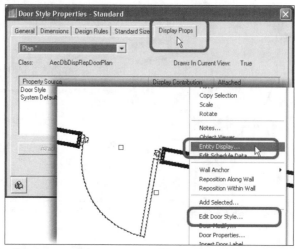

Figure 2–3 *Accessing Display properties*

Data Properties are assigned to objects through Schedule Data and Property Sets (see Chapter 6). This type of information is used to add data to a model that is then reported directly to a Schedule object or to an external database. Data properties can be "automatic," where they link directly to a specific parametric property. They can also be "non-automatic" in which case the data will require manual input. Data properties are edited from the Property Sets button on the General Tab of object Properties dialog boxes (see Figure 2–4), or by using the SCHEDULEDATAEXT command.

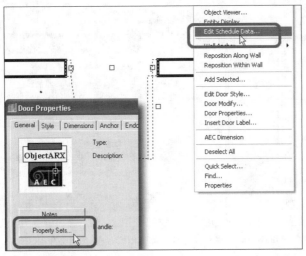

Figure 2–4 *Accessing property data*

IS IT A PARAMETRIC, DISPLAY OR DATA PROPERTY?

It may be all three. Sometimes it is difficult to determine exactly which properties are global (parametric) and which affect only a certain view (Display) or simply the schedule data (property data).

Take the example of Door louvers. This is an item that would be nice as a parametric property, since it is a common physical characteristic of doors and it needs to be shown in elevation and model as well as on the schedule. However, for whatever reason, louvers are not *fully* parametric. Therefore, we must construct them.

To do this, we begin in the Door Style properties. Here we can assign a profile to shape the Door and give it a "hole" for the louver to occupy. This is a parametric property, as it is assigned directly to the Object or Style properties and it will be seen in all views. When you add a profile to a Door in this way, the "hole" defaults to glass within the Door. To change this, we need to change the Display properties of the Door to show a louver in place of the glazing. This must be done for each view where we would like to see the louver using a custom display Block.

Finally, simply displaying the louver (regardless of the number of views) does not make it appear in the Door Schedule. Therefore, our final step is to add property data to the Door that indicates that the louver is there in the Schedule (see Figure 2–5). Refer to Chapter 9 for a detailed tutorial for creating this type of Door style. Please note that this type of style can be built once, and reused repeatedly from then on without the need to repeat these steps.

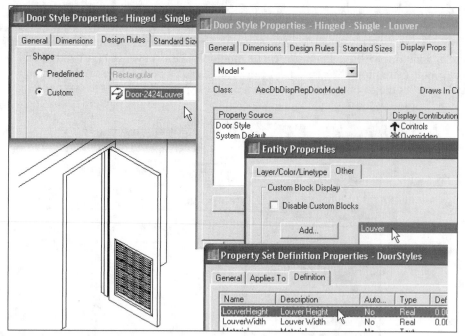

Figure 2–5 *Understanding parametrics, Display properties and property data*

UNDERSTANDING THE TERM "MODEL"

We see many uses of the term "Model" throughout both ADT and traditional AutoCAD. Let's assign definitions to the three most common uses of this term.

▶ **Building Object Model** – Also Building Model; a virtual model of a complete building composed of a collection of ADT objects and XREF files. This is very close to being the digital equivalent of the traditional foam-core presentation model.

▶ **Model File** – An AutoCAD file used to contain objects (ADT or AutoCAD) that represent some actual physical piece of a Building Model. Model files are created at full scale (life size) and may contain either two-dimensional or three-dimensional data or both.

▶ **Model Space** – The limitless and scaleless three-dimensional "working" space in the AutoCAD/ADT environment intended for developing building model components. Model space is three-dimensional space, but it may contain either two-dimensional or three-dimensional model data or both.

In general, when the term *model* is used, it refers to something real and full-scale. Model space is our full-scale working environment in which we build full-scale models of "real" things. Those real things may be buildings or portions of buildings, or even components placed within or around buildings. The *content* of the model does not distinguish as much as "*real*" and "*full-scale*" do. A Building Object Model is comprised of one or more model files. How many model files there are and exactly what they contain is dependent on a variety of project specific factors. These will be discussed below in detail.

SHEET TERMINOLOGY

Anything added to your ADT drawing files strictly for the purposes of enhancing plotted output can be thought of as a sheet component. Items such as Title Blocks, General Notes and Legends belong to this category. Consider the following terms:

- ▶ **Sheet File** – An AutoCAD file configured especially for the purpose of printing an individual sheet from the document set. Sheet files are sometimes refered to as document files or paper files. These files should contain those XREFs and other items necessary to guarantee reliable and consistent plots.

- ▶ **Annotation** – The category of all scale dependent text-based data, dimensions, schedules and symbology used on drawings to clarify and convey design intent.

OBJECT DIMENSIONALITY

The dimensionality of an object is an important consideration in its use and understanding. Objects in ADT may be two-dimensional, three-dimensional or hybrid. A 2D object remains "flat" even in a 3D view. A 3D object remains 3D even when viewed in plan. A hybrid object dynamically "switches" from a 2D Display Representation to a 3D one as the viewpoint in the drawing changes. The hybrid object is by far the most powerful and versatile of all ADT objects. This is because hybrid objects offer the best of both 2D and 3D objects and drafting/modeling techniques while being more powerful and useful than either type alone. (All standard AutoCAD objects are either 2D or 3D; AutoCAD does not have native hybrid objects.)

Not all ADT objects are hybrid. Walls, Doors, Windows, Stairs, Railings, Curtain Walls and many other ADT objects are examples of hybrid objects. All of these objects, and many more, offer both 2D Display Reps (such as Plan) and 3D Display Reps (such as Model.) In this way, it is possible to work exclusively in 2D while in plan view without worrying about "messing up" something in the 3D view. The same is true of working in Model. This point cannot be stressed enough.

 GUIDING PRINCIPLE: Autodesk Architectural Desktop is a hybrid two-dimensional and three-dimensional software package. It is neither a 2D drafting tool nor a 3D modeling engine. It has the ability to do both simultaneously!

Please do not forget that statement as you work through the remainder of this book. Although not every object in ADT is a hybrid object, most of them are. Therefore, we should structure our projects, drawing files and external references to take the fullest advantage of this hybrid structure wherever possible (see Figures 2–6 and 2–7). This single factor will drive many of the recommendations made throughout this book. The hybrid nature of ADT is one of the software's biggest strengths. This is especially important in architectural design and documentation. The architect is frequently required to "think" in 3D to understand, solve and convey architectural ideas and construction, while simultaneously being required to document those conditions and solutions in two-dimensional documents.

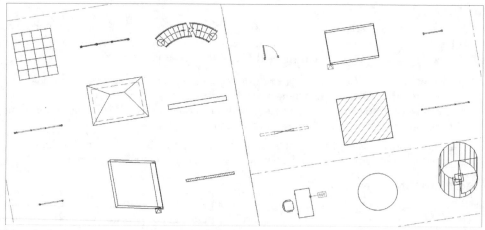

Figure 2–6 *Architectural objects shown in Plan representation*

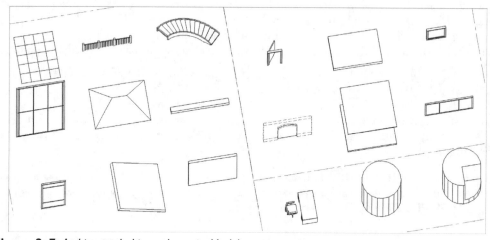

Figure 2–7 *Architectural objects shown in Model representation*

Figure 2–6 shows the collection of Architectural objects and a few of the Multi-Purpose ones in their Plan representation. Figure 2–7 shows the same objects in their Model representation. Notice that the Roof, Roof Slab, Slab and Mass Group objects all appear to display three-dimensionally despite the fact that each one has its "Plan" Display Representation active. There are some subtle differences that we will explore later in Chapter 5, but the one conclusion that can be drawn now from these two illustrations is that the Slabs and Roofs are more or less "3D only" objects.

Following the same type of logic, observe the Ceiling Grid object in both figures. Notice that there is no difference between the two. The Ceiling Grid is a "2D only" object. The remaining objects in these illustrations are hybrid objects—they have *true* 2D and 3D Representations. To be a *true* 2D object, the Plan representation must be completely "flat" and ideally will show information unique to that view.

For instance, a Wall object in Plan understands the architectural concept of a "cut plane." Therefore, the Plan representation of a Wall has the capacity to indicate through an "Above" and "Below" sub-component that portions of the Wall have variance in height. To be a true 3D object, the object must use 3D geometry to represent itself in its Model representation, and not simply change its Z elevation. There are several reasons why this is significant, as we will see in the coming passages. The remaining objects from Figure 2–1 are not shown here, as they do not offer as significant an influence on the overall structure of the Virtual Building Model. The Annotation objects are largely 2D-only objects used in 2D drawings. The Helper objects are also mostly 2D, but some are 3D depending on their designated purpose.

ARCHITECTURAL DESIGN APPROACH

The way to use ADT and the specific tools that are deployed will differ from firm to firm and possibly even project to project within the same firm. To start with:

- How does your firm approach design?
- What drives the design solution?
- What are the primary concerns?
- Is the look and character of the building's exterior the most important issue, or are meeting the square footage requirements, adjacencies and stacking the overriding principles?
- Are the plans developed as an overall layout and slowly refined to detail each of the individual spaces, or are typical unit configurations worked out in detail first and then assembled to form a complete building plan?

The answers to these questions and the collection of issues they raise lead us to two prevailing types of design: "Outside – In" and "Inside – Out."

CONCEPTUAL DESIGN

Although there can be found many similarities, all architects approach the design process in their own unique way. The critical thing to understand about creating projects with Architectural Desktop is that there is no required or "correct" way to approach design. There is no required set of tools or correct set of procedures. The point at which you begin using ADT tools in your design and production and how extensively you employ them are not as important as how *well* you use them and being sure that you are using them in the most *effective* way.

There are many ways to execute your conceptual design explorations. You can use traditional means of sketches and trace paper or study models, or you can use digital tools such as Autodesk VIZ, Autodesk Architectural Studio or ADT. The most important issue to address at this phase is:

- How will early design data be transitioned into the production phases of a project?

In Outside – In design, the 3D modeling capabilities of Autodesk VIZ and the Mass Modeling in ADT are far superior to traditional study models. With the mass modeling features of ADT, you can quickly create almost any three-dimensional form you wish and build complete 3D study models of your project for design exploration and client presentation. These models may then be transitioned into the DD and CD phases with minimal effort. Autodesk Architectural Studio also has a complete suite of excellent study model tools as well as the most complete digital 2D sketching toolset made available to architects to date. Working in a vector-based digital sketchpad, you can generate sketches with the same ease and freedom that traditional tools offer. The added bonus is Architectural Studio's ability to "talk" directly to ADT. This allows design and production to come together in a nearly seamless fashion.

Once one has acquired an understanding of the respective toolsets, models can be generated much more quickly than with traditional means. You can perform shadow studies and generate quick perspective images placing your proposed building directly on the site.

If your work is mostly Inside – Out, then again Architectural Studio is likely the tool to consider for freeform sketching and quick studies. When you are ready to make your designs more "hard-line," switch to either Spaces and Boundaries or Areas and Groups in ADT.

 Note: For more information about interoperability to Autodesk VIZ, refer to Chapter 13, "Using ADT Models in Autodesk VIZ." This chapter is included on the CD in PDF format. Also, visit the Autodesk web site at *www.autodesk.com.*

DESIGN DEVELOPMENT AND CONSTRUCTION DOCUMENTS

Although ADT can be used quite effectively in conceptual design with or without additional software products, design development and construction documentation are the phases in which ADT truly shines. Much of the effort in setting up office standards and building an implementation plan for ADT will concern these two phases.

OUTSIDE – IN DESIGN

When a building is designed from the outside in, presumably the overall form of the building and its relationship to surrounding context are the primary concerns driving the design. An architect may choose to explore these relationships primarily in plan or three-dimensionally in model. When the overall form, shape and context of the building are the driving factors in design, the building is first thought of in terms of its mass and shape. There can be several design issues to consider in the Outside – In approach— the qualities of the site: urban, rural, hilly, flat, open, closed, wooded, clear, and the possibility of existing construction. There may also be strict site setback requirements and other building code or municipal ordinances to consider. None of these factors necessarily dictates an approach to take in design or delineates a choice of tool for the job.

Traditional models and sketches may be used; software tools like Autodesk VIZ and Autodesk Architectural Studio offer a wonderful suite of tools ideal for these types of conceptual design studies. You may also choose to begin modeling directly in ADT using the Mass Modeling tools or sketch perimeter Walls. You may also use any combination of these methods to achieve your best results, utilizing whichever toolset allows you and your design team the most freedom and flexibility. Once a form is solidified, floor plates can be extracted from the form directly in ADT with the Slice object. Slice objects can be used to generate Walls and other architectural building components. These components must then be inserted into a series of pre-determined files based on office standards. Many file structure and naming schemes are addressed below in the "Project Types and File Structures" topic.

There are some key points to stress about the Outside – In approach:

- It is not necessary that a Mass Model be built first to follow an Outside – In approach with ADT.

- Begin with whatever tools you and your team find most comfortable. If you wish, you may design the building mass outside of ADT and begin your ADT files with the exterior perimeter Wall.
- If you do build a Mass Model of your building in ADT or another application, keep the 3D Mass Model in a file separate from any DD or CD drawings extracted from it.

Commercial office buildings are an obvious example of the Outside – In approach. However, almost any kind of building could use this approach. When following this approach in ADT, you will usually see the best results when a separation between the elements of the building's core (elevators, stairs, toilets and utility spaces) and shell (exterior skin and outer enclosure) and any of the interior spaces is maintained. This separation is often made quite literally with the use of external references (XREFs). As the project progresses, the core, shell and any interior build out should remain in separate files. The complete building model will be an assemblage of all of these components.

INSIDE – OUT DESIGN

When the overall form and shape of the building are a result of the internal layout of spaces as driven from programmatic requirements, the spaces themselves become much more critical. Projects of this type may begin with the individual rooms and departments as dictated by a building program. This is typical in medical or school design. In such projects, the needs of the spaces to both adhere to a required square footage and to maintain critical adjacencies to other spaces often drive the entire project. Therefore, in projects like this, design needs to begin on the inside, and work its way out to the building skin. Cores fit in where structure, code requirements and available space allow. There is less need for modeling the building mass, and more need to work efficiently with the plan layout. An equal reliance on ADT Wall, Space and Area objects is likely in these project's early phases.

Another form of Inside – Out design shares some features of the core and shell (Outside – In) design scenario. In "unit-based" design projects such as multi-family housing and hotel design, the governing factor is also the building program like the medical and educational; however, the need to maintain a few well-designed units is critical to these projects for a variety of reasons, including managing budgets. Getting a 500-bed hospital or a 100-apartment housing project completed where every unit was unique would likely be prohibitively expensive. If the number of unit configurations can be minimized, costs of production and construction can be decreased as well. As the unit plans and the building floor plans they generate become more refined, an active exchange between inside and outside becomes more frequent. Decisions made regarding the building skin design will influence the overall plan and unit plan design, while maintaining the fewest number of unique units remains the critical factor.

 GUIDING PRINCIPLE: It is recommended that the *dimensional* nature of ADT objects be considered in all of your file separation decisions.

DIRECTORY AND NETWORK STRUCTURE

In the previous chapter, we set up three network shares for office standard resources, two of which we have already populated. The third, the P Drive, we set aside for storage of project files, and it has yet to be discussed. Unless your firm is embarking on CAD for the first time, you likely have a project directory on the server and a sub-folder structure already in existence. If this is the case, you can simply use this existing network location in lieu of creating a new network share. The drive letter does not need to be "P." However, it will continue to be referred to as "P" throughout the remainder of this book. You may also choose to continue using the existing folder structure. If it meets your firm's needs, then there may be no need to change it. Whether or not you have an existing folder structure, consider these criteria:

- The structure should be easy to understand.
- Does the structure comply with the U.S. National CAD Standards 2.0 (NCS2) recommendations (or another industry standard if not in the United States)?
- Files should be easy to locate.
- Folders and files ought to be accessible by all project team members.
- The structure should make the external reference process simple.

Resist the temptation to build overly complex folder structures and cryptic naming conventions. Users should be able to quickly locate any file required for a project with a minimum of hassle. In order for files to be easily accessible to all users, all project data must be stored in the network project folder (see Figure 2–8). Network file permissions need to be established appropriately to allow all users the proper level of access. **CAD Software Users** will naturally require Read and Write access, while **CAD Data Consumers** will require only Read access. (For more information on these two user designations, please refer to the "How to Use This Book" section of the Preface.)

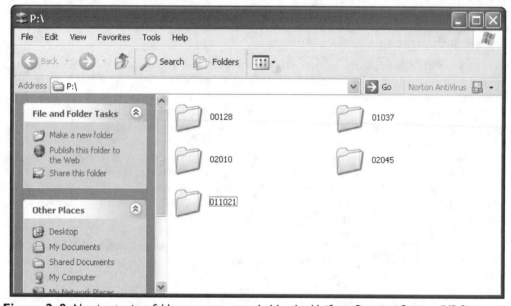

Figure 2–8 *Naming project folders as recommended by the Uniform Drawing System (UDS) published by CSI and incorporated into the National CAD Standards 2.0*

EVALUATING YOUR EXISTING FOLDER-NAMING SCHEME

If you already have a folder-naming scheme and directory structure in place, ask yourself the following questions:

- Does it meet the criteria listed above?
- Do people use it?
- Do people like it?
- Do they comply with the industry standard?

If the answer to the questions above was "yes," then there is no reason to change your system. You and your users are already familiar with it, and that is one less thing you have to worry about building at this time. If, however, when you analyze the current directory structure, it does not meet one or more of the above criteria, you may want to consider revising it or abandoning it altogether. Consider the recommendations of an industry standard such as the U.S. National CAD Standard. (Refer to Appendix A for information on acquiring these resources.)

Two folder and file naming schemes are common. The first suggests using project descriptors such as job numbers only on the folder names and not in the naming of the CAD files themselves. The rationale in this recommendation is the assumption that the CAD data will live on in the form of facilities management documentation long after the original construction project is complete. Therefore, eliminating the job number from the file names avoids the hard coding of legacy project descriptors. However, in the NCS, an equally compelling argument is posed for the inclusion of project numbers directly in the file names as a way of preventing accidental file corruption and deletion that is possible when files from different projects share the same file names.

Both arguments are sound, and this text will not make a firm recommendation other than to stress the importance of consistent file naming. Regardless of whether job numbers are included or not, it is important that the directories and file names used in a firm remain consistent from one job to the next. This will make it easier for all parties that use your files to reliably predict and utilize their contents.

 GUIDING PRINCIPLE: Make sure your file- and directory-naming structure makes it easy to understand file relationships and locate and utilize the files.

That is the only meaningful criterion from which to develop your standard. As a final point, whatever your directory structure is, be sure to perform regular backup and archival of all project data. A well-conceived disaster recovery procedure is critical to any firm's success.

PROJECT FOLDER NAMING

There are many theories regarding folder naming and structure. Some firms use the project number for the names; others use the project name. If you do not currently have a file-naming and folder-naming convention or, if you are not satisfied with your existing one, consider using both. Create a folder on the P Drive for every active project within the firm in the following format: <Project Number> - <Project Name> (see Figure 2–9). The National CAD Standard explains its recommendations for naming folders to sort projects, phases and file types on page UDS-01.30.

Figure 2–9 *Sample project folder names*

A scheme like this can satisfy proponents on both sides of the issue. Project folders can easily be organized by number in a list, yet including the name makes it much easier to spot your project from a long list quickly. Just be careful not to make the name portion too long. This can create very long absolute path strings when models are XREFed to sheets. See the "Understanding XREF Paths" section below for more information.

SUB-FOLDER STRUCTURE

Next, we have the issue of sub-folders. There are many theories here as well. Some firms create folders for every phase of the project; others organize by discipline, and still others by each type of drawing and so on. It is very common to see some combination of all of these.

None of these systems is intrinsically wrong or right—there is no such thing in this context. Sometimes these structures have evolved over time. If the system you currently have works for the majority of your users, then do not change it. However, if you have several pages in your CAD manual describing the folder structure, it may be a bit too complex. As you analyze your current structure, ask yourself the following question:

 Note: Is there a compelling reason why all of the DWG files for a particular project (or phase) could not be all stored in the same folder?

The first argument against this approach is usually the quantity of files. Granted, this can be a compelling argument. Once you have more than two or three dozen files, it becomes a bit cumbersome to locate them in Windows Explorer. However, this is where a well-thought-out file-naming scheme will prove invaluable. The next section discusses this in detail, but let us assume for the moment that a consistent and logical file-naming scheme has been adopted by your firm. Long lists of files can be sorted in a variety of ways in Windows, such as by name, date modified, type and file size. In addition, we can quickly use the search capability in Windows to locate a small list of files that have similar names, thereby shortening the scroll bar and shortening the list to a more manageable length. We can also do this directly in ADT using wildcards like the asterisk and the question mark (*,?). Simply type them with a partial name directly in the **File Name** field of the Open dialog box. (See Figure 2–10.)

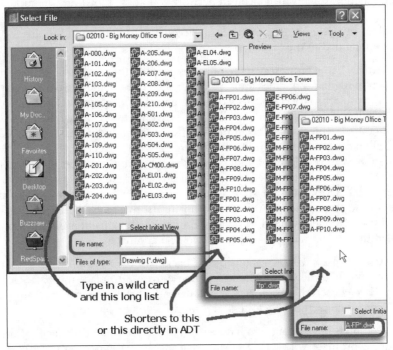

Type in a wild card
and this long list

Shortens to this
or this directly in ADT

Figure 2–10 *Type into the File Name field using wildcards to shorten the list*

The benefits of using a single folder include ease of use and simplified XREF paths. When XREFs are attached outside of their own folder, the name must include the path in order to be located when the drawing re-opens. Although it is possible to use relative path and **ProjectName** system variables to help manage complex paths, the easiest way to manage them is not to have them in the first place. (See the section below: "Understanding Files, XREFs and Blocks.") When files are sent to consultants outside the firm, there will be no issues related to missing XREF files when all the files are contained in the same folder. This is because the current folder is always included in the AutoCAD search path. Ultimately, the choice of folder structure is one you and your firm will need to make based on a variety of criteria. As such, you may find that a single folder does not meet your needs. In fact, many firms organize their CAD drawings into two main folders, *Models* and *Sheets*. This does introduce paths into the sheet file XREFs, but it helps users to better understand the model/sheet metaphor and can assist network administrators with managing user permissions for viewing and editing rights. Regardless of whether you agree with or decide to follow the suggestions here, rest assured that ADT will still function properly. The recommendations here are meant only to foster simplicity in project setup, not as a moral imperative.

 GUIDING PRINCIPLE: Even if you follow none of the recommendations here, just remember to keep it *simple* and *consistent*. Even a very complex file scheme can be easy to use if it is implemented consistently.

BACKUP STRATEGY

Do not overlook this critical component of your implementation plan. Regular backup and archiving are functions of a well-conceived network topology. However, the topic is so critical to productivity that it bears mention as a separate heading. The easiest way to make regular backup simple is by designating a single easily accessible root location for all office resources. This was recommended in the last chapter.

In the drive letter scheme proposed in the previous chapter, there could be three separate back-ups performed on regular schedules appropriate to the type of data the drives each contained. For instance the project share (Drive P) would require daily incremental backup with a supplemental Friday full backup, while the install share (Drive I) would likely only need backup after it had been modified, which would occur mostly after product updates. The library share (Drive L) would likely have a schedule similar to the project share. Keeping the folders within the shares simple and distinct, such as a folder for each active project, makes it very easy to archive entire projects to CD at the completion of critical phases and at project completion. You may prefer to create folders for each phase of a project within the project root folder, and keep all project files on the server until a project is complete. In this case, the backup may take a bit longer, but if users frequently need to reference archived versions of a file from previous submissions, then the backup may prove the most logical approach.

Don't forget that the users will be responsible for backup of their own Local Content folder (see the previous chapter). Some CAD Managers choose instead to place these user folders on the server. If you have chosen this approach, it will naturally have an impact on the steps taken in the Client Deployment setup. Regardless, the specifics will be left to you.

If your organization is large enough to have a separate IT group that manages disaster recovery and archiving, coordinate with IT to ensure that all the bases are properly covered and an effective policy and plan are currently in action.

DECIDING HOW TO STRUCTURE FILES

Architectural Desktop projects can be structured in several ways. Deciding how many model files to create and what they each should contain is an important first step in developing a process for your firm. Many factors will contribute to this decision, including the type of project, the size of the project team and the user's overall comfort with the ADT tool set.

Let's first consider the single model file approach. Due to the nature of several of the individual ADT architectural objects, one could make the case for building a single model file that contains all of the project data. In an approach such as this, a combination of display configurations and layer settings are used to control which objects are viewed at any given time. The inter-relationship between ADT objects is also easy to maintain in the single model file approach. Spaces, Boundaries and Walls may all maintain their linkages and relationships. Ceiling Grids may be linked to Space objects. Locations of tags and schedule files present no issues since they are together in the same file.

However, the benefits of the single model file approach end here. Although potentially viable as an approach for very small projects and/or small or single person project teams, the single model file presents many challenges not easily overcome. For instance, multi-story building models are not easily created in a single ADT file due to the object's hybrid 2D/3D nature. Customized Display Representations, saved User Coordinate Systems, Clip Planes and elaborate Layer States can be built in an attempt to manage the display of a multi-story single model file project. However, the effort expended on this task and the complexities it introduces to the file can often outweigh any benefits derived from the single model file.

In other than the simplest of single-story projects, it is highly recommended that you adopt and implement the *Separated Building Model* approach. In fact, there is little reason not to seriously consider adopting the Separated Building Model for all projects and teams, large and small. The Separated Building Model approach is defined and described next.

THE CASE FOR THE SEPARATED BUILDING MODEL

So far, in this chapter, we have discussed parametric design and defined much of the terminology required for a complete understanding of the ADT Building Model approach. The one component needed for a complete understanding of the virtual Building Model is what we will term the Separated Building Model (or XREF Building Model). The term *Separated Building Model* is used to describe the process of "separating" portions of a project into individual drawing files and then using external references (XREFs) to join them all together in a composite model. Think of it as building a model in small portions. See Figure 2–11.

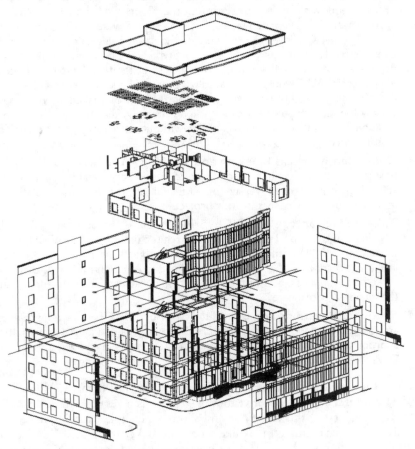

Figure 2–11 *Example of a Separated Building Model*

There are several reasons to construct the Separated Building Model in ADT:

- **File Size** – By separating the model into smaller files, we can keep sizes of individual files down to a minimum.

- **Project Team File Sharing** – Since the Windows operating system locks files when opened, separating the data into smaller files allows us to share the workload among several team members at once, without being limited by file locking.

- **Project data is more secure** – Separate files are safer. Should a file become corrupt, it can become a huge problem. If your entire project is contained in a single file, then the potential inconvenience or potential devastation from file corruption or loss or incorrect versions is magnified many times.

- **Superior Project Data Organization** – The file structure becomes an integral part of the project's organization. Entire XREFs can be unloaded (**Insert>Xref Manager**, select the XREF, click **Unload**). The combination of file, display and layer control gives us unparalleled control in managing our data.

- **Makes best use of Auto-Layering** – ADT automatically adds layers as you go. In a Separated Building Model, each XREF Layer will include its file name as a prefix. In a Single File Model, you will have to manually add all of the other layers or layer modifiers required to separate other floors while you work. ADT does have a Layer Key Override feature, but the tool is flawed and not very user friendly. It merely allows you to override the name of a particular layer. It does not override the Layer properties (Color, Linetype and Lineweight) of the "overridden" layer. Therefore, if you used overrides to generate a demolition layer for Wall objects, ADT would automatically generate the layer "A-Wall-Demo" (if using AIA Layering), but this new layer would not be gray or dashed. It would have exactly the same properties as "A-Wall." With an XREF structure, you can add additional layers in any file as you please without causing any grief because of the global control you have with XREF **Unload**, VISRETAIN and **Viewport Freeze**.

- **Conveniently checks blocking and stacking** – You can use XREF Overlay (instead of Attach) within any floor plan to check stacking alignments and other relationships. If you use Overlay, the XREF will occur only in that file and not nest its way through any others in your project.

- **Takes full advantage of ADT hybrid objects** – This approach takes fullest advantage of the hybrid nature of most Architectural objects in ADT (see earlier discussion). Within each file, users work with objects inserted at a zero elevation in the Z direction. In this way, users are not saddled with the need to consider the three-dimensional aspects of their objects or constant worries of adding data at the wrong Z coordinate.

- **Sections, Elevations and Schedules support XREFs** – Sections, Elevations and Schedules all work through XREFs. Simply gather the individual files at the appropriate heights into a separate composite model file and generate Sections, Elevations and Schedules there.

The Separated Building Model is not a new concept. In the days before CAD software, separation of data onto different sheets of vellum or mylar was a very common practice. Pin-bar drafting was the pinnacle of this technique and allowed a complex drawing to be broken into simplified "layers" of data. In fact, it is likely that the CAD term "layer" evolved from this practice. In pin-bar drafting, the separations were usually made between the various building trades. For each floor of the building, each discipline would draft on their own sheet, yet being able to see through the translucent pages allowed them to "reference" the drawings of the other disciplines in their own work.

When thought of in those terms, it seems as though things have not really changed all that much. In fact, many firms today still follow this practice. Certainly, it is prevalent among the large multi-discipline AE firms. Even projects generated in small and one-person firms frequently make use of this technique, albeit with the members of their "virtual" project team. When an architectural drawing file is sent to an engineer, they XREF it into their own file, and work "on top" just as in the old pin-bar days.

For firms using standard AutoCAD, it has been possible to simply utilize the digital equivalent of the venerable pin-bar system as just described with great success. With ADT however, the building model concept has evolved. Although it is possible to continue to separate our building models by discipline and floor, there are other issues to now consider when determining what data to separate and where to place it. Consider the following list of issues:

- How extensively will ADT objects be used in the project?

Some firms have found great success in the "slow and steady" approach to ADT deployment. There is a paradigm shift at work when migrating to ADT from AutoCAD and for some, this shift is disruptive and difficult. Some of the culture shock can be eased by choosing to begin with a small subset of ADT tools on the first project and adding more to the mix incrementally with each future project.

- Is the project new construction or renovation?

It is not terribly important which it is, but the approach to new construction is a bit simpler than renovation. If your firm does a large amount of renovation work, you will need to devise a suitable way to differentiate new construction from existing to remain and demolition. There are several strategies to this. Please refer to Chapter 4 for complete details.

- Does the project utilize any legacy AutoCAD drawings as backgrounds?

The extent to which legacy drawing data and standard AutoCAD entities will be used is an important consideration. Always remember that although ADT offers many new and often superior tools to generate architectural components, these in no way preclude us from using AutoCAD entities when necessary and/or appropriate. Recognizing this reality and planning for it with procedures and standards will make the transition to model-based tools much more seamless.

- Is the project large (greater than 25,000 SF [2,323 SM]) or small?

Large and small are of course relative terms. A large project for firm A may be small for firm B. The point is raised here simply because it is very common to see variance in the typical office standard based on the size and scope of a project. It is not necessary to have one standard for small projects and another for large ones; however, the practice is very common.

- Is the project simple or complex?

This is again, another relative distinction, but also worthy of consideration. There is no virtue in complexity for complexity's sake. Remember to always seek the simple solution before resorting to a more complex one. At the end of the day, our job is to get buildings built, not generate CAD drawings. Therefore, if the procedure seems overly complex, it probably is. Strip off a few layers of complexity wherever you can, and you and your users will be much happier. They may even follow the standard.

ADT OBJECT CHARACTERISTICS

After you address these points at the project level, there are a couple of additional considerations at the object level to take into account before finalizing a project file composition.

- Is the object 2D, 3D or hybrid?

This was discussed in the "Parametric Design" topic earlier. Dimensionality is an important consideration in organizing project data. 3D-only data ought to be kept separated from 2D and hybrid data. This is because the tendency is to work with 2D and hybrid objects extensively in plan view. If 3D-only objects are included in the same file, they could prove deceiving when viewed from a plan view. Even though the object may look as though it is positioned properly in plan view, it may in fact be skewed when viewed in 3D. You should always insert 2D and true hybrid objects at Z=0. 2D objects will not impact the third dimension, so a zero elevation is appropriate. Hybrid objects, even though they are inserted at a zero elevation, can "appear" at different heights in 3D because of their hybrid nature. The parameters for the three-dimensional Display Rep are separate from those of the 2D Plan Rep. A 3D object does not have separate parameters for plan and model, and therefore it often must be inserted at a Z elevation other than zero. These are the reasons for recommending the separation between 2D and 3D components. The two will be reconciled when they are XREFed together into a single composite building model.

- Is the object anchored or free?

Anchors enable two objects to be linked together in a logical, physical manner. Usually, they are used to build intelligent links between the two objects. As one object moves or transforms, the anchored object will move and transform along with it. The way in which the anchored object moves and transforms is governed by the rules established within the anchor's parameters. ADT objects can be either anchored or free. If an object is anchored, its position, rotation, scale and/or relationship to property data (in the case of schedule tags) are governed by the Anchor. If an object is not anchored, it is "free." As the name implies, the object will not be constrained in any way when it is free. The implication of Anchors to project setup is simple. If an object is anchored, it cannot be separated from its parent object (the object it is anchored to). This is the only potential limitation currently imposed by ADT on the structure of project file composition.

- Is the object or assembly predominately horizontal or vertical (spanning)?

As stated above, traditionally (both in CAD and in hand drafting), project data has been separated by floor and then by discipline. In hand drafting, you then had front and back of the sheet to work with after that. With traditional CAD files, we had layers. ADT objects introduce additional parameters and Display properties. It can be a very daunting task to try to understand the myriad of possibilities offered by the combination of File, XREF, Layer, Parameters and Display on an object-by-object basis. Fortunately, we can approach the task from another angle. As a general rule of thumb, if the object is predominately created and edited in a horizontal orientation, then we can treat it as such in the overall project structure. Likewise, if the object is thought of and edited predominately in the vertical, the same applies.

Tables B2–1, B2–2 and B2–3, (in Appendix B on the included CD ROM,) detail the characteristics of each ADT object with respect to dimensionality, Anchor state and orientation. The information contained in Table B2–1 and accompanying notes can be summarized as follows and as shown in Figure 2–12:

- Walls, Doors, Openings, Windows and Window Assemblies will all be in the same file named with an "FP" (Floor Plan) drawing type code. A separate "CP" (Ceiling Plan) file will contain the Ceiling Grid object and potentially the Space to which the Grids are clipped.

- Place Column Grid and any Anchored Structural Members in a separate "GD" (Grid) file.

- Curtain Walls and Curtain Wall Units should be contained in a separate model file containing only elements of the building skin. Files of this type could also use the "FP" designation, but this may foster confusion. In the predecessor to this book, (*Mastering Autodesk Architectural Desktop*), the code "SH" was used to indicate "Shell." However, according to the National CAD Standards' UDS Module 01, "SH" ought to be used for Schedules. Refer below to Table 2–2, "NCS 2.0 Model File Types" for more information and potential resolutions to this issue.

- Roofs, Roof Slabs and Slabs should be kept separate from the other building model components. Depending on the needs and complexity of the project, these items could all be contained in the same file, or a separate "Slab" or "Horizontal Surface" file could be made for each floor of the building.

- Stairs and Railings should be kept in separate "Core" files in most building types. If the project is very simple or has very few floors, the Stairs could be placed directly in the "FP" file.

- Space objects can be saved in many locations depending on how they are being used in the project. Possible locations include the FP and the CP. Potentially they could be stored in their own file as well. Structural Members that are not anchored Columns should be kept in a separate "FP" (Framing Plan) file.

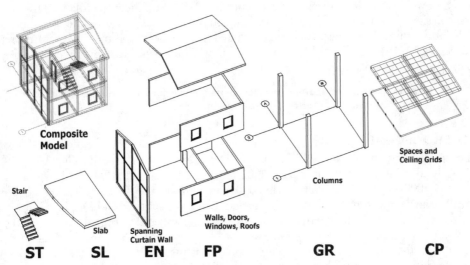

Figure 2–12 *A typical Architectural Desktop file structure*

The information contained in Table B2–2 and accompanying notes can be summarized as follows:

- Tags and AEC Dimensions must be included in the file with the items to which they are linked or anchored.
- Area Objects and Area Groups ought to be created and maintained in a file separate from the Building Model.
- Building Section and Elevation Line objects need not be contained in the same file as the model; however, they must be in the same file as the Section or Elevation object that they generate.

FILE NAMING CONVENTIONS

The topic of file naming is covered in detail in the National CAD Standard (*NCS*). The file-naming convention should be logical, easy to understand and consistent. For purposes of the tutorials in this book, we will adhere as closely as possible to the *NCS* recommendations (see Figure 2–13). The most critical issue regarding file naming involves the file's role in the project. It should be evident at a glance what the purpose of the file is.

NCS 2.0 MODEL FILE NAMING											
A	-	A	A	U	U	U	U	.	A	A	A
Discipline Designator											
A	-	A	A	U	U	U	U	.	A	A	A
Placeholder											
		A	**A**								
Type of Model											
				U	**U**	**U**	**U**				
User-defined Model Type Modifiers											
								.	**A**	**A**	**A**
File Extension											

Figure 2–13 *Anatomy of an NCS model file name*

Model files will begin with a discipline code such as "A" for Architecture, followed by a model file type designator such as "FP" for Floor Plan, and then two digits for the floor number such as "03" for the third floor. Sheet files will be named for the sheet type/number that they represent, such as "A-101" for sheet **A1.01**. Feel free to use your own file name descriptors if your firm uses an alternate naming scheme.

 Note: As suggested above, remember to use the Windows file search capability to search for a partial file name and shorten the list of available files.

XREF PROCEDURE

Architectural Desktop files use the AutoCAD DWG file format. Specifically, Architectural Desktop R3.3 is built on the AutoCAD 2002 drafting engine, and ADT 3.0 is based on the AutoCAD 2000i engine. Both AutoCAD 2000i and 2002 use the **AutoCAD 2000** file format. Internally, the data within ADT files is organized with Layers, Styles and Display Control settings. Externally, folders and external references (XREFs) are used to organize and manage the collection of files that make up an architectural project. Projects of different building types will differ most in the structure of these organizational constructs. The type of building being designed may influence the XREF and folder structure. Different types of projects as well as project and team size may also influence the choice of objects and procedures employed on a project.

Once you have settled upon a folder-naming and file-naming scheme, the next thing to do is to figure out which files will be XREFed together. XREFs allow two files to remain separate yet be linked together. One file becomes a virtual "underlay" to the other. There are many good uses for XREFs in a well-thought-out project structure. Universal to any use of XREFs is the need to have a clear understanding of how XREFs function. There are two types of XREF: an XREF Attach and an XREF Overlay.

Attached XREFs are useful when you want to build a hierarchical, nested XREF structure. If file "**A**" is attached to file "**B**," and then file "**B**" attached or overlaid to file "**C**," then file "**A**" will automatically appear within "**C**" as well (see Figure 2–14).

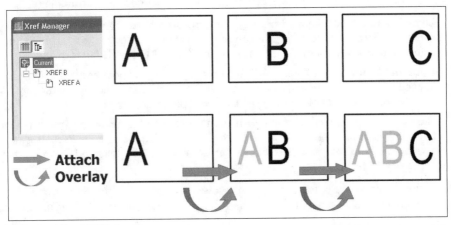

Figure 2–14 *Using XREF Attach*

On the other hand, an overlaid XREF operates only one level deep, meaning that if file "**A**" is overlaid to file "**B**," and then file "**B**" overlaid or attached to file "**C**," file "**A**" will *not* automatically appear within "**C**." If it were desired that "**A**" appear within "**C**," it would need to be separately overlaid directly to "**C**" (see Figure 2–15).

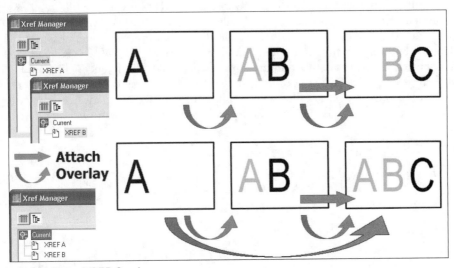

Figure 2–15 *Using XREF Overlay*

Consider the nesting question carefully when you establish your standards for XREFing. There are advantages and disadvantages to both approaches and often both types of XREF can be used within the same model.

UNDERSTANDING FILES, XREFS AND BLOCKS

There is ongoing debate in the Architectural Desktop community regarding if, when and how much to use external references when working with ADT. The simple answer to the "if" question is an unequivocal *yes*. XREFs should most definitely be used in Architectural Desktop projects. The answers to the "when" and "how much" questions are naturally a bit more involved. Let's address each of the three questions in detail, beginning with some more terminology:

▶ **Drawing File** – A collection of related information saved together under a particular name to a particular location on a computer hard drive. Files are formatted by the software that saved them. ADT writes to the AutoCAD DWG file format.

▶ **External Reference (XREF)** – Everything you need to know about an XREF is implied by its name: "external reference." It "references" something "externally." What exactly does it reference externally? A drawing file.

▶ **Block** – A collection of related information saved together under a particular name, but in the case of the Block, it is *part* of a file rather than being a file itself. It is not saved directly to the hard drive but rather is one of those "collected" items within some larger file.

The distinction between these items comes down essentially to where they are saved. This is the only tangible difference between a file, XREF and Block—besides the way they are used, of course.

In simple terms, an XREF is simply a Block that automatically updates itself. This may be surprising to some because the two objects, Blocks and XREFs, at first seem so different. First, you cannot just create something in a drawing and then run a simple command to make it into an XREF as you can with a Block. In addition, different commands are used to add them to the drawing, remove them from the drawing and update them. However, these differences aside, there remains only one difference between a Block and an XREF that matters. A Block's definition lives inside another file, and the definition of an XREF *is* another file. As a result, the definition of an XREF lives "externally." It really is that simple. If you would like some proof, open a file that contains both an XREF and a Block. At the command line, type LIST and press ENTER. Select both the Block and the XREF and press ENTER again. When you read the listing for either object, you will note that they are both listed as "Block Reference." The only clue that one of them is an XREF is five lines into the listing where it reads "External reference" (see Figure 2–16). Therefore, yet another way to think of the difference between these two reference types is that an XREF is an *external* reference, while a Block is an *internal* reference.

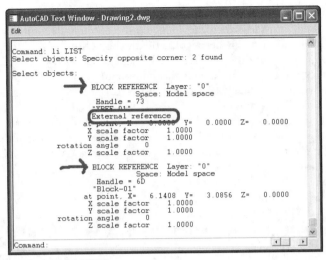

Figure 2–16 *Listing Blocks and XREFs*

At this point, the last two paragraphs may seems a bit contradictory. After all, in the first passage we said that an XREF was a file, and in the next we said it was basically a Block. So how can something be both file and a Block at the same time? The answer is simple. This is exactly the point of an XREF. You get the features and benefits of both files and Blocks at the same time.

Let us add two more terms to our repertoire: Definition and Reference.

- Block (or XREF) **Definition** – The actual data of the Block. Think of it as the recipe of the Block. It is the "instructions" for how to make a reference to a particular Block.

- Block (or External) **Reference** – An "instance" of the Block that exists within the drawing. This "instance," refers to the Definition of the Block. Thus the word "reference."

A definition exists only once and is saved either internally within a file in the case of a Block, or externally in the case of an XREF. There can be several references to a single definition within a particular file.

To understand some of those benefits, consider the reasons you create separate files and the reason you use Blocks. We place information in separate files for managerial purposes. Those reasons include keeping the size of files reasonable, adherence to accepted norms and standards, and personnel issues (having multiple users work on different files at the same time). We place information in Blocks also for managerial purposes. However, the specifics tend to be a bit different. For instance, Blocks are often used to manage repetitive drawing components and office standard components. The reason Blocks are ideal for these purposes is their ability to be globally updated. This is, of course, the same reason why XREFs are used. The real difference, from a usage point of view, tends to involve size (and update procedure). When we would like to manage and globally update a large portion of the project, we favor an XREF. This is true for a complete drawing like a Floor Plan or a Ceiling Plan. When we are looking for an easy way to update many smaller items, like fixtures, furniture, Unit Plans and Layouts or drafting symbols globally within the same drawing, we favor a Block.

Once you have a good understanding of XREFs and Blocks, consider the myriad of tools available to work with them. To add either reference, use the commands found on the **Insert** menu (**Insert>Block** or **Insert>External Reference**). To edit the definition of either, you can use the **Reference Edit** command (**Modify>In-Place XREF** and **Block Edit>Edit Reference**). You can also just double-click a reference to launch this command. You can "crop" off a portion of a reference using the XCLIP command (**Modify>Clip>Xref**). Few people realize that this command works on both Blocks and XREFs (see the tutorial below). The process of updating Blocks and XREFs is a little different. In order for a Block to update, the process must be manually initiated by either editing the Block directly with the **RefEdit** command, exploding and redefining or, if a version of the Block exists externally as a "Wblock," it can be re-inserted manually into the drawing to update the definition and all references. With an XREF, update occurs automatically each time the file is opened. It may also be manually updated any time using the XREF **Reload** command.

ACTUALLY, THERE IS NO SUCH THING AS A "WBLOCK"

It may come, as a bit of surprise to a veteran AutoCAD user to hear that there really is no such thing as a *Wblock*. It is common among AutoCAD users to interchange the terms Block and Wblock. In reality, Wblock is not a *thing* it is an *action*. Wblock is a verb, not a noun. You *Wblock* information from a drawing to create another separate file on the hard drive. Wblock is simply the name of the command used in this process, not the name of the thing it creates. The online help defines Wblock this way: "Saves objects or converts a block to a file." Users often use the term Wblock to describe drawing files created using the **Wblock** command, or files contained within Block Library folders. It is not necessarily bad to think in these terms as long as you realize that AutoCAD and ADT make no such distinction in files. Nothing about a drawing file saved in a Block Library folder prevents it from being used in exactly the same way as any other drawing file.

BLOCK AND XREF USAGE

Let's be sure that we have a clear understanding of the previous passage. To do this we will work through a small hands-on exercise using a dataset file located on the CD ROM. A few general tips are in order before beginning the tutorial.

Install the CD Files and Open the Sample File

1. Install the files for Chapter 2 located on the Autodesk Architectural Desktop: Advanced Implementation Guide CD ROM.

Refer to the "How to Use the CD" section in the Preface for instructions on installing the sample files included on the CD.

2. Open the file named *Chapter02-01.dwg*.

There is a series of jack arches in the center of the screen two levels high.

Redefine a Block

1. List one of the arches, and note that each is a Block with the name "Arch."

If you have an alternate version of a design component, such as the arches in this drawing, you can insert one version into the drawing to replace the instances already in use. This type of "swapping" of Block references can be very useful when you consider various design schemes.

2. From the **Insert** menu, choose **Block**.

3. Click the **Browse** button and in the *Chapter02* folder, locate the file named *Arch-02.dwg* (see Figure 2–17) and click **Open**.

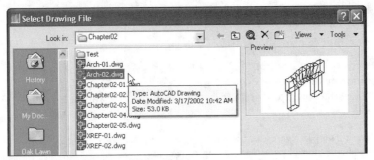

Figure 2–17 *Browse to the file named Arch-02.dwg*

4. In the **Name** field, type **Arch** in place of the existing "Arch-02" text (see Figure 2–18).

Figure 2–18 *The path to the file remains even if the name is changed*

Note the path to the file we are inserting remains unchanged: *C:\ADT_AIG\Chapter02\Arch-02.dwg*. This is the key to being able to "swap" one block definition for another. Remember, there is only a single definition for each Block name within a drawing. By redefining it, we are updating all references simultaneously. A reference simply "looks" to the definition for instructions on what to draw. Therefore, if we change the definition, the references must also update to reflect this change.

5. In the Insert dialog box, click **OK**.

6. In the AutoCAD dialog box that appears asking if you wish to redefine the Block, click **Yes** (see Figure 2–19).

Figure 2–19 *To redefine the Block click Yes*

Note the change to the drawing. All of the arches have changed.

7. Press ESC to terminate the Block insertion.

We can cancel the insertion since the update that was desired has already taken place and we do not need another instance of the block. The previous exercise was effective, but perhaps we only want the top tier to be arched and leave the bottom tier as jack arches. The problem is that right now, both tiers reference the same block definition name. To allow for more variation, we need to reference the top tier to a different block definition.

8. **Close** the file without saving.

Working with a Two-tier Scheme

1. **Open** the file named *Chapter02-02.dwg*.

Note the different color on the top tier. In this file, the same Block was inserted twice as two different definition names: Arch-Bottom and Arch-Top. For additional organizational control, they were inserted on two different layers: Bottom-Tier and Top-Tier. You may list an arch on each tier to verify this.

 Note: The naming conventions here are strictly for convenience in this exercise and not a recommendation of preferred naming convention.

2. From the **Insert** menu, choose **Block**, locate the file named *Arch-02.dwg* and then click **Open**.
3. In the **Name** field, type **Arch-Top** in place of the existing "Arch-02" text and then click **OK**.
4. When prompted about redefining the Block, click **Yes**.

Note the change to only the top tier.

5. Press ESC to terminate the Block insertion.

The arches can be returned to the original jack arch configuration by repeating the same steps and browsing to the file named *Arch-01.dwg* instead. Using the technique outlined here, you can save as many variations of a design element as required in separate files and then redefine them on the fly by inserting one of these files to the appropriate Block name within the current file.

6. **Close** the file without saving.

Block/XREF Insertion and User Coordinate Systems (UCS)

1. **Open** the file named *Chapter02-03.dwg*.

It is very important to pay attention to the orientation of the User Coordinate System (UCS) when defining and inserting Blocks and XREFs. Whatever coordinate system is active when a Block is created will become the "World" Coordinate System (WCS) for that Block. The WCS is the default X, Y, Z coordinate system in AutoCAD. As a general rule of thumb, when building Blocks or saving files that will be used as XREFs, it is best to work in (or save in) WCS exclusively, unless a particular task specifically requires working in another coordinate system. Some AutoCAD functions such as changing views with the View toolbar also change the UCS. This is not necessarily a bad thing; just remember to set the UCS back to "World" before saving a Block. To do this, at the command line type UCS and then press ENTER. Type W (for

World) and then press ENTER again. In this next sequence, we will explore the effect of the UCS on both defining and inserting Blocks.

2. From the **Insert** menu, choose **Block**.
3. From the **Name** list, choose **Arch-01** and then click **OK**.
4. At the "Specify insertion point" prompt, click a point within the gray rectangle on screen.

Note that the arch is standing "upright."

5. At the command line, type **UCS** and then press ENTER.
6. Type **X** and then press ENTER again.

 Note: Although "X" will not appear as an option at the command line, it is still a valid option.

7. At the "Specify rotation angle about X axis" prompt, press ENTER again.

This will accept the default and rotate the coordinate system 90° about the X axis.

8. Insert the same Block again. (Follow the previous steps.)

Note the change. (see Figure 2–20)

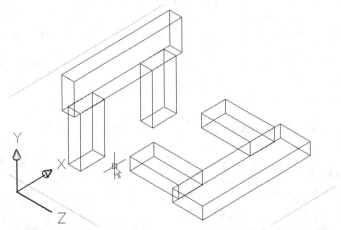

Figure 2–20 *The current UCS affects the orientation of inserted Blocks and XREFs*

9. Rotate the UCS about the Y axis **20°.**
10. Insert the Block one more time at **0,0,0** insertion point.

Note the orientation of this Block.

Redefine a Block after Rotating the UCS

1. **Explode** the last Block inserted (the one rotated 20°).
2. Rotate UCS on X axis **90°.**
3. From the **Draw** menu, choose **Block>Make**.
4. From the **Name** list, choose **Arch-01**.

Verify that in the **Base point** area that each of the **X, Y** and **Z** fields are set to zero and that the **Objects** area reads: "3 objects selected." If the **Objects** area reads: "No objects selected," verify that you have chosen the **Arch-01** Block from the **Name** list (see Figure 2–21). If it still shows "No objects selected," click the **Select Objects** button and click each of the three elements that make up the exploded jack arch and then press ENTER.

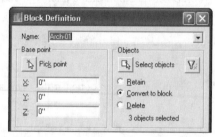

Figure 2–21 *Redefine the Block with the same base point and objects*

5. Click **OK** in the Block Definition dialog box.

Note the change. Clearly, the current UCS has a significant impact on the orientation embedded within a block definition. The easiest way to ensure predictable orientation results when inserting a Block is to always build your block definitions relative to the World Coordinate System. The same is true for a file being used as an XREF. Do not build a model in an XREF file in a UCS other than World unless you have very definite reasons for doing so. It is always possible to rotate a Block or an XREF after it has been inserted, rather than attempt to build the rotation into the definition.

6. **Close** the file without saving.

UNDERSTANDING XREF PATHS

Unlike Blocks, which refer to a definition saved within the same drawing file, XREFs refer to a separate drawing file. As a result, the path to the external drawing file is critical to maintaining the link to the file. XREF paths can be absolute, relative or unpathed. You can also use a "ProjectName" system variable if you wish (see "The Project Name Search Path" sidebar).

- **Absolute Path** – A complete path including the drive letter, or server name (for UNC). *P:\ProjectFolder\Plans\Plan01.dwg* and *\\MyServer\ProjectFolder\Plans\Plan01.dwg* are examples of absolute paths.

- **Relative Path** – Includes only the file name and one or more leading folder names not including the drive letter (or server name). *ProjectFolder\Plans\Plan01.dwg* and *Plans\Plan01.dwg* are examples of relative paths. With a relative path, the XREFs can be moved freely from one server to the next so long as the relative folder structure remains in tact.

- **No path** – XREFs with no path omit all drive letter, server names and folder names from the path and save *only* the file name. An example would be simply *Plan01.dwg*. These XREFs must be saved in the same folder as their hosts (or in a folder along the AutoCAD search path).

THE PROJECT NAME SEARCH PATH

For a Project Name system variable to be used, a path must be added to the **Project File Search Path** on the Files tab in the Options dialog box. You can add several if you wish. Each Project name will contain a name of the project and a path beginning with a drive letter or server name. This is the "root" path of the project. Individual XREFs may use relative paths (or no path, just the file name) that are appended to this root. Once you have set up a Project Name system variable in the Options dialog box, you must then "assign" each drawing file to that project. This is done by typing **projectname** at the command line and then inputting the name of the project as saved in the Options dialog box. Keep in mind, however, that AutoCAD will always search the entire Support File Search Path as well as the Project Search Path. Project Name variables are most useful for firms sharing files with teams that use drive letters different from those in use in their own firm or department.

Swapping XREFs

Just as we were able to "swap" one block definition for another in the sequence above, so are we able to do the same with XREFs. In some ways, it is easier to do with XREFs than with Blocks.

1. **Open** the file named *Chapter02-04.dwg*.
2. **List** any one of the jack arches (see Figure 2–22).

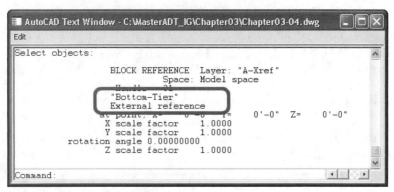

Figure 2–22 *List an XREF file to see its Block Name*

Notice that the entire ring of arches highlights. The complete collection of arches is part of an XREF file. The name of this file is *XREF-01.dwg*; however, the name of the Block reference within this file is actually "Bottom-Tier." Another way to say this is that this Block called "Bottom-Tier" references an external file named *XREF-01.dwg*. We do not see that it references *XREF-01.dwg* in the **List** command. For that we need to visit the Xref Manager.

3. Select the XREF again, right-click and then choose **XREF Manager**.

Here in the Xref Manager we see the above Block Name and reference more clearly. In this dialog box, we can change the Block Name of the reference, and choose a completely different reference file for that Block by changing the saved path to the file. The most important thing to understand is what has been already stated about XREFs in the earlier passages. An XREF is simply a Block that automatically updates itself, because its definition points to an external file and it is *not* saved within the current drawing. If this file is changed and saved, the Block that references it must update as well. The other critical thing that this passage illustrates is that although the Block Name and the reference file name are often the same, they do *not* need to be the same.

4. Highlight **Bottom-Tier** in the XREF list (see Figure 2–23).

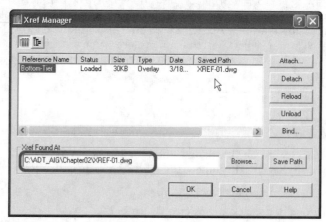

Figure 2–23 *The XREF Manager shows the Reference (Block) Name and path to the reference file*

Note that the "Saved Path" for this reference is simply its name; in this case *XREF-01.dwg*. In the **Xref Found At** field, it lists the location where ADT has found this file including its complete path; in this case *C:\ADT_AIG\Chapter02\XREF-01.dwg*. As was mentioned above, we can save this complete path to the XREF or simply the file name or even include a relative path that includes some part of the folder structure, but not the complete path to the drive letter or server name.

5. In the **Xref Found At** field, type **Test\XREF-01.dwg** and then click the **Save Path** button (see Figure 2–24).

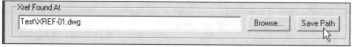

Figure 2–24 *Change the path to a relative path*

6. Click **OK** to view the result (see Figure 2–25).

There is a copy of the *XREF-01.dwg* located in the *Test* folder. By inputting this *relative* path to that file, ADT will seek to locate the reference file for this Block within a subfolder called test. However, if the file is not located there, it will begin searching the current folder and then the normal Support File Search Path. (Refer to Chapter 1 for more information on the support file search path.)

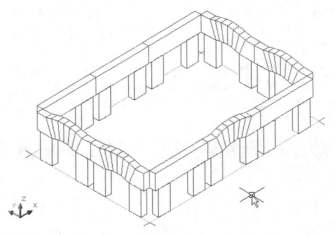

Figure 2–25 *A different scheme loads even though the file names are the same*

7. Reopen the Xref Manager, select the **Bottom-Tier** listing and view the path listed in the **Xref Found At** field.

Note that it is now *C:\ADT_AIG\Chapter02\Test\XREF-01.dwg.*

8. In Windows, browse to the *C:\ADT_AIG\Chapter02\Test* folder and rename the *XREF-01.dwg* file to **XREF-00.dwg**.

9. Return to ADT with the Xref Manager still open, select the **Bottom-Tier** listing, click the **Reload** button and then click **OK**.

Notice that the XREF has reverted to the previous configuration. This is because the required file no longer exists in the location indicated by **Saved Path**. Therefore, ADT begins looking for the file elsewhere. It will first look in the current folder, where in this case, a file with the correct name does exist—even though it is the wrong scheme! ADT has no way of knowing which scheme is "correct." It looks only for the files based on name and location. If, after searching the entire Support File Search Path, ADT is unable to locate the file, the reload will fail and ADT will display nothing but a piece of text with the name of the file that should be there (see Figure 2–26).

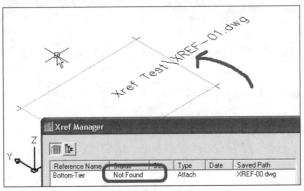

Figure 2–26 *Only a placeholder will display when an XREF file cannot be found*

There are several other ways that the **Block Name, Saved Path** and **Xref Found At** variables can be manipulated. Deciding which if any of these techniques to use depends on what you are trying to accomplish. Let's look at one more example.

10. Reopen the Xref Manager and select the **Bottom-Tier** listing.

11. Next to the **Xref Found At** field, click the **Browse** button.

12. In the *C:\ADT_AIG\Chapter02* folder locate the file name *XREF-02.dwg* and then click **Open** (see Figure 2–27).

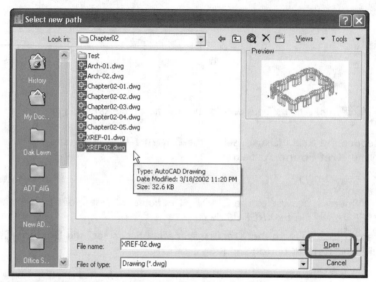

Figure 2–27 *Locate another file to reference to the Bottom-Tier Block Name*

13. Click the **Save Path** button.

 Note: this saves the entire path to the XREF, making it an example of an absolute path as defined earlier (see Figure 2–28).

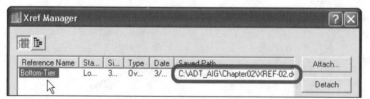

Figure 2–28 *Using an absolute XREF path*

14. Click **OK** to view the results.

This technique can be used to quickly swap out a different scheme for the current one. The original file is still intact on the hard drive, ready to be swapped back into the reference should the design change again.

Cropping an XREF

Both Blocks and XREFs may be "cropped" using the xclip command. The XClip establishes the cropping boundary for the Block or XREF. Any part of the reference that is not contained within the Clip Boundary will not be displayed. This can be very useful way to display only part of a Block or XREF.

1. From the **View** menu, choose **3D Views>Top**.
2. From the **Modify** menu, choose **Clip>Xref**.
3. When prompted, select the XREF and then press ENTER.
4. At the command line, choose **New boundary**, and then **Rectangular** (see Figure 2–29).

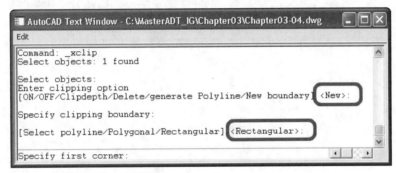

Figure 2–29 *Create a Rectangular XCLIP*

5. Following the command prompts, draw a rectangular boundary around approximately half of the XREF (see Figure 2–30).

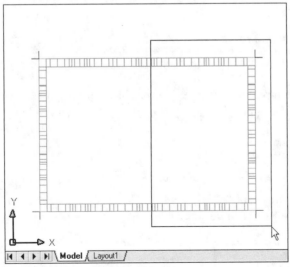

Figure 2–30 *Draw the clip boundary around half the plan*

Notice that half of the XREF is cropped away. This does not affect the reference file in any way; it displays this way in this file only. If you wish to verify this, you can open the XREF file and see it first hand.

6. From the **View** menu, choose **3D Views>SW Isometric**.

The clipping boundary is three-dimensional. We can adjust it so that it crops away portions of the file along the Z axis as well.

7. Repeat the XCLIP command, select the XREF when prompted and then press ENTER.

8. Right-click and choose **Clipdepth**.

9. At the "Specify front clip point" prompt, right-click and choose **Distance**.

10. Type **20'** and then press ENTER.

11. At the "Specify back clip point" prompt, right-click and choose **Distance**.

12. Type **10'** and then press ENTER.

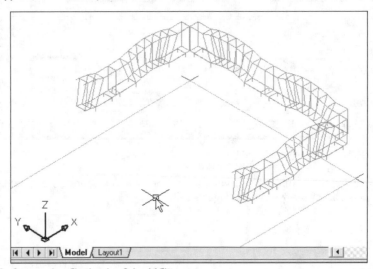

Figure 2–31 *Setting the Clipdepth of the XClip*

Notice how the bottom portion of the XREF has been cropped away (see Figure 2–31).

13. **Close** the file without saving.

Cropping a Block

1. **Open** the file named *Chapter02-05.dwg*.

Contained in this file are a table and four chairs. Using XCLIP, we can crop off the portion of the chair that would be concealed under the table.

 Note: If you use XCLIP, be sure to read the next section on XREF performance options.

2. From the **Modify** menu, choose **Clip>Xref**.

 Note: Even though the command says "Xref" it can be used on a Block.

3. Select the chair on the left and press ENTER.
4. Right-click and choose **New boundary**, and then **Select polyline**.
5. When prompted, click the polyline surrounding the back of the chair (see Figure 2–32).

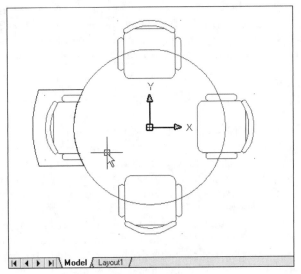

Figure 2–32 *Using XCLIP on a Block*

If you wish, you can repeat the steps on the remaining chairs, or simply delete them and polar array the clipped one in their place. The copies will each have their own clip applied.

 Note: XCLIP does not work on Multi-View Blocks, only on XREFs and AutoCAD Blocks.

6. **Close** the file without saving.

ADDITIONAL BLOCK CREATION CONSIDERATIONS

Now that we have seen many of the features common to both Blocks and XREFs, there are likely situations when you could justify the use of either tool. There are no hard rules regarding the use of Blocks and XREFs. Their functionality and features speak for themselves. As a general rule of thumb, Blocks tend to be used for small components such as furniture, equipment and other symbology, or even small repetitive unit plans and restroom layouts. XREFs tend to be used for organizing entire drawing sets. Floor plan information may be in a file that references a separate core and shell file, a column grid file and possibly even separate files for the furniture and/or ceiling plan information. Many factors will weigh in here, however, as discussed in the previous section. The next section will also explore some these situations in more specific terms.

- Always create Blocks while the World Coordinate System (WCS) is active, and later insert them into an active UCS. This guarantees that you will not get an unexpected orientation when inserting a Block.

- Always consider how the Block will be used before creating it and assigning the insertion point. This can save lots of rework and unnecessary editing later.

- If the Block is being used as a "View Block" within a Multi-View Block, set the Color, Linetype, Lineweight and Plot Style of all components within the Block to "ByBlock." This will allow those View Blocks to take on the properties of the Multi-View Block when inserted. (See Chapter 8 for more information.)

- If the Block is being used as a "View Block" within a Multi-View Block, use AutoCAD Nodes on the Defpoints layer within the Block as alternate insertion points. (See Chapter 8 for more information.)

DRAWING INDEXES

As a final note before moving into specific file, object and project type considerations, let us consider the issue of performance. In general terms, performance will be enhanced by separating information into separate files and XREFing them together. However, several external factors can then come into play and affect performance as well. We will not discuss network topology and hardware considerations here, as they are out of the scope of this book. However, these issues will have an impact, so discuss these issues with your network support personnel. In addition, some items within the AutoCAD environment are also worthy of consideration. First, a drawing file can be indexed by AutoCAD. An index is simply a list of information that AutoCAD can refer to when loading a drawing to speed up the loading process. Rather than having the entire file loaded when an XREF is loaded, an index enables AutoCAD to load only those portions required to display the portions of the file needed. There are two types of indexes: Layer and Spatial; both are set by the INDEXCTL command.

- **Layer Index** – Keeps a list of all objects organized by their layer. In this way if a layer is frozen, the XREF simply ignores it and all of its objects when loading.

- **Spatial Index** – Keeps a list of all objects organized by their location in 3D space. In this way, if a portion of the drawing is excluded by the use of an XClip, the XREF simply ignores all of the objects within that portion of space when loading.

Indexes are saved with the drawing and can be enabled within your office standard template files if you wish. There are two ways to set up indexing. Either type **indexctl** at the command line and input the appropriate value from Table 2–1 or access it from the SaveAs dialog box under **Tools>Options** (see Figure 2–33).

 GUIDING PRINCIPLE: Even though indexes can make the file size a bit larger, it is recommended that you enable both types of indexes in your template file. (Set INDEXCTL to "3.")

Table 2–1 *Index Control* (INDEXCTL) *Settings*

Value	Result
0	No indexes are created
1	Layer index is created
2	Spatial index is created
3	Layer and spatial indexes are created

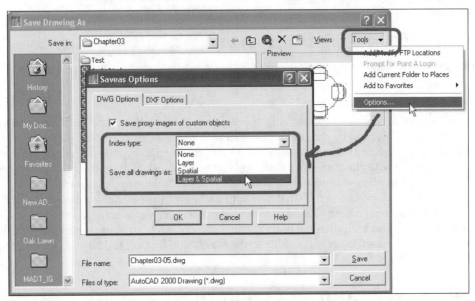

Figure 2–33 *Setting Indexes in the SaveAs Option dialog box*

PRESERVING XREF LAYER VISIBILITY

Another important system variable to consider is the VISRETAIN setting. VISRETAIN, or **Visibility Retain**, controls the configuration of layers in a reference file. If this setting is turned on, then any layer states configured within the main drawing will be retained the next time the file is opened, regardless of how they were saved in the reference file itself. The opposite is true if VISRETAIN is turned off. VISRETAIN is saved in the drawing file and should be made a part of the template file (see "Template Checklist" in this book's Conclusion). By default, it is on in all new files. It is highly recommended that you keep this setting on in all of your template files and in all drawings within your firm. There is little good reason to ever turn it off. To turn it on, go to **Tools>Options**, click the Open & Save tab and place a check mark in the **Retain changes to Xref layers** check box (see Figure 2–34). You can also type **visretain** at the command line and set it to **1** (on).

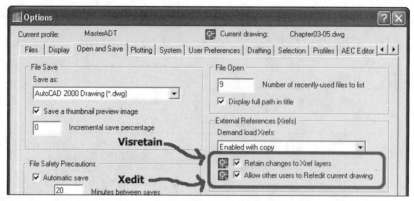

Figure 2–34 *Turn* VISRETAIN *on in all template files and drawings*

XEDIT SYSTEM VARIABLE

In the same location in the Options dialog box is the **Allow other users to Refedit current drawing** control. This is on by default and is saved with the drawing. (You can also set this with the XEDIT system variable at the command line.) With this setting on, the current drawing may be edited remotely with the **Reference Edit** (REFEDIT) command. How you set this will depend on the XREF procedures and strategies you have or would like to put in place in your firm. REFEDIT can be a very powerful tool that can help users save time switching between multiple drawings. If you do not have any strong feelings against its use, please leave it turned on in all of your template files and let individual users decide how and when to make use of it. Regardless of how this variable is set, users will be able to use REFEDIT on Blocks, and should be encouraged to do so.

PROJECT TYPES AND FILE STRUCTURES

Throughout this chapter, we have tried to explain the virtues of the Separated Building Model. We have explored the influence of Blocks, XREFs, file naming and ADT object parameters. Let's try to incorporate all of these factors and see the Separated Building Model concept in practice with a few common project types. Throughout these examples, we will look specifically at the interaction of the following issues:

- Design phase (SD, DD, CD etc.)
- Design type (Inside – Out, Outside – In etc.)
- Drawing Type (Plan, Section, Elevation etc.)
- ADT Architectural and Annotation object relationships

Clearly, the Schematic Design phase will require the fewest drawings, while the later design phases will require progressively more files. The following examples will use an expanded version of the NCS 2.0 Model File Types list shown here in Tables 2–2 and 2–3. The rightmost column indicates the ADT and AutoCAD objects most likely to populate each of these files. Other objects can often be included as required on a project-by-project basis.

Table 2–2 *NCS 2.0 Model File Types*

Discipline used	Code	Description	Objects Typically Used
All	FP	Floor Plan	Wall, Door, Opening, Window, Window Assembly, Tags, Stair, Railing, Space, AEC Dimension, Tags
	FP	Roof Plan	Wall, Roof, Roof Slab
All	SP	Site Plan	Drafted Entities
All	DP	Demolition Plan	Wall, Door, Opening, Window, Window Assembly, Tags, Stair, Railing, Space, AEC Dimension
All	QP	Equipment Plan	Multi-View Blocks, AutoCAD Blocks
All	XP	Existing Plan	Wall, Door, Opening, Window, Window Assembly, Tags, Stair, Railing, Space, AEC Dimension
All	EL	Elevation	XREF, 2D Section/Elevation, Bldg Section, Bldg Elevation Line, Bldg Section Line, Tags, Blocks, Dimension Labels
All	SC	Section	XREF, 2D Section/Elevation, Bldg Section, Bldg Elevation Line, Bldg Section Line, Tags, Blocks, Dimension Labels
All	DT	Detail	Drafted AutoCAD Entities
All	SH	Schedules	XREF, Schedule Table
All	3D	Isometrics/3D	XREF, Any
All	DG	Diagrams	Any
A, I	CP	Ceiling Plans	Ceiling Grid, Mask Block, Space
A, I	EP	Enlarged Plans	XREF, Wall, Door, Opening, Window, Window Assembly, Tags, Stair, Railing, Space, AEC Dimension
A, I	RP	Furniture Plans	XREF, Multi-View Blocks, AutoCAD Blocks, Tags
A, I	NP	Finish Plans	XREF, Multi-View Blocks, AutoCAD Blocks, Text, Dimensions, Tags, Blocks, Dimension Labels

Notes:

1. The code "FP" is used for any horizontal Floor Plan view. Although an argument could be made that a special code is needed for Roof Plans, here we are simply using the FP designation with the number following the final Floor Plan. Therefore, if it were a six-story building, the Roof Plan would be in file *A-FP07.dwg*. Another option: consider using RP for the roof plan designation in the "A" discipline and the again using the RP for furniture plan designation with the "I" discipline. The latter method is used by The CADD/GIS Technology Center of

the Department of Defense in their A/E/C CADD Standard release 2.0. This standard is compliant with the National CAD Standard 2.0 and available free for public download. See Appendix A for complete URL information.

Table 2–3 *Additional Recommended Model File Type Designators*

Discipline used	Code	Description	Objects Typically Used
A	GR	Column Grid	Column Grid, Column Grid Bubbles
A	CR	Building Core	
A	EN	Building Enclosure (Skin or Shell)	Curtain Wall, Curtain Wall Unit, Wall, Door, Window, Window Assembly, Tags
A	CM	Composite Model	XREF
A	SL	Floor Slabs	Slab
A	ST	Stairs	Stair, Railing
A	AN	Annotation File	Multi-View Blocks, AutoCAD Blocks, Text, Dimensions
A	SK	Sketch File	Any
A, I	AA	Area and Space Plan	Space, Space Boundary, Area, Area Group
A	MM	Mass Model	Mass Element, Mass Group, Slice
A	FR	Framing Plan	Structural Member
A	UN	Unit Plan	Wall, Door, Opening, Window, Window Assembly, Tags, Stair, Railing, Space, AEC Dimension
I	BB	Base Building	Any (Existing Building Permanent Conditions)

CHOOSE A PILOT PROJECT

The most common approach to ADT deployment is to choose a pilot team of users and a project typical of your firm's body of work. The project ideally is starting up new at the time of migration. There will be fewer hurdles to overcome if you begin with a new project rather than attempt to convert an existing one mid stream. The exact type of project or scope of work is not important, only that it is typical of the work commonly performed by your team. If your firm has made its reputation designing commercial office buildings, it would not be wise to have your pilot project be the firm's first high-profile school project.

The project schedule pace is also important to consider when choosing the pilot. Projects with longer project schedules can more easily absorb the learning curve associated with the adoption of ADT Building Model methods. For example "High Tech" fabs are extremely fast paced and do not provide a good environment in which to deploy ADT for the first time. In contrast, a hospital project that may last well over a year (and sometimes several) with lengthy DD and CD phases is much more likely to be able to sustain the learning curve and "ramp-up" time needed to deploy ADT. Choosing the wrong project for your pilot could seriously compromise upper management's perception of the software and its potential to improve productivity and process workflow.

There are far too many different types of architectural projects to effectively count or quantify. The purpose of this section will be to reverse engineer a few common project types with the aim of drawing practical conclusions that we can use in our implementation plan. To assist in this endeavor, gather up the following resources for reference as we go through the exercise:

- List the most common project types practiced by your firm with percentages relative to total project load.
- Draw a diagram of a typical file structure used by your firm. (Include folders, files and XREF relationships.)
- Gather the complete sets of drawings for half a dozen recently completed projects.

As you work through this section, keep the items listed above close by for comparison and use the following questions to help you assess what your firm's needs are.

SMALL COMMERCIAL BUILDING EXAMPLE

This is a Model/Sheet example where Annotation is included in the model files. The project is a 30,000 SF four-story commercial office building (see Figure 2–35). The project is mostly core and shell with some build out occurring on one of the tenant floors. The project structure shown here is for the construction documents phase.

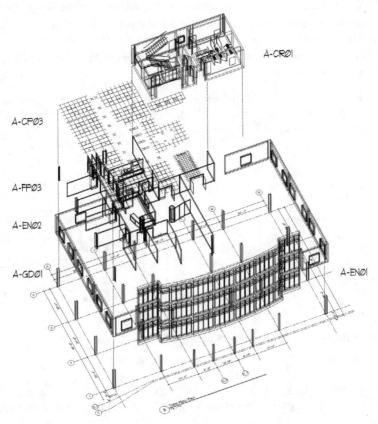

Figure 2–35 *A small commercial office building model structure*

> ## THIS MAY LOOK FAMILIAR
>
> If you have read Mastering Autodesk Architectural Desktop, this project will be familiar. This is the structure of the commercial project from that book. The structure as presented here has been modified slightly, particularly in the inclusion of two additional composite model files. One of these new composite models, *A-CM01.dwg*, includes only the exterior shell components, the roof and the site plan. This model is used to generate full building elevations and will generally perform better than the model that includes all of the files (*A-CM00.dwg*). The other new composite model is for generating Door schedules and based on the same assumption, fewer files will increase performance. These composite models have also been renamed to use the "CM" designation rather the "COMP."

The project is for new construction and includes four floors of "core and shell" speculative office space. Much of the design is in the overall massing and form of the exterior of the building and its main public spaces.

The separation of files is a bit more granular than typically recommended by the *NCS 2.0*. The primary concerns involved in devising the structure were as follows:

- Maximize the number of team members who could work in the files at the same time.
- Maximize the performance of each file for the generation of Elevation, Section, Schedule and Plot files, while at the same time minimizing individual file size.
- Make best use of ADT object parameters and strengths (covered previously in this chapter), while minimizing the effect of any object's individual weaknesses.

Figure 2–36 shows the file relationships, while the complete file structure chart is presented in Appendix B (on the CD). Table B2–4 details the structure of the sheet files and Table B2–5 details the structure of the model files that rely on XREFs.

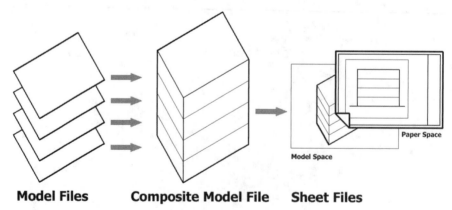

Model Files **Composite Model File** **Sheet Files**

Figure 2–36 *Composite model file relationship to models and sheets*

It is recommended that these file structures be built as early in the project as possible. The purpose of that recommendation is two-fold: First, by building the file structure early, you have created a digital "Cartoon Set" that can be used by the Project Manager for planning and staffing. Secondly, all of the files will already be created, named, XREFed and saved in their respective folders at the start of the project. This makes it very easy for users to get right to work on the project without worrying about needing to stop and create files. The initial project structure setup ought to be performed by the Project Data Coordinator. For a complete description of the role of the Project Data Coordinator, refer to the "How to Use This Book" section of the Preface.

LARGE COMMERCIAL OFFICE TOWER

The following is a file structure for a twenty-five-story core and shell office tower. The lobby is a double-height space with a Curtain Wall entrance and solid shell on the remaining three sides of the building. As such, the skin for the first three levels is separated into three files. The first one is for the solid portion of the first floor (indicated in Table B2-7 in Appendix B on the CD as "Horizontal Skin 1st"). Another is for the Curtain Wall entrance on the first floor, (indicated in Table B2-7 as "Curtain Wall 1st") and the final file is used for the entire skin of both the second and third floors. (Indicated in Table B2-7 as "Horizontal Skin 2nd - 3rd"). Additional Curtain Walls are used to enclose the next twenty floors (these are in separate files as well), and the final stories of the building have an enclosure similar to the first floor with a horizontal solid enclosure on three sides and a separate Curtain Wall on the front. The structure of this project makes use of a separate Annotation file. The Annotation files here are built much like the composite model files. They use XREF Attach to gather all of the individual files required to properly represent a complete floor plan. Annotation (text, notes, symbols and AutoCAD dimensions) is added in this file, in model space directly on top of these XREF files.

 Note: Due to the behavior of the tools, AEC Dimensions and Tags may not be added to the Annotation files. (Refer to the "ADT Object Characteristics" topic earlier and Tables B2–1, B2–2 and B2–3 in Appendix B for more information on this issue. Appendix B is located on the CD in PDF format.)

Some CAD Managers prefer to add annotation directly to the sheet files, as is the case in the "Small Commercial Building Example," while others prefer to separate the annotation into its own file, as will be shown in this example. The main advantage of the separate Annotation file is that it allows the complete separation of any "scale dependent" information from the full scale model files without requiring users to *work* in the sheet files. While working in a sheet file, if you change the Layer, Linetype Scale, Viewport Scale, Display Control Settings or forget to save the file with the correct paper space layout active, it can have an adverse affect on plotting. The Annotation file seeks to remedy this situation by making the sheet file a "set it and forget it" affair. The sheet file is configured at the start of the project by the Project Data Coordinator and then opened *only* for printing the document set. In its purest form, this process should never require any edits to be made directly in the sheet file. An additional bonus of the separate Annotation file approach is the ease with which "clean" model files can be sent to consultants. If you do not include your Annotation files in the package you send to your engineers, they will not have any of your notes, dimensions and other annotative items that they would likely freeze anyway.

Issues such as the need to place AEC Dimensions and Tags directly in the model file tend to erode the philosophical underpinning of this technique. In addition, the added complexity that this approach builds into the project file structure can render the use of a separate Annotation file undesirable. However, minimizing the amount of actual work that needs to take place in the sheet file is a worthwhile goal, and the separated Annotation file is one way to achieve this (see Figure 2–37). Regardless of the approach you take on this matter, it is important once again to maintain consistency. However, it is not necessary to mandate the firm's position on the use of Annotation files as a global office standard. This particular topic is typically best decided on a project-by-project basis.

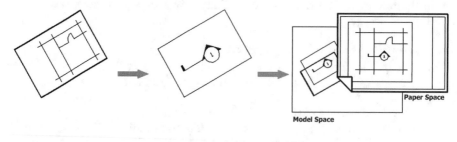

Model File   **Annotation File**   **Sheet Files**

Figure 2–37 *Using separate Annotation files*

It can potentially become confusing when the numbering scheme does not necessarily synchronize with the floor numbering, particularly in this example, where several floor plates are being used as "typical" floor conditions. If you prefer, you could substitute a lettered scheme (remaining NCS compliant) for the numbering used in Tables B2–7 in Appendix B and in Figure 2–38. Two possible letter schemes are shown in Tables B2–8 and B2–9 in Appendix B.

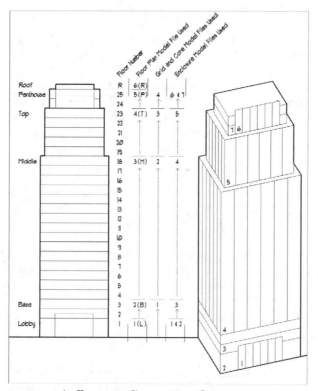

Figure 2–38 *Large commercial office tower file usage per floor*

Figure 2–38 shows that the height of the building is divided into six sections: Lobby, Base, Middle, Top, Penthouse and Roof. You could use the Letters: **L, B, M, T, P** and **R** to represent each of these divisions. The advantage would be that the names would connote the function of the file, making it easier for users to find the file they needed for a given task. The disadvantage to a scheme such as this is that the files would not sort correctly in lists. Therefore, you could instead use the letters **A, B, C, D, E** and **F** where A = Lobby, B = Base, C = Middle, D = Top, E = Penthouse and F = Roof. In this scheme, the files would sort correctly, but the users would still need to become familiar with the lettering scheme to understand the content of each file. However, in this scheme, numbers could be reserved for items that were "floor specific." Regardless of the scheme you ultimately choose, it is important to reiterate that files be set up at the beginning of the project, that the naming scheme (whether it use numbers or letters) be used consistently throughout the entire project, and that all team members understand the structure completely so that they can use it effectively.

There are some additional issues regarding projects of this type. Information on how to work with multi-story Stair and Curtain Wall files can be found in Chapter 5. Information on how to deal with schedules on repetitive (typical) floors is found in Chapter 6.

INSTITUTIONAL PROGRAM DRIVEN PROJECTS

The NCS does not specify a separate model file type for Space Plans. In Table 2–3 above, the code "AA" has been recommended for this purpose. Likely the intention of the NCS is for Space Plans to be akin to Floor Plans and therefore to use the code "FP." If used with the "I" discipline code, this can be a viable approach. Regardless of how you decide to name the file(s), it is recommended that your space planning schemes be maintained in a separate file or files. One reason for this recommendation is the way that Space objects and Space Boundary objects function. A Space Boundary object is set to "manage" the Space objects. This means that you will be able to manipulate the edges of the Space Boundary and have the shape and area of the included Spaces react in real time. As powerful as this feature is, it does not work with Wall objects, only Space Boundaries. Thus when it is decided to generate Walls (usually when beginning DD or CD phases), the Walls ought to be moved to an "FP" (Floor Plan file), while the Spaces and Boundaries remain in the "AA" file. One other item that can make this transition between phases a bit of a challenge is the relationship of Ceiling Grids to Spaces. If in your CD phases you opt to clip the Ceiling Grids to the Space objects, you will likely want to have those Space objects in the same file as your Ceiling Grids. This may necessitate moving those Space objects to the "CP" (Ceiling Plan) file.

The ADT's space planning tools (Spaces and Space Boundaries) are not limited to conceptual design space planning tasks. They can also be used throughout the life of the project for area analysis and code compliance checking. For all of these reasons, it is recommended that your Space objects and Space Boundaries be maintained in their own file throughout the life of the project (see Figure 2–39).

In addition to Spaces and Boundaries, there are also Areas and Groups. Areas are 2D objects that provide robust area calculation analysis potential. Since they are drawn like polylines or converted from existing objects, they can also provide a good tool for space planning. A Space object is a hybrid object that includes a floor and ceiling plane in 3D and Section. In plan, it is useful for tracking area. Due to the potential overlap in functionality it is not necessary to use both Area objects and Space objects in the same project; however, there is no reason they cannot both be used. This topic is covered in detail in Chapter 10. Regardless of your ultimate choice of Areas or Spaces, they should be maintained in separate files.

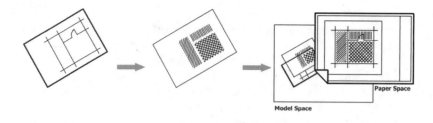

Model File   **Area or Space File**   **Sheet Files**

Figure 2–39 *File Structure with separate Area or Space Plan files*

Presented in Tables B2–10, B2–11 and B2–12 in Appendix B (on the included CD ROM) are some sample project structures that incorporate the use of separate Space/Area files. The remainder of the projects that are heavily dependent on space planning and programmatic requirements will be set up in a fashion similar to the two commercial projects illustrated in the preceding examples.

HOTEL AND MULTI-FAMILY HOUSING

Sharing similarities to each of the other examples, the distinguishing factor in hotel and multi-family housing projects is their "unit-based design" approach. In the unit-based design process, several typical units are designed and then fit together into an overall Floor Plan scheme. For a hotel, these units would be the typical guest rooms. In housing, they would be the individual apartment layouts. To develop a file composition strategy, first create a file for each unique unit plan. For this task, we could use the "FP" designation, but it will likely be more desirable to use a different code such as "UN" for Unit. (See Table 2–3.) A variety of factors will influence the particular design items included in the Unit Plan. However, as a general rule of thumb, include all of the fixed, interior architectural components such as Walls, Doors and millwork. Columns will likely be in a separate Grid file just like the projects previously shown in the above passages. Each of these Unit Plans will XREF Attach to the FP (Floor Plan) file. Attach is important here because we want the FP file to serve its typical function as the overall floor plan for a given floor of the building. When we XREF it to other files such as the sheet file, it will need to call the Unit Plans along with it; otherwise we would see an empty file when referenced. You will also need to make use of XCLIP (discussed above) and in some cases, invisible Wall Endcaps to make the union between the various Unit Plans "seamless."

The exterior building skin must be addressed next. Just as in the commercial project examples above, it will be desirable to keep the building's exterior enclosure in its own separate file or files. These can use the "EN" (Enclosure) drawing type code used in the examples above. Therefore, the "FP" files of this type of project will be comprised of several XREFs. There will be an XREF for each Unit Plan, copied, mirrored and rotated several times as appropriate. There will also be the XREFs for the exterior enclosure (EN) and the Column Grid (GR.)

The Column Grid and the building skin enclosure could be built from units as well if design needs warranted it. If column position is driven by the Unit Plan design, then you could consider building a small Column Grid file for each Unit Plan and then building the overall building floor grid from XREFs of these smaller grids. Likewise, the exterior building enclosure could be broken into smaller sections of building skin matched up to each Unit Plan's exterior exposure. Naturally incorporating either of these techniques will increase the quantity and complexity of the XREF structure of your floor plans. There is nothing inherently wrong with having a large quantity of XREFs. However, many users tend to resist this level of complexity. Recall the "Understanding Files, XREFs and Blocks" topic earlier in this chapter and consider using Blocks within the Enclosure and Grid files to achieve the same general result (see Figure 2–40). When doing so, you will need to use "invisible" Wall Endcaps and XClips to "hide" the seams between portions of the exterior Skin Wall. Editing the individual Blocks can be accomplished easily with the REFEDIT command.

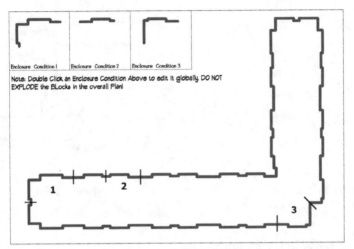

Figure 2–40 *A Building Enclosure Plan comprised of Blocks of several typical conditions*

The same could be done for the Unit Plans as well, but in general, you will achieve better results from a staffing perspective if you use separate Unit Plan XREFs (see Figure 2–41). Several users could be working each in their own Unit Plans simultaneously. With the Block approach, only one user can access the file at a given time. This is, however, acceptable in the case of the building skin in most circumstances. This is because the skin is not likely to change as often as the Unit Plans themselves.

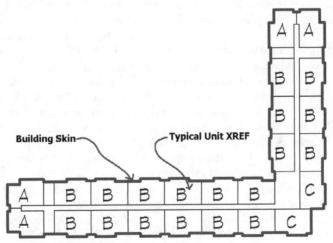

Figure 2–41 *Unit-based project building Floor Plan diagram*

For this example, we will assume individual Unit Plan XREFs and a single Column Grid and single building enclosure file for the entire floor plate (see Figure 2–42). Details can be found in Tables B2–13 and B2–14 in Appendix B (on the included CD ROM).

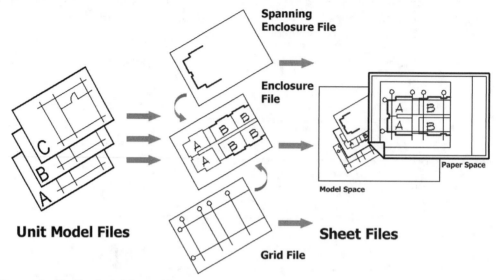

Figure 2–42 *Unit-based Design file structure*

TENANT BUILD-OUT (INTERIOR ONLY) PROJECTS

Many of the file structures that have been already suggested will work well for an Interiors project. If you plan to use ADT space planning tools, keep in mind the recommendations above in the "Institutional Program Driven Projects" example. Keep Space Plan files separate. Often in Interiors projects, a Base Building file is provided by the building owner or client. These files are often AutoCAD drawing files saved in previous DWG or DXF formats. It is not necessary to convert these files to ADT objects, particularly if you do not intend to change anything in the existing file. This is especially the case when these existing files show core and shell conditions only. If existing interior build-out is included in the files that you have been provided, separate the geometry into two files—one will become the Base Building and the other will become the Existing Conditions file. The Base Building file will use the "BB" (Base Building) file designation and should contain all of the "permanent" building components. This will usually include the Core, Shell Column Grid and Columns. In a tenant improvement, these are items not likely to be disturbed. All of the remaining construction, which would include all of the existing interior partitions and build-out, should go into the Existing Conditions file with the designator "XP" (Existing Plan). As you work in the Existing Conditions file, redraw only those items that are required by the design of the new work. Again, there is little benefit to attempting to convert existing geometry to ADT objects, particularly if there is little to no new work being done on them. To designate items in the existing background files as Demolition and Existing to Remain, simply change the layers of the AutoCAD geometry. Refer to the recommendations below in the "Remodeling and Renovation" topic for more information on working with Demolition and Existing to Remain.

Composite models are used to facilitate the creation of overall building sections, elevation and schedules. In addition, if 3D drawings of a complete building design are desired, then the composite model serves this purpose as well. Since most interior projects will not likely have these types of "overall" views, it may not be necessary to generate a composite model file. Even if the project spans many floors of a building, there is still little need for the composite model file. If you plan to include all floors of a project on the same schedule, then you should create the composite model. Otherwise, it may be safely omitted from your project structure. Details of a sample file structure can be found in Tables B2–15 and B2–16 in Appendix B (on the included CD ROM).

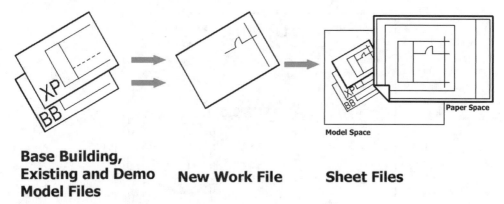

Base Building, Existing and Demo Model Files   **New Work File**   **Sheet Files**

Figure 2–43 *Plan structure for interiors or renovation type projects*

REMODELING AND RENOVATION

In addition to all of the other items outlined in the various project types, Demolition, Existing to Remain and New Construction must be properly articulated. Naturally any building type may be remodeled; therefore start by determining the overall file structure based on the size, scope and type of project. Use the previously outlined guidelines to assist in your decisions. Demolition and Existing to Remain can be incorporated into any of the project structures. The *NCS 2.0* makes provision for separate model files for Demolition and Existing Construction in the codes "DP" and "XP." There is also the Status modifier for Layer names, which would be used in each of these files. One complication to this approach is the way in which ADT objects receive their layers. Layering is handled automatically by ADT and can be customized by editing the Layer Standard used and the accompanying Layer Key Style. The topic of Layers and Layer Key Styles will be discussed in detail in Chapter 4.

If you place items in separate files based on Status, this situation is easily remedied. Build a Layer Key Style for Demolition and make it the default in the DP file. Do the same for the XP file. Save both of these as template files. Leave your standard Layer Key File assigned to the FP file. There is a disadvantage to working in separate files for purposes of Status, at least with respect to the Demolition and Existing to Remain. Both of these items exist when the project begins. Throughout the course of the project, we decide to demo certain portions of the existing construction while leaving others. If you keep them in separate files, one would have to physically remove those items to be demolished from the existing file and place them in the demolition file. This may not prove to be the best workflow. Regardless of your choice for Demolition and Existing to Remain, New construction ought to be created in a separate file in most cases. By doing this, you maintain a clear separation and understanding of the new work scope.

As mentioned above, Chapter 4 is devoted to working with Layers and Layer Key Styles. For our discussion here, we will simply summarize the options available for dealing with the Status of ADT objects should you decide to place two or more them in the same file. Details of a sample file structure can be found in Table B2–17 in Appendix B (on the included CD ROM).

- ADT and AutoCAD objects alike can be manually re-layered to a Demo or Existing layer (see Figure 2–44).

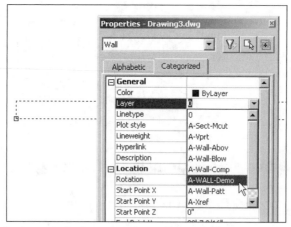

Figure 2–44 *Manually changing the layer to indicate Status*

- Layer Key Overrides may be used to temporarily add the Status field to Layers names as new objects are added to the drawing (see Figure 2–45).

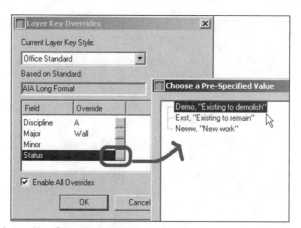

Figure 2–45 *Using Layer Key Overrides to create or use layers to indicate Status*

- Separate Layer Key Styles can be devised for each Status: **Demolition**, **Existing** and **New**. The appropriate Layer Key Style can be made active before objects are added (see Figure 2–46).

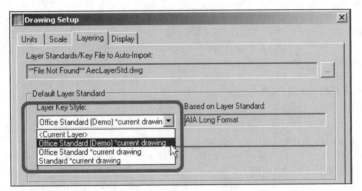

Figure 2–46 *Using alternate Layer Key Styles to set the drawing Status*

- Demolition and Existing to Remain can be designated by object Style. (Wall Style, Door Style, etc.) Sub-components within a particular Style are hard-coded to a particular pre-defined layer. In this way, choosing this Style assigns the correct graphic properties to the object (see Figure 2–47).

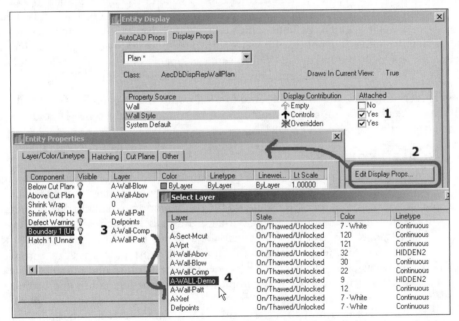

Figure 2–47 *Using Entity Display to create a "Demolition" object Style*

Although there are several different ways to accomplish the delineation of Status, none of them completely outshines all of the rest. Layer Key Overrides merely change the Layer name and not the Layer Properties, and swapping Layer Key Styles can be counter-intuitive and error-prone. Object Styles have the disadvantage that the technique solves only the layering of the object's sub-components and not the layer of the object itself. However, since it is unlikely that you will ever need to freeze the Wall layer, this technique can prove very viable. Once a Style has been added to the library, you may simply use it like any other. This Style will in turn assign the proper sub-component Layer, and any other settings such as a Demolition Cleanup Group. It is recommended that you use Demolition Styles for items like Walls and simply edit the layer manually for all other objects.

LARGE HORIZONTAL PLAN

For projects with a very large floor plate, it may be worthwhile to cut the plan into sections. If a single floor plate is larger than 50,000 SF [4,600 SM] and reasonably complex, this is a valid consideration. The NCS makes the distinction based on what is able to fit on a sheet at 1/8" = 1'-0" scale. A number of factors will play into this decision. Wherever possible it is advisable for the entire floor plate to be continuous in a single file. However, the larger the project gets, the more likely it is that a large floor plate will slow system performance and file load times. In addition, only one person can be in the file at the same time, so on large projects, this could prove a significant limitation.

Whatever criteria you use to decide when to break a model up, do follow the NCS recommendation regarding where the break should occur. Specifically: "The seams between the partial floor plan models need not align with the match lines of the sheets, but are typically located where there is minimal geometry crossing the seams." When breaking files that contain Wall objects, you will need to add invisible Endcaps where the Walls are broken. To do this, create an Endcap from a single segment polyline with a non-zero width. It is not important what the actual width is, only that it is not equal to zero. Create an Endcap Style from this polyline and when applied to a Wall, the end of the Wall will appear "open." This will allow you to splice two XREFs of different sections of the building plan together without a line showing at the seam.

The remainder of the file configuration setup can follow any of the guidelines outlined here as appropriate. Details of a sample file structure can be found in Tables B2–18 and B2–19 in Appendix B (on the included CD ROM).

SMALL-SCALE PROJECTS

These recommendations apply to any small-scale project type, including single- and double-family residential and small-scale commercial projects. A popular strategy among CAD Managers is to have a variation of the office standard file configuration for small-scale projects. The value in this approach comes in the simplicity that can usually be attained in file structure for small projects. Often, only one or two individuals are assigned to small projects and as a result, this fact tends to encourage fewer XREF files. Simplifying project setup is always a valid and desirable goal. However, simply having XREFs does not necessarily make a project more complex. There are some very elaborate and complex layering schemes in use and promoted by many of the very same people who shy away from XREFs. Be careful that you are not simplifying one aspect of the project only to make another more complex. It is important to strike a balance. Follow these guidelines when working on small-scale projects:

- Create a separate FP file for each floor (or level change) in the project, including a separate file for site plan and another for the roof.

- Separate Grid, Enclosure, Ceiling and Furniture files are often *not* required. All model geometry may be placed in the FP file for a small-scale project.

- If there is relatively little demolition, place all construction regardless of Status in the FP file and use one of the techniques mentioned in the "Remodeling and Renovation" example to manage existing construction.

- If using Slabs, Spaces or Areas, do consider placing those items in separate files. It is easier to unload an XREF than to build elaborate Layer Manager and Display Manager schemes to control these items.

- Create a composite model file as indicated in the other examples. However, place each XREF in it on its own layer. These may have names like A-Xref-0 (basement) A-Xref-1 (first floor) and so on.

- Use the composite model file to build your sheet file. In this way, and following the previous recommendation, you will be able to display several Plans on a single sheet of paper (see Figure 2–48).

The final point requires a bit more elaboration. To set up a sheet that shows both the first and second floor plans, create a sheet file with two viewports in the printing layout. In the first viewport, set the view direction to top and set the scale as desired (for example 1/4"=1'-0" [1:50]). Float model space in that viewport and freeze in the current viewport all of the XREF layers except the ones you want to see. In this case, to see the First Floor Plan, freeze in the current viewport A-CM00|A-Xref-0, A-CM00|A-Xref-2 and A-CM00|A-Xref-3. Details of a sample file structure can be found in Tables B2–20 and B2–21 in Appendix B (on the included CD ROM).

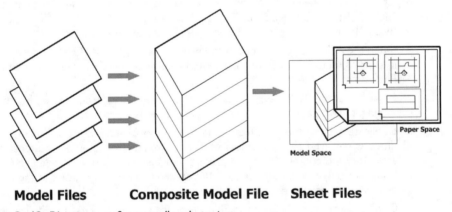

Model Files　　**Composite Model File**　　**Sheet Files**

Figure 2–48 *File structure for a small-scale project*

MASS MODELING

The needs of a Mass Model built in or imported into ADT are similar to those of the Space Plans as outlined above in the "Institutional Program Driven Projects" topic. Mass Models have specific Display Control and procedural needs. It is best to build your Mass Models in their own file. You can use the "SK" (Sketch) designation above or devise your own scheme for file naming. Once a Mass Model is substantially complete, the entire model or data extracted from it can be exported to the other files in the project set.

PROJECT SETUP CHECKLIST

Now that we have taken a comprehensive look at a variety of project file structures, the final thing remaining is a list of steps to perform at the start of a new project. The earlier you perform these steps in the life of a project the better for all persons involved. The following tasks are the province of the Project Data Coordinator. This point has been raised before, but the value of developing a complete set of drawing files early in the life of a project cannot be overstated. Take the time to do this early, and you will be glad you did.

Start with the files that have no XREFs and work your way forward. The following steps assume that template files have already been developed and are in use. Please refer to the Conclusion for a checklist of traits for drawing template files.

1. Using your standard *PlanSectElev.dwt* template file (basic model template), create a file for the Site Plan. Insert any existing geometry into this file and save it as *A-SP00.dwg*.

2. Open any Existing Conditions files (from outside sources, previous projects, clients, etc.) and run AUDIT and PURGE.

   a. For Interiors projects, separate the "permanent" items from the non-permanent ones. Using CUT and PASTE or WBLOCK, move the permanent geometry into a Base Building file (*I-BB##.dwg*), and move the rest to an Existing Conditions file (*I-XP##.dwg*).

   b. For other projects, simply rename the files using the EX designations.

3. Create new drawings based on the *PlanSectElev.dwt* template file for Building Grids (GR), Core (CR), Stair (ST), Enclosure (EN) and Plan (FP.) Create only those files required by your project type. These are your "Base" files.

4. Open the *A-FP01.dwg* (created in the previous step) and XREF the other files to it using **Overlay**. The Floor Plan should reference the Grid, Stairs, Core and Enclosures if present.

   a. Add "dummy" geometry to each of these files for reference at early project stages. This geometry will be deleted later in the project.

5. Duplicate each of the Base files (Grid, Stairs, Core and Enclosures) as many times as required by the quantity of floors and anticipated number of design variations and complexity in the project. For instance, you may have five Core Plans, but only two Column Grids. (If you are unsure at this point in the project, make an educated guess.)

6. Double-check all settings in the First Floor Plan. Save the First Floor Plan as the Second Floor Plan (*A-FP02.dwg*) and rename and re-path the XREFs as required. For instance, if the Column Grid is the same as the First Floor, but the Core is different, the Grid XREF will require no change, while the Core must be renamed and re-pathed.

7. Follow similar steps to create the Ceiling Plans and if necessary, any other specialty plans such as Furniture, Finish, etc.

8. Create a composite model file. If you are using a "Shell Only," "Plans Only" or "Stairs Only" composite model, create it first. Begin with the template file and XREF each required file into it. Pay close attention to the Z elevation as you insert your XREFs. They should be "stacked" at the correct height based on true "floor-to-floor" elevations.

9. Create your composite model (complete) file. Follow a similar process as the previous step. On small projects where you anticipate printing more than one plan per sheet, use separate layers to insert each XREF; otherwise they can all be on the same A-Xref layer.

10. Using a sheet file template, create your First Floor Plan sheet file. All XREFs should be inserted as Attachments, usually at 0,0,0 with no rotation, full scale and in model space. The Title Block file should be XREFed into the layout that you will plot from, also at 0,0 and full scale.

11. Set the scale of the viewport(s) in a paper space layout. (You can set up more than one layout if desired and appropriate.) Configure all Layer and Display Control settings within each viewport. When complete, set the viewport scale and lock the viewport. Save the file with the main layout active.

12. Double-check all settings in the sheet file. Using SAVEAS, create the remaining sheet files by re-pathing XREFs as required.

13. All files should have some sort of "dummy" geometry in them. Run a test batch plot on all sheets to see if everything is configured properly. Make any required adjustments.

14. Back up all files.

The keys to success in this setup task are paying close attention to detail and being very methodical about each step. Depending on the size of the project, setting up all files can be a long process, but your efforts will be greatly rewarded as users reap the benefits of a well-conceived project file structure. Make sure to double-check all settings at each critical step, particularly before saving a file or duplicating a file.

SUMMARY

ADT object composition influences best practices regarding their use and placement within specific files.

ADT objects have Parametric, Display and Data Properties.

The term "model" refers to the environment where work is performed in ADT as well as the result of those efforts.

ADT object dimensionality (2D, 3D and hybrid) influences best practices regarding their use and placement within specific files.

Architectural design strategy and phases should be considered when deciding how to structure an ADT project file structure.

Keep folder and file naming simple and easy to understand.

The Separated Building Model is the recommended structure for ADT projects.

Blocks and XREFs share many similarities and provide crucial benefits to the Building Model Structure.

Project and building type exert a significant influence on project file structure.

Be sure to set up all project files at the very beginning of a project for maximum benefit to project planning and scheduling.

CHAPTER 3

CAD Standards Assessment

INTRODUCTION

When you attempt to integrate new methods into an existing process, it is important to perform a thorough assessment of the existing system and practices. Through this process, it can be determined which aspects of the current process work well and which ones will need rework. Successfully implementing Architectural Desktop begins with this very strategy. Therefore, we will begin to analyze the way you currently work and on what tools you currently rely. This chapter is intended to be prescriptive in nature, with the goal of generating a complete checklist of action items you can use to work through the remainder of the book.

OBJECTIVES

The main objective of this chapter is to create an action plan for your firm's ADT implementation. With this goal in mind, we will explore the following topics:

- Set up plotters
- Decide whether to use Named or Color Dependent plotting
- Migrate existing or set up new AutoCAD Format items (text, dimensions, etc.)
- Identify and understand the key components of a successful ADT migration
- Understand staffing issues and training needs
- Build the checklist

WHO SHOULD READ THIS CHAPTER?

This chapter is intended for the **CAD Manager,** the **Project Data Coordinator** and the **Project Manager.** The **CAD Software User** should also be engaged in the discussions that this chapter would prompt. Therefore, this chapter is also important to that group as well.

CAD STANDARDS

Long before computers were used to produce architectural drawings, there were graphic drafting standards established governing everything that appeared on a drawing. Many firms adopted a rigorous methodology that dictated hand lettering to be as identical amongst drafters as possible. Accepted drafting techniques, lead weights and proper symbol templates were all accepted industry best practices.

Within these industry standards, there is plenty of room for variation and as such, nearly every architectural firm has its own stylized symbology and methods. A CAD standard includes the digital equivalent of the old drawing standards binders that many firms had on their shelves. CAD standards, however, reach beyond simply outlining the type of symbols and title blocks used by your firm. Computer aided design goes further than drafting, by enabling design exploration and coordination to happen more readily and efficiently. Therefore, while drafting standards are only a part of a CAD standard, they are an important part of a total CAD standard, and we will include in our discussion of CAD standards many issues related to drafting and drafting standards. However, since Architectural Desktop is much more than a drafting tool, it is impossible to separate the modeling and design aspects of ADT from the drafting functions. All features are intertwined and inherent throughout.

The essential argument for good CAD standards is simply the increased productivity attained by using consistent methods, and thus saved time. That is the goal of standardization, CAD standards included. A well-thought-out CAD standard should make everyone's job easier. If it does not, it is broken, plain and simple.

A CAD standard can solve many common issues by providing users with an easily accessible and organized library of frequently used components, to be reused repeatedly from one project to the next. More importantly, it can give staff procedures to follow when creating items that are not "standard" or quite as repetitive. CAD standards also suggest best practices to help eliminate wasted efforts on items that have already been solved by others in the firm or items that yield little overall benefit to the final project set. A well-conceived Office standard is essential to overall firm-wide productivity because it eliminates redundancy, improves quality and enhances everyone's capabilities.

For a simple example, let's consider something as pervasive as text. In architecture school, all "would be" architects were taught to letter the same way. This was a form of standardization. If two draftspersons worked on the same sheet, it was important to ensure that the drawing did not "look" like it was done by two or maybe even several people. This need remains an important issue today. Computers offer us many more choices in fonts and other text characteristics, but the desire to have drawings "read" consistently from one project or staff member to the next is still an important goal. Text that is drawn on screen uses a Style definition, which establishes the shape, orientation, height and, of course, the font used to draw text. There are different types of font technologies available. Factor in drafting symbols, title blocks, dimensions, and schedules, all of which use text, and we begin to see how one "small" part of a larger package can have a dramatic impact on the whole process.

The point of this little exercise is to illuminate the "spider web" of connections between the various pieces of a well-planned CAD standard. Therefore, we will take two basic approaches to this type of problem throughout this book.

- First, we will often start small, beginning with the smallest component of a topic and branching out to topics that are more complex.
- Second, we will often start with the desired final output and work backwards to the details of achieving them.

Although these two approaches may seem at odds with one another, they are actually quite complementary. A frequent switch from macro to micro scales will help us simultaneously address the "big picture" and the "small" details. In the next topic, we will begin by building on whatever existing standards, CAD or drafting, your firm may already have.

PLOTTING

In many ways, plotting is the most critical aspect of a well-conceived CAD standard. Although this is the final step in CAD production, it should be considered as part of the first step, since virtually all of the work we produce will ultimately be printed for distribution to the extended project team. In addition, the complexity of the drawings and our own sensibilities about the way our ideas are communicated on paper impose other demands on plotting. This includes not only the actual content of that printed information, but the specific qualities with which it is rendered. Therefore, items such as lineweights, colors, tones, drawing organization, symbology and clarity are critically important. For these reasons, plotting is the first issue we will address in this chapter (see Figure 3–1).

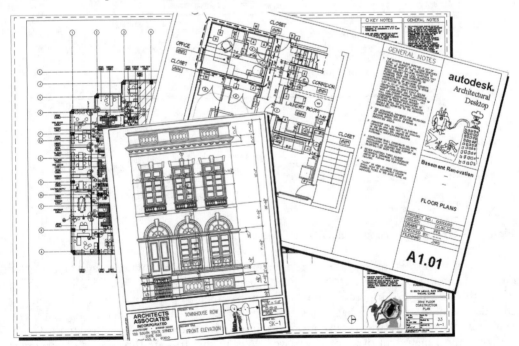

Figure 3–1 *Plotted ADT drawings*

BASIC PLOTTING

Our primary concern with plotting is determining which objects will display and how they should be rendered. Layer settings and Display Control within the drawing itself typically determine which objects are visible and invisible. Visibility rarely becomes a function of the plotting parameters. In contrast, plotter settings usually influence the rendition of objects at the time of plotting. Nearly every physical characteristic of an object, such as Color, Linetype and Lineweight, can be assigned within the drawing directly or "interpreted" while plotting. In prior releases of AutoCAD/ADT, many of these physical characteristics of a plotted object were determined exclusively by the interpretation of their interaction with a pre-built pen table. The modern "pen table," named more appropriately the "Plot Style Table," offers far more flexibility and functionality than its PCP and PC2 ancestors. In today's AutoCAD/ADT, pen tables still play an important role in the plotting process, but they are no longer the only way to control graphic quality and output.

As with any standard piece of Windows software, you call **Page Setup** from the **File** menu to establish your page size, scale, rotation and the printer you wish to use. The only obvious differences between printing and plotting are the size of the output and the use of a Plot Style Table in AutoCAD/ADT plotting. Plotters tend to be much larger devices, but many use the same ink-jet technology of many standard office printers. For these reasons, we will focus this discussion on understanding Plot Style Tables and plotter configuration.

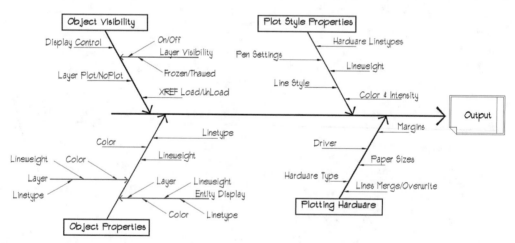

Figure 3–2 *Factors influencing AutoCAD/ADT output*

As you can see in Figure 3–2, there are four major categories of influence affecting plotted output. Beginning at the top left arm of the diagram is Object Visibility. This is controlled within the drawing by Display System and Layer parameters. Some overlap occurs when you consider the next arm of the diagram. Object Properties presents the potential for Layer and Entity Display settings, (which can control visibility in the first arm), but may assign the Color, Linetype and Lineweight of an object as well. Within this arm are three levels of hierarchy: Entity Display is furthest away, while explicit Object Properties is closest. This means that Entity Display can optionally assign a Layer, which can in turn assign a Color, Linetype, Lineweight and Plot Style. There can of course be other combinations, but the flow moves "downstream." For example, Entity Display can assign a Layer, but Layers cannot assign Entity Display Properties. Moving to the top right arm, Plot Style Properties occur next. This means that the settings within our Plot Style Table are applied next. This has the potential to influence or override the rendition dictated by the other two arms. Finally, the Plotting Hardware arm shows settings unique to certain printing devices that are applied by that device after all of the others at the time of plot.

INSTALLING PLOTTERS

ADT uses the plotting functionality of AutoCAD. Therefore, for a more thorough explanation of the features of plotting, consult a book on AutoCAD 2002, such as *AutoCAD 2002: Complete* by Bill Burchard, Dave Pitzer and Art Liddle, published by Autodesk Press. Plotters are installed in AutoCAD in much the same way as printers are installed in Windows. Once you have completed the hardware installation as per manufacturer's instructions, you will need to install the proper driver files. Many drivers for a large collection of standard plotting devices are included with AutoCAD upon installation. This makes installation of many of the most popular plotters very simple. However, it is always a good idea to check the Web site of your plotter manufacturer for the latest version of the driver before installation. Also, check to see if the manufacturer provides drivers specifically for AutoCAD.

Another important issue to understand is that there are three types of printer drivers available in AutoCAD. They are the **Heidi®** driver, the **Windows system** driver and the **Optimized Windows system** driver. Heidi drivers are specifically designed to interface directly between AutoCAD and print devices. Other Windows software cannot use Heidi drivers. They must print using Windows system printer drivers. The two forms of Windows system printer driver make it possible to install your device in the Windows Printers folder, where it will be accessible to all of your applications, including AutoCAD. AutoCAD will not require any additional installation, since by default all Windows system printers are accessible directly within AutoCAD. The only downside to this approach is a potential loss of certain functionality that is unique to AutoCAD. The Optimized Windows system printer driver is designed to alleviate this problem; however, currently HP is the only manufacturer creating optimized drivers.

So how do you decide which driver to use? This can be a tricky question, particularly since new hardware and drivers are being released all the time. The easiest way to decide is to consult your manufacturer's documentation or Web site. In addition, many Autodesk Resellers, Consultants and reprographic service bureaus have staff who understand the devices and can assist in making recommendations. You can also run some test plots to see which installation gives you the best results. If you can get the results you require without any loss of functionality or features by using the Windows system printer driver, then it is recommended that you choose this route. You will only need to install your printers once in the Windows Printers folder, and you can skip altogether the following tutorial on installing plotters. However, if the quality or features available from the Windows System drivers offered by your printer manufacturer do not give you acceptable results, then use the Heidi drivers as outlined below. (If you have an HP Plotter, the Optimized Windows system driver is your best choice.)

When you first install the printer/plotter, a collection of default settings will be established. These are likely the settings configured by the manufacturer, but you can change them to suit your needs. This will be the "default" version of this plotter when used within AutoCAD. You may, however, have a need to use the same plotter in several different ways. For instance, you might wish to configure the same hardware device to use Bond paper for check plots and Vellum or Mylar for final submission plots.

ISN'T A PC3 JUST A NEW VERSION OF A PC2?

If you have been using AutoCAD for a while and are familiar with plotting in Releases 14 and prior, it is important not to confuse a PC3 file with the legacy PC2 and PCP files. These legacy files had either pen settings or a mixture of pen settings and hardware settings saved within. Today the only real similarity between the PC3 files and their legacy ancestors is the name. ("PC" in all cases refers to Plotter Configuration.)

There is one other type of configuration file associated with plotting: it is called a PMP file. PMP stands for *Plot Model Parameter*. A file of this type can be created to store custom plotter calibration and custom paper size information associated with a particular plotter configuration file. There is more on this below.

Using the Plotter Installation Wizard

This is a short tutorial on installing plotters. Performing the steps in this tutorial will not permanently affect your system in any way. The Add-A-Plotter Wizard simply creates a file with a PC3 extension in your *Plotters* folder. To undo the results of this tutorial, simply delete the resulting PC3 file later.

1. From the **File** menu, choose **Plotter Manager** (see Figure 3–3).

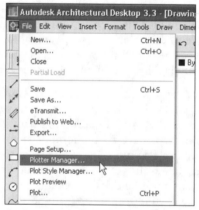

Figure 3–3 *Access the Plotter Manager window from within ADT*

This opens a folder in Windows that contains all of your Plotter Configuration files. These files have the file extension of PC3. Each file contains the instructions on a particular way to configure a plot device. There may be more than one PC3 file with different instructions for the same plot device. For example, you may want to set the default paper for one PC3 to Vellum and another of the same device to Bond, so that you can use one for check plots and the other for final plots.

2. Double-click the icon labeled **Add-A-Plotter Wizard**.

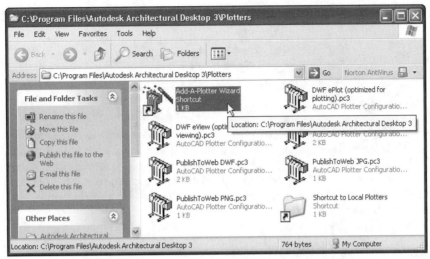

Figure 3–4 *Add a plotter with the Add-A-Plotter Wizard*

If an Autodesk Hardcopy System dialog box appears with the following message: "There are multiple software programs installed on this computer that use the Autodesk hardcopy system. Please choose the software program you want the edited file to work with from the list below:" choose **Autodesk Architectural Desktop 3.3** and then click **Continue.**

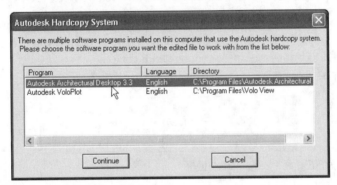

Figure 3–5 *Multiple Autodesk Products Alert*

This message appears if there is more than one Autodesk product installed on your system, such as Architectural Desktop and Volo View, as shown in Figure 3–5.

The next screen, the Add Plotter - Introduction Page, will have a message explaining how a PC3 file will be created at the completion of the wizard.

> 3. Click **Next.**

The Add Plotter – Begin screen presents the first choice needed to install a particular plotter configuration. You should have a good understanding of each of the three choices presented: **My Computer, Network Plotter Server** and **System Printer.** Let's explore each.

▶ My Computer

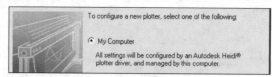

Figure 3–6 *Install plotter directly to My Computer*

Use this option only if there is a plotter (or printer) attached directly to the back of your computer (see Figure 3–6). This option does not make the plotter available to anyone else on your network. If you are on a network and other people use the printer attached to your machine, you will likely use one of the next two options instead. If you are running ADT stand-alone without a network, you may choose this option, or the **System Printer** option below. This option uses an Autodesk Heidi driver.

▶ Network Plotter Server

Figure 3–7 *Install a printer to access a network plot server*

This option is used when the printer or plotter is installed on a network print server (see Figure 3–7). A network print server provides a dedicated interface between the printer and any machine that needs to access it. The server can govern and log usage and negotiate traffic to the printer. The print server can be a dedicated server machine, or a special network device installed directly in the printer. Check with your IT support person to see if this is the way your printers are configured. This option uses an Autodesk Heidi driver.

▶ System Printer

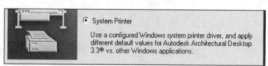

Figure 3–8 *Install a printer as a System Printer*

As stated above in the section on drivers, this is the best option if the hardware you are using supports it, and if the quality output meets your firm's standards (see Figure 3–8). In fact, if you are using system printers successfully, you may not even need to run the Add-A-Plotter Wizard. You only need to configure a system printer as a separate PC3 file if you need your AutoCAD defaults to be different from those already established in the system printer itself for other Windows applications.

 Note: You cannot choose the **System Printer** option if you have not first installed the printer in Windows using the Add New Printer wizard in the Printers folder. If you are unsure about how to do this, consult a resource on your version of Windows.

 Tip: As a general rule of thumb, always try the **System Printer** option first. If you are not able to use this option for whatever reason, the **Network Plotter Server** is your next best choice in a network environment, as it does not place the print server burden upon a user's workstation. If neither option is workable, then use the **My Computer** option above.

To understand a bit more about each option, we will go through the installation of each plotter type. Before going any further, be certain that you have at least one printer or plotter installed in your Windows Printers folder. To check for this, click the Windows **Start** button and choose **Printers and Faxes** (or something like that depending on your version of Windows). See Figure 3–9.

Figure 3–9 *Open the Windows Printers folder*

 Note: Verify that at least one device has been installed in the Windows Printers folder and then close the folder. If there are no printers, install one before continuing.

4. In the Add Plotter – Begin screen, click the **System Printer** option and then click **Next**.

5. Choose a Printer from the list of installed System Printers and then click **Next**.

On the Add Plotter - Import Pcp or Pc2 screen, you have the option to import settings from a legacy PC2 or PCP file. If you do not have any legacy data that you wish to import, skip this screen by clicking **Next**, or else click the **Import File** button and locate the legacy file to import (see Figure 3–10).

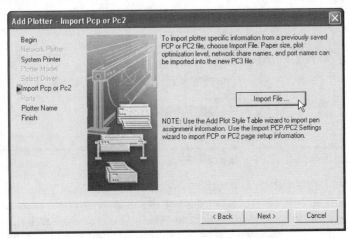

Figure 3–10 *Optionally import settings from a legacy configuration file*

6. Click **Next** and type a name for the new Plotter Configuration.

7. To edit the **Device Specific** parameters of the plotter, click the **Edit Plotter Configuration** button.

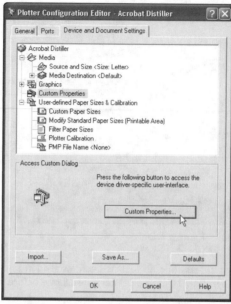

Figure 3–11 *Edit the Device Parameters in the Plotter Configuration Editor*

8. Make any edits you require and then click **OK**, then **Finish**.

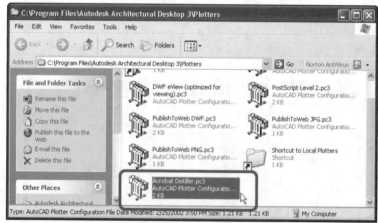

Figure 3–12 *Your new plotter in the Autodesk Plotter Manager folder*

If you return to the Autodesk Plotter Manager folder, you will have a new plotter listed with the name you assigned (see Figure 3–12). Notice the PC3 extension. (If you have extensions turned off in Windows, you will not see the extension.) If you double-click this new Plotter, it will return you to the Plotter Configuration Editor for that Plotter (see Figure 3–11). You can change the default settings for this plotter any time this way. If you are the CAD Manager, you should consider installing all of the plotters required in normal production for your firm. Once these have been created, place them on the server in to the *Plotters* folder on the L Drive. Make this folder Read Only to all but yourself and any other IT– or CAD Manager–level employees. More information on this topic is covered below.

Note: If you use the Batch Plotting utility frequently, you may wish to create separate PC3 files for each unique Batch Plot configuration. The reason for this recommendation is that when PC3 files are shared across the network, more than one user may potentially edit the parameters of a single Plotter Configuration, which could produce an adverse result on an ongoing batch plot.

Much of the steps for creating plotters that use a network server and connect directly to My Computer are virtually the same. Repeat the steps above, and choose **My Computer** from the Add Plotter - Introduction Page this time. On the Add Plotter - Plotter Model screen, note the list of manufacturers and their plotter models.

9. Choose a **Manufacturer** on the left, and then a **Model** on the right and then click **Next** (see Figure 3–13).

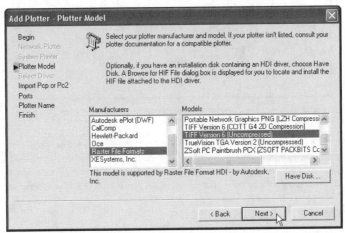

Figure 3–13 *Choose a Manufacturer and Model; the Raster File Formats was chosen for example purposes only*

Again, you will have the option to import legacy Plot Parameter files.

10. Make your choice and then click **Next**.

11. On the Add Plotter – Ports screen, choose an available port for this plotter and then click **Next**.

Since we chose **My Computer**, port options may include the various output ports on your PC. There is also the option for **Plot to File** and **AutoSpool**. If you chose a file format such as Postscript, or Raster File Formats in the Manufacturer screen, then the port option will be gray, and **File** will automatically be chosen. Using these options allows you to create high-resolution raster and vector images from your AutoCAD data. To learn more about AutoSpool, click the **What is AutoSpool** button.

12. Click **Next** again to name the plotter, and then click **Finish**.

As before, you may optionally edit the default configuration before clicking **Finish**.

Note: Plotters created this way are available only to the workstation on which they were created.

The final way to configure a new plotter is through a network plot server. The process is nearly the same, except for the choosing of the plotter from a network location on the Add Plotter - Network Plotter screen. Click the **Browse** button, or type in the path to the network plotter (see Figure 3–14). The remainder of the configuration is largely the same as the others.

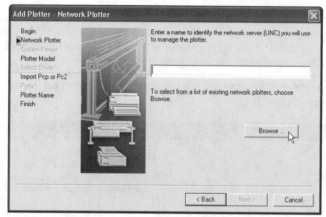

Figure 3–14 *Click Browse to access a network installed plotter, or type in the path*

Note: It is not necessary to use the **Network Plotter Server** option to access a network printer/plotter in AutoCAD. If a Windows system printer has been installed over the network, then it will also be available to AutoCAD at its network location. Therefore, the only option that does not make use of the potential of network printers is the **My Computer** option.

TAKING STOCK OF PLOTTING NEEDS

Now that we have learned how to add new plotters to our office, we need to look at exactly how many and what types we will actually need. Take an inventory of the types of output generated by members of your firm. Include all conceivable types of output desired from ADT. At this level however, we are only concerned with things that may impact the plotter configuration. To complete this task, you will need to understand what those parameters are. Take some time with one of the plotters you just added and look at all of the settings in the Plotter Configuration Editor. In most cases, you will find that a single configuration per piece of hardware is often sufficient for the firm at large. However, there may be those situations that compel you to set up the same device a few different ways. It is easy to get carried away. In the world of CAD standards, "*less is most definitely more.*" Do not add additional plotter configurations for settings that are easily controlled within Page Setup or a Plot Style Table. For instance, you could configure a special plotter for black and white plotting, and another for color. However, these settings can also be controlled with Plot Style Tables and Page Setup. So, which is the best place to set these two options? If the only change to the configuration is the color, then having a separate plotter configuration may not be necessary as it would be just as easy for the user to select their choice of color in the Page Setup. However, if the color plots were always going to use high gloss paper and the black and white ones vellum, then having two plotter configurations is the only way to go. Just keep in mind that configuring all these settings in a PC3 does not guarantee that someone restocked the ink or changed the roll of paper. Some things still need the human touch.

Tip: As a general rule of thumb, always try to limit additional plotter configurations (PC3 files) to items that cannot be controlled elsewhere.

Consider Figure 3–15, which shows the Merge Control setting. This is a hardware feature of only certain plotters. It may be desirable to have a PC3 that merges lines and another that does not (blends the colors).

Figure 3–15 *Merge Control is a hardware feature of some plotters*

As you devise your list of plotter output needs, consider the following typical architectural output types:

- Full-sized check plots (Bond)
- Full-sized production quality plots (Vellum or Mylar)
- Half-sized check plots (Bond)
- Half-sized production quality plots (Vellum or Mylar)
- Letter [A1] sized sketch plots (ASK)
- Tabloid [A2] sized check plots
- Full-sized presentation plots (high quality Bond)
- Color renderings (high gloss paper)
- Color presentation plots (high gloss paper)
- Faxes (letter or A1 size)
- PDF Files
- EPlots (DWF format)
- Web sites (Publish to Web Wizard)
- Raster images and Postscript files (TIFF, BMP, EPS, PS, etc.)
- Other

Keep this list handy. You will use it now to determine which PC3 files you require, but later on, you will use this list again to help you build all of your required title block files if you do not already have them in DWG format. If possible, try to streamline this list to the files types used 90 percent of the time by 90 percent of the people in your firm. You will probably find that the remaining 10 percent on both counts is very hard to "pin" down.

UNDERSTANDING PLOT STYLE TABLES

In a departure from the way things were done in legacy releases of AutoCAD (R14 and prior), the tools used in plotting have been neatly divided into items that are "hardware driven" and items that are "software driven." A Plotter Parameter File (PC3) governs the configuration of hardware-specific parameters. As such, you will often find a different set of options available for each type of plotting device within the Plotter Configuration Editor. In contrast, the features of a Plot Style Table are identical from one plotter to the next, regardless of manufacturer or printer format. This is because the Plot Style Table is the collection of all the software settings available in plotting that do not rely on a particular printer device. This can be said to be the "device independent" settings, while the PC3 file controls the "device dependent" settings.

The Plot Style Table contains a collection of settings that determine how each line in a plotted drawing is rendered. Specifically, there are two or more plot styles in every Plot Style Table. Each plot style can control a variety of settings at the time of printing. This includes the following features: color, dithering, grayscale, pen and virtual pen number, screening, linetype, lineweight, line end style, line join style and fill pattern.

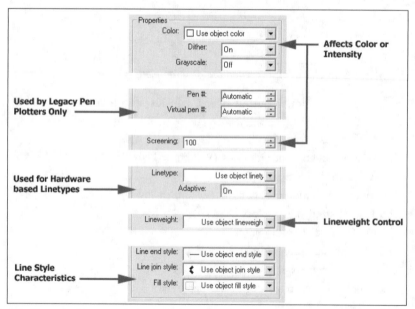

Figure 3–16 *Plot style parameters*

This list includes quite a few very useful features. Figure 3–16 helps to segregate them into categories. The features most commonly used are color, lineweight, line end style and line join style.

- **Color** – The actual color of ink used to draw objects of this style. Final color is also affected by the Dither, Grayscale and Screening settings.

- **Lineweight** – Determines the thickness of the linework for this plot style. Thickness values are in final plotted units.

- **Line End Style** and **Line Join Style** – Determine the way that the ends of lines and the corners of lines will be rendered.

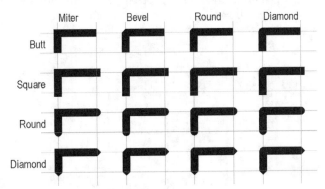

Figure 3–17 *Combinations of line join and end styles*

Depending on your specific situation, you may use more or fewer of these features in practice. Most people will have little use for the legacy pen plotter and hardware linetype settings. The various color and intensity settings (see Figure 3-16) offer many possibilities for color plots and halftone output. Plot styles can be applied to objects in the drawing in a few ways. If an object is assigned to a plot style, its plotted rendition will be governed by that plot style and potentially override any drawing level settings. Plot styles come in two varieties: Named and Color Dependent.

LINEWEIGHTS

One of the most important aspects of achieving high-quality plotted output is having good control over object lineweights. This is one of the most potent graphical tools we have as architects to help make drawings read. Gather up several completed project sets, lay out the drawings on a large table, and study them; are you satisfied with how they read? Do the lineweights provide the necessary contrast to help the most important parts of the drawings "pop" graphically? If you cannot answer yes to both of these questions, it may be time to reassess you firm's use of lineweight.

In the days of AutoCAD R14 and prior, there was one way to assign lineweights: through object colors at plot time. In today's ADT/AutoCAD, you can choose to assign the lineweights directly in the drawing and view them on screen prior to plot time. However, if your firm is like most, you are likely still using the "object color equals lineweight" method. This is not necessarily a bad thing, but it is worthwhile to take a moment here to discuss your options.

Using the Lineweight Object Property

If you choose to assign lineweights directly in the drawing, the lineweight information becomes a part of the drawing. You will no longer need to worry about convincing your consultants to use your pen table when they print. Regardless of how they print your data (screened or not) your lineweights, and therefore the "read" you intended, will be preserved. Since the lineweight is another object property like color and linetype, users will already understand how to apply it, which you will likely want to do ByLayer. You will also appreciate that since it is not tied to color as the legacy system is, you will be free to use color for whatever other purposes you find appropriate. The Lineweight property may also be viewed directly on screen. In model space, it shows proportional to other lineweights, so a .020 is always twice as thick as a .010, but it has no relationship to the objects themselves and will therefore not give a very good read of the overall drawing. However, when toggled on in paper space layouts, lineweights display properly in the true thickness that they will have relative to the printed sheet of paper. This feature alone will save you many wasted trips to the plotter and scrap plots (not to mention saving money and a few trees).

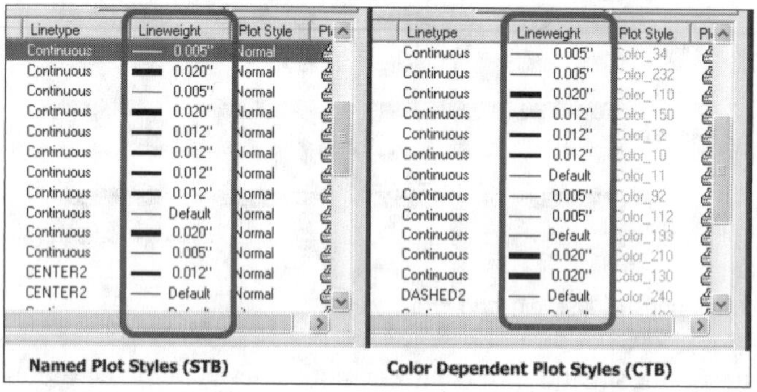

Figure 3–18 *Using layers in the drawing to assign the Lineweight property in both a STB and CTB*

If you use a Plot Style Table, your lineweights are referenced from a single table used by all drawings in your document set. Therefore, you have the ability to make a global change to the lineweight settings and update the entire project at once. If necessary, you can build variations of your table for each of your consultants, to take into account each of your combined preferences and standards. You can also use this "swapping" technique as an alternative to the **Scale Lineweights** option in Page Setup to vary the relative scale of lineweights for a half-size plot. Plot styles can be previewed on screen, but only in paper space layouts (not model space).

If you wish, it is possible to use both the Lineweight object property and a Plot Style Table, but this is generally not recommended, as it is likely foster confusion among your users. Therefore, consider your choice carefully. Although you can change your mind later, it will require some effort to do so.

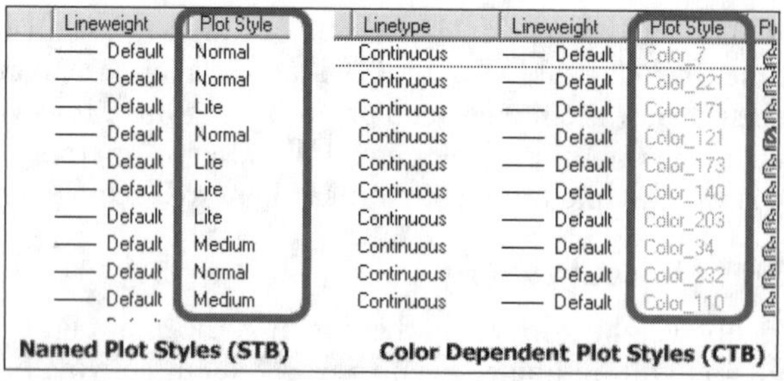

Figure 3–19 *Using layers in the drawing to assign plot styles in an STB and automatically assigned in a CTB (plot styles control the lineweights in this example)*

Once you have decided how you intend to apply lineweights to your drawings (by color, by Named plot style or ByLayer through the Lineweight property), you will want to standardize their application as much as possible. If you use the Lineweight object property, it is possible to apply lineweight directly to a selected object. However, this is not recommended. Rather, you should pre-assign lineweights to all of your office standard layers (see Figure 3–18) and Display Control settings. Standardizing layers is covered in Chapter 4 and standardizing the Display System in Chapter 5.

If you have chosen to use Plot Style Tables to assign your lineweights, the specific way they are actually applied to the objects within your drawing will vary depending on the type of Plot Style Table you choose to use. There are two types, Named and Color Dependent. Both are covered in the next topic.

With Color Dependent, the layers will show the plot style assignment (in gray), but you will be unable to change the assignment; they are linked directly to the color number. With Named, you can assign them to the layers in the same way you assign the Lineweight property. (Examples of both are shown in Figure 3–19.) If you do not assign a lineweight in the layers or the plot style, the drawing default lineweight (as established by clicking the **Lineweight Settings** button on the User Preferences tab of the Options dialog box) will be used. An example of this is shown at "1" in Figure 3–20. Also, be careful of assigning both the Lineweight property and a Named plot style (that also assigns a lineweight) to a layer at the same time. At "2" in Figure 3–20, both a lineweight and a plot style are assigned, but the plot style used does not assign its own lineweight. Therefore, the one assigned by the layer will be used. However, at "3" in Figure 3–20, both plot style and layer assign a lineweight. Although this is permissible and in some cases desirable, if a lineweight other than **Use Object Default** is assigned in the Plot Style Table, the Lineweight property assigned to the layer *will* be ignored in favor of the assignment in the Plot Style Table. This is true of all duplicate properties that can be assigned both in the drawing and in the plot style. The plot style *always* "wins."

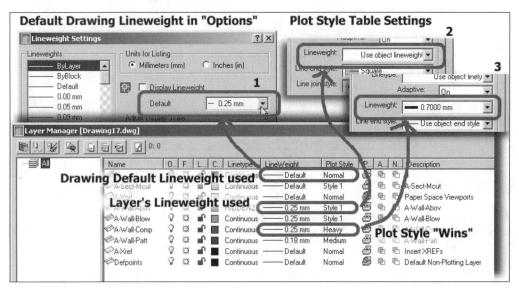

Figure 3–20 *Example of assigning Named plot styles by layer*

Choosing Your Office Standard Lineweights

Once you have decided how you will assign your lineweights, you ought to choose what they will be. If you already have a collection of lineweights that works well for your firm, there is no reason to change it now. If you do not have a standard set of lineweights, or are unhappy with the one frequently used, now would be a good time to consult the National CAD Standards for their list of recommended lineweights (UDS-04.14), which is also ISO compliant.

To assist you with in this endeavor, open the file named *Lineweights.dwg* located in the *C:\Program Files\Autodesk Architectural Desktop 3\Sample* of your Management Install.

AutoCAD Standard Lineweights

mm	inch		ISO
0.00			
0.05	.002		
0.09	.003		
0.13	.005		
0.15	.006		
0.18	**.007**		X
0.20	.008		
0.25	**.010**		X
0.30	.012		
0.35	**.014**		X
0.40	.016		
0.50	**.020**		X
0.53	.021		
0.60	.024		
0.70	**.028**		X
0.80	.031		
0.90	.035		X
1.00	**.039**		
1.06	.042		
1.20	.047		X
1.40	**.056**		
1.58	.062		
2.00	**.078**		X
2.11	.083		

Figure 3–21 *The default lineweights provided with AutoCAD/ADT*

This file has a series of lines in it, each assigned a different lineweight. The sizes are given in both Metric and Inches, and ISO standard lineweights are indicated by an "X" in the ISO column directly in the DWG file. (Most of the NCS recommended lineweights correspond directly to an equivalent ISO lineweight, as shown in Figure 3–21.) Print this file on each of your plotters and printers. Use it as a guide to help determine which lineweights work best on all of your devices and give you the best contrast. Try similar experiments with some typical drawings (preferably some including ADT objects if possible).

For most architectural drawings, you will only require a handful of the lineweights actually available. If possible, within the corporate culture of your firm, consider choosing the lineweights listed by the NCS and ISO. Hard code them into the Layer Key files, Display Control settings and object styles you will build later in Chapters 4, 5 and 9. Everyone in your firm can benefit from a definitive decision on this and many other similar matters. Your success at implementing ADT can be slowly eroded by indecision in these seemingly trivial matters. It is here that adopting an industry standard can prove efficient and beneficial.

NAMED VERSUS COLOR DEPENDENT PLOT STYLE TABLES

There is no difference in feature set between these two types of tables. The difference is in how they are applied to the objects within the drawing (see Figure 3–22). Color Dependent Plot Style Tables (CTB) offer compatibility with legacy plotting in AutoCAD. There are 255 styles within a Color Dependent Plot Style Table. Each style in the CTB maps directly to a single color in the AutoCAD Color Index (ACI.) Therefore, Color 1 maps to Style 1, Color 2 to Style 2 and so on. This mapping cannot be changed.

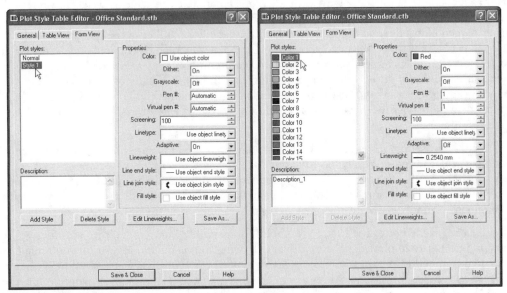

Figure 3–22 *Both CTBs and STBs offer the same features*

A Named Plot Style Table (STB) contains a minimum of two plot styles: Normal and Style 1. Style 1 may be renamed, but Normal may not be renamed or edited. Normal is a "pass through" style. It simply "passes" an object through to the plotter without changing any of the drawing-level settings. An STB may have as many styles as are required to render all objects in your drawing appropriately. When Named plot styles are used, a plot style within an STB behaves like any other AutoCAD object property. In other words, STBs are assigned directly to objects in the same manner as other object properties like color, linetype and layer and can be assigned to objects ByLayer as well. This offers some interesting possibilities not available with Color Dependent styles.

What makes the Named Plot Style Table compelling is its ability to correspond directly to the intended function and to be named accordingly. For example, if your primary concern was assigning lineweights, and you preferred not to use the Lineweight property, you could have plot styles named accordingly. For instance, the NCS names Extra Wide, Wide, Medium, Thin and Fine could be used directly as plot style names. If you wanted to have pens for half-tone plots, you could create styles named 50% Halftone and 30% Halftone. Even more interesting, the same table could be used to house color pens as well: Pantone Red, and Logo Blue.

Note: Colors can be assigned in RGB values, but not Pantone PMS numbers; therefore to create the pens listed above, you would need to dial in the closest RGB equivalent to the Pantone value.

Your Named plot style could be made even simpler if you use the Lineweight property. In that case, you would not require a plot style for each lineweight in your standard, rather you would only need two plot styles initially: Black and Halftone.

You could achieve all of these pens with a Color Dependent pen table as well. However, you would still need to apply the values to one of the AutoCAD ACI colors (rather than a descriptively named plot style) and then assign objects to that color number in the drawing in order to use the pen at plot time. As stated above, the difference in Color Dependent and Named plot styles lies not in their respective feature sets, but rather in how they are named and applied. In general, Named plot styles will seem more intuitive and user-friendly to users, assuming you choose a straightforward and easy to understand naming scheme (highly recommended). Color Dependent plot styles will have the benefit of familiarity and ease of interoperability with outside consultants.

With either type, do not forget the potential of having several Plot Style Tables, each designed for a particular purpose. You could have one for plotting Construction Documents that complies with the NCS Tri-Services Plotting Guideline. Another could be built for color presentation drawings, where colors are used rather than black and white as well as different lineweight assignments. Simply swap out the one you need before plotting the drawing.

Even if you decide not to use Plot Style Tables for assigning lineweights, they are still important and necessary to plotting. For example, a drawing in ADT uses many colors on screen. If you do not intend to plot in those colors, you will need the Plot Style Table to address this. In Named Plot Style Tables, you simply assign all objects to a plot style or styles whose color (the ink color it will use to plot) is set to Black. In Color Dependent, you must assign all 255 colors to Black to achieve the same goal. Another area where Plot Style Tables are critical is Line Join and Line End Style. Neither of these can be changed from the default "round" shape within the drawing. You can only achieve the shapes shown in Figure 3–17 by using a plot style. This is true regardless of the type of Plot Style Table you choose. Therefore, making the decision of whether or not to use the Lineweight feature of Plot Style Tables is only a small part of the work required to set these plotting resources up for your firm. Take the time now to build a list of requirements and set up one or more Plot Style Tables to meet those needs. Save these Plot Style Tables in the *Plot Styles* folder on the L Drive. This way they will be available to all members of the firm.

Tip: If you did not do so in Chapter 1, create a Windows shortcut to *C:\Local Content* and the *Plot Styles* and *Plotters* folder. In this way, users may create their own custom versions of plotting resource files if necessary. These custom versions will be available only to them.

Tip: To give all members of a project team access to a project-specific Plot Style Table used only by members of that team, place an additional Windows shortcut to the project folder within each team member's *C:\Local Content* folder.

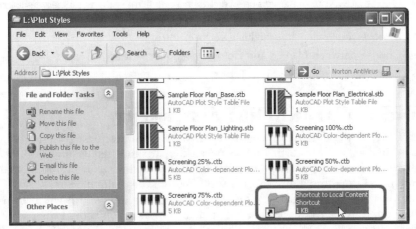

Figure 3–23 *A shortcut to the Local Content folder allows users to create custom plotting resources as project needs dictate*

Lineweight and Plot Style Recommendations

Remember that although our discussion of lineweights and plot styles involves several issues, two questions stand out as most critical: Which lineweights will our "standard" list include, and how will they be applied to objects? The following two recommendations summarize the issues:

- Use the Lineweight object property to assign lineweight, not plot styles.
- Use plot styles to assign the final color, intensity, join style and end style of all objects (for example, Black ink color, Square end and Miter Join Style).

Following these two recommendations will give the greatest overall benefits while maintaining maximum simplicity and flexibility in the drawings.

WORKING WITH SERVICE PROVIDERS

Many of the recommendations made regarding plotting in this section have not taken into account any issues associated with using third-party plotting applications such as Repro-Desk. If your firm uses one of these packages to print from AutoCAD/ADT, please consult that manufacturer's documentation for specific recommendations regarding any of the topics in this section.

If you use an outside plotting service to plot your Architectural Desktop drawings, it is highly recommended that you choose a service provider that owns its own copy of ADT R3.3 or at a bare minimum, AutoCAD 2002 with the ADT Object Enabler for R3.3. This point cannot be stressed enough. If the service bureau you employ does not use ADT, the hassles you will encounter while trying to achieve acceptable plotting results cannot be understated. If the service bureau wishes to keep your firm's business and the business of its other clients doing similar work, it is not unreasonable to require them to procure a copy of the software you use. Refer to Chapter 12 for more information on issues related to sharing files with persons not using ADT.

PLOTTING CHECKLIST

1. Install plotters and printers in the Windows Printers folder.
2. Install your plotters and printers within AutoCAD/ADT (build PC3 files).
3. Decide how you will assign lineweights in your firm: using the Lineweight property or Plot Style Tables.
4. Choose your office standard lineweight palette.
5. Choose between Named and Color Dependent Plot Style Tables.
6. Build one or more Plot Style Tables.

EXISTING CAD STANDARDS ASSESSMENT

Since AutoCAD is the underlying engine upon which Architectural Desktop is built, it is important to understand the distinction between the two tool sets (AutoCAD and ADT). There are several items of office standardization that rely solely on core AutoCAD functionality. For instance, formatting and styling of text is within the realm of AutoCAD. This topic will explore those items that utilize AutoCAD core functionality, and it will help you determine what modifications from your current standards, if any, you should consider when migrating these settings to ADT. We will consider best practices in migrating AutoCAD Text, Dimension and Formatting settings to ADT. If no current standard prevails in your firm for these items, this section will assist you in establishing one.

Do you have CAD standards in place already? Are you migrating from a previous release of ADT or AutoCAD or are you new to CAD? Perhaps you are migrating from a CAD package other than AutoCAD or ADT. Regardless of your current standing and level of standardization, we will use the topics in this section to help you answer the following questions:

- Are your methods logical, consistent and efficient?
- Are there areas that require improvement?
- Do your users understand the procedures that are in place?
- Are the standards being followed by users?

In the previous two chapters, we have already discussed network shares, folders, files and XREFs In this chapter so far, we have covered plotting standards, lineweights and plot style settings. Let's do an inventory of the remaining items related to CAD standards that you may already have established. As you read this section, keep a list of items that need to be created or improved.

LAYOUT (PAPER SPACE) USAGE

Architectural drawings are printed in scale. However, unlike the days when they were drawn by hand, the drawings are not actually drawn to scale, but rather they are drawn at "full size" and then plotted to scale. The layout tabs of ADT provide the means to establish plotting scales for and compose the layout of a printed sheet. Layouts can be used in other ways, but their use as a plotting tool is by far the most common use of the layout tabs in both standard AutoCAD and ADT. If you use sheet files, as recommended in the last two chapters, the layout tabs will be set up to include a title block with sheet information and at least a single viewport set to a particular scale. The goal of the sheet file is to create a single "ready to print" file that represents a single sheet in a document set.

In addition to the usage outlined above, layouts are also very useful for setting up alternative sheet layouts such as a half size or color presentation version of the same sheet, or for creating an ASK from the same portion of the total sheet that will be printed on letter size paper and faxed. This is possible using multiple layout tabs, Named Page Setups and the ability of viewports to each show their own display configuration and layer settings.

Note: Prior to adopting multiple layout sheets, ensure that the teams that will accept your ADT files have the capability to read this AutoCAD sheet layout standard. Many smaller firms and vendors may be using older versions of AutoCAD/ADT or other CAD packages that do not support multiple layouts. Coordinate with those consultants, clients and vendors on a project-by-project basis and have a contingency plan for this possibility. Refer to Chapter 12.

Using layouts

For clarification purposes, paper space is the two-dimensional "paper" scaled environment that contains one or more layout tabs (not including the Model tab, which is model space). Having multiple layout tabs allows for the presentation of model space data in a variety of sheet configurations. Although not exactly the same, you will frequently hear the terms "paper space" and "layouts" being used interchangeably. When using layouts, keep the following points in mind:

- A drawing must have at least one layout and may have several. This is true even in Model files (see Chapter 2). The distinction between model and sheet files is procedural and not a distinction recognized by the AutoCAD file format.

- When a drawing is saved to R14 file format, the active layout at the time of saving will become the "paper space" of that R14 drawing file. The other layouts in the file will *not* be available to the R14 user.

- Build your most common layout configurations in a template file. They will be easily imported into the current drawing by right-clicking on one of the layout tabs and choosing **From Template** (see Figure 3–24). This will load a list of template files in your default template folder (see Chapter 1), which will allow you to import any layout tab saved in any of the template files listed.

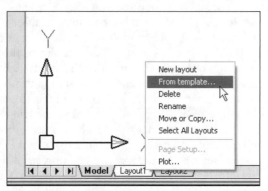

Figure 3–24 *Import a layout from an existing template file*

- In the same way that you are able to use the **Freeze in Current Viewport** (VPLAYER) feature to control layers independently in each viewport, you may also assign a different display configuration to each viewport. In this way, the same model may be viewed quite differently in each active viewport (see Figure 3–25).

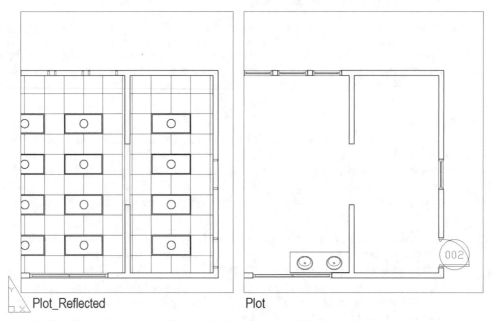

Plot_Reflected   Plot

Figure 3–25 *Showing a model in Plan and Reflected in two different viewports*

- All of the settings in the Page Setup dialog box may be saved and named as a "reusable" Page Setup configuration. In many cases, this can be even more useful than setting up multiple layout tabs.

Consider items like choice of plotter, page size and plot scale. Frequently we have need to print a sheet from a document set to our tabloid-sized laser printer at half-size scale or perhaps scaled to fit. Before you build a new layout to accommodate this need, consider using a saved Page Setup instead. When you build a new layout, you must duplicate the title block, title block information and viewport configuration. If the layout of the sheet, including title block, information input into the title block, general notes and viewport organization, are identical in both the full-size configuration and the tabloid version, then duplicating all of these items to a separate layout will prove counter-productive. This is because the two layouts maintain no link to one another. If information in the original title block changes, it will need to be changed manually in the second copy as well. However, if you build two saved Page Setups instead, they both will relate to the same layout tab. Therefore, you will only need to maintain one copy of your title block and all of its information and viewports. Before plotting, the user simply chooses the appropriate Page Setup from a list. If the Page Setup you need is not listed, you can add one by clicking the **Add** button (see Figure 2–26).

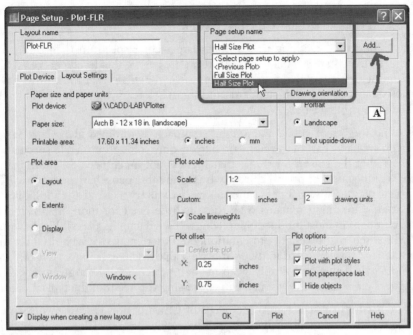

Figure 3–26 *Choose a saved Page Setup from the list*

LAYER NAMING AND USAGE

No discussion of CAD Standards would be complete without a look at layers. Layers are often one of the first things addressed by a CAD Manager when building new standards. In Architectural Desktop, layers continue to play an important role in our CAD standards and everyday production. However, they are not as pervasive as they once were. Display Control now addresses many of the tasks that heretofore were exclusively the province of layers. If you have been using AutoCAD for a while, you almost assuredly have a list of standard layers already in place. For the purpose of this discussion, simply get a copy of that list and keep it handy. The next chapter is devoted entirely to the subject of layers, and specifically how to integrate layer strategies into your standardization and usage of ADT. We will therefore postpone any specific discussion of layers until that chapter.

COLORS

In societies the world over, color is imbued with all manner of meanings. For example, Red means stop, Green means go, Blue invokes calm, Red can signal fire, put us on "Red Alert" and invoke a sense of aggression. On the other hand, Red may simply be the color of our dimension lines. Within AutoCAD, many very elaborate color-coding schemes have been built relating to the "look" of certain colors on screen, color's relationship to legacy Rapidograph® pens, or a variety of other justifications.

In AutoCAD, there are 255 colors. However, in looking at most AutoCAD drawings, you would never know this. Most CAD Standards tend to favor the use of the first six (the primaries) or perhaps the first sixteen colors in the AutoCAD palette. There are reasons for this that are not necessarily within the scope of this text. However, in the following paragraphs we will make the case for color usage above the first sixteen.

By default, Architectural Desktop uses a much larger color spectrum than the first eight to sixteen colors. In fact, there is a deliberate effort on the part of the developers of ADT content to avoid using the first eight colors. The simple reason for this is that most firms have already assigned those colors within their standards and the ADT team wanted to avoid conflict with these standards "out-of-the-box."

If you are using the "color equals lineweight" approach as outlined above in the lineweight discussion, you are likely part of the eight-or-sixteen-only color group. Wild speculation would have us believe that the reason for this is that there is not much need for more than eight to sixteen pens within your Plot Style Table, so why set up more pens than are required? There are other legitimate reasons as well, but as stated above, we will try to avoid any additional "groundless" speculation. One thing remains true: there is little need for more than eight, let alone sixteen different pens in most production drawings. If you factor in all possible drawings (color, black and white, schematic design, presentation, etc.) there may well be a need for more than sixteen pens, but as was indicated in the Plot Style Table discussion above, you can build more than one Plot Style Table and simply swap in the one required for a given task. Therefore, current practices seem perfectly reasonable in addressing the needs of the typical document set.

Let us limit our current discussion to the needs of Construction Documents, for the sake of simplicity. In this case, you will likely need approximately five pens (fine, medium, wide, extra wide, etc.) that correspond to your standard lineweights. You may then need some halftone pens and perhaps one or two other special-purpose pens or combinations of any of these. This is how we came up with the need for between eight and sixteen pens.

The problem with approaching color this way is that it focuses too much attention on the final output, the drawing, rather than the screen view of your drawing, where users spend most of their time. If you were able to quantify the amount of time the average user spends looking at the screen and compare that to the average amount of time they spend generating plots, it is a near certainty that the large majority of their time is spent simply working in the CAD drawing and not plotting. This is not to say that plotting is unimportant; on the contrary, it is critical. It is simply the opinion of this author that plotting issues should not be the only driving force in building a color scheme. The human eye is capable of discerning even the subtlest variation in color. This is evidenced by the fact that it takes over 16 million colors on our computer screen to fool our eyes into seeing "photo-realistic" color. Therefore, do not underestimate the eye's ability to clearly distinguish between each of the 255 colors in AutoCAD's small palette. In summary, there are 255 colors; use them!

For instance, NCS assumes the use of Color Dependent plotting and recommends eight Black pens and seven Half Tone pens. These pens are then dispersed throughout the entire AutoCAD Color Index (ACI) so that all 255 colors have a recommended NCS pen assignment (see Figure 3–27). If you make use of this pen table, or one similar, you will be able to have two different objects and/or layers that look different on screen, yet plot in exactly the same way. This allows your colors to convey much more meaning than a simple one-to-one mapping (where most colors go unused) would allow.

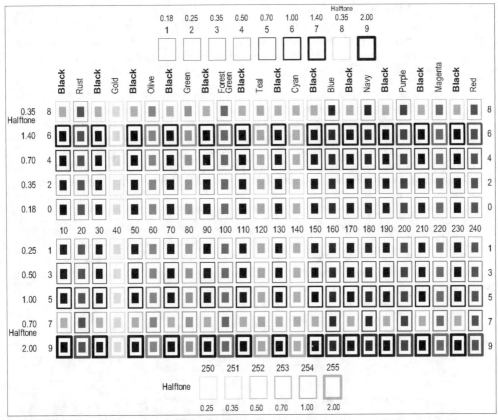

Figure 3–27 *The complete ACI color list assigned as recommended by the NCS Tri-Services Plotting Guidelines*

The drawing shown in Figure 3–27 is located in the *Chapter03* folder and is called *NCS Color Map.dwg*. Also included is a sample CTB Plot Style Table named *ADSK_NCS3.ctb* (CTB file provided by Autodesk, used with permission). A PDF version in color of Figure 3–27 is also saved to the same folder named *ADT_AIG-Figure3-27-NCS Color Usage.pdf*. You can print this file on a color printer for reference if you wish.

Consider the following uses of color:

- Color can be used to help distinguish unique information on screen. Two lines drawn in the same color look the same. If they are in two different colors, there is probably some difference between them.

- Make each layer you create a unique color. This makes it easy to immediately distinguish one layer's information from others. Remember, two different colors can be assigned the same plotting parameters if it is appropriate that they look different on screen while plotting the same.

- Make each ADT object type a different default color. This can be done ByLayer, or by object or even down to the sub-component level in Display Control, and it affords benefits similar to those of the previous point.

The NCS Color Table has the advantage of being part of the complete NCS CAD Standard. However, although comprehensive in its approach to provide "consistent black-and-white and limited color plotting," the NCS Tri-services pen table can still prove a challenge to master, as it relies on a number-matching scheme to map colors to lineweights. As an alternative to the NCS table, consider the following scheme:

- Use different shades of the same color to indicate similar information, such as the same lineweight, new or existing construction or the same object type belonging to different construction project phases. This will give you visual cues on screen to allow you to quickly spot these connections in similar types of information.

Figure 3–28 shows the entire AutoCAD Color Index (ACI) divided into six distinct pen types. Colors 1 to 5 (Red, Yellow, Green, Cyan and Blue) are used for Lineweight 1 through Lineweight 5 respectively. Pen 6 (Magenta) is reserved for "special" use pens. This includes items like no plot items where plotted pen weight is not important, or the three "Extra Wide" pens recommended by the NCS. In addition, the group of six gray colors are mapped to the same pens, but are assigned to 50% halftone. The drawing shown in Figure 3–28 is located in the *Chapter03* folder and is called *ADT_AIG Color Map.dwg*. A PDF version in color of Figure 3–28 is also saved to the same folder and is named *ADT_AIG-Figure3-28-Color Map.pdf*. You can print this file on a color printer for reference if you wish.

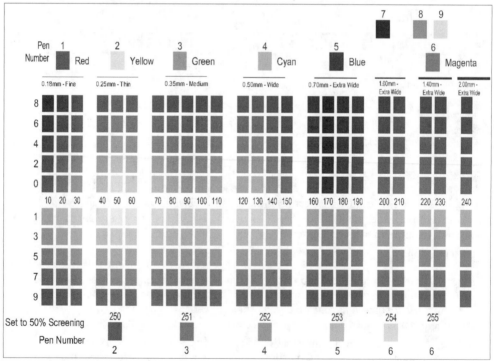

Figure 3–28 *The complete ACI color list assigned to six pen types*

Keep in mind that the main purpose of this pen table is to foster simplicity. The ADT default color scheme used in many of the template files does not match these (or the NCS) mappings. If you adopt either color scheme, or if you invent your own, you will need to adjust layers and display settings within template files, Layer Key Styles and the Display System. This is one of the reasons that the topics of plot styles, lineweights and colors have been discussed first. Layers and the Display System will be discussed in the next two chapters. If you wish to build your own color scheme, you can also create your own version of the color tables shown in Figures 3–27 and 3–28. Open either of these files form the *Chapter03* folder, or load the original drawing file located in *C:\Program Files\Autodesk Architectural Desktop 3\Support\chroma.dwg* of your Management Install. If you prefer to work with a color wheel version, use the *C:\Program Files\Autodesk Architectural Desktop 3\Sample\colorwh.dwg* file instead.

- With the ability to use the lineweight and plot style properties discussed above, consider allowing your colors to be "colorful." In other words, color can be used to evoke design intent rather than a color "code." This is why the files corresponding to Figure 3–28 do not include a Plot Style Table. Regardless of the type of plot styles you are using, you can simply assign the lineweights to the layers and use the *Monochrome.ctb* or *Monochrome.stb* Plot Style Tables to plot in black and white.

- With the Display System, you can achieve all of these points and more at the same time. For instance, in the Plan Display Rep, color could be used traditionally to indicate object lineweight (as recommended by the NCS Tri-Services Plotting Guidelines or the other color mapping scheme as noted above), and it can be assigned ByLayer. In the Model Rep, color can be assigned explicitly and used to convey design rules for Sections and materials for rendering.

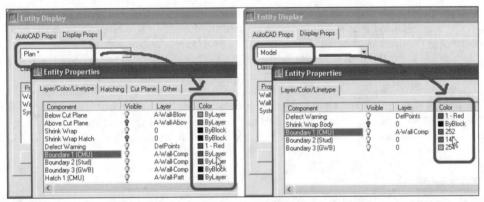

Figure 3–29 *The Plan and Model Display Reps of a typical Wall style each using color differently*

All ADT objects have a default color assigned through their respective Layer Key Styles. The complete list of pre-assigned ADT colors (for all layer standards included in the default *AecLayerStd.dwg* file), sorted by Layer Key is included in Table B3–1 in Appendix B on the CD ROM and is also provided as Table XL3–1 in an Excel spreadsheet file named *Layer Key Colors.xls* located in the *Chapter03* folder. The same information is also sorted by color in Tables XL3–2 and XL3–3 in the same Excel file. Figure 3–30 shows a portion of this file.

(The default *AecLayerStd.dwg* file was copied to your L Drive in Chapter 1. Its original location in your Management Install is: *C:\Program Files\Autodesk Architectural Desktop 3\Content\Layers*).

Note: Layer Key Styles and Layer Standards are covered in detail in the next chapter.

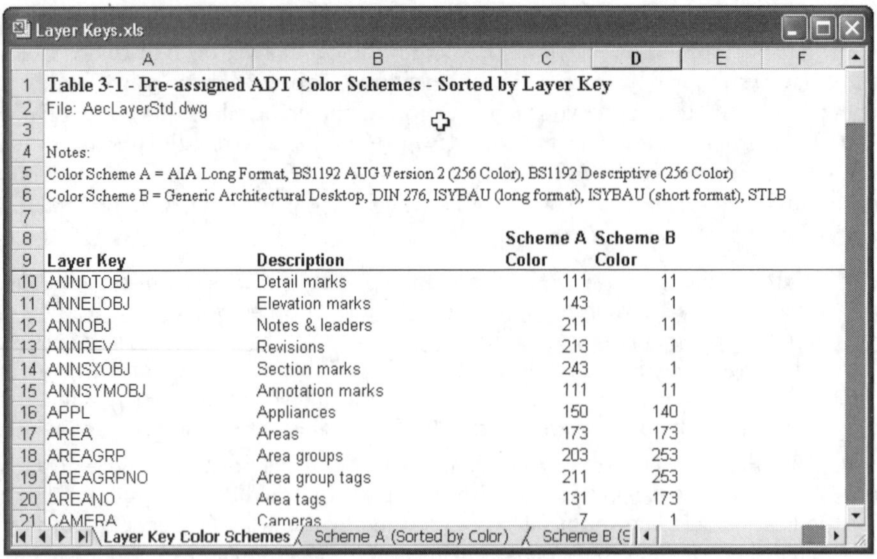
Layer Keys.xls

Table 3-1 - Pre-assigned ADT Color Schemes - Sorted by Layer Key

File: AecLayerStd.dwg

Notes:

Color Scheme A = AIA Long Format, BS1192 AUG Version 2 (256 Color), BS1192 Descriptive (256 Color)

Color Scheme B = Generic Architectural Desktop, DIN 276, ISYBAU (long format), ISYBAU (short format), STLB

Layer Key	Description	Scheme A Color	Scheme B Color
ANNDTOBJ	Detail marks	111	11
ANNELOBJ	Elevation marks	143	1
ANNOBJ	Notes & leaders	211	11
ANNREV	Revisions	213	1
ANNSXOBJ	Section marks	243	1
ANNSYMOBJ	Annotation marks	111	11
APPL	Appliances	150	140
AREA	Areas	173	173
AREAGRP	Area groups	203	253
AREAGRPNO	Area group tags	211	253
AREANO	Area tags	131	173
CAMERA	Cameras	7	1

Layer Key Color Schemes | Scheme A (Sorted by Color) | Scheme B (S

Figure 3–30 *Tables XL3–1, XL3–2 and XL3–3 located in an Excel spreadsheet on the CD ROM; Table XL3–1 is also in Appendix B as Table B3–1*

In *Mastering Autodesk Architectural Desktop*, an image similar to the ones shown in Figures 3–27 and 3–28 was included that showed all colors used by the *AEC Arch (Imperial - Intl).dwt* file. The drawing file used to generate that image (*MasterADT Color Usage.dwg*) has been included on the CD, as has a PDF (*MasterADT-Figure13-28-Color Usage.pdf*) containing the resulting image. Regardless of what decisions you make regarding your firm's use of color, you will want to keep track of which colors you use and for what purpose you are using them. In this way, you will be able to make good decisions regarding specific color choices and be certain that you have not assigned two opposing meanings or uses for the same color. Print one or both of these files to assist you in this endeavor or, if you wish, work directly in the drawing file (*MasterADT Color Usage.dwg*) within ADT.

Color Recommendations

Regardless of your choice of Color Dependent or named Plot Style Tables, with ADT you can begin to consider using color a bit differently than has been typical in the past.

- Assign colors ByLayer in Plan Display Reps for ADT objects and to all 2D-only AutoCAD entities.
- Use explicit color assignments in three-dimensional Display Reps such as those used for Sections and Elevations and for export to Autodesk VIZ. (See Chapter 11 for information on Section and Elevations and Chapter 13, located on the CD ROM in PDF format, for information on VIZ.)
- Do not limit color usage to the first eight to sixteen colors. Use lots of color!
- Develop a color scheme that is simple to understand and use or adopt an industry standard such as NCS Tri-Services Plotting Guidelines.
- Remember, color does *not* have to equal lineweight anymore!

LINETYPES

Linetypes should be one of the simpler items in your implementation plan. The default list provided in the *acad.lin* should be suitable for most users. However, if you decide to comply with the NCS, it is very easy to create custom Linetype definitions that match those listed in Module 6 of the National CAD Standard. Most ISO compliant linetypes have been included with the standard AutoCAD linetype *acad.lin* file. In the "Set Up Network Resources" topic in Chapter 1, we discussed the preparation required to set up a customized version of the Linetype Definition File (.LIN). If you need more information on the procedure required to build your own linetypes, pick up a copy of *Customizing AutoCAD 2002* by Sham Tickoo, published by Autodesk Press.

Linetypes must be loaded into a drawing file before they can be used. Like most similar resources in AutoCAD/ADT, importing another item such as a Block or style that references a particular linetype will also import that linetype. Therefore, it is not critical to anticipate the needs of every user and bulk up the drawing with a long list of linetypes. Seek to identify the most common linetypes, used about 90 percent of the time, and make these readily available. Users will be able to load any exotic linetypes they need on the fly during production with a minimum of effort. Loading linetypes can be done from the **Linetype Manager** available on the **Format** menu, or through a variety of shortcut methods and dialog boxes. Once a linetype has been loaded, it becomes part of the current file and no longer needs to reference the external file from which it came.

Linetype Scale

There are three ways to scale the proportion of dash to space defined in a linetype.

▶ **LTSCALE** (Linetype Scale) – The global scale factor used to set the proportions of all linetypes within the entire drawing. The value of the LTSCALE is multiplied by the values built into each Linetype definition.

For example, a linetype defined to draw a dash at .25 and then a gap at .125 would draw a 12-unit dash and 6-unit gap at a LTSCALE of 48. At a LTSCALE of 50, the dash would be 12.5 units and the gap 6.25.

There can only be one value set for LTSCALE at any given time. If you need to have two different values, use either the PSLTSCALE or the CELTSCALE variable to adjust the global scale for either the current viewport or the current element. Linetype definitions are typically built to display at a scale of 1. Therefore, the proper LTSCALE for a given drawing ought to be that drawing's "scale factor," the number that represents the proportion between the size the model is drawn (usually full life size) and the size it will likely be plotted. Therefore, in a building model which we intend to plot at 1/4"=1'-0" [1:50], the scale factor would be 48 [50]. This is because the building model is 48 [50] times larger in ADT than it will appear on paper when plotted. In this scenario, the LTSCALE of the drawing in model space ought to be set to 48 [50] to preview the correct linetype proportions on screen. In sheet files, the LTSCALE should always be set to 1. This is because the viewports will already be scaled in the drawing and the drawing containing those viewports will be plotted at a scale of 1=1. Therefore, it is consistent with the formula given here that the LTSCALE ought to be set to **1**.

 Tip: Be sure to also use the PSLTSCALE setting of 1 as well (see below).

There are firms that advocate additional multipliers to these values. For instance, it is common to see firms using LTSCALE values of .5 or .33 times the scale factor of the drawing. Although there is nothing inherently wrong with this approach, it is recommended that you not engage in this practice. If you are dissatisfied with the proportions yielded when using the unmodified scale factor of the drawing, consider editing the Linetype Definition File instead. There is one justification for this recommendation: *keep it simple*. There are enough hurdles for users and CAD Managers alike to overcome when adopting new technology. Let us not add additional artificial complexity to the mix.

> ▶ **PSLTSCALE** (Paper Space Linetype Scale) – A toggle setting whose possible values are 1 and 0. When set to 1, PSLTSCALE is turned on, and the global LTSCALE of the drawing will be multiplied by each of the viewport scales in paper space layouts. When this value is turned off (set to 0), the viewport scale will have no effect on LTSCALE and the single global LTSCALE value will be applied to all viewports.

There is little reason to ever set the PSLTSCALE to anything other than 1. When this value is on, it will enable all viewports regardless of scale to display and print properly. There will be no visible effect on model space, nor will there be any detriment. For best results in all drawings, set PSLTSCALE to on (value of **1**) in all template files.

> ▶ **CELTSCALE** (Current Element Linetype Scale) – An object property used to change the LTSCALE of a single object. The CELTSCALE multiplies by the current global LTSCALE of the drawing for a selected object(s). The initial value of the CELTSCALE is 1; therefore most objects simply use the global LTSCALE "as is."

The effects of each of these commands are cumulative. Therefore, be certain to set the LTSCALE first. Changing the LTSCALE later could have a detrimental effect on all other settings that you have made to both the PSLTSCALE and especially the CELTSCALE. For instance, if the current drawing has a global LTSCALE of 48 [50], and you have since edited the CELTSCALE of several objects to a value of .5 (making their cumulative LTSCALE equal to 24 [25]), changing the LTSCALE to 96 [100] will also change the individual objects, to the net result of 48 [50]. They will still have a CELTSCALE of .5, but since the current global scale is now 96 [100], their net result changes to 48 [50]. This may be your intended result, but if it is not, you will need to reselect all of the previously modified objects and set their new value of CELTSCALE to .25, a tedious process to be sure.

To edit the CELTSCALE, it is recommended that you first create the object, and then edit the CELTSCALE. This is recommended to prevent accidentally leaving the current setting of CELTSCALE for the drawing set to something other than 1. To change the value of AutoCAD objects such as lines and circles, use the Object Property Manager, and for ADT objects, use Display Props in Entity Display, as shown on the right in Figure 3–31.

The situation will arise most often when using "complex" linetypes (linetypes containing shapes). For instance, suppose you wanted to show Batt insulation within a Wall style. The *acad.lin* file contains a "batting" Linetype definition. By importing it into your template file (or just the drawing that will use it), you can assign it to the Boundary lines of a component representing Batt insulation. There are a few tricks to make this work properly: First, you must add a component to your Wall style for the Batt insulation. The width of this component must be set to zero. Next, as we are discussing in this topic, the CELTSCALE of this component must be adjusted for the Batt insulation to display at the proper size. As you can see in the diagram at left in Figure 3–31, the height of the shape used within the Batting Linetype definition is .8 units tall at an LTSCALE of **1**. Therefore, if you consult Table 3–1, you can perform a few calculations to arrive at an acceptable value for CELTSCALE. For instance, if you had a Wall 6" thick, and the drawing was going to plot at 1/4"=1'-0", you would multiply the LTSCALE (48 in this case) by .8 (the height of the Batting) to arrive at the Calculated Height of **38.4** in Table 3–1. Next, divide the Desired Batt Thickness by this number to arrive at the correct value for the CELTSCALE, in this case **0.13**. This example was given in inches, but as you can see in Table 3–1, it works the same way in millimeters.

Table 3–1 *Calculating the Celtscale for Batting*

LTScale	Calculated Height (multiplied by .8)	Wall Thickness	Desired Batt Thickness	CELTScale
48	38.4	6	5	0.13
96	76.8	6	5	0.07
50	40	150	120	3.00
100	80	150	120	1.50

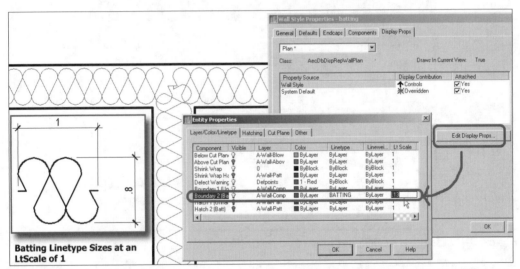

Figure 3–31 *Editing the value of the Current Element LtScale to create Batt insulation in a Wall style*

 Tip: As a general rule of thumb, use of the CELTSCALE variable should be avoided if the same result can be achieved in another way. Leave the drawing default CELTSCALE set to **1**.

TEXT

All text in AutoCAD/ADT is controlled by a style. The style determines the font used by the text as well as the height, width factor and other special effects. It is highly likely that you already have a collection of Text styles in use in your firm. Nothing about ADT implementation requires that you rework your existing Text styles. Therefore, if you are satisfied with the styles you currently have, continue to use them in ADT. Consider the following issues and options:

- If you use custom fonts not installed by default in ADT, make sure that these fonts are available on the common support path for all users (refer to Chapter 1 for more information on setting up support paths).

- It is recommended that you do not "hard code" text height into your Text styles. In other words, the height field of all Text styles ought to be set to 0. This will allow text height to be set "on the fly" (see Figure 3–32).

There are two reasons for this recommendation: first, tools that reference the current Text style may yield unpredictable results if the text height is set explicitly. This is because the value will not be able to be changed from the value set in the Text style. The second reason is again the desire to "keep it simple." If you do not hard code text heights, you will only need one Text style per type rather than several. For instance, if you wish to have a Text style for "Notes," you will only need one with the height set to 0, as it will be able to rescale to any height as required by the scale of the drawing. If you do not set the height to 0, you will need to create as many copies of this style as the anticipated quantity of drawing scales for which it will be used. This could be dozens.

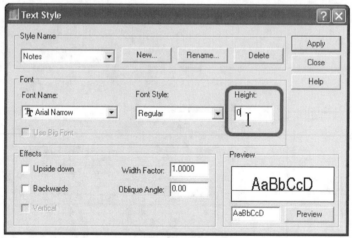

Figure 3–32 *Setting a variable text height in a Text style*

- Use descriptive names. A Text style and a font are not the same thing. Name your styles for how you intend them to be used. If you are creating a style for typing notes, call it "**Notes**" not "Simplex." Nothing about a font name indicates to a user *how* they ought to *use* the Text style in production.

- Do not be afraid to use True Type fonts. There was a time when True Type fonts were very slow—that is no longer the case on newer hardware. Use True Type fonts; they look better than AutoCAD SHX fonts.

- If you use "non-standard" fonts, (fonts not shipped with AutoCAD/ADT), you will need to provide those fonts to anyone with whom you share your ADT files. The eTransmit tool (covered in Chapter 12) can be used to make this easier to do, but just be aware that there is no way to "force" your external partners to actually install and use your fonts.

DIMENSION STYLES

All dimensions in AutoCAD/ADT are controlled by a Dimension style. The style determines the size of all Dimension components, the Text style and the unit of measurement used by dimensions. It is highly likely that you already have a collection of Dimension styles in use in your firm. Nothing about ADT implementation requires that you rework your existing Dimension styles. Therefore, if you are satisfied with the styles you currently have, continue to use them in ADT. Consider the following issues and options:

- When sizing a Dimension component within the style, you can input the actual size of the component as it will appear in the final plot, and then input an "Overall Scale" (DIMSCALE) factor by which these values will all be multiplied in your model, or you can "pre-scale" the values to a particular plot scale and not use a DIMSCALE.

Both methods will achieve the same net result; however, the first method will generally be easier for all parties to edit and use. The point is raised here because this method is not used in the default Dimension styles, used by the AEC Dimension objects in default ADT template files. It is recommended that you customize the AEC Dimension styles to use your office standard Dimension styles. More information on AEC Dimensions can be found in Chapter 7.

- As indicated in the "Text" topic, it is recommended that you use a Text height of 0 for Text styles referenced by Dimension styles. In this way, the Dimension style and not the Text style will determine the height of the text used in dimensioning.

- If you wish to build a custom Dimension arrowhead Block, make sure that it is inserted into all of your template files. The Block used for a "User Arrow" must be resident in the current drawing to be used (see Figure 3–33).

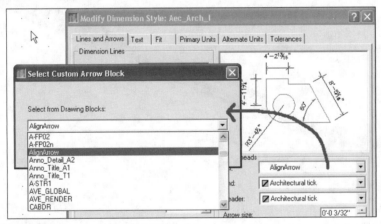

Figure 3–33 *Accessing a user defined arrowhead*

- Associative dimensions *do not* work on ADT objects. To get the associative features on ADT objects, you must use AEC Dimensions. Associative dimensions *can* be used on AutoCAD objects within ADT drawings. However, the system variable that controls this is turned off by default in ADT. To turn it on, type **dimassoc** at the command line and set the value to **2**. You can also set it in the Options dialog box on the User Preferences tab. For more information on this variable, look up DIMASSOC in the online help.

Note: Dimensions created with the QDIM command are not Associative by default. If you are part of the ADT Subscription program, an Associative QDIM Extension is available that corrects this.

SCALE SETTINGS

The same recommendations made in the "Linetype Scale" topic above will be made for the drawing as a whole. Again, our goal is to simplify the process of creating a model and its documentation from ADT. Therefore, it is recommended that all scale-dependent information (linetypes, text, dimensions, hatching, and Blocks used as symbols) use the same strategy for scaling. Simply determine the scale factor for the drawing (as covered in the "Linetype Scale" topic) and use this value as the basis for all scale-dependent annotation.

Few people stop to consider that there have always been at least two scales (possibly many more) involved in every architectural drawing. There are, at a minimum, the scale of the building itself (termed the model in ADT terminology) and the scale of the annotation. In the days before CAD, we used to draw the model (the building) at a scale factor, and the annotation at "full size." This approach has simply been reversed. We now model at full scale and create annotation using a scale factor. In hand drafting, we had our architectural scale ruler that allowed us to switch scales effortlessly (see Figure 3–34.) We need a tool within CAD that is just as simple.

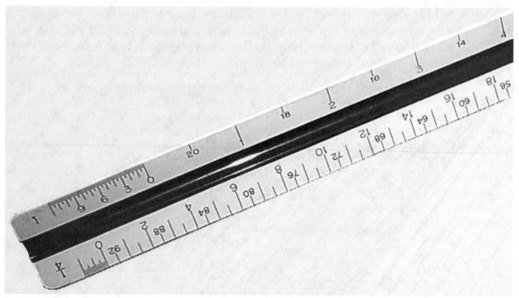

Figure 3–34 *The original "high tech" piece of equipment for architects*

So how do we use the "keep it simple" strategy to scale management? Calculate the scale factor of the drawing (see Table 3–2), and then apply it directly to each annotation type.

Table 3–2 *Common Scale Factors*

US Architectural Scales		US Engineering Scales		Metric Scales	
1.1.1 Plot Scale	Scale Factor	1.1.2 Plot Scale	Scale Factor	1.1.3 Plot Scale	Scale Factor
1/64" = 1'-0"	768	1" = 10'	120	1 : 1	1
1/32" = 1'-0"	384	1" = 20'	240	1 : 5	5
1/16" = 1'-0"	192	1" = 40'	480	1 : 10	10
3/32" = 1'-0"	128	1" = 50'	600	1 : 20	20
1/8" = 1'-0"	96	1" = 60'	720	1 : 50	50
3/16" = 1'-0"	64	1" = 70'	840	1 : 75	75
1/4" = 1'-0"	48	1" = 75'	900	1 : 100	100
3/8" = 1'-0"	32	1" = 80'	960	1 : 200	200
1/2" = 1'-0"	24	1" = 100'	1200	1 : 500	500
3/4" = 1'-0"	16	1" = 150'	1800	1 : 1000	1000
1" = 1'-0"	12	1" = 200'	2400	1 : 2000	2000
1 1/2" = 1'-0"	8	1" = 250'	3000	1 : 5000	5000
3" = 1'-0"	4	1" = 500'	6000		
1' = 1'-0"	1	1" = 1000'	12000		
		1" = 2000'	24000		
		1" = 5000'	60000		

The preceding scale factors can be used to properly scale all forms of annotation in a drawing file. Apply the scale factors to each type of annotation as follows:

- **Linetypes** – Use the scale factor as the LTSCALE. In a 3"=1'-0" [1:5] drawing, set the LTSCALE to 4 [5].

- **Text** – Multiply the scale factor by the desired plot height of the text. To create text that is 3/32" [2.5 mm] tall in a drawing printed at 1/8"=1'-0" [1:100], multiply 3/32 [2.5] by 96 [100] to arrive at a text height of 9" [250].

- **Dimension Scale** – Use the scale factor as the DIMSCALE. In a 1/4"=1'-0" [1:50] drawing, set the DIMSCALE to 48 [50].

- **Hatch patterns** – Not quite as straightforward: starting with the scale factor as the hatch scale, modify up or down as necessary to achieve a desirable result. For standard hatch patterns used in construction documents, you should build a table of typical hatch patterns and scale factors. Note: user defined hatch patterns (**User Single** and **User Double** in the ADT Entity Display dialog boxes) do not use a scale factor; rather they use a true value in drawing units for the spacing between lines.

- **Blocks used as symbols** – Build the Blocks at the size they should appear on the final printed sheet, and insert them into drawings using the scale factor as the X, Y and Z Block scale. For example, a section marker might be drawn with a 5/8" [16 mm] diameter circle. When inserted into a drawing with a 1/8"=1'-0" [1:100] plot scale, the size of this target would be 5'-0" [1,600 mm].
- **Multi-view Blocks** – Can use **Drawing Scale** or **Annotation Scale** (in the Drawing Setup dialog box) and can scale themselves automatically. Therefore, in most cases, the user will not need to worry about inputting scale factors at all. For more information, refer to Chapter 8.

BLOCK LIBRARY

Block libraries have long been a staple among AutoCAD users. Since ADT is built on and around the AutoCAD 2002 engine, all of your existing blocks will be usable "as is." However, for a variety of reasons you may choose not to continue using them as is. Take stock of your current collection of AutoCAD Blocks.

- What is the primary purpose of your existing Blocks?
- Do they provide symbology for construction documents annotation needs, or do they serve a broader need?
- If your firm has decided to adopt the NCS or other industry standard, do your Blocks comply with that standard?

Most of us have been dragging the same ever-growing collection of Blocks and resources around for years and rarely take the time to clean out the "dead wood." Now is a good time to clean house. Divide your existing library into three categories: "keep," "edit" and "delete."

You will find that most of the symbology required on a typical set of construction documents has been provided in one form or another in the "out-of-the-box" ADT offerings. This fact may influence your decisions of what to keep and what to delete. The issue will likely involve your level of satisfaction with the Content that has been provided compared to your own versions. For our discussion in this chapter, we will focus on the issue of Block library access. The details of assessing the quality and usefulness of your existing Blocks and symbols and building new ones will be left to Chapters 6, 7 and 8.

The AutoCAD DesignCenter is the primary interface for accessing all of the ADT Content provided. Your existing Block library is accessible from the DesignCenter, but not in the same capacity as the ADT Content provided. In order to be accessible from the ADT-enabled portion of the DesignCenter, Content items must be saved in a special way, using the AEC Create Content Wizard. However, the remaining two nodes of the DesignCenter, Desktop and Open Drawings, allow access to *any* AutoCAD or ADT drawing file and the Blocks, Dimstyles, Layers, layouts, linetypes, Textstyles and XREFs contained within them (see Figure 3–35).

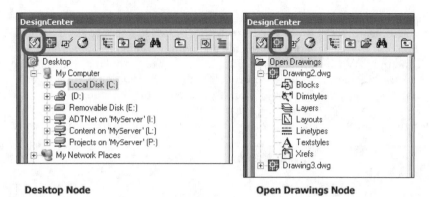

Figure 3–35 *The Desktop and Open Drawing nodes give access to any drawing file*

For the purposes of this chapter, create a folder on the L Drive named *Legacy Content* and copy all of your existing Block library content to this folder. Add as many sub-folders as necessary. The goal is simply to get all resources in one easily assessable location. Users will be able to access this folder from the DesignCenter right away. Later in Chapters 6, 7 and 8, we can begin to convert them as appropriate to AEC Content.

DETAIL LIBRARY

Follow the same recommendations for existing Detail libraries as were made for Block libraries. Create a folder on the L Drive named *Details* and copy all existing Details to this location. Again, you may include sub-folders. If you wish to perform some housecleaning first, go ahead and do so. Details can be left as standard AutoCAD Blocks, or you can convert them to AEC Content to enable more intelligent preview features and layer keying on insertion.

TITLE BLOCKS

As with most of the other resources covered in this chapter, your title blocks likely will need little to no revision to be used with ADT. Consider the following issues and recommendations when migrating existing or building new title blocks.

- Create a title block for each sheet size commonly used in your firm and save each one to its own drawing file. If compliance with an industry standard is a goal, check for compliance. Standard sheet sizes and title block organization are provided with the National CAD Standard.

- Draw the title block in model space at full size. A 36" x 48" [914 x 1,219] title block will be drawn 36" x 48" [914 x 1,219].

- Determine the printable area of each sheet size you use for each of your plotters as configured at the start of this chapter.

For example, the printable area of an "ARCH E (36.00 x 48.00 Inches [914 x 1,219])" sheet might have a printable area of 47.50 x 34.50 inches [1,149 x 826 millimeters]. This means that the there will be a 0.75 [44] margin on both the top and bottom of the sheet and a 0.25 [37.5] margin on the left and right. When you build your title block, you need to build it relative to the printable area size and *not* the paper size. Furthermore, the 0,0 origin for plotting purposes will be the lower left corner of the **printable area,** *not* the sheet itself. (See Figure 3–36 on the left.) If you wish to adjust the margins of your printed sheet, input values into the **Plot Offset** fields of Page Setup. This will actually shift the origin for plotting purposes as shown on the right of Figure 3–36.

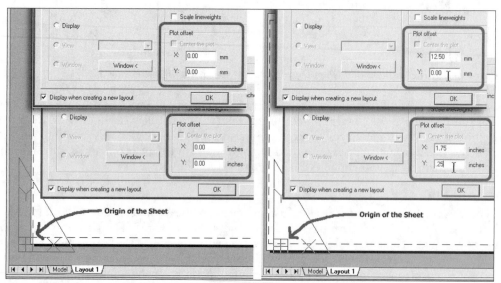

Figure 3–36 *The printable area and the Plot Offset determine the origin for plotting*

- The lower left corner of your title border (not the sheet size boundary) should be drawn at 0,0 in the title block file. Do not include margins in your title block file. Use the Plot Offset feature to adjust the position of the origin in your layout tabs so that this title border falls in the correct location relative to the paper size.

Using the "ARCH E (36.00 x 48.00 Inches [914 x 1,219]" sheet example from above, let's assume a 2" [50] binding strip on the left, a 1" [25] right, top and bottom margins. This means we would draw a rectangle to represent our outer border starting with a lower left corner at 0,0 and an upper right corner at 45,34 [864,1144]. 48- 2- 1= 45 and 36 - 1-1=34 [1,219 - 50 - 25=1,144 and 914 - 25 - 25=864]. Assuming the same printable area of 47.50 x 34.50 inches [1,149 x 826 millimeters], we would then set the Plot Offsets in Page Setup to: X=1.75" [12.5] and Y=.25" [0] (see Figure 3–37).

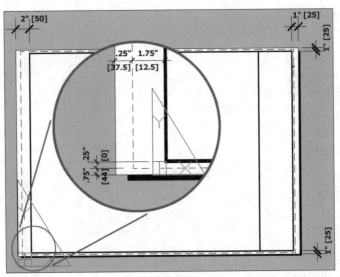

Figure 3–37 *Setting the dimensions of the the title block to match the printable area of the sheet*

- Draw the remaining "non-changing" graphical items such as borders, lines, general note columns, company logos and fixed text (like firm name and address). Draw each of these items on an appropriate layer and apply Color, Linetype and Lineweight properties.

- Save the file to a folder on the L Drive named *Title Blocks*. Use a name such as: *48x36 Title Block.dwg* [*914x1219 Title Block.dwg*]. (This file will XREF into the paper space layouts of your sheet files.)

- Add all of the variable text like drawing name, author, dates, etc. as attributes. Establish the Text style characteristics desired for each of these items and position them relative to the fixed graphical components.

- **Wblock** these items out to a separate file (being sure to delete the originals from the title block file) with the same name as the title block and append the word "info" to the end, such as *48x36 Title Block_info.dwg* [*914x1219 Title Block_info.dwg*]. (This file will be inserted into the paper space layouts of your sheet files on top of the XREF title block file.)

Once you have completed all of your title blocks and the accompanying "info" files, document the procedure for their use using the following recommendations when building this procedure:

1. At the start of each project, copy the entire *Title Blocks* folder from the L Drive to the project folder on the P Drive.

2. Rename the title blocks by adding the project name and/or number to the beginning of the file names within the folder.

3. Open the files containing the fixed graphical information and input any fixed text required for the project, such as the project name, number, client name and address.

4. When using the title blocks, XREF the fixed graphics from the project-specific copies made in these steps, and insert the "info" portion as a Block and edit their attributes accordingly. This is done in the paper space layouts of the sheet files.

CUSTOM COMMANDS

If you have existing LISP or VBA routines in wide use in your firm prior to your ADT implementation, they will need to be tested for usefulness and compatibility with ADT. It is likely that some of your routines will be rendered obsolete by ADT functionality. Others that still serve useful purposes may not be compatible with the latest release of AutoCAD/ADT. Be sure to test these routines thoroughly on your Management Install prior to rolling them out to general firm use. Be sure to document the intended uses of these items in your CAD Manual and Procedures documentation.

HATCH LIBRARIES

If you have any non-standard hatch pattern files (*.PAT), you can share them in one network folder on the L Drive. This method would be useful for hatch patterns not used very frequently by all members of the firm. Each hatch pattern can be saved to its own PAT file in this folder. Users can then simply browse to them with the DesignCenter, and drag and drop. Several files of this type are installed with the ADT Details Package. You can make them available to all users by moving them to the network. If you have custom hatch patterns that are used by everyone on a regular basis, you can also customize the standard *acad.pat* file and save it to the network in the I Drive Common Support folder as recommended in Chapter 1.

DOCUMENT SHARING PROCEDURES

If you currently have procedures and tools in place for conversion and sharing of drawings with outside consultants, you may need to amend those procedures to work with ADT. Refer to Chapter 12 for complete details on this topic.

ADT STANDARDS

In addition to the collection of items covered thus far, you will also need to build standards and resources for each of the following ADT items:

DRAWING SETUP

The **Drawing Setup** command has several functions in ADT. Separated into four tabs: Units, Scale, Layering and Display, the Drawing Setup dialog box configures many of the settings required proper usage of ADT objects. The Units and Scale tabs work together. Users are cautioned to use the Units tab of the Drawing Setup dialog box when working in ADT rather than the **Units** command on the AutoCAD **Format** menu. Drawing Setup Units *will* set the AutoCAD Units, but AutoCAD Units will *not* set the ADT Units. Therefore, do not use the **Units** command on the **Format** menu to set drawing Units. Always choose a Unit first, as this will dynamically change the choices available on the Scale tab. The Scale tab is used to automatically scale any ADT Annotation objects dragged into a drawing from the DesignCenter. The topics of Scale and AEC Content are covered in detail in Chapter 8.

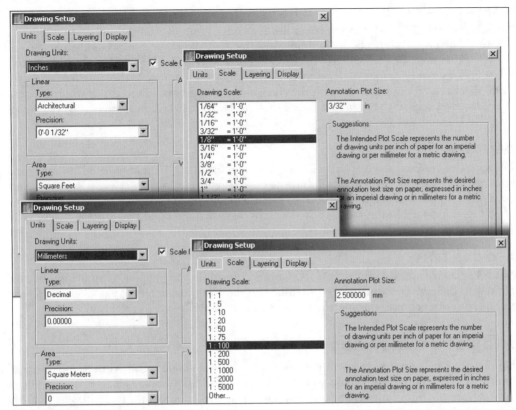

Figure 3–38 *The Drawing Setup dialog box*

In Chapter 1, we visited the Layer tab of the Drawing Setup dialog box and chose a Layer Key File from which to "Auto-import." In the next chapter, we will discuss layering and Layer Key Files in detail. Finally, the Display System is one of the most powerful and complex tools in ADT. Chapter 5 is devoted to all the details of this topic. Therefore, the two remaining tabs of the Drawing Setup dialog box will be saved for Chapters 4 and 5 respectively.

TEMPLATE FILES

Most ADT R3 configuration settings are saved with the drawing file. Therefore, drawing template files provide an excellent tool for promoting and maintaining office standards. Within our template files, we can build in all of our office standard ADT settings and configurations.

We will begin building our office standard template file in the next chapter. Our template file will evolve throughout the course of this book. The Conclusion will recap all that we have built regarding templates in the entire book with a checklist to help us finalize our templates and make them ready for production. A well-conceived template file is worth its weight in gold in ADT production.

DISPLAY SETTINGS

The Display System in ADT brings the rules and conventions of architectural drafting into the software. No longer is it necessary to draw plans, sections and elevations as completely separate drawings that must be manually coordinated in order to remain current. The Display System introduces a series of unique parameters to all ADT objects that enable the "intelligent" switch from plan to elevation to section with the change of viewing direction in the drawing window. All parameters related to the proper use of the Display System will become part of your complete CAD standards. Refer to Chapter 5 for more information.

OBJECT STYLES

Object styles are used to create classes of object types that can be manipulated from a single interface and that globally update all drawing objects referring to that style. Styles offer a very powerful design and management tool. An extensive collection of styles available for all manner of objects is included in the software. You will likely wish to edit and tweak these styles to suit your firm's unique needs. This topic is covered in Chapter 9.

AEC CONTENT

Much has been discussed regarding AEC Content already in this chapter. Chapter 8 is devoted entirely to the understanding of the potential and the customization of AEC Content. Chapters 6 and 7 offer coverage of specific types of content as well.

PROJECT STANDARDS

Despite your best efforts, you will not be able to build a single standard that applies to all projects all of the time, nor should you try. The goal of CAD standards is to make everyone's job easier by providing answers to common questions and resources for typical tasks. A CAD standard will provide proven and useful solutions to typical issues and problems that arise within the firm. Since it will not be possible to provide solutions to all problems, project teams should be encouraged to have a "production coordination meeting" at the start of each new project. The goal of this meeting should be to assess those unique requirements and issues of the current project that would prevent the adherence to accepted office practices and standards. The meeting should also expose items that might not have been addressed yet within the overall office standard. The meeting should not become a gripe session, where current processes and procedures are challenged, criticized and undermined. If that sort of sentiment is prevalent among the members of the firm, then there is clearly a need for a broader office-wide discourse.

In the "How to Use This Book" section of the Preface, we identified the following six personnel types: **IT Support Person, CAD Manager, Project Manager, Project Data Coordinator, CAD Software User** and the **CAD Data Consumer**. We also acknowledged that a particular individual might fill many of these roles simultaneously. If you have not read that section of the Preface, do so now. Choose your **Project Data Coordinators** carefully at these early stages. They will facilitate these project kick-off meetings and should hold regular follow-up meetings throughout the life of the project.

 Note: Having an individual with a particular "axe to grind" or a less than open-minded outlook on existing office procedures in this role could prove a detriment to all team members and the project as a whole. The person in the **Project Data Coordinator** role ought to be a true champion of the Building Model process and a good facilitator of people and resources.

All project-specific standards must be documented and made available to all team members. If appropriate, these standards may also be submitted for consideration as additions and/or replacements to office-wide procedures. Regardless of the procedures adopted for a specific project, it is critical to get "buy-in" from all team members, not just one or two visionaries. Again, it is critical to remember that the goal is the completion of an architectural project, not the creation of a brilliant CAD standard. Today success will be judged by how successful the architecture is and whether the project came in on budget, not in how clever we were at using the CAD software.

Do keep in mind, however, that the downstream application of data is becoming increasingly important to clients. Providing that data to clients is an area where architects can begin prospecting for additional services. The ongoing maintenance and sharing of that data makes compliance with industry standards such as the NCS useful and attractive.

AGENDA FOR PRODUCTION COORDINATION MEETING

It may not be possible to complete all of these items in a single meeting, but all of them should be completed within the early weeks of a new project.

- **Roles and Responsibilities** – Establish each team member's role in the project as it pertains the above referenced personnel types. Use the "How to Use This Book" section of the Preface as a guide, and remember that the roles describe areas of responsibility and not necessarily actual job titles.

- **Project Specific Issues** – Discuss unique characteristics of the project and any deviations from accepted office standards that these might impose. Begin the discussion of establishing border sizes, text styles and other graphic considerations, and any variances that may be required to meet the specific needs of the project.

- **Collaboration** – Develop a method for sharing base files with consultants/clients. Not all will be willing to accept Object Enablers. If you wish to work toward a true "Virtual Building Model," you are encouraged to set a baseline software requirement directly in your contracts with consultants. Working with consultants who are still using Release 14 should be strongly discouraged (see Chapter 12 for more information).

- **Approval Process** – Develop an approval process for items of deviation from office-wide standards.

- **Submission Process** – Develop a submission process for items to be considered for addition to the office master library.

- **Build a Digital Cartoon Set** – Prior to the meeting, the Project Data Coordinator ought to create all files required by the project. This may only be a rough draft for discussion purposes, but some attempt to determine all drawings required and build those skeleton files should be made at the start of the project. Refer to the "Project Types" section of the previous chapter (and Chapter 4 of *Mastering Autodesk Architectural Desktop*) for more information on this critical task.

TRAINING

Training of the staff is a critical component of any successful implementation. There are many types of training available in many formats and durations. The exact method or duration of training is not as critical as the actual need for some form of training. Autodesk Architectural Desktop requires a different way of thinking and approaching problems than standard AutoCAD does. For this reason, it is not advised to simply treat ADT as any other upgrade to AutoCAD. Take the training component seriously, because it will definitely pay for itself in productivity and shortened learning curve. Many firms have found success in training their staffs in shifts as project loads warrant and permit. There is never a good time to have people away from the office in training, but there is also never a good time to have people's productivity crippled due to lack of understanding of the software. When you build your budget for ADT implementation, factor training costs into this budget for a realistic picture of implementation costs. ROI strategies need to be carefully considered when training budgets are established. Consider using an annual percentage of time devoted to training as compared to total annual percentage of time spent in front of the computer. For the person who spends most of their day working in CAD, twenty hours of classroom instruction is only 1 percent of available billing time. Improvements in methods and procedures gained in training can be shown to management that will make up the twenty-hour differential in a short period of time. Management must be made to see that, in the end, the cost of not doing training far exceeds the cost of doing training.

Ways to Keep Down Training Costs

Most training providers offer different levels of training. It may not be necessary to train all users in all levels. Consider training only the project team just before the start of the ADT project. Everyone will likely benefit from some sort of overview course, but perhaps only certain individuals will have need for the advanced courses. In addition, if your firm has figured out exactly how they will deploy ADT on projects, this will allow for more focused customized training delivered in less time.

What to Look For

Make sure you interview the various training vendors in your area carefully and speak with the actual instructors who will administer the training. Not all training providers and trainers are equal. Shop around and know your options, including content covered, cost and location of training. It sometimes sounds really good to have the instructor come to you, but if people are constantly being pulled out of class to take phone calls and check email, they will not get the benefits from the training that they need and for which your firm is paying. It is sometimes better to have the attendees leave the office for the training and turn off their cell phones while in attendance. Make sure the instructor and training center are willing to accept "off topic" questions pertinent to the way your firm works. Ask about their policy on custom training.

Training Options

- **Dealer Training** – The reseller that sold you your copies of ADT most likely offers training courses and other consulting services.
- **Autodesk Training Centers** – Autodesk sanctions a small group of training centers worldwide to provide intense, hands-on, instructor-led training. To become an ATC, an organization must meet rigorous requirements established by Autodesk for the quality of the instructors and the center itself. A complete list of ATCs is available at the Autodesk Web site.
- **Colleges** – Many colleges and universities offer courses in Autodesk software. They are usually offered over the course of a semester rather than in an intense weeklong session like the Dealer or ATC training. If the intense weeklong pace does not appeal to you, consider these alternatives.

- **Autodesk Consulting Services** – Autodesk Consulting Services (ACS) provides standard and custom consulting offerings that help you apply Autodesk software products, tools and technologies so you can realize the full potential of Autodesk technology. ACS has domain expertise in Building Design and can deliver a wide range of services, from needs analysis through implementation, training and custom application development.

- **Independent Consultants** – There are many independent consultants offering training and other services. Some of the best Autodesk trainers are independent consultants; however, be sure to check their credentials carefully. Your best bet is to use an Autodesk Authorized Independent Consultant—use word of mouth to find a reputable one.

- **In House Expert** – If you are fortunate enough to have an in-house expert who truly understands the software, that person can save you a lot of money. However, it can sometimes be difficult to objectively assess the credentials of a "self proclaimed" in-house expert. Make sure that the individual truly understands the software and the best practices to its use. In addition, the same issues apply to in-house training regarding the potential for interruptions. As a rule, people do not tend to take in-house training as seriously as they do when going to an outside training center. In addition, the in-house expert may also be responsible for the support of the CAD environment or billable project work. Having them unavailable to the rest of the staff may not be an acceptable practice.

- **Self-Taught** – Although it can work for some, most individuals will greatly benefit from even a short introductory class. If cost is the limiting factor, consider the cost of not receiving training. In the end, a lack of productivity and the frustration users will feel will cost much more than a good training course.

ONGOING EVALUATION AND REVISION

With the completion of these first three chapters, we have addressed most of the global issues facing your firm and its ADT implementation. If you have been keeping a checklist of action items as you read these three chapters, it has likely grown quite long. Do not despair—most of the tasks that you face are not difficult; they simply require careful thought and a methodical approach. Use your lists to track your progress toward the completion of required tasks and as a rough draft for your CAD manual of standards and procedures. For the person or persons responsible for setting up corporate-wide resources, the tasks include the methodical execution of each of the items and recommendations made so far in the chapters of Section I. Your users will need to utilize on a daily basis all of the items you create. Therefore, it is critical that you provide them with proper documentation and training on each of the items you create and set in place.

Finally, do not become complacent. Your job is far from over with the completion of your standards and procedures and accompanying documentation. On the contrary, users will bombard you with questions, requests and criticisms immediately upon publication of your standards. Treat your implementation plan as a living thing, with room to grow and evolve as the needs of your firm and personnel dictate. Regular evaluation and revision should be an integral part of any successful implementation plan.

CAD STANDARDS CHECKLIST

Now that we have taken a comprehensive look at your current CAD standards, it is time to build a list of "action items" that you can use to begin making improvements. All of the topics covered in this chapter are summarized here in the following checklist. On the CD, the list is included in more detail (including additional sub-headings) in an Excel file for your convenience. Feel free to edit this list as required to make it most useful to you as a tool. The name of the Excel file is *Chapter03 Checklist.xls*, and it is located in the *Chapter03* folder.

I Plotting

   1. Install Plotters

   2. Import Legacy PCP or PC2?

   3. Configure Plotter Parameters

   4. Output Types/Sheet Sizes

II Plot Style Tables & Lineweight Control

   1. Decide What Parameters the Plot Style Table will control

   2. Set Up Lineweights

   3. Choose Plot Style Table Type

III Assessment of Existing CAD Standards

   1. Usage of Layouts and Page Setup

   2. Layer Names and usage (Chapter 4)

   3. Color

   4. Linetypes

   5. LTScale standardization

   6. Text

   7. Dimension styles

   8. Use a Common scale factor for all Annotation items

   9. Block Library

   10. Detail Library

   11. Title Blocks

   12. Custom Commands and Macros

   13. Hatch Libraries

   14. Document Sharing

IV ADT Standards

   1. Drawing Setup

   2. Display System Setup (Chapter 5)

   3. Property Set Setup (Chapter 6)

   4. Annotation Setup (Chapter 7)

   5. Content Library Setup (Chapter 8)

   6. ADT Object Styles (Chapter 9)

SUMMARY

Many items of your existing CAD standards can be transitioned directly to ADT.

Set up plotters and plotting resources for all types of output.

Lineweights are one of the most important aspects of plotting. Set these up carefully and consider using the Lineweight property rather than pen tables for lineweight.

Consider using Named Page Setup configurations rather than multiple layout tabs.

There are 255 colors, don't be afraid to use them!

Text and Dimension styles can be migrated directly from existing AutoCAD standards.

Try to standardize all scale-dependent information on a simple use of scale factors.

Existing Detail and Block libraries can continue to be used as is or "upgraded" to use ADT functionality.

Use Drawing Setup rather than the AutoCAD Units command to setup drawing units.

ADT Layer, Display, Object style and Content configuration are all required by a complete CAD standard.

Allow individual project standards where appropriate.

Set out a clear set of project team expectations and procedures at the start of each project.

Be sure to budget training cost and time for all users.

SECTION II

Display

This section is devoted to the display of objects in Architectural Desktop. Object Display is a function of the combined effects of Layer Control and the ADT Display System. Such advanced topics as Layer Standards, Layer Key Styles, Display Configurations and the Display Manager are among the topics covered in this section.

Section II is organized as follows:

Chapter 4	Layers
Chapter 5	The Display System

WHO SHOULD READ THIS SECTION?

This section is intended for the **CAD Manager,** the **Project Data Coordinator** and the **Project Manager**. The **CAD Software User** should also be engaged in the discussions that these chapters would prompt. Therefore, this section is important to that group as well.

CHAPTER 4

Layers

INTRODUCTION

Layers have provided the means for organizing data in AutoCAD drawings since the earliest releases. Devising "proper" ways to name and use layers has occupied the efforts of CAD technicians, CAD managers and professional organizations alike. In 1990, the American Institute of Architects established their first version of a CAD layer naming and usage guideline for U.S. Building Professions. This guideline was later updated and published as the "CAD Layer Guidelines" in 1997. Today this publication, along with other industry standards, has been incorporated into the U.S. National CAD Standards. In the United Kingdom, the yield of a similar progression of events and publications is the "BS 1192 Layer Standard." Similar efforts in Germany and throughout the world have fostered similar results.

Architectural Desktop provides many tools to help us organize our drawing data and our building models, including the Display System, object styles and layers. Deciding which of these topics to cover first in a manual such as this is a bit like deciding what came first, the chicken or the egg. Layers were chosen to be first *not* because they are more important than the Display System or object styles, but rather because display and style settings *can* control layers, but layers *cannot* control the Display System or object styles. Regardless of this distinction, layers remain an important aspect of the successful implementation of Architectural Desktop. ADT includes a robust set of layer management tools not provided in the base AutoCAD 2002 package; these include Layer Standards, Layer Key Styles and automatic layering, all of which support the NCS and other worldwide industry standards. This chapter will help you gain a full understanding of layer usage and techniques in order to fully implement them with the other drawing management and organizational tools of ADT.

OBJECTIVES

Many layering functions are performed automatically by the software based on a set of predefined rules and templates. In this chapter, we will explore all of these tools in detail and understand when and how the software generates layers and how they are maintained. Finally, it is critical to understand when ADT does not provide automated tools and what techniques may be employed to implement a layering standard successfully.

- Understand layering
- Understand features of auto-layering

- Understand Layer Standards and Layer Key Files
- Understand which features of layering do not occur automatically
- Customizing layer Setup

LAYERS VERSUS DISPLAY CONTROL SETTINGS

As was stated in the introduction above, layers and Display Control settings are both critical to maintaining and managing the display of objects in ADT drawings. Layers have been with us since the earliest releases of AutoCAD. The Display System is unique to Architectural Desktop. In the simplest way, they perform similar functions; they both help us decide what items to display on screen and print by making some items visible and others invisible. However, they achieve this task in very different ways. Layers provide a global drawing-wide management tool. Think of layers as a set of **categories** for the geometry within the drawing. Every object in an AutoCAD/ADT must belong to a single Layer category. That layer may in turn control the other properties of its objects such as color, linetype, lineweight and visibility.

The Display System shares many of the same functions, with a few distinct differences. The Display System is both much broader and more specialized. Its reach applies only to ADT Objects—AutoCAD entities are unaffected by the Display System. Like layers, Entity Display Props, which are the specific controls for a particular class of object such as Wall or Door, also control many of the common object properties like color, linetype and visibility. However, they exert this control based on a specific set of criteria, including the type of object, the current viewing direction of the drawing and the type of drawing required, such as Plan or 3D Model. Layers cannot achieve this level of granularity. On the other hand, the Display System cannot easily achieve global drawing-level control independent of specific object or drawing type. As a result, both tools offer critical benefits to ADT production and should both be well understood for successful implementation.

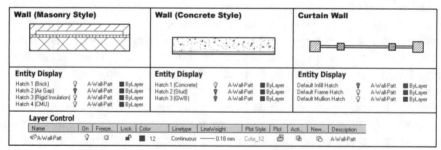

Figure 4–1 *AutoCAD layers and the Display System offer an overlap in functionality and some distinct traits as well*

Consider Figure 4–1. Each ADT object has Entity Display Props that can control the fine details of its display. These display properties can be controlled separately for each ADT object by class of object (system default), style of object or even object by object. However, you cannot "link" the properties of two object styles or two different classes of object with Display Control. As you can see from Figure 4–1, layers are required to achieve this control. Notice that even though the first two images are Walls and the third is a Curtain Wall, we can choose to assign the hatch components of both objects to the same layer, and thereby control them together.

In addition to the standard AutoCAD functionality of the layers, ADT offers Layer Standards, Layer Keys and Layer Key Styles. These are defined as follows:

▶ **Layer Standards** – Rules for a standardized layer naming convention.

▶ **Layer Key** – A "hard coded" tag within each ADT object that is mapped to a particular layer when used in a drawing (by the active Layer Key Style).

▶ **Layer Key Style** – A table mapping each Layer Key to a specific layer name designation, including its associated properties.

Each of the ADT layering tools can be found on the **Desktop>Layer Management** sub-menu and the Layer Management toolbar, as shown in Figure 4–2.

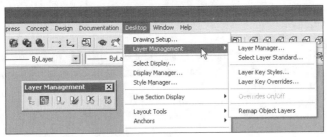

Figure 4–2 *ADT layer management tools*

LAYER MANAGEMENT DIALOG BOXES

The Layer Key Style includes all of the *potential* layers that *may* be created automatically by ADT within a drawing. In addition, every drawing contains a list of layers that *actually* occur within that drawing. These two lists are rarely identical. To see the list of Layer Keys and their associated layer definitions, choose **Layer Key Styles** from the **Desktop>Layer Management** menu. See "Layer Keys and Layer Key Styles" below for more information. To view the list of layers actually resident within a drawing, use the AutoCAD Layer Properties Manager or the ADT Layer Manager. In both these dialog boxes, you can see the complete list of layers defined within the current drawing, as well as the Color, Linetype, Lineweight, Plot, Lock and Visibility states of each layer. Both dialog boxes also offer a mechanism to filter and sort the list. The AutoCAD Layer Properties Manager uses Layer Filters for this purpose; the ADT Layer Manager uses Layer Groups. The features of Layer Groups are a bit more extensive than Filters, but the functionality is very similar. You can also save and name a collection of layer settings (Layer States) in both dialog boxes; the AutoCAD version uses Layer States (Figure 4–3) while the ADT version uses Layer Snapshots (Figure 4–4).

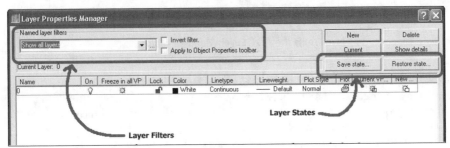

Figure 4–3 *AutoCAD Layer Properties Manager*

178

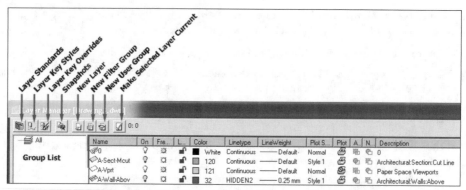

Figure 4–4 *ADT Layer Manager*

Upon exploring both of these interfaces, it can be a bit difficult to decide which one to use. When making your decision, consider that the ADT Layer Manager contains a few features specific to ADT that are lacking in the AutoCAD Layer Properties Manager. Layer groupings are expressed in a tree view in the left pane of the dialog box, layer names may be given an optional description and Layer Standards are accessed through this dialog box.

The first icon on the toolbar at the top of the ADT Layer Manager dialog box is used to access Layer Standards (see Figure 4–4). The details of working with Layer Standards are covered below in "Working with Layer Standards." When you click the New Layer icon (see Figure 4–4), a dialog box will appear allowing you to "build" a new layer name from the predefined descriptions defined by the Layer Standard (see Figure 4–5).

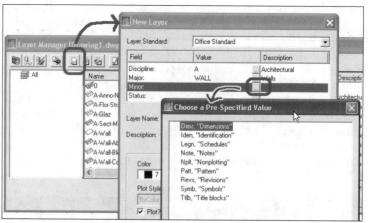

Figure 4–5 *Create a new layer from the ADT Layer Manager dialog box*

LAYER STATES, FILTERS, SNAPSHOTS AND GROUPS

There are many apparent similarities between the various tools used to group and filter a drawing's list of layers. (If you have the Autodesk Express Tools installed, you will have even more choices at your disposal.) Let us take a brief look at the four tools built into ADT by default. Two of them, Layer State and Named Layer Filter, are part of the core AutoCAD functionality and are accessible through the standard AutoCAD Layer Properties Manager. The other two, Snapshot and Layer Group, are available through the ADT Layer Manager.

- **Layer State** – Available in the standard AutoCAD Layer Properties Manager. A Layer State is used to capture and record the current state of all layers in the drawing. You can then use the saved Layer State to restore the drawing to this state whenever required. Specifically, the Layer State records the complete list of layers and each of their settings for On/Off, Freeze/Thaw, Plot/NoPlot, etc. If a new layer is added after the Layer State is captured, its settings are not included within the state and therefore will be unaffected by the Layer State as it is restored. The Layer State will need to be redefined to include settings for the new layer. Create and restore Layer States from the two buttons located in top right of the AutoCAD Layer Properties Manager (see right side of Figure 4–3) and edit the properties the state controls with the tools shown in Figure 4–6.)

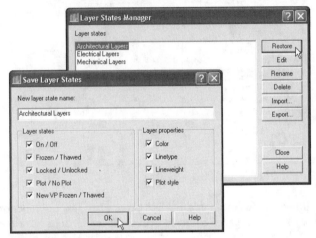

Figure 4–6 *Working with Layer States*

- **Layer Snapshot** – Available in the ADT Layer Manager, the functionality is nearly identical to a Layer State. It is also useful for saving the current state of all layers, thus the name "Snapshot" (see Figure 4–7). Create, restore and edit Layer Snapshots from the Layer Snapshots icon in the ADT Layer Manager (see Figure 4–4).

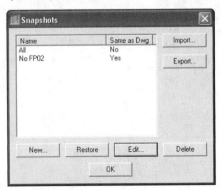

Figure 4–7 *Create and restore a Layer Snapshot*

Tip: In the **Same as Dwg** column in the Snapshots dialog box, you can quickly tell if the Snapshot matches the current state of the drawing.

- **Named Layer Filter** – This is available in the standard AutoCAD Layer Properties Manager. Scrolling through a long list of layers can become tedious. Layer filters shorten the list to those that share a common set of criteria. This does not eliminate or purge the excluded layers; it simply "filters" them temporarily from the list. Some layer filters are created automatically, such as filters to show only the layers belonging to XREFs. Create, restore and edit Named Layer Filters from the drop-down list at the top left or from the right-click menu of the AutoCAD Layer Properties Manager (see Figure 4–8).

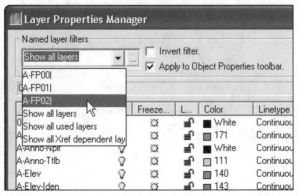

Figure 4–8 *Accessing the list of Named Layer Filters*

- **Layer Groups** – Available in the ADT Layer Manager, the functionality is nearly identical to a Named Layer Filter. However, Layer Groups have the ability to be both dynamic (like a Layer Filter) and static (similar to a Snapshot.) Static Layer Groups differ from Snapshots in one significant way: With a Snapshot, you must first manually configure all the layer settings the way you want them in the Snapshot, and then save the Snapshot to record and name the condition. With static Filter Groups, you establish criteria, as you would with dynamic Filter Groups, but once applied, they remain fixed like a Snapshot unless you edit them. A dynamic Filter Group will "remember" the criteria and continue to re-apply it to the list of layers as they change. Layer Groups differ from Named Layer Filters in one other way: they are aware of ADT Layer Standards (see Figure 4–9) and whether or not a layer complies with a layering standard may be used as a grouping criterion.

Figure 4–9 *Building a Layer Group*

REPLACING THE STANDARD LAYER COMMAND WITH THE AEC LAYER MANAGER

The AEC Layer Manager is available from both the Desktop>Layer Management menu and the Layer Management toolbar. If you would rather have the users use this dialog box exclusively in lieu of the AutoCAD Layer Properties Manager, you can redefine the AutoCAD LAYER command and its command alias to point to this dialog box rather than the AutoCAD Layer Properties Manager. To do this, add the following lines to a text file named *Acaddoc.lsp* and save it to the *I:\Support* directory. You can create this file in a text editor such as Windows Notepad.

```
(command "undefine" "layer")
(defun C:LAYER ( / cmdsave )
(setq cmdsave (getvar "cmdecho"))
(setvar "cmdecho" 0)
(command "AEClayermanager")
(setvar "cmdecho" cmdsave)
(princ) )
```

Note: If you already have an *Acaddoc.lsp* file, simply add the code to your existing file.

The Acaddoc.lsp file (if present along the AutoCAD search path) loads automatically when AutoCAD/ADT is launched. With these lines of code present, the standard LAYER command will be undefined and replaced with the ADT version. From that point on, it will not matter which method is used to execute the LAYER command (menu, keyboard or icon), the AEC Layer Manager will appear.

As an alternative, you could simply add an alias to your *acad.pgp* file instead of undefining the actual default LAYER command. In that case, you would need to type your custom alias at the command line to call the ADT Layer Manager.

WORKING WITH LAYER STANDARDS

Earlier in this chapter, we defined ADT Layer Standards as "the rules for a standardized layer naming convention." What this means is that we are able to establish one or more distinct portions of a preferred layer naming convention and assign rules to each of those components. The software will then be able to both assist us in quickly choosing a "standard" layer name when creating a new layer, and also to immediately flag those layer names that do not conform to the standard. It is important to mention that although a Layer Standard is very useful tool, it is not required in order to use ADT automatic layering. The Layer Standard functionality serves as a guide for users and a check for CAD Managers. (If you prefer, you can skip the Layer Standard and go directly to a Layer Key Style file.)

Getting to the correct dialog box to view and edit a Layer Standard is a bit tricky. The **Select Layer Standard** command, available on both the Desktop>Layer Management menu and the Layer Management toolbar, calls on the Layering tab of the Drawing Setup dialog box. We visited this dialog box back in Chapter 1 where we assigned a temporary "Office Standard" Layer Key Style for all new drawings. See Figure 4–10.

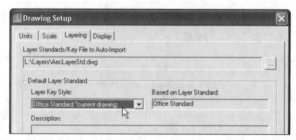

Figure 4–10 *The Office Standard Layer Key File was set to the default in Chapter 1*

 Note: You can launch your Office Standard ADT icon that was built in Chapter 1 and choose the **Desktop>Layer Management>Select Layer Standard** command to verify this.

Nothing in this dialog box speaks to establishing rules for a standardized layer naming convention. The actual Layer Standard in use in particular file is meant to be a "set and forget it" type of tool. The best place to do this is in the file that is designated to Auto-Import (as shown in Figure 4–10), in this case *AecLayerStd.dwg* (or whatever file you set here back in Chapter 1).

Install the CD Files and Open the Layer Standards/Key File

1. Install the files for Chapter 4 located on the Autodesk Architectural Desktop: Advanced Implementation Guide CD ROM.

Refer to the "How to Use the CD" section in the Preface for instructions on installing the sample files included on the CD.

A folder named *Chapter04* was created containing a copy of the original *AecLayerStd.dwg* provided with ADT. If you followed the installation procedures covered in Chapter 1, you mapped your default Layer Standards/Key File to a copy of this file in the *L:\Layers* folder. (Refer to Chapter 1 for more details.) The name will be *AecLayerStd.dwg*, or whatever you renamed it in Chapter 1. Regardless of whether you created your own file in Chapter 1 or used the default, for this tutorial, you should work with the copy of *AecLayerStd.dwg* that you just installed from the CD. Use this one for practice, thus preserving your original until you are ready to perform your actual edit. Later you can either use it to replace your original copy or repeat the steps on your own version.

2. Launch ADT 3.3 and open the *AecLayerStd.dwg* file (from the *Chapter04* folder).

3. From the Desktop menu, choose **Layer Management>Layer Manager**.

This loads the ADT Layer Manager. As stated above, the ADT Layer Manager is an alternative to the AutoCAD Layer Properties Manager. The first icon in that Layer Manager dialog box calls the Layer Standards dialog box (see Figure 4–11).

 Note: This is the only way to open this dialog box.

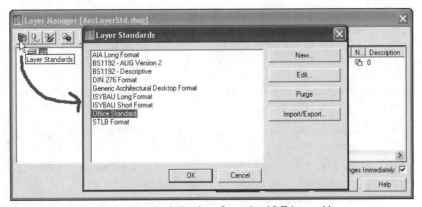

Figure 4–11 *Open the Layer Standards dialog box from the ADT Layer Manager*

Contained in the default file are the following Layer Naming Standards: AIA Long Format, BS1192 - AUG Version 2, BS1192 - Descriptive, DIN 276 Format, Generic Architectural Desktop Format, ISYBAU Long Format, ISYBAU Short Format, Office Standard and STLB Format.

 Note: The Office Standard entry is the one we added in Chapter 1; it is not a part of the default ADT offering.

Understanding a Layer Standards/Key File

Let's reverse engineer a few of the entries in this dialog box to understand the features of an ADT Layer Standard.

1. Select **AIA Long Format** and then click the **Edit** button.

The Layer Standard Properties dialog box opens revealing four tabs: Component Fields, Edit Descriptive Fields, Edit Descriptions and Description Specification. The "AIA CAD/Layer Guidelines" defined four layer component fields. The **Discipline** field designates the author of the information on a particular layer and includes codes like "A" for Architectural, "M" for Mechanical and "C" for Civil. The **Major** field designates a major building component, such as

Wall, Door, Flor (flooring) and Glaz (glazing). **Minor** and **Status** fields are also specified, but are not required.

    2. Select the **Discipline** Field Name and look at each of the settings (see Figure 4–12).

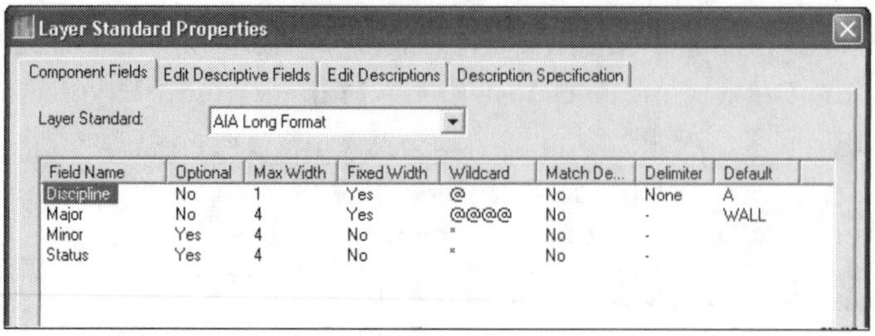

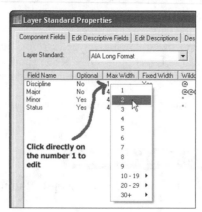

Figure 4–12 *Study the settings for each of the fields*

The name of a field can be anything you wish. Notice that the **Discipline** field is not optional and is required by the AIA guideline. The width of this field is also fixed at one character, which must be an alpha character and it (unlike the other three fields) will not be preceded by a delimiter. The **Major** field is also required, has a fixed width (four characters this time), uses only alpha characters and is preceded by a delimiter. The other two fields are not required and may be as many as four characters, but are not required to be four. They can use any character, including numbers and symbols.

The AIA guideline allows for a variation in the **Discipline** field to include two leading characters rather than one. To edit this standard to allow the double-character discipline code, do the following:

    3. Next to the **Discipline** field, click on the number **1** in the **Max Width** column.

    4. In the menu that appears, choose **2** as shown in Figure 4–13.

Figure 4–13 *Choose 2 characters for the Discipline field*

The **Discipline** field will now be required to have two characters. If you prefer to allow the field to be *either* one or two characters, then click the word **Yes** in the **Fixed Width** column and choose **No**. In addition, a literal reading of the AIA guideline dictates substituting the first delimiting dash (-) for the second discipline character. It is the opinion of this author that the layer name becomes more difficult to read when you follow that recommendation. However, if you wish to follow the "letter of the law" rather than interpret the intent, you can change the delimiter of the **Major** field to **None** (see Figure 4–14). This means that the "dash" will not be added automatically. Therefore, when using single character codes in this system, you will need to add the dash manually to the discipline code.

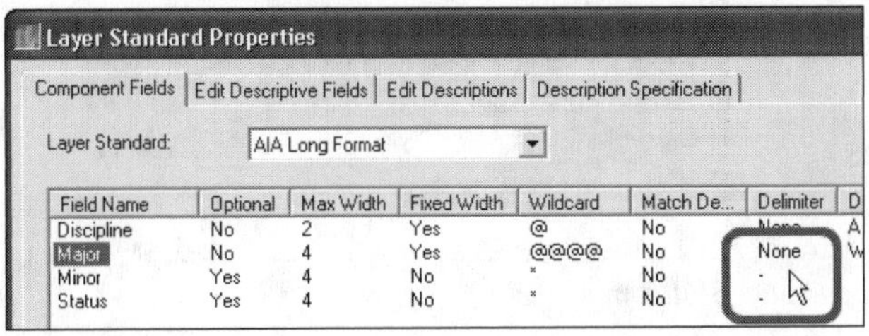

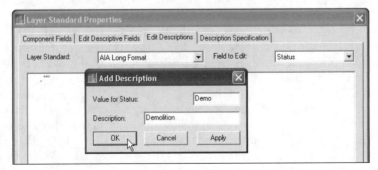

Figure 4–14 *Removing the delimiter between the Discipline and Major fields*

The Component Fields tab is perhaps the most important part of the Layer Standard. It establishes the rules that will be used by ADT to determine if a layer complies with the standard. By adding descriptions, you can make your Layer Standard easier to use and conform to.

    5. Click the Edit Descriptions tab.

    6. Scroll through the list and read the entries.

Notice that this list includes most of the common Discipline codes. If you changed the Discipline code to two characters above, then you will need to change each of the Discipline codes in this list to their two-character equivalent and/or add and delete entries as required.

    7. In the top right corner of the dialog box, choose Major from the **Field to Edit** list.

Notice that this list includes all of the common Major field codes for AIA.

    8. Choose Status from the list.

Note that there are no Status codes defined.

    9. Click the **Add** icon at the bottom of the dialog box.

  10. For **Value for Status** type **Demo**.

  11. For **Description** type **Demolition** (see Figure 4–15).

Figure 4–15 *Add a Status field Description for Demolition*

12. Add others if you like, such as **Exst**, **Phs1** and **Phs2** (see Figure 4–16).

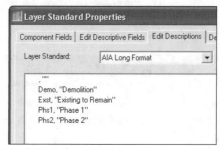

Figure 4–16 *Add additional Status modifiers*

13. Click **OK** or **Cancel** to exit the Layer Standard Properties dialog box.

Explore the BS1192 - AUG Version 2 Layer Standard

1. From the Layer Standard list, choose **BS1192 - AUG Version 2**.

This layer naming scheme, developed by Autodesk and the U.K. Autodesk Users Group (AUG), relates to the BS1192 – Part 5 guideline used in the United Kingdom. The Layer Naming Convention uses up to eight fields of distinct information. The **Discipline** field is required and indicates the author of the information contained on that layer. The **Category** field is required and uses the CI/SfB common referencing system. The remaining six fields, **Graphics**, **Grade**, **Level**, **Status**, **Scale** and **Time** are all optional.

2. Return to the Layer Standards dialog box if you are not there presently.
3. Select **BS1192 - AUG Version 2** and then click **Edit** (see Figure 4–17).

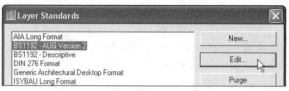

Figure 4–17 *Edit the BS1192 - AUG Version 2 standard*

The first thing you will note is that there is a single-digit **Discipline** code defined to include mostly letters. Next, rather than a single three-digit Category code as actually indicated by the BS 1192 guideline, there are three separate **CI/SfB** code fields. This is because the CI/SfB codes are hierarchical, reading from left to right. If the second digit were **2**, it could represent any number of categories depending on the value of the first CI/SfB code. For instance, code 200 is used for Carcass, Primary Elements. This is a general overall category that could include Walls, Stairs, and Roofs. Code 210 is used to indicate Exterior Walls. Should even more detail be required to differentiate Walls further, codes such as 212 – Outer Leaf of Cavity Wall, and 213 – Inner Leaf of Cavity Wall can be used. When the level of detail is not broken down to all three characters required by the **Category** field, zeros are used as placeholders. Notice that zero is the default of both **CISfB-B** and **CISfB-C** (see Figure 4–18).

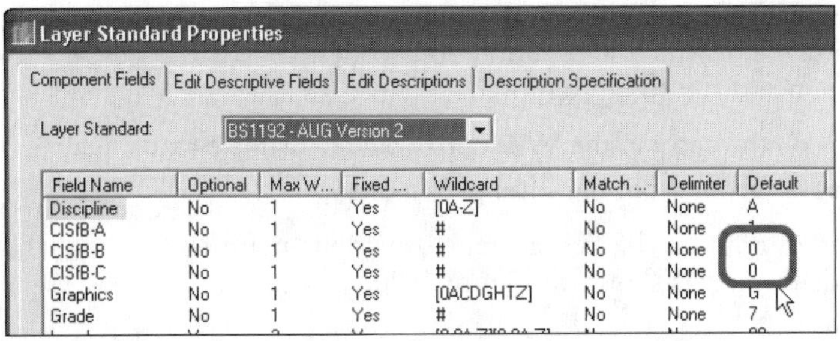

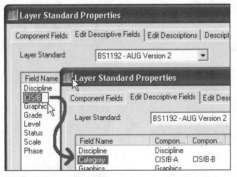

Figure 4–18 *Category fields use a zero (0) as their default value*

The hierarchical relationship between each of the three **CISfB** fields is established on the Edit Descriptive Fields tab.

    4. Click the Edit Descriptive Fields tab.

Descriptive fields match the field definitions as defined by the Layer Naming Convention (BS-1192 in this case), while the Component Fields are the actual fields as used by the ADT Layering Standard. In the previous example where we looked at the AIA CAD/Layer Guideline, the Component fields and the Descriptive fields were the same. However, in the BS-1192 guideline, three ADT **Component** fields (**CISfB-A**, **CISfB-B** and **CISfB-C**) are required to build the single **Category** as required by the BS-1192 standard. A **Component** field cannot be used twice in a layer naming convention; therefore, when building a hierarchical relationship between **Component** fields, begin at the top of the dialog box and work your way down. As a point of minor technicality, the BS-1192 guideline refers to Field 2 as the "Category" field rather than **CISfB** as found here. If you wish you could change this.

    5. Click on the **CISfB** text in the **Field Name** column. (It should go into a rename mode.)

Figure 4–19 *Rename the CISfB field to Category*

    6. Type **Category** in place of **CISfB** (see Figure 4–19).

    7. Click back to the Component Fields tab.

    8. Take a close look at both the **Wildcard** and **Delimiter** columns.

BS-1192 sometimes uses a very limited number of codes in a particular field. Take, for instance, the **Status** field, which allows only four codes: **0** – Not Used, **N** – New Work, **X** – Existing to remain and **R** – to be Removed.

9. Click on the entry in the **Wildcard** column for the **Status** field.

10. Choose **Other** from the small pop-up menu.

Setting this kind of limit on the field is done by simply inputting the legal characters into the Default Value dialog box as shown in Figure 4–20.

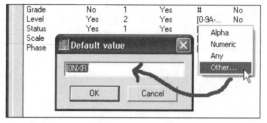

Figure 4–20 *Use the Other option to set a specific set of legal codes*

11. Click the Edit Descriptions tab.

Note the plus (+) signs next to each of the Major Categories (CISfA).

12. Expand the plus (+) sign next to **2, Carcass, Primary Elements** and then again next to **21, Ext walls.**

When the ADT Component Fields are set up in the hierarchical fashion on the Edit Descriptive Fields tab, the entries on the Descriptive Fields tab follow this organization.

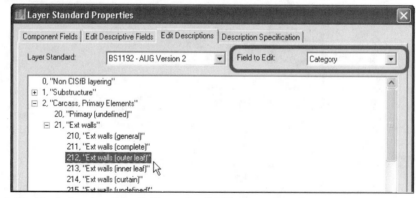

Figure 4–21 *The heirarchy of the Fields on the Edit Descriptions tab*

 Note: The **Field to Edit** on the top right now appears as Category rather than CISfB.

13. Click **OK** to complete your exploration of BS1192 - AUG Version 2.

Feel free to scan through some of the other standards included in this file as well. Several countries and jurisdictions are represented by the standards included in this file. Take the time to explore each one, and you will note that there is much potential in the way that Layer Standards can be configured.

BUILD YOUR OFFICE STANDARD LAYER STANDARD

By now, you have seen enough examples to begin to set up your own layering standard. If you are using one of the industry-standard layering schemes included in the *AecLayerStd.dwg* file, congratulations, your work is already done for you. If you are not using one of these standards, you will likely wish to build your own based on the standards used in your office.

Tip: Before you embark on the path of building a custom layering standard, take the time to critically analyze the standard that you currently use or are conceiving. Does it meet needs not met by one of the Layer Standards included? Could one of the ones provided meet your needs for organizing drawing data and categorizing information? If so, you should seriously consider adopting one of those standards. The work of building Layer Standards and Layer Key Files has already been completed, and you will have a much easier time in sharing your files with outside firms that use the same standard.

If you have decided to build your own Layer Standard, use the following tutorial as a guide. (If you have decided to use one of the Layer Standards provided, you may skip to the "Layer Keys and Layer Key Styles" section below.)

First, try to diagram the structure you wish your Layer Standard to take. Determine how many fields you will need and what rules should apply to them. If your layering scheme is very similar to one of those in the Layer Standard file, you may copy it and edit it to save time; otherwise, you may need to create your standard from scratch.

Build a Custom Layer Standard

Let's build a very simple Layer Standard as an example. Suppose we want to build a three-field layer scheme, where the first field indicates the phase of construction (like New, Demo or Existing), the second field indicates the type of component (like Walls, Doors or Annotation) and the final field is an optional description field (like Interior, Exterior or Special). A diagram for the Layer Standard might look like Table 4–1, while Table 4–2 shows the sample descriptions.

Table 4–1 *A Sample Layer Standard Specification*

Field Name	Optional	Max Width	Fixed Width	Wildcard	Delimiter
Phase	No	1	Yes	@	None
Component	No	6	No	Any	- (dash)
Modifier	Yes	6	No	Any	_ (underscore)

Table 4–2 *A Sample Layer Standard Descriptions List*

Code	Field Name	Description
N	Phase	New Construction
D	Phase	Demolition
E	Phase	Existing to Remain
Walls	Component	Walls
Doors	Component	Doors
Annot	Component	Annotation
Intr	Modifier	Interior Construction
Extr	Modifier	Exterior Construction
Spcl	Modifier	Special Construction

 Note: This tutorial was not intended as a recommendation of Layer Standard, but rather to illustrate the steps required while building a custom standard. It is highly recommended that you adopt an industry standard, preferably one already included in the *AecLayerStd.dwg* file. The steps above are intended to serve as a guide for those firms that have already developed their own in-house layering standard.

1. If you are still in the Layer Standard Properties dialog box, click **OK** to exit.
2. In the Layer Standards dialog box, click the **New** button.

If you wish to base your standard on an existing one, place a check mark in the **Based on** check box, choose the standard to use as a basis for your new one and type a name in the **Name of Standard** field.

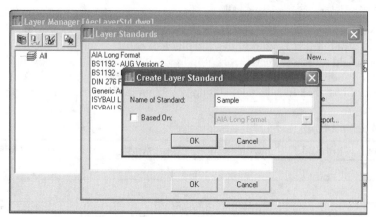

Figure 4–22 *Create a new standard not based on an existing one*

3. For this example, do *not* choose **Based on** and type **Sample** for the name (see Figure 4–22).
4. Click **OK** to create the Standard and then click the **Edit** button.

5. Based on Table 4–1, edit the values of the default field to look like Figure 4–23.

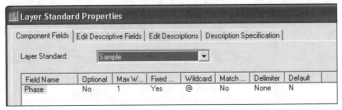

Figure 4–23 *Add the Phase field and its parameters*

6. At the bottom left of the dialog box, click the **Add below the selected item** icon (see Figure 4–24).

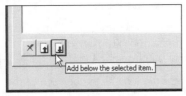

Figure 4–24 *Add a new field below the first*

7. Change the values of this field to match the Component field from Table 4–1.
8. Add the final **Modifier** field based on Table 4–1 (see Figure 4–25).

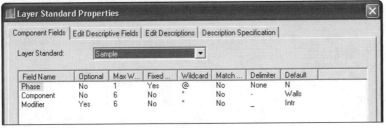

Figure 4–25 *The completed Component field definitions*

Add Descriptions

To add the descriptions from Table 4–2, we must establish the fields' relationships to one another. They can have either a one-to-one relationship to the components of the layer name, or a hierarchical relationship. In this simple layer example, we will build a one-to-one relationship.

1. Click the Edit Descriptive Fields tab.
2. Edit **Default Value** to read **Phase**.
3. Next to **Phase**, click in the **Component 1** column, and choose **Phase** from the menu that pops up.
4. Click the **Add below the selected item** icon (see Figure 4–24) and then type **Component**.

5. Choose **Component** from the pop-up menu in the **Component 1** column.
6. Add the Modifier field next. Your screen should look like Figure 4–26.

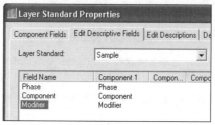

Figure 4–26 *The completed Edit Descriptive Fields tab*

7. Click the Edit Descriptions tab.

The top of the dialog box will show that we are still editing the Sample Layer Standard and that the Phase field is the **Field to Edit**. Choose the correct items from the lists if this is not the case. If you added a Default value such as N on the Component Fields tab, it will be listed here.

8. Select **N** and click the **Edit** icon at the bottom of the dialog box.
9. Type a description and click **OK**.
10. Click the **Add** icon at the bottom of the window, and add the remaining Phase codes from Table 4–2. (See Figure 4–27.)

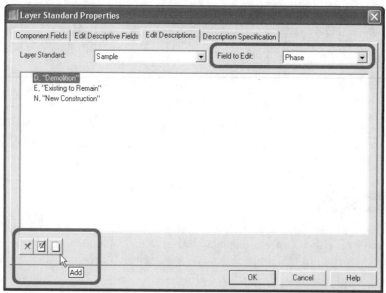

Figure 4–27 *Add all the Phase codes*

11. Following the same procedure and using Table 4–2 as a guide, add the remaining codes. Remember to choose the correct **Field to Edit** before adding.
12. When finished, click **OK** to complete the standard.

THE FINAL WORD ON LAYER STANDARDS

Layer Standards are only half of the ADT "auto-layering" picture. In fact, as has already been mentioned, Layer Standards serve only to flag inconsistencies in a layer name. In other words, if a "non-standard" layer designation is encountered, a message will typically appear. Furthermore, this functionality applies only to the ADT tool set. Layers created manually within the AutoCAD Layer Properties Manager will produce no such alert. However, your chances of catching inconsistencies will be much greater with a Layer Standard in place, and all ADT objects that use Layer Keys will conform to the standard automatically.

LAYER KEYS AND LAYER KEY STYLES

Definitions for Layer Keys and Layer Key Styles were offered in the "Layers versus Display Control Settings" section at the start of this chapter. Most Layer Keys are hard-coded into the software. We are able to add additional Layer Keys for use by custom content (see Chapter 8 for information on creating custom content), but we cannot edit the name or delete the hard-coded Layer Keys built into the software. However, we can freely edit the layer mapping and properties of *any* Layer Key. This is the point of the entire Layer Key system. By hard-coding in the keys, but providing the means to map those keys to any layer definition we wish, we gain the power of automatic layering with the freedom to customize layer mapping to meet our own office standard needs. Combine this with the Layer Standard tools covered in the previous topic and you gain the ability to flag layer inconsistencies across your organization.

WORKING WITH LAYER KEY STYLES

In the previous topic, we worked within a copy of the *AecLayerStd.dwg* file that was installed from the CD. You may continue to work in this file now if you wish, or you may open your own Office Standard default file from the L Drive (or wherever you saved it).

Understanding a Layer Key Style

1. Open the **AecLayerStd.dwg** file (or your renamed version of it).

In Chapter 1, we built an OfficeStandard profile, which mapped all of the standard ADT resources across the network to the individual user workstations. Every attempt was made to build this profile with future flexibility in mind. This flexibility will be guaranteed as long as you refrain from unnecessarily changing the names of any of the resources built and mapped to in Chapter 1. In other words, in the specific instance of layering, a Layer Key Style name has already been chosen for your office default, as was the name of your Layer Key/Standards file. You may freely edit either of these resources now, but it is advisable that you *not* rename either of them. Doing so would require that you also update your OfficeStandard profile to reflect these changes. For obvious reasons, this ought to be avoided.

2. From the Desktop menu, choose **Layer Management>Layer Key Styles.**

If you have been using ADT for a while, you have likely encountered the Style Manager. The Style Manager is the single interface used for browsing, editing and creating new styles of all types. (See Chapter 9 for more information.) In this case, since we have chosen specifically the **Layer Key Styles** command, the Style Manager has opened in a filtered state revealing only the Layer Key Styles of the current drawing. The filter button can be easily toggled on or off to switch the view within the Style Manager from a single type of style to all style types and back (see Figure 4–28).

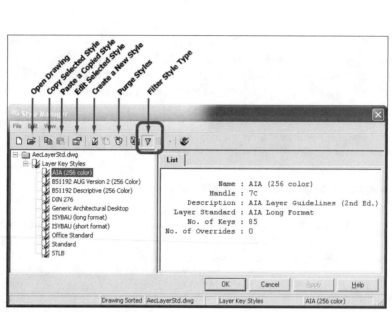

Figure 4–28 *Icon bar in the Style Manager*

Setting up a Layer Key Style is very similar to creating layers in the traditional way. The only difference is that you are not actually creating a layer when you work in a Layer Key Style; rather, you are establishing the parameters that the layer ought to have when it is created automatically by ADT later.

3. Double-click any Layer Key Style in the list, such as AIA (256 Color). You may also select it and then click the **Edit Selected Style** icon (see Figure 4-28).

4. If necessary, in the Layer Key Style Properties dialog box, click the Keys tab.

5. Enlarge the dialog box and each of the columns as much as your monitor will allow.

Regardless of the specific Layer Key Style you are viewing, the first several columns will be the same. In fact, the only columns that vary are those in the **Allow Overrides** section to the right (see Figure 4–29). The Allow Overrides settings tie into the Layer Standard that is assigned at the bottom left corner of this dialog box, and they are used to indicate which fields can be overridden when the **Layer Key Override** command is used.

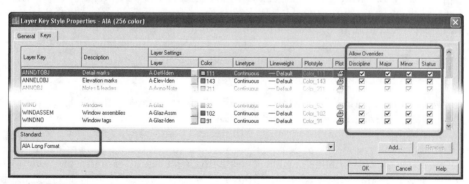

Figure 4–29 *The Layer Key Style Properties dialog box*

You may recall that a Layer Standard established a series of component fields to which layer names must conform. In some instances, it may be desirable to override the pre-set value of a layer name for a particular purpose within the drawing. For instance, sometimes Layer Key Overrides are used to designate the status of an object as demolition or existing to remain. This is not the only way to designate status in ADT; more information on this topic can be found below in the "Customizing Layer Settings" topic.

The first column is the **Layer Key**. As has already been stated, Layer Key names are fixed, however, any of the other parameters assigned to a Layer Key can be edited. In addition, we can also add new Layer Keys that can later be referenced by custom AEC Content. The process to create custom AEC Content is covered in Chapter 8.

6. Select the Layer Key named **Door**.

The Description for this key is Doors because this key is used by ADT Door objects. Following that is the actual layer name that will be used or created in the drawing when the **Add Door** command is used. In the AIA (256 color) Style, the layer is **A-Door**; in BS1192 AUG Version 2 (256 Color), it is **A315G2**. The layer A_____NTTU will be used when ISYBAU (long format) is active, and so on.

The **Layer** name field can be directly edited, but the best way to edit a layer name is to click the small browse (...) button next to the **Layer** name.

7. Click the browse (...) button next to the **Door Layer** name.

The Layer Name dialog box will appear, and depending on the Layer Standard, will break the layer name into its component fields. Each component will also have its own browse (...) button from which a standard description may be chosen. Even though it was stated above that Layer Standards are not required for ADT auto-layering to function, you can see here the benefit that they provide. Each of the possible choices in the Choose a Pre-Specified Value dialog box comes from the Layer Standard, as defined in the previous topic. For instance, in the AIA and BS1192 standards, you can designate the Status of work by adding a code to the Layer name.

8. Click the browse (...) button next to the **Status** field (or any field) to see your list of choices (see Figure 4–30).

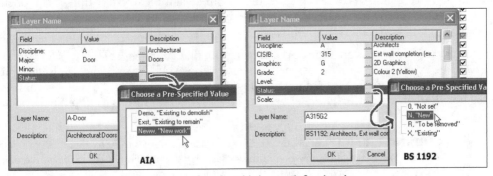

Figure 4–30 *To designate the Status of work, add the predefined code*

The remaining columns in the Layer Key Style Properties dialog box are identical to those in the standard Layer dialog boxes. In the previous chapter, we discussed Color, Linetype, Lineweight and Plot Style. This is where you can assign all of your office standard default values for those items. You can also designate a layer as NoPlot by default and decide whether it can receive overrides (see below). If you wish, edit other Key Styles to compare them. There is a series of tables in an Excel spreadsheet file named *Layer Keys.xls*, located in the *Chapter04* folder. Each of the Layer Key Styles from the *AecLayerStd.dwg* file is included in this Excel file.

> **Note:** The Layer Key Style named "Standard" is *not* based upon a Layer Standard and simply uses the Layer Key for the Layer name. All Layer properties of "Standard" are identical as well. "Standard" should not be used.

9. Click **Cancel** twice to dismiss all dialog boxes without saving the changes.

Adding Additional Layer Key Styles

In the last chapter, we discussed some techniques to manage the status of work (Demolition, Existing to Remain, and so on). One of the suggestions was to create separate Layer Key Styles for each status. This technique can be very viable in a Separated Building Model structure, where each status is kept in a separate file. For instance, if you keep Demo in one file and New Construction in another, you could consider creating a separate Layer Key Style for Demolition and making it the active Layer Key Style in the Demolition file.

To do this, create a new file from your standard template and choose **Layer Management>Layer Key Styles** from the Desktop menu. Make a copy of your Office Standard Layer Key Style and edit each Layer Key to include the proper layer name and properties for demolition. In the Drawing Setup dialog box, make this Layer Key Style current and save the file as a new template file for demolition. You can repeat the steps for an Existing to Remain if you wish, but it might be better to have a single template for both kinds of existing construction (demo and to remain). If you keep both in the same file, you will need to create two Layer Key Styles, and users will "swap" when they switch status during drawing.

THE AUTO-IMPORT FEATURE

The most important thing to understand is that Layer Key Styles are like all other ADT styles and are part of the drawing file. Unlike other ADT styles however, Layer Key Styles can be set to import into the drawing automatically. This is why it is so critical to get the default file to Auto-Import (see Figure 4–10) set up correctly. If your default file to Auto-Import is properly set up, you can be certain that layers will always be generated by ADT based on the proper standards.

The file to Auto-Import is an AutoCAD drawing file that contains ADT Layer Key Styles. A Layer Key Style may in turn be assigned to reference an ADT Layer Standard. Refer again to Figure 4–10 and note that we are not able to choose a Layer Standard directly within this dialog box. Rather, we must choose a Layer Key Style, which in turn is based on a particular Layer Standard.

> **GUIDING PRINCIPLE:** Important: Do not build multiple files to Auto-Import in your organization. There should be only one of these files for the entire firm. If required, include more than one Layer Key Style and Layer Standard within the single file.

Although it is possible to create several files to Auto-Import, there is little benefit in doing so. Presumably, you would consider such a technique to manage varying Layer Standards used by certain clients and/or projects. However, if you decide to build multiple Auto-Import Layer Key/Standard files, you will then need to build separate AutoCAD profiles on the users' work-stations that require the standards within these files. Furthermore, the file to Auto-Import *cannot* be changed through the profile, and therefore, if these users need to sometimes work on a project requiring the client-based standards, and at other times work on projects using the office standards, there will be no easy way to allow them to switch between the two.

Place all Layer Key Styles and Layer Standards within the single *AecLayerStd.dwg* file placed in the *L:\Layers* folder in Chapter 1. As the following tutorial will show, there can only be one file to Auto-Import per workstation, but you may have template files that point to different Layer Key Styles/Layer Standards within that file.

Using Client-based or Project-based Layer Standards

1. Create a new drawing file using the Start from Scratch option.
2. From the Desktop menu, choose **Layer Management>Select Layer Standard**.

If you followed the recommendations of Chapter 1, your dialog box should look like Figure 4–10. The file to Auto-Import is *L:\Layers\AecLayerStd.dwg* and the Layer Key Style is Office Standard *current drawing. (If you changed the paths or file names in Chapter 1, yours will match those changes instead.)

3. Click **OK** to accept these settings.
4. From the **File** menu choose **Save As**, and in the dialog box that appears, set the **Files of type** to **AutoCAD Drawing Template File (*.dwt)**.
5. Type **Test - Office Standard** for the file name and then click **Save** (see Figure 4–31).

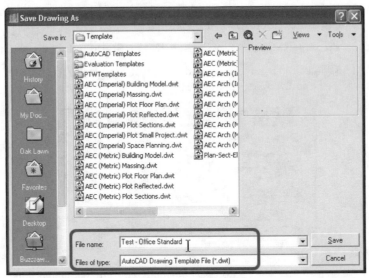

Figure 4–31 *Save a test template file referencing the Office Standard Layer Key Style*

6. Click **OK** to dismiss the Template Description dialog box.

7. From the Desktop menu, choose **Layer Management>Select Layer Standard** again.

8. Change the Layer Key Style to any other entry on the list, such as **AIA (256 color)**.

9. Click **OK**, and save this file as another template named **Test – Client Standard**.

We now have two "dummy" templates that we can use to test the effects of the layering controls on Drawing Setup.

10. From the **File** menu choose **New**, and then using the Template option, choose **Test - Office Standard.dwt** (see Figure 4–32).

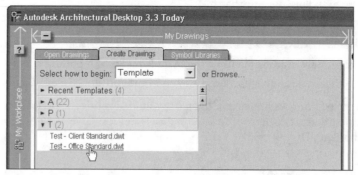

Figure 4–32 *Create a new drawing from the Test – Office Standard template*

11. From the Desktop menu, choose **Layer Management>Select Layer Standard** again.

Note that the Office Standard Layer Key Style is active in this file. Also, note that the **Layer Standards/Key File to Auto-Import** is *L:\Layers\AecLayerStd.dwg*.

12. Repeat the same steps using the **Test – Client Standard.dwt** template this time.

Note that the AIA (256 color) is now active and the **Layer Standards/Key File to Auto-Import** remains *L:\Layers\AecLayerStd.dwg* (see Figure 4–33).

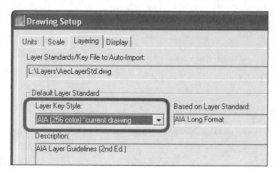

Figure 4–33 *A file created from the Test – Client Standard template will reference the AIA (256 color) standard*

This simple exercise shows that it is easy to create template files that reference different layering standards, providing those layering standards are contained within the same file to Auto-Import. If this concept is not completely clear, try the following:

1. On the Layering tab of Drawing Setup, change the **Layer Standards/Key File to Auto-Import** to another file.

 Note: You can use any file that has Layer Key Styles/Layer Standards saved within it, such as the one installed above from the Autodesk Architectural Desktop: Advanced Implementation Guide CD ROM in the *C:\ADT_AIG\Chapter04* folder.

2. Choose a Layer Key Style from the list available (one you haven't used yet such as DIN 276) and then click **OK**.
3. Save this file as another template drawing.
4. Create a drawing from this template and test it in the same way as the others.

This test should reveal that although the Layer Key Style did remain DIN 276 for this template, the change in the file to Auto-Import was ignored (see Figure 4–34).

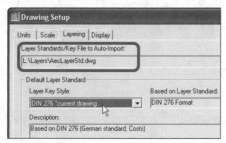

Figure 4–34 *The alternate Layer Standards/Key File to Auto-Import was ignored*

As a result, note that next to "DIN 276" it reads, "*current drawing." This may prove unimportant if you have no plans to edit the Layer Key Style. However, if you do edit the Layer Key Style in any way, those edits will *not* be automatically read into new drawings based on this template. In fact, this will also be the case with the above "test" templates as well. This gets back to this issue raised at the start of this topic: Layer Key Styles are like all other ADT styles and are part of the drawing file. Therefore, if a change is made to a Layer Key File or Standard, it must be re-imported into a drawing or template to be used. This can be accomplished manually using the Style Manager, but this is why the Auto-Import feature for Layer Key/Standard files exists. If the Layer Key Style named in the Drawing Setup dialog box does not exist in the drawing, it will automatically be imported from the file to Auto-Import.

Manual Layer Key Style Update with the Style Manager

This will be a simple tutorial to show you how to manually update a Layer Key Style (or any ADT Style) using the Style Manager. This approach is practical if only a few drawings are involved.

1. Create a new drawing file using the Start From Scratch option.
2. Save it as **Layer Test 1.dwg**.
3. Go to the Layer Key Styles (as above) in this drawing.

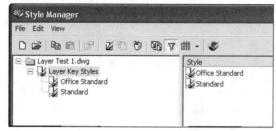

Figure 4–35 *Both Standard and Office Standard Layer Key Styles have been created automatically*

Notice that two Layer Key Styles have been automatically created in these drawings (see Figure4–35), even though they were created from scratch: Standard and Office Standard. (Office Standard may be another style if you designated a different one as the default in Chapter 1.) This is because of the Auto-Import functionality.

4. Open the *AecLayerStd.dwg* drawing and **Edit** the **Office Standard** Layer Key Style.

5. On the Keys tab, at the bottom right of the dialog box, click the **Add** button.

6. In the Name dialog box, type **Test** and then click **OK** (see Figure 4–36).

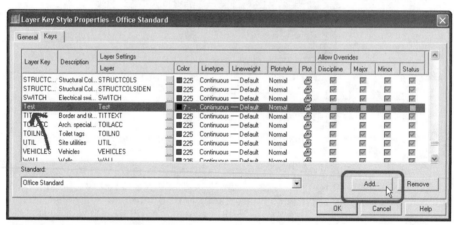

Figure 4–36 *Add a Layer Key called Test*

7. Click **OK** twice more to exit all dialog boxes and then close and save the *AecLayerStd.dwg* drawing.

8. Close, save and re-open the *Layer Test 1.dwg* file.

9. Return to the Layer Key Styles (in *Layer Test 1.dwg*) and **Edit** Office Standard (see Figure 4–37).

Notice that the **Test** Layer Key has not been created in this file. Even though the layer settings designate *AecLayerStd.dwg* and the Office Standard Layer Key Style as Auto-Import, they did not overwrite the existing version of Office Standard within this file.

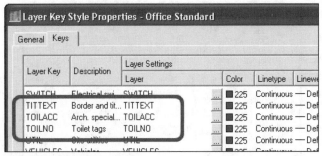

Figure 4–37 *The Test Layer Key did not appear*

10. Click **OK** to close the Layer Key Style Properties dialog box.

11. In the Style Manager, click the **Open Drawing** button.

12. Browse to the *C:\ADT_AIG\Chapter04* folder and open the *AecLayerStd.dwg* drawing.

13. Expand the Layer Key Styles tree and drag **Office Standard** from the *AecLayerStd.dwg* file listing to the *Layer Test 1.dwg* listing (see Figure 4–38).

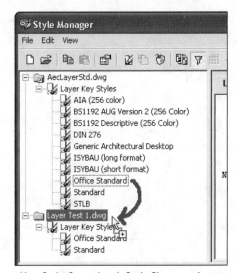

Figure 4–38 *Drag the Layer Key Style from the default file to update it in the Style Manager*

14. In the Import/Export - Duplicate Names Found dialog box, choose **Overwrite Existing** and then click **OK** (see Figure 4–39).

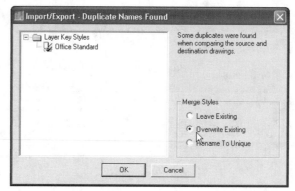

Figure 4–39 *Overwrite the Existing Layer Key Style with the newly updated one*

15. Double-click **Office Standard** beneath the *Layer Test 1.dwg* entry and scroll the list to the **T's**.

Notice that the **Test** Layer Key is now present. This is because we chose to overwrite the existing Layer Key Style with the version within the *AecLayerStd.dwg* file (see Figure 4–40).

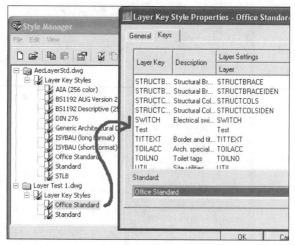

Figure 4–40 *The Test Layer Key is now present in Layer Test 1.dwg*

16. **Close** the Style Manager and **Close** and save *Layer Test 1.dwg*.

Forcing Automatic Update of Layer Keys

The preceding procedure is effective in updating any ADT Style from one drawing to the next. The procedure can be used for any ADT style type. The trouble with this procedure is that it is manual and must be performed drawing by drawing. However, because we have the Auto-Import feature for layers, we can actually force the correct Layer Key Style from the file to Auto-Import to be used without need for manual update. In order to do this, we must first purge all Layer Key Styles from the current drawing. In this way, the correct style will Auto-Import when the file is opened, since it does not exist in the current file. This technique will be effective if you have decided to update several drawings to a new version of your standard, but more importantly, this technique will be very valuable when building drawing template files. The recommended approach for template files is to purge Layer Styles from the template immediately prior to saving, so that new drawings from that template will always use the settings specified in the Drawing Setup dialog box. The command to do this, AECPURGELAYERSTYLES, must be input at the command line.

1. Open the template file named *Test - Office Standard.dwt* built earlier.

2. At the command line, type: **AECPURGELAYERSTYLES** and then press ENTER.

3. At the "This will remove all Layer Key Styles and Layer Standards from the drawing. Do you wish to continue?" prompt, type **y** and then press ENTER.

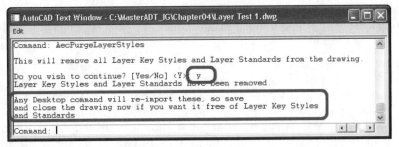

Figure 4–41 *Purge all Layer Key Styles from the drawing*

 Note: A message notifies you regarding the re-import of Layer Key Styles upon execution of any ADT command (see Figure 4–41).

4. **Close** and **Save** the *Test – Office Standard.dwt* template file.

5. Open the *L:\Layers\AecLayerStd.dwg* file.

6. **Add** a Layer Key to the Office Standard layer Key Style named **Test2** and then **Close** and **Save** the file.

7. Create a new drawing based on the *Test - Office Standard.dwt* template file.

8. **Open** the Layer Key Styles (see Figure 4–42).

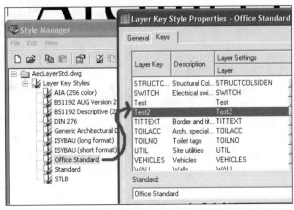

Figure 4–42 *The template now imports the latest Layer Key Style from the library every time*

Notice that Office Standard has been Auto-Imported and that the **Test2** Layer Key has been created. By not including a local version of any Layer Key Styles or Standards in your template files, you will be assured that all new drawings will always Auto-Import the latest Layer Key Styles from the office standard file to Auto-Import.

 GUIDING PRINCIPLE: Be sure to purge all Layer Key Styles from your template files each time you edit and save them.

CUSTOMIZING A LAYER KEY STYLE

The Layer Key Style brings the issues raised in the last chapter (regarding color and lineweight) and the rules set by the layer naming convention within a Layer Standard together. For this reason, taking the time required to build your default Layer Key Style correctly will prove well worth the effort in the end.

It is actually a very simple task to build a custom Layer Key Style. Many of the critical issues that will arise because of building it have already been discussed in the previous chapter. For instance, by now you should have a good idea of how you intend to assign lineweights in your drawings. You should also have decided how you will use colors in your drawings and what role Layer colors will play. If you are still uncertain on some of these issues, you may want to review Chapter 3 again. In order to complete this task you must have the following issues in order:

- You will need to know which Layer Standard you want to use.
- You will need a list of colors and their intended meanings.
- You will need a list of lineweights and the preferred way of applying them (Color, Lineweight property or Named Plot Style).
- You will want to begin thinking of any custom Layer Keys you may need later.

Build an Office Standard Layer Key Style

1. **Open** your actual *AecLayerStd.dwg* drawing file (the one located in the *L:\Layers* folder, *not* the copy from the Autodesk Architectural Desktop: Advanced Implementation Guide CD ROM).
2. From the **Desktop** menu, choose **Layer Management>Layer Key Styles**.
3. If you already have an Office Standard Layer Key Style, right-click it and choose **Purge**.

If you chose another name for your Office Standard Layer Key Style in Chapter 1, then Purge that one instead. We are purging this file so that we can build a new clean copy of it. If you have already made edits to the file that you wish to maintain, then you may choose not to Purge it and skip the next two steps instead. If you are using an industry-standard style that is already included in this file, you may simply edit it directly in the steps that follow, rather than building it anew.

It has already been recommended in this chapter and in Chapter 1 that you either use one of the existing Layer Standards as is or copy it to use as a basis for your own customized version. It is simply a practical matter—it is easier to edit an existing file than to build one from scratch.

4. Right-click the Layer Key Style in the list that is closest to the one you wish to build and choose **Copy**.
5. Right-click again and choose **Paste**.

A copy of the Layer Key Style will be created with "(2)" appended to the name (see Figure 4–43).

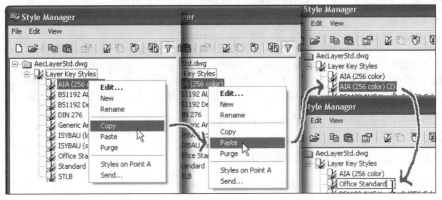

Figure 4–43 *Make a copy of one of the existing Layer Key Styles*

6. **Rename** the copy to the same name as you chose in Chapter 1.

"Office Standard" will be used in this example.

 Caution: It is very important to use the same name as you chose in Chapter 1, or your default resources that you built there may not function properly.

Plot Style Setup in the Layer Standards File

A Layer Key Style contains the complete Layer properties for each layer that ADT can auto-create. Along with Color, Linetype and Lineweight, this includes the Plot Style that is assigned to each layer. In Chapter 3, we discussed the differences between Color Dependent and Named Plot Style Tables. You were also encouraged to decide at that time on which type to standardize your firm. At this point, it is assumed that you have made this decision and have built at least one Plot Style Table to meet your firm's general plotting needs. You may in fact need additional Plot Style Tables for special plotting needs, but you ought at least to have the basics covered in an Office Standard Plot Style Table. In Chapters 1 and 3, we named that file *Office Standard.stb* if you used Named plot styles and *Office Standard.ctb* if you chose Color Dependent plot styles.

If you have chosen to use Named plot styles as your office standard, you will assign them in the Layer Key Style when you assign the other object properties. To do this, we must convert the default *AecLayerStd.dwg* drawing file, which is based on Color Dependent plot styles by default, to a Named Plot Style drawing. If you chose Color Dependent Plot Style Tables as your standard, skip to the "Edit the Office Standard Layer Key Style" below and do not perform these steps.

Convert *AecLayerStd.dwg* to Named Plot Styles

1. At the command line, type CONVERTCTB and then press ENTER.

I DON'T HAVE THE CONVERTCTB COMMAND...

If you are using ADT 3.0, you must install the Migration Assistance tools to use this command. (It is built into ADT 3.3, so a separate install is not required for 3.3.) The tools are located in the *Migration* folder on the ADT CD. Install them from that location and then enter APPLOAD at the Command prompt. In the Load/Unload Applications dialog box, navigate to the *Migration* folder and select the *accvps.arx* file to load.

You will be prompted in a Select File dialog box to select an existing Color Dependent Plot Style Table.

2. Choose **Acad.ctb** and click **Open**.

Note: You can choose another CTB file if you wish.

A Create File dialog box will appear prompting you to type the name of the new file you wish to create from the *Acad.ctb* file you selected.

3. Type **Convert.stb** in the **File Name** field and click **Save**.

A dialog box like the one shown in Figure 4–44 will appear to confirm the successful creation of the file. This step is necessary because AutoCAD will use this Plot Style Table to map the current drawing from Color Dependent to Named in the next step.

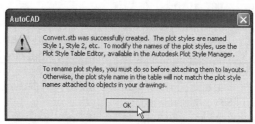

Figure 4–44 *A message confirming the successful creation of the new Plot Style Table*

4. At the command line, type **convertpstyles** and then press ENTER.
5. Click **OK** in the AutoCAD dialog box that appears.
6. In the Select File dialog box, select the **Convert.stb** Plot Style file and click **Open**.

At the command line, "Drawing converted from Color Dependent mode to Named plot style mode" will appear confirming that the drawing has been successfully converted.

7. Save the drawing.

Edit the Office Standard Layer Key Style

At this point, you have a new Layer Key Style named Office Standard (or the name chosen in Chapter 1) or you have decided to work with one of the existing ones, like AIA, or BS1192.

Note: From here on in this tutorial, the style will be referred to as "Office Standard" for simplicity.

1. From the Desktop Menu, choose **Layer Management > Layer Key Styles**.
2. Double-click **Office Standard** to edit it.
3. Click the Keys tab of the Layer Key Style Properties dialog box.
4. Enlarge the Layer Key Style Properties dialog box as much as your monitor will allow.
5. Double-click the border between each column header to size the columns to fit their contents (see Figure 4–45).

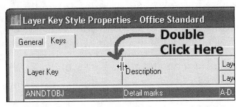

Figure 4–45 *Double-click the border between columns to size to fit*

Layer Key names do not always clearly indicate their purpose. Therefore, you will likely have better luck using the **Description** field for this. You can edit the text in the **Description** fields if you wish.

 Tip: You can sort the list of Layer Keys by any column; to do so, simply click the column heading you wish to sort by. Click again to reverse the sorting in that column.

6. Optional: If you built your own Layer Standard above, choose it from the list of standards at the bottom of the dialog box.
7. Work your way through the list of Layer Keys. For each one, edit the Description, Layer Name, Color, Linetype, Lineweight, Plot Style (if using Named Plot Styles) and Plot/NoPlot settings as required.

 Tip: If you need to rename a layer, use the small browse (**...**) button next to the Layer entry to edit the name based on the chosen Layer Standard.

If you know of any additional Layer Keys that you will need when building custom AEC Content in Chapter 9, you can add them now. However, it is not necessary—you will have the opportunity to add them later as required.

8. When you have finished editing the Layer Keys, click **OK** twice to exit all dialog boxes.
9. **Save** the drawing.

If you have identified the need for additional Layer Key Styles (refer to the "Using Client-based or Project-based Layer Standards" topic above), you may add them now. Repeat the same steps covered here to add as many additional Layer Key Styles as required. Just make certain that they are all being added to the *AecLayerStd.dwg* drawing file. If you are not sure which additional Layer Key Styles you may need, don't worry, you can always add them later.

DOCUMENTING YOUR LAYER KEY STYLES

As you have been working your way through this book, you have been encouraged to keep lists of "to do" items and to document choices you have made along the way. You will likely want to have a list of Layer Keys and their layer mappings for use in your office CAD Standards documentation. This is a very simple task to achieve in the Layer Key Styles dialog box.

1. Open your Office Standard Layer Key Style to the Keys tab.
2. Right-click anywhere in the list and choose **Copy All**.

You now have a complete list of all Layer Keys on your Windows Clipboard in Tab-Delimited format, ready to paste into Excel, Word or any other word processor or spreadsheet application.

3. **Open** the other application and choose **Paste**.
4. **Save** the file.

You now have all the settings from the Layer Keys Style dialog box in a file that you can include in your CAD Standards documentation manual.

UPDATING EXISTING DRAWINGS

In spite of your best efforts, a time will come when you realize that the office standard Layer Key Style must be edited. This will naturally raise the questions of how and if you should attempt to update existing project drawings to this new standard. There are not always easy answers to these questions. The need to update will depend on a variety of factors, beginning with how extensively the edits to the office standard file have been. If existing projects do not require the changes made to the file, it may not be worth the effort to update them.

However, if you do decide to update existing files, you can use the techniques covered above in the "Forcing Automatic Update of Layer Keys" tutorial. If you have only a few files to update, use the manual process through the Style Manager. If you have many files, you can write a script file to batch process several drawings at once. Since the AECPURGELAYERSTYLES command is command line based, it will be easy to script. You will only need to generate a list of drawings to perform the script on.

However, keep one thing in mind before performing a batch update of Layer Keys. The update will be of only the Layer Key Style, which will affect all future objects generated in those drawing files. To change the layers of existing objects requires a few additional steps. ADT objects can be easily "remapped" through the **Remap Object Layers** command. For AutoCAD entities, you will need to use the **Layer Translator** command.

Remap Object Layers and Layer Translator

1. In the C:\ADT_AIG\Chapter04 folder, open the file named A-FP01.dwg.
2. Open the Layer dialog box and scan the Layer list.

Note that the layers in this drawing use the AIA Layer Standard.

3. From the Desktop menu, choose **Layer Management>Select Layer Standard**.

Again, note that the Layer Standard used for this drawing is AIA (256 Color).

4. From the Layer Key Style drop-down list, choose another Layer Standard such as **ISYBAU (long format)** and then click **OK**.

5. From the Desktop menu, choose **Layer Management>ReMap Object Layers**.

6. When prompted to "Select objects to re-layer" type **all** and then press ENTER twice.

7. At the "Enter a Layer Key or [?/byObject]" prompt, right-click and choose **byObject**.

8. **Open** the ADT Layer Manager (**Desktop>layer Management>Layer Manager**).

9. Scroll through the list of Layers.

Notice that all of the original layers remain in the drawing, but several new ones have been created. Further, the small icon before the Layer name in the ADT Layer Manager indicates whether the layer contains objects. Although all of the original layers are still present in the dialog box, you can see that these layers no longer contain objects and may now be purged. (See Figure 4–46.)

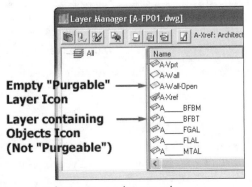

Figure 4–46 *Original Layers remain but may now be purged*

10. **Close** the file.

Remap Object Layers works well when all objects are ADT objects, but when there is a mixture of ADT and AutoCAD entities or when two standards do not match exactly, you can use the **Layer Translator** to remap the layers.

11. Open the file named *A-FP02.dwg*.

12. From the **Tools** menu, choose **CAD Standards>Layer Translator**.

13. On the right side, click the **Load** button, locate the file named **Layer Translator Sample.dwg** and click **Open**.

You now have a list of layers from the current drawing on the left, using BS1192 AUG Version 2 (256 Color) in this case, and the remote file on the right, using BS1192 Descriptive (256 Color) in this case. You can scroll through the list and choose a layer on the left and then another on the right. Click the **Map** button to map the two together. You can save the layer mappings to a drawing standards file (*.DWS) for use with other drawings.

14. Map several Layers and then click the **Translate** button to execute the translation.

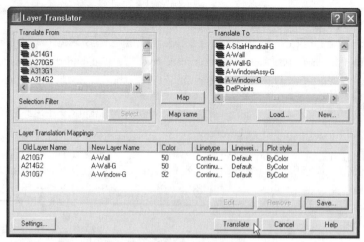

Figure 4–47 *Using the Layer Translator to remap layers*

15. **Close** the file when you are satisfied with the results.

This same procedure can be used to map layers contained in consultant or client files to your office standard.

USING NON-KEYED LAYERS

The Layer Key system is a great tool to aid in standardization of company's layer usage. Unfortunately, it does not solve all of our layering needs. Layer Keys are built into all ADT objects and are used specifically by the "Add" commands that place objects in the drawing. You can also build custom Layer Keys that are used by AEC Content when it is dragged and dropped from the DesignCenter. Those are the only two instances when layer keying will be applied. Therefore, we are left with the following layer issues: First, unlike ADT objects, AutoCAD entities do not use Layer Keys even when they are being added to the drawing. Second, it is not possible assign a Layer Key to ADT object sub-components. "Hard" layer assignments must be used with Entity Display Props instead. Finally, editing of objects: ADT or AutoCAD never invokes layer keying; users will be free to move objects to any layer they choose. Therefore, in all three instances, we need procedures to ensure that layers created and used during production will follow an acceptable office standard.

- **AutoCAD Entities and Layer Standards** – There are two ways to help maintain standard layering when using AutoCAD entities in a drawing: pre-create the layers within your template files, or use the ADT Layer Manager. CAD Managers have long pre-defined layers within office standard template files in an attempt to encourage standard layer usage. This technique will still prove valuable for common drawing types like Section, Elevation and Detail sheets. On occasion, non-ADT objects will be required in Plans as well, and those layers can be included in the templates used to generate Plan drawings. In addition, if you review the "Layer Management Dialog Boxes" topic earlier in this chapter, you may recall that the ADT Layer Manager prompts you to choose from predefined fields when adding a new layer to a drawing (see Figure 4–5). This will help ensure that layer names are based on proper usage of the layer naming convention built into the

Layer Standard. However, you will still need to refer to a chart or other resource for guidance in assigning the appropriate color, lineweight and other properties to the new layer. The New Layer dialog box shown in Figure 4–5 does not automatically assign or suggest the Color or other properties of the layer. You will still need to assign these manually according to office standards.

- **ADT Objects and Sub-Component Layers** – All ADT objects are comprised of one or more sub-components. We can access the graphical characteristics of these sub-components within the Entity Display Properties of each object (see the next chapter for more information), but we cannot assign a Layer Key to these sub-components; rather, we must assign a layer directly. Furthermore, this layer must already exist within the drawing before we can assign it through Entity Display.

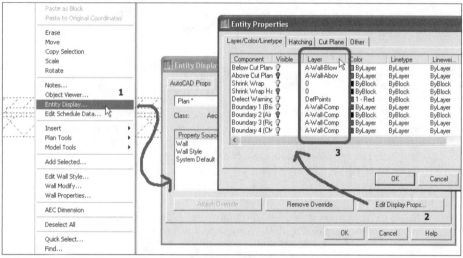

Figure 4–48 *Adding Layer references to ADT Object Display Props*

- **Editing Layers Manually** – There is not much that can be done to keep users from changing object layers after the fact. In this situation, you must rely on well-conceived procedures with complete and easy-to-understand documentation. If your CAD standards make sense to your users and they can find answers to questions easily and quickly, they will follow them—it is that simple. If you work in a small firm and do not have a CAD manual or CAD Manager, it will still behoove you to have a simple procedure and strategy for dealing with layer inconsistencies. For instance, you may revisit an old project that used a different set of layers than you use today, or you may receive drawings from outside firms that use a different layering scheme than you do. As suggested above, the Layer Translator can be used to help correct discrepancies that may materialize throughout the course of the project.

THE ROLE OF TEMPLATE FILES

Drawing template files are a very important piece of the implementation puzzle. A drawing template file (or simply template) is really nothing more than an AutoCAD/ADT drawing file with a DWT extension. DWT files can be opened in the drawing editor in the same way as DWG files. One of the only differences between the two file formats is that a DWT contains a small bit of header information with a name and description of the intended use of the template file. Users can see this description in the traditional startup dialog box. The other differ-ence is that when a template file is chosen when creating a new drawing, a copy of the template file is used to begin the new drawing. Users could achieve nearly the same thing with a SAVEAS, but a template has this functionality built in. Template files are so critical to productive use of ADT that, starting with this chapter and continuing through the remainder of the book, we will slowly build an office standard template file for your use in implementation.

As useful as Templates are, we must address a few limitations.

- You *cannot* INSERT a template file as a Block into another drawing.
- You *cannot* XREF a template file.
- You *cannot* access a template file from the DesignCenter.
- Template files are useful only for starting new drawings. Updating and saving them later will have no effect on the drawings created from them.

When you consider this list, it can seem to mount a convincing case against the use of template files. Regardless of the apparent limitations to template files, they are still the best way to *begin* new ADT drawing files and help enforce the collection of settings and standards contained within them.

 Note: If drawing is begun without a template or with the incorrect template, you can correct this quickly. Start a new drawing with the correct template. Insert the incorrect drawing into the new drawing using the **Insert>Block** command using the **Explode** check box in the Insert dialog box. This will insert the contents of the file as separate entities, each of which will take the properties of the correct template.

You can build a single template or, if you wish, you can create a collection of template files designed to begin each of your most common drawing types. For instance, you could have a single template for plans, sections and elevations or a separate one for each type. At a mini-mum, you ought to create a Model file template and a Sheet file template. (See Chapter 2 for more information on Model and Sheet files.) In addition to the collection of template files, it will also prove useful to build a collection of "Library Drawing" files.

Library Drawings are DWG files that contain much of the same information as the template files do, and often much more. A Library Drawing can be accessed at any point in the produc-tion process, not just at the creation of the drawing. Use them as repositories for your library resources such as object styles, Display Control settings, layers, Blocks and more. This content can be imported into project files as the project needs dictate. This helps to keep the size of template files small, and consequently all drawings created from them small. A good balance between file size and overall productivity can be achieved with a good collection of lean tem-plate files, for creating the initial project files, and a robust set of Library Drawings from which additional Content resources may be quickly imported.

BEGINNING A TEMPLATE FILE

To be certain that we are creating compact template files with only the items we require, and to ensure that you fully understand each of the components contained in a template file, we will build our template file from scratch. In Chapter 1, we created our first template file (*Plan–Sect–Elev.dwt*) as a copy of the basic AutoCAD template file. This gave us the same result as start-ing from scratch would. This chapter is devoted to understanding layers. Therefore, we will expand this basic template to include the layers that we need.

Remember, the purpose and function of Layer Keys is to map the creation of ADT objects to a set of predefined standard layers. As was discussed in the previous topic, this does not include layers required by AutoCAD entities or ADT sub-components. Therefore, at this stage, it will be useful to open the *Plan-Sect-Elev.dwt* template file and add any layers that will be required "*most* of the time." Do not add every layer you can dream up. Remember, it is desirable to have the template files remain small. If you are uncertain what your layer needs will be, it is not critical to get them all in the template now. We will be modifying the template files throughout this book, so you will have plenty of opportunities to add layers later. In addition, Display Control will likely reduce your layer needs considerably. This is because Display Properties can achieve many of the functions traditionally requiring layers. Therefore, you may wish to complete the next chapter before adding too many layers to your template files.

Consider some of the following layers as possible candidates for the template file. These have been gleaned from the template files provided with ADT. If the specific names do not comply with your current standards, you may simply rename them as appropriate.

Table 4–3 *Layer Candidates for Template Files*

Layer Name	Purpose	Usage in AEC Templates provided with ADT
A-Anno-Note	Notes and General Annotation	Graphics on the Template-Overview tab
A-Glaz	Windows and Glazing	Used for the Glass component in Elevation view in some Door styles
A-Sect-Mcut	Cut line of Sections	Used for defining Line component in Sections
A-Wall-Abov	Wall Components Above Cut Plane	Used for Above component in Plan view of Wall styles
A-Wall-Blow	Wall Components Below Cut Plane	Used for Below component in Plan view of Wall styles
A-Wall-Comp	Wall Components General	Used for Boundary lines in Plan view of Wall styles
A-Wall-Ftng	Wall Footing Lines	Used for Below component in Plan view of Footing Wall styles
A-Wall-Patt	Wall Hatching	Used for Hatch components in Plan view of Wall styles
Viewport	Paper Space Viewports	Used for paper space viewports

The list of layers in Table 4–3 is mostly required by the object styles saved within the file. Most of the template files shipped with ADT include this same list of layers. The exception is the templates from the German speaking countries. In those templates, the only layer included is a layer for paper space viewports. The layer names listed above conform to the AIA Layer Standard. If you are using another standard, you can change the names to suit that standard. Another point to consider is that when a style is imported into a drawing, any layers that it requires (as determined by the style's Display Props) will automatically be imported into the current drawing as well. Therefore, if you do not intend to include many styles in your template files, you can also choose to omit the layers.

Table 4–4 shows some additional layers to consider adding to your template files.

Table 4–4 *Additional Layer Candidates for Template Files*

Layer Name	Purpose
A-Xref	Insert All XREFs on This Layer
A-Vprt	Paper Space Viewports
A-Ttlb	Title Blocks
A-Sect-Lite	Lite Section Lines
A-Sect-Medm	Medium Section Lines
A-Sect-Heav	Heavy Section Lines
A-Sect-Patt	Section Hatching
A-Sect-Hidd	Hidden Section Lines
A-Elev-Lite	Lite Elevation Lines
A-Elev-Medm	Medium Elevation Lines
A-Elev-Heav	Heavy Elevation Lines
A-Elev-Patt	Elevation Hatching
A-Elev-Hidd	Hidden Elevation Lines

Before saving your template files, be sure to run the AECPURGELAYERSTYLES command (see "Updating Existing Drawings" above) and remember to choose **Drawing Template File (*.DWT)** as the file type when saving.

Regardless of the number of layers you add to your template files, you can create a Layer Standards Library file, which is a normal AutoCAD/ADT drawing file that contains every possible layer you can conceive. Set up each layer with all of the proper settings for Color, Linetype, Lineweight, etc. and save this file to the L Drive. Users can access this file through the DesignCenter and drag and drop layers from it into their drawings as required. In this way, you can ensure that new layers will be created based on proper office standards.

 LAYERS CHECKLIST

1. Build or adopt a layer naming convention (ADT Layer Standard).
2. Build or adopt a Layer Key Style (layer name, colors, lineweights, etc.).
3. Document Layer Keys and Layer Standards.
4. Begin your template files by including common non-keyed layers.
5. Build a Layer Standards Library file containing all office standard layers.

SUMMARY

Although substantially similar, the ADT Layer Manager dialog box offers features specific to ADT not available in the standard AutoCAD Layer Properties Manager.

Layer Groups in the ADT Layer Manager can be set up to flag layers non-compliant with the current Layer Standard.

An ADT Layer Standard is used to establish a fixed layer naming convention for all layers created by ADT.

You can adopt one of the Layer Standards provided or build your own custom one.

Layer Keys are used by ADT objects to map predefined layers to ADT objects automatically as the objects are created.

You can only have one Auto-Import Layer Key/Standards file per workstation; create a single file for office-wide use.

A single file to Auto-Import can contain multiple Layer Key Styles if required.

Purge all Layer Key Styles from template files prior to saving to force them to Auto-Import the latest Layer Key Styles into all new drawings.

Customize your Layer Key Styles to include all layer settings such as Color, Linetype, Lineweight and Plot Style.

To update or translate drawings to your office standard, use Remap Object Layers and the Layer Translator.

Add non-keyed layers to drawing template files and Layer Standards Library files for easy use and retrieval.

CHAPTER 5

The Display System

INTRODUCTION

If you are like most users of ADT, you have likely found the Display System a bit baffling, if you have given it that much consideration. The Display System offers us immense power and flexibility to control all aspects of our drawing's graphical display, and yet most of us have done little more than perhaps purge the configurations for which the firm saw no immediate use. This is a shame, since the Display System is by far the most powerful and flexible tool that ADT has to offer. Unfortunately it is also perhaps the most complex and difficult to master. If you want to reap the fullest benefits from your ADT implementation, Display Control must play an integral role.

OBJECTIVES

The primary objective of this chapter is to demystify the Display System and make it accessible to you in everyday use. To that end, the following specific goals have been set:

- Understand the Display Manger
- Understand Display Configurations
- Understand Display Sets
- Understand Display Reps
- Learn what, why and when to customize

EXPLORING THE DISPLAY SYSTEM

Early in his or her career, an architect is taught the traditional rules of architectural drafting. These rules govern such things as what a Plan or Elevation drawing represents, how to create one, and most importantly, what to include and what not to include in order to make a drawing "read." Although there are accepted universal rules in place, a large part of the process involves personal style. Therefore, the rules need to be consistent enough to allow a drawing to convey information reliably, and flexible enough to allow for variation. Amazingly enough, although CAD software such as AutoCAD revolutionized the way in which design drawings were created, prior to Architectural Desktop, the software offered no specialized tools to assist the architect in achieving the unique graphical look required by architectural documents. Rather, lines were still painstakingly laid out one at a time following the internalized prescriptions learned "on the boards" in architecture school.

If a plan, section and elevation were required to convey design intent, three completely separate drawings needed to be created and, more importantly, coordinated. The Display System in ADT addresses this situation by incorporating the rules of architectural drafting directly into the software. Plans, Sections and Elevations can now be generated directly from a single building model (see Figure 5–1). This reduces rework and redundancy by requiring one set of objects, with three different modes of display. The tools are flexible and fully customizable, so we may fully benefit from this powerful tool and still introduce the nuances of our own personal style into the process.

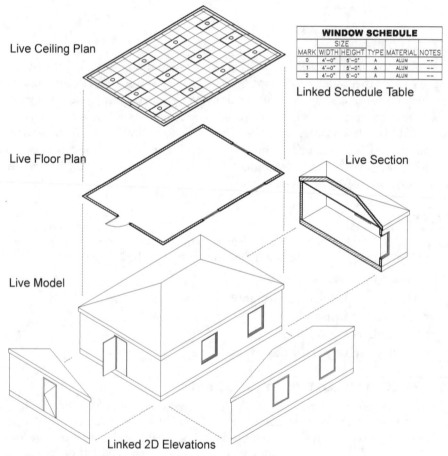

Figure 5–1 *Generating Plans, Sections and Elevations from a single ADT model.*

DISPLAY CONFIGURATIONS

Display Configurations establish which display *condition*, specifically referred to as a "Display Set," will appear on the screen as determined by a particular viewing direction. Loading a configuration can effect subtle changes over a selected group of ADT objects, or it can completely transform the graphical display of the entire drawing. In traditional AutoCAD, the display of objects is controlled by careful configuration of layers. As we saw in the previous chapter, layering schemes can become very complex. This is because layers are merely categories that are assigned to objects and have no inherent awareness of the type of objects they manage or for what purpose the objects are designed.

When a Display Configuration is loaded, it activates a "set," which determines which object types should be visible and in what modes they should be displayed. These modes, called Display Reps, are the means by which ADT objects receive layers and other display properties. Therefore, in the typical Display Configuration, many changes to the drawing will occur before the layers are even taken into consideration.

There is another configuration/layer comparison worth mentioning here. The process of loading a Display Configuration is very similar to loading a Layer State or Snapshot. However, Layer States and Snapshots fail when new layers are added to the drawing. This is not an issue for configurations since they are tied directly to object types, and the list of ADT object types is fixed. The downside of the Display System is that it applies only to ADT objects. AutoCAD entities are not affected.

UNDERSTANDING ADT DEFAULTS

There are two basic approaches to understanding and customizing the Display System to meet your firm's needs. The first is to begin with one of the sample drawing template files provided and then remove or edit the portions you choose not to use. The other approach is to start from scratch and build or import everything yourself. Both methods are valid and useful. However, before you can appreciate the extent of the pre-configuration afforded you by working with Architectural Desktop's in-the-box template files, it is instructional to see what is included in drawings that are started from scratch.

Install the CD Files

1. Install the files for Chapter 5 located on the Autodesk Architectural Desktop: Advanced Implementation Guide CD ROM.

Refer to the "How to Use the CD" section in the Preface for instructions on installing the sample files included on the CD.

A folder named *Chapter05* was created containing a collection of sample files that we will use throughout this chapter.

Begin a New Drawing from Scratch

It is important to note that "scratch" drawings do not actually contain any Architectural Desktop content or display settings. However, immediately upon execution of an ADT command (which usually happens automatically upon drawing creation), some basic Display Configurations are automatically created based on "hard-coded" defaults. The same configurations are added regardless of whether you choose Imperial (feet and inches) or Metric for units. However, the Display Configuration that is set active by default will vary based on your choice of units.

Think of a *Display Configuration* as the glasses through which the drawing is viewed. You choose a Display Configuration in the drawing to force all the ADT objects to display in a particular way. A scratch drawing includes four configurations by default: Plan 1-100, Plan 1-50, Reflected and Standard. In a scratch drawing generated using Imperial units, the active Display Configuration is set to Standard. When Metric units are chosen, the active Configuration will be Plan 1-100. The differences between Plan 1-100 and Standard are minor and will be elaborated upon further in the next topic.

1. Create a new drawing using the Start from Scratch option.

You may choose either Metric or Imperial units; it is not important for this exercise.

Loading Display Configurations

There are four Display Configurations automatically generated in a scratch drawing. Display Configurations can be loaded from either the **Desktop>Select Display** menu or the **Set Current Display Configuration** icon on the AEC Setup toolbar (see Figure 5–2).

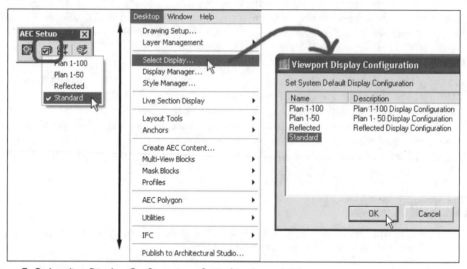

Figure 5–2 *Loading Display Configurations from the menu or icon*

1. Try both methods in the current drawing.

Naturally, there will be no change to the drawing since there are no objects. However, do make note of the presence of the four above-noted Display Configurations.

 Note: Throughout the remainder of this book, you will simply be instructed to "load a Display Configuration," and the exact method and individual steps will be left to your preference.

Exploring the Default Configurations

To make the exercise of exploring the four default configurations more meaningful, a test file, *Display.dwg*, has been provided in the *Chapter05* folder. This file contains one of every ADT object type organized on screen in the categories shown in Figure 2–1, Chapter 2. We will use this file throughout this chapter as a way of testing the various Display Configurations and settings that we build and explore (see Figure 5–3).

1. **Open** the file named *Display.dwg*.

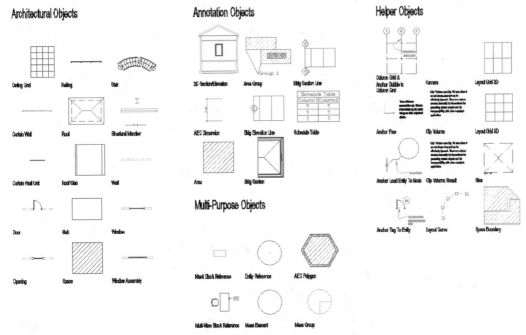

Figure 5–3 *The Display.dwg test file includes one of every ADT object*

Every ADT object has been inserted free from other objects wherever appropriate. In some cases, such as Doors, Windows, Openings and Window Assemblies, which are typically anchored to Walls, a Wall has been included to help visualize context. These Walls have been placed on a temporary layer with the color assigned to number 254 (light gray.) Similar techniques were used for objects that rely on other objects for their shape and form such as Space Boundaries, Area Groups and Mass Groups.

Many of the Helper objects have also required special attention. For instance, Anchors are typically important to drawings because of the governance they place on relationships between two objects. Rarely are anchors expressed graphically and in fact, many of them cannot be. The exceptions to this are the Leader and Column Bubble Anchors, which are displayed by default (see Figure 5–4). *Anchor Free* is not a separate object type, but rather is a sub-component associated with all ADT objects. It is used for graphically marking an object's insertion point within the drawing. Anchor Free is turned off by default in all Display Configurations. The one included in this file is actually an exploded version for illustration purposes only. Tag Anchor Leaders can be very useful as a diagnostic tool for locating mismatched tags. This will be covered below. Tag Anchors are off by default in all four configurations.

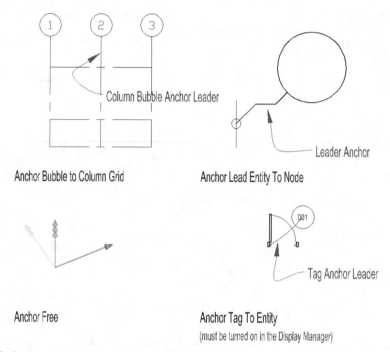

Figure 5–4 *Anchors that can be displayed graphically*

 Note: Clip Volume and Clip Volume Result are not shown in this file and can be effectively ignored. These two objects are used internally by the software for generating section objects and for interoperability with other Autodesk applications. Unfortunately, they cannot be filtered out of the list of objects shown in the ADT Display Manager.

2. **Zoom** and **Pan** the drawing to become acquainted with its contents.
3. **Load** each Display Configuration (using your preferred method from above) as you explore and note the differences.

Using Layout Tabs to Switch Display Configurations

Each viewport in a layout tab can be assigned its own Display Configuration. This makes switching configurations even easier. It also allows us to set up a View/Scale/Configuration combination and save it within the drawing. Several layouts have been included with this file to demonstrate this point.

I. Click in succession through each of the first four layout tabs: Standard, Reflected, Plan 1-50 and Plan 1-100 (see Figure 5–5).

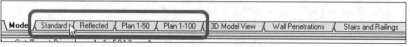

Figure 5–5 *Use layout tabs to switch Display Configurations*

This will reveal the objects in the drawing in the same way that loading each Display Configuration manually did in the previous step sequence.

 Note: The first time you switch layouts, the following message will appear at the command line: "Regenerating model - caching viewports." Subsequently switching to the same viewport will be faster.

Not surprisingly, the most dramatic differences can be seen comparing the Reflected display to the other three. However, if you take a close look at some of the objects, particularly Wall penetrations, Stairs and AEC Dimensions, you can see some distinct differences in other configurations as well.

2. Click the 3D Model View layout tab.

This view shows the dimensionality (see Chapter 2) of objects well (see Figure 5–6). Notice that certain objects do not appear in 3D, such as AEC Dimensions, Areas and Area Groups.

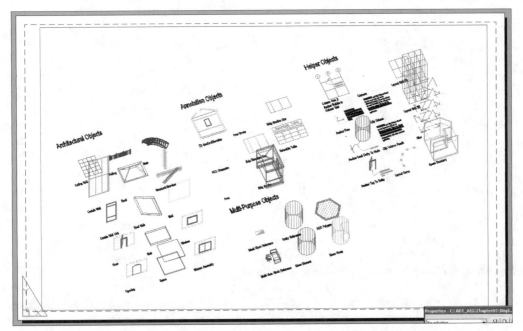

Figure 5–6 *Each of the objects shown in 3D Model View*

3. Observe the Desk and Chair as you click from Standard through 3D Model View.

Notice that the desk, chair and tag (all Multi-View Blocks) show the same in Standard, Plan 1-50 and Plan 1-100. In Reflected, they all disappear, but the small leader remains. This is because the leader is actually an AutoCAD Leader and therefore does not use Display Control. In 3D, the desk and chair remain but the tag disappears. Similar behavior can be observed in the Door Tag (Anchor Tag To Entity), the Section and Elevation markers and the Column Bubbles—these are all Multi-View Blocks. Multi-View Blocks do not use Display Properties in the same way that other ADT objects do, so they do not always follow the same behavior as the other objects. However, they can be easily customized to suit any need. For instance, if you wished you could edit the furniture Multi-View Blocks to include a "dashed" or "half-tone" view in the Reflected Display Configuration. For more information on Multi-View Blocks, refer to Chapter 8.

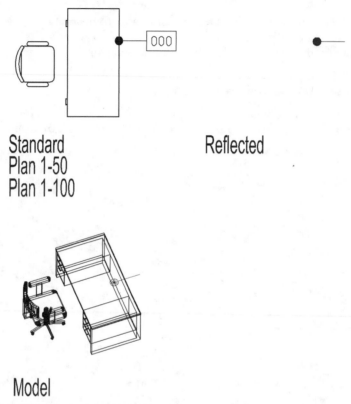

Standard
Plan 1-50
Plan 1-100

Reflected

Model

Figure 5–7 *Multi-View Blocks viewed in each configuration*

There are two layout tabs remaining in this file: Wall Penetrations and Stairs and Railings. They focus more directly on the object types that exhibit the most differences between the four Display Configurations being explored here. Pay particular attention to the variations between Doors and Windows in each of the Plan Configurations (see Figure 5–8). Too often, the potential of the Plan 1-50 and Plan 1-100 Display Reps goes untapped.

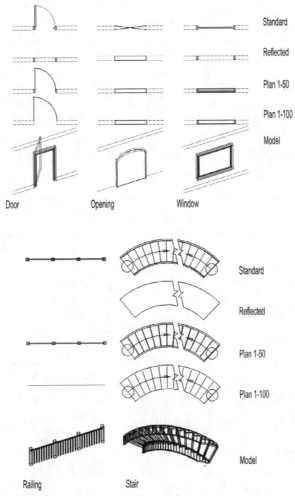

Figure 5–8 *A sampling of object types shown in several views*

    4. **Close** the *Display.dwg* file without saving.

IN-THE-BOX TEMPLATES

Let's continue our exploration of Display Configurations by creating a few drawings from some of the templates provided and seeing the effect on our sample *Display.dwg* test file. The template files provided with the software present several variations on the basic Display Configurations explored so far. In these templates, much of the Display System has already been configured for typical design situations.

Architectural Desktop ships with eighteen AEC template files. These will be found in the *Templates* folder at the root of your Management Install. If you did not include the AEC templates in your L Drive, you will need to access them from that location. Table 5–1 shows the distribution of these eighteen templates by use, units and purpose. Eight of the templates (the first eight listed, including "AEC Arch" in their name) are the ones you will find most useful for exploration. The first four of these are generic "multi-purpose" template files, and the remaining four are localized to needs of particular regions.

Table 5–1 *AEC Template Files Included in the Box with ADT*

Template File Name	Units	Purpose
Generic		
AEC Arch (Imperial).dwt	Imperial	Multi-Purpose Template
AEC Arch (Metric).dwt	Metric	Multi-Purpose Template
AEC Arch (Imperial - Intl).dwt	Imperial	Multi-Purpose Template including settings for Int. Ext.
AEC Arch (Metric - Intl).dwt	Metric	Multi-Purpose Template including settings for Int. Ext.
Localized Generic		
AEC Arch (Metric) Austria.dwt	Metric	Multi-Purpose Template specific to Austrian needs
AEC Arch (Metric) D A CH.dwt	Metric	Multi-Purpose for Germany, Austria and Switzerland
AEC Arch (Metric) Germany.dwt	Metric	Multi-Purpose Template specific to German needs
AEC Arch (Metric) Switzerland.dwt	Metric	Multi-Purpose Template specific to Swiss needs
Task Specific		
AEC (Imperial) Building Model.dwt	Imperial	DD & CD Building Model
AEC (Imperial) Massing.dwt	Imperial	Mass Modeling
AEC (Imperial) Space Planning.dwt	Imperial	Space Planning
AEC (Imperial) Plot Floor Plan.dwt	Imperial	Plotting Floor Plans
AEC (Imperial) Plot Reflected.dwt	Imperial	Plotting Reflected Ceiling Plans
AEC (Imperial) Plot Sections.dwt	Imperial	Plotting Sections and Elevations
AEC (Imperial) Plot Small Project.dwt	Imperial	Plotting Plans, Sections, Elevation and RCP
AEC (Metric) Building Model.dwt	Metric	DD & CD Building Model
AEC (Metric) Massing.dwt	Metric	Mass Modeling
AEC (Metric) Space Planning.dwt	Metric	Space Planning
AEC (Metric) Plot Floor Plan.dwt	Metric	Plotting Floor Plans
AEC (Metric) Plot Reflected.dwt	Metric	Plotting Reflected Ceiling Plans
AEC (Metric) Plot Sections.dwt	Metric	Plotting Sections and Elevations
AEC (Metric) Plot Small Project.dwt	Metric	Plotting Plans, Sections, Elevation and RCP

AEC Arch (Imperial).dwt and *AEC Arch (Metric).dwt* are the basis of each of the other template files. The international version of these two (the versions with "Intl" in their names) include minor adjustments that account for the presence of the International Extensions.

Note: As was mentioned in the Preface, it is assumed (and recommended) that you have the International Extensions installed.

Each of these is a "generic, multi-purpose" template file. This means that each includes styles, layers and Display Control settings for a variety of drawing types, including Plans, Sections, Elevations, 3D Models and Plot sheets. The style content and units settings vary between the Metric and Imperial versions, but the Display Control settings and layout tabs are the same in all four generic multi-purpose templates (see Figure 5–9).

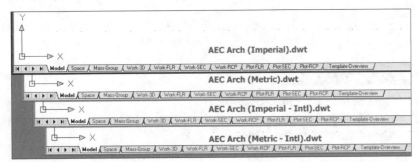

Figure 5–9 *All four Generic templates share the same Display Control settings and layout tabs*

AEC Arch (Imperial - Intl).dwt is the template file that we will explore here.

The four Localized Generic template files are customized to the needs of the German-speaking nations. They are also "generic, multi-purpose" templates, which can be used for a variety of drawing types, including drawings required by building code and planning departments in those countries. All of these files are Metric. Most of the style content is the same for all four files. All four templates contain the same basic group of layouts and Display Control settings as well. *AEC Arch (Metric) D A CH.dwt* includes a few additional Layout tabs and Display Configurations not included in the other three (see Figure 5–10).

Figure 5–10 *The D A CH template includes all of the configurations of the other localized plus a few others*

AEC Arch (Metric) D A CH.dwt will be explored here.

Feel free to look at the Task Specific template files on your own later. We will not look at them here. Each of them includes a smaller sub-set of the contents of the associated Generic version. Therefore, looking at the contents of the *AEC Arch (Imperial - Intl).dwt* and *AEC Arch (Metric) D A CH.dwt* will give us a complete cross-section of Display System potential, as represented by the template files included.

Create Two Sample Files Based on AEC Templates

1. Create two new drawings, one based on the **AEC Arch (Imperial - Intl).dwt** Template and the other on the **AEC Arch (Metric) D A CH.dwt**.

2. **Save** them as drawing files (DWG extension) with the same name as the template that they were created from in the *Chapter05* folder.

For example, you should now have: *AEC Arch (Imperial - Intl).dwg* and *AEC Arch (Metric) D A CH.dwg*.

In order to explore the Display System as it is configured in each of these files, we will insert the *Display.dwg* test file, used earlier, into each of these new drawing files.

Insert the Display Test File

1. From the **Insert** menu, choose **Block**.

2. Click the **Browse** button, navigate to the *Chapter05* folder, click the file named *Display.dwg* and then click **Open**.

3. Clear the **Specify On-screen** check box for **Insertion Point** and place a check mark in the **Explode** check box (see Figure 5–11).

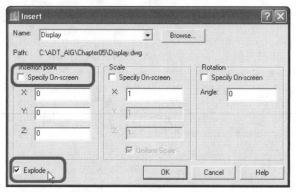

Figure 5–11 *Insert an exploded copy of the file into the current drawing*

4. **Zoom Extents** and **Save** the file.

5. Repeat these steps on the other file.

Explore the AEC Template Configurations

Throughout the remainder of this tutorial, the following abbreviations will be used when each of the Templates is referred to:

* **Imperial** – *AEC Arch (Imperial - Intl).dwg*
* **D A CH** – *AEC Arch (Metric) D A CH.dwg*

1. In the Imperial file, click the Space tab.

2. **Zoom** and **Pan** within each viewport.

Note that several object types are not visible. The intention of this Display Configuration is to show only those objects typically used while Space Planning: Space, Space Boundary, Door, Window and Annotation objects.

3. In the *D A CH* File, click the Space Planning tab.

4. **Zoom** and **Pan** within each viewport.

The purpose of this tab is similar, but it is more restrictive concerning which objects are visible. Only Space, Space Boundary and Annotation objects can be seen in this tab.

We should note here an important variation between these Space Planning configurations and the configurations seen in the "scratch" files above. In the scratch files, nearly all objects were shown in all views; they were simply displayed in different ways. Here the configuration is being used to exclude entire classes of objects from displaying at all.

5. In the *D A CH* file, click the Massing Tools tab.

6. In the Imperial file, click the Mass-Group tab.

These two tabs use configurations designed for optimal use of the Mass Modeling tools of ADT. In Imperial, there are two configurations: Concept_Mass and Concept_Group. The reason there are two separate configurations is that both Mass Elements and Mass Groups occupy the same physical three-dimensional space. Therefore, if both were shown together, they would appear directly on top of one another. (You can see this if you switch to model space.)

7. In the Imperial file, **Zoom** in on the Mass Group object in both the right and left viewports (see Figure 5–12).

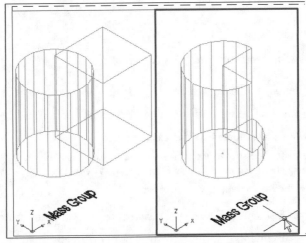

Figure 5–12 *Only Mass Elements show on the left, while Mass Groups appear on the right*

Notice that two Mass Elements appear on the left, while a single Mass Group, representing the result of the Box being subtracted from the Cylinder, appears on the right. (You may have to REGEN to get the actual display shown in Figure 5–12.)

The *D A CH* template takes the Mass Modeling tab a bit further. In it we have four viewports, one showing only Mass Elements in 3D, one showing Mass Elements only in 2D, one showing only Mass Groups (2D or 3D will depend on viewing direction) and the last one for Slices.

8. In the *D A CH* file, **Zoom** into the Mass and Mass Group objects in all viewports (see Figure 5–13).

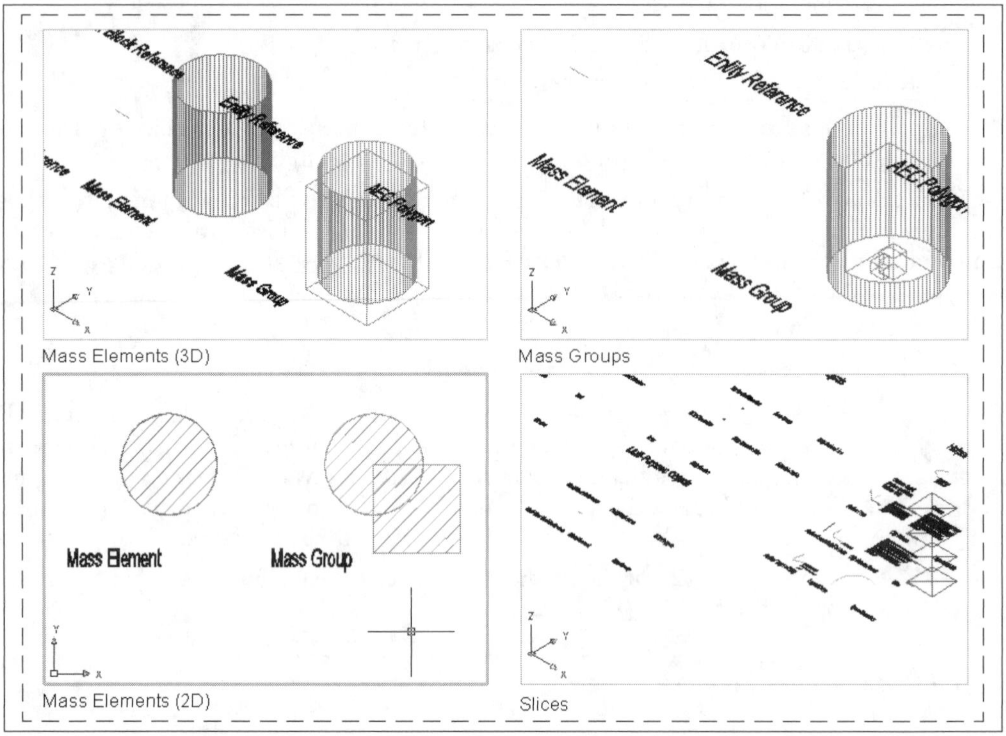

Figure 5–13 *The Massing Tools tab of the D A CH template*

Notice that the neither Mass Elements nor Mass Groups show in the Slices viewport.

There are several compelling features of the *D A CH* file not present in the Imperial version.

- The viewports are labeled so it is easier to use.
- Mass Elements are shown exclusively in 3D in the Mass Elements (3D) viewport and exclusively in 2D in the Mass Elements (2D) viewport.
- Slices are shown in their own viewport.

9. In the Mass Elements (2D) viewport, switch to an axonometric view.

Notice that the Mass Elements continue to display two-dimensionally even though we are viewing them from a 3D vantage point. This illustrates an important feature of Display Configurations: They determine what "set" of Display Control parameters will be displayed on screen. They may also be configured to react to a particular viewpoint when doing this. In this case, the configuration has been designed to *not* consider viewpoint when determining what to display. It simply shows 2D graphics all the time. We will explore exactly how this is accomplished in the next section.

10. Click the Space Planning and Structural Members tabs.

Similar techniques are used on each of these to limit the type of objects displayed and control the specific display modes used to show them.

11. Click through the remaining tabs in both files.

Upon completion of your exploration of the Display Configurations built into each of these template files, one distinct difference comes through. In the Imperial file, there are separate Work and Plot modes, while in the *D A CH* file, there is no distinction made between working and plotting, but rather between the various scales at which the drawings typically plot out (see Figure 5–14).

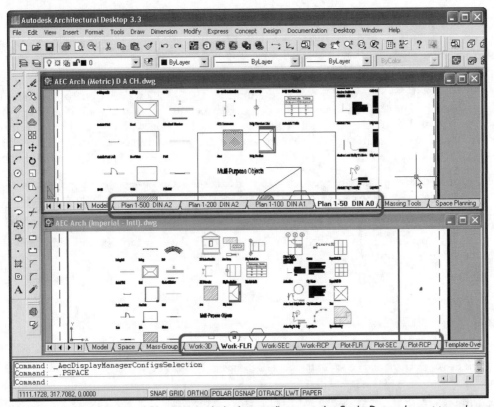

Figure 5–14 *The Work and Plot approach (in Imperial) versus the Scale Dependent approach (in D A CH)*

Working mode (Work) is used to display the drawing in a mode suitable for creation and editing of the model. The items shown in Work are not necessary displayed the way they will when plotted. Plot mode is the complementary display mode to Work. It shows objects the way we want them printed. Plot is used in sheet files and any layout tab that is intended for printing.

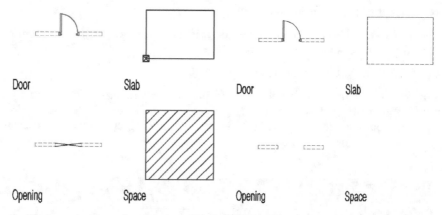

Door   Slab   Door   Slab

Opening   Space   Opening   Space

Figure 5–15 *Several objects shown in both Work (left) and Plot (right)*

Figure 5–15 shows some examples of this. For instance, an "X" appears through Opening objects in Plan while in the Work Display Configuration. The "X" allows you to select the object to move or edit it. The "X" disappears in the Plot display because we do not want it to print. The pivot points of Slab objects are treated the same way. The Space object, while displayed in Work, disappears altogether in Plot. Explore some of the other object behaviors in each of these displays as well to get a full sense of the differences between the Work and Plot configurations.

 Note: Work displays by default in color in a "not-to-scale" viewport. Plot displays in black and white in a "locked scale" viewport. These are features of the layout tabs independent of any Display Configuration. Refer to Chapter 3 for additional information.

The Scale Dependent tabs vary the level of detail and display characteristics of individual objects depending on intended plotting scale (see Figure 5–16).

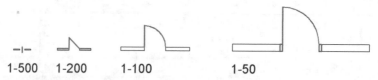

1-500   1-200   1-100   1-50

Figure 5–16 *Door shown in each of the Scale Dependent D A CH tabs*

Explore Configurations Not Attached to Layouts

Several configurations in the Imperial file are not assigned to any of the included layout tabs. Let's take a moment to look at some of those as well.

1. In the Imperial file, click the Model tab (model space).
2. From the **Desktop** menu, choose **Select Display**.
3. Load the **Design Development Poche** Display Configuration.

Notice the change to the drawing. Many of the objects, including Walls, Mass Elements, Area Groups and Slabs, have filled with a Solid Hatch Poché.

4. Try **Contract Documents Large Scale** and **Contract Documents Small Scale** next.

These are very similar to the Scale Dependent tabs in the D A CH Template. They too adjust the level of detail shown in each of the objects. However, these two configurations also "flatten" the model to a two-dimensional drawing.

5. With either **Contract Documents Large Scale** or **Contract Documents Small Scale** active, change to a 3D view.

Notice that the drawing remains "flat" even though we have changed to a three-dimensional viewpoint.

6. Try the remaining configurations and note their effects.

Table B–1 in Appendix B shows the complete list of Display Configurations in the Imperial and D A CH files, with a brief description of the intended purpose.

Now that we have explored the collection of Display Configurations pre-configured in the Imperial and D A CH templates, it is highly likely you would benefit from having portions of both. As we continue to work on our Office Standard template file in this chapter, we will borrow pieces from both of these sample template files.

THE DISPLAY MANAGER

The Display Manager is the primary interface for building and editing Display System components. Access the Display Manager from the Desktop menu or the AEC Setup toolbar. The Display Manager is organized in a Windows Explorer–type interface with three folders on the left, one for each of the major Display System components. See Figure 5–17.

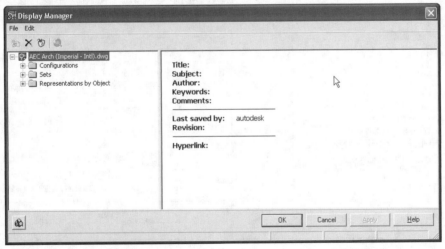

Figure 5–17 *The Display Manager*

Using the Display Manager

1. In the *Chapter05* folder, **Open** the file named *Display.dwg*.

Do not close the Imperial and D A CH files.

2. From the **Desktop** menu, choose **Display Manager** (see Figure 5–18).

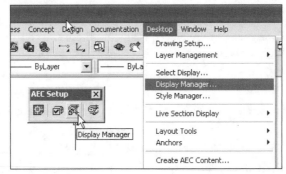

Figure 5–18 *Open the Display Manager from the menu or toolbar*

The Configurations Node

1. Click on the **Configurations** folder and expand the plus sign (+) below it.

There are several icons and symbols used in the Display Manager. Bold is used to indicate the active configuration or set. A check mark on an icon indicates that the item is being used somewhere in the drawing (cannot be purged). The small cubes on the right side indicate the view direction(s) to which a setting is applied (see Figure 5–19).

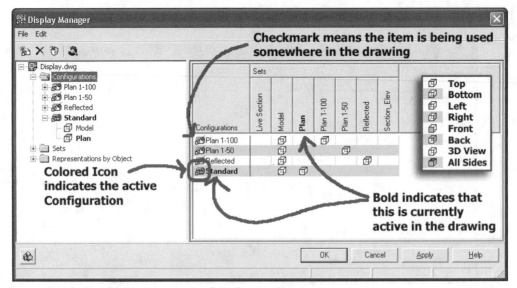

Figure 5–19 *The Configurations node of the Display Manager*

2. Beneath the Configurations node, select **Standard** and then click the Configuration tab on the right.

The Configuration tab allows you to edit the settings of the selected Display Configuration. Display Configurations may be built in two ways: View Direction Dependent and Fixed View. A View Direction Dependent configuration will swap a new Display Set on screen simply by changing the viewpoint in the drawing. For instance, the Standard configuration selected here defaults to a 3D Model Display Rep unless the current viewpoint of the drawing is Top view. In that case, a Plan Display Rep is shown (see Figure 5–20).

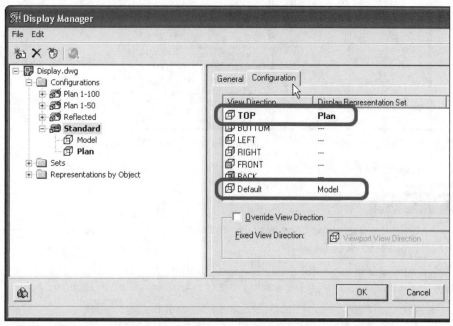

Figure 5–20 *Standard is an example of a View Direction Dependent Display Configuration*

Earlier, we looked at the Mass Modeling tab of the **D A CH** Template file and noted that the Mass Element viewports, both the 2D and 3D versions, did not adjust with the change in view direction. These were examples of Fixed View configurations.

3. **Close** the Display Manager and switch to the **D A CH** file.

4. **Open** the Display Manager.

5. Under Configurations, select **Mass Elements (2D)**.

What makes this a Fixed View configuration is the check mark in the **Override View Direction** check box. Notice that on the Configuration tab, **Default** is set to load the **Mass Elements (2D)** Display Set (see Figure 5–21). Furthermore, when **Top** is selected from the **View Direction** list, this configuration will display only 2D graphics as if viewed from the Top regardless of the actual viewing direction active in the drawing.

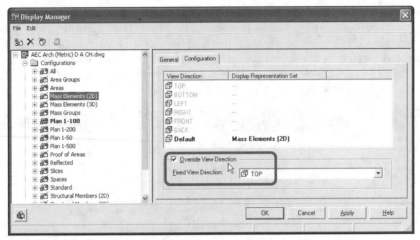

Figure 5–21 *Mass Elements (2D) is an example of a Fixed View configuration*

6. **Close** the Display Manager.

The Sets Node

1. Make the *Display.dwg* drawing active and open the Display Manager.
2. Expand the **Sets** folder.

Notice that since Standard is still the active configuration, and we have not changed the view direction in the drawing, **Plan** is the active Display Set (as indicated by its being bold).

3. Click on **Plan** on the left side tree view beneath the Sets folder.

This expands a detailed matrix on the right showing all of the ADT objects in a column at the left and the total list of Display Reps available in this drawing along the top.

SETS AND REPRESENTATIONS BY OBJECT

A *Display Set* determines which object types should be visible and in what mode(s) they ought to be displayed. A Set closely approximates a particular type of drawing such as Floor Plan, or Building Section.

A *Representation (Display Rep)* controls the behavior of objects under various drawing situations such as Plan, Elevation and Reflected. Each Rep contains one or more object sub-components, which are in turn assigned a collection of traditional AutoCAD properties, such as layer and color.

A Set answers two basic questions: *what* is displayed and *how* is it shown? For instance, 2D Section/Elevation is the first object type listed. Scrolling horizontally reveals that it is turned on in this Set. This makes 2D Section/Elevation objects part of the answer to the "what" question. The answer to "how" it is displayed is "General." It turns out that for 2D Section/Elevation objects, General is the only option, so this is an easy one.

4. Scroll to the **Door** listing.

Doors are also visible in this set; however, Door objects have ten Display Reps in this drawing. This makes the "how" question is a bit less obvious in this example. In this case, the check mark is in the **Plan** check box (see Figure 5–22). Therefore, in the **Standard** Display Configuration, the **Plan** Display Set is shown when the drawing is viewed from the **Top** view. Moreover, **Door** objects are displayed in the **Plan** Set using their **Plan** Display Rep.

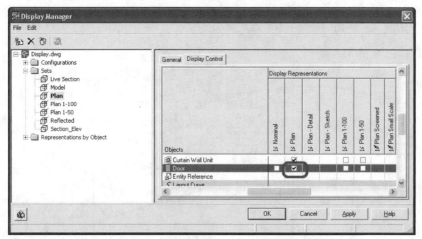

Figure 5–22 *The Plan Rep is used to display Doors in the Standard configuration*

5. Continue to explore the other objects and their accompanying Display Reps in the **Plan** set.

 Hint: While exploring the Display Manager, increase the window to the largest size your monitor will allow.

6. Beneath the Configurations folder, expand the **Reflected** entry and click on **Reflected** beneath it (see Figure 5–23).

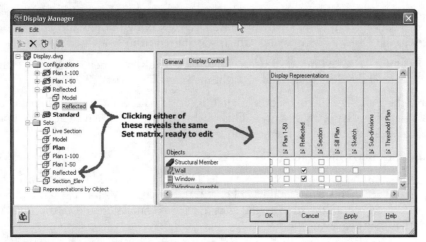

Figure 5–23 *Explore Reflected*

Notice that the same matrix appears on the right when the Reflected set is highlighted this way. It makes no difference whether you edit sets this way or directly from the Sets folder. Use whichever method works best for you. Using this technique can sometimes make it easier to keep track of where you are in the hierarchy.

If you study the matrix of Reps to Objects, you will note that in each set, a particular Display Rep is predominant for most of the objects. Figure 5–24 shows how the column of check marks appears to shift across the matrix as you compare Model with Plan and Plan 1-100.

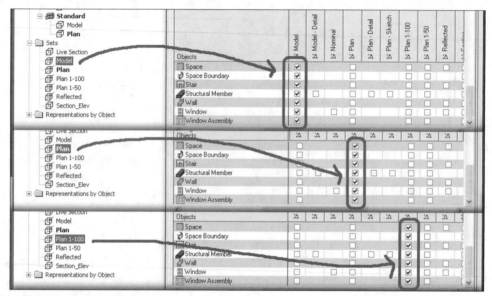

Figure 5–24 *Each set favors a dominant Display Rep*

Explore the AEC Templates

Let's return to our two files based on the Imperial and **D A CH** template files. There is no difference in concept, but there are many more Display Configurations and Display Sets than there were in the *Display.dwg* drawing, which was started from scratch.

1. In the **D A CH** drawing, select the Massing Tools tab, open the Display Manager, expand the **Sets** folder and highlight the **Mass Elements (2D)** Set.

Notice that there is only a single check mark in this set. Mass Elements are visible in Plan 1-100 Rep. All other object types are invisible. This is why the Mass Elements (2D) viewport in the **D A CH** file showed only Mass Elements in the exercise above.

2. Click **Mass Elements (3D)**.

The same situation occurs here; only the Model Display Rep is used rather than Plan 1-100. Spaces (2D), Spaces (3D), Structural Members (2D), Structural Members (3D) and Structural Members (Logical) all follow similar strategies in their composition and in the small quantity of objects that they display.

UNDERSTANDING DISPLAY REPS

The Display System is hierarchical. A configuration points to one or more sets. A set points to a collection of Display Reps. A Display Rep is an assemblage of graphics that represent an object specifically for a certain type of drawing or from a particular vantage point. Display Reps are really the heart of the Display System. Before you can build or edit configurations or sets, you must understand which Display Reps are available for each type of object. Furthermore, while nearly all sets and configurations are user defined, the majority of Display Reps are built into the software. Table B5–2 in Appendix B shows all of the Default Display Reps included in the software.

The Representations by Object Node of the Display Manager

1. In the *Display.dwg* drawing, open the Display Manager and click on the **Representations by Object** folder.

A matrix will appear, identical to the one viewed above in the Sets node except that in place of the check boxes there are now small "properties" icons (see Figure 5–25 at "1"). A few gray property icons appear as well (see Figure 5–25 at "2"); these indicate that although the Display Rep is available for that object type, there are no user-editable properties.

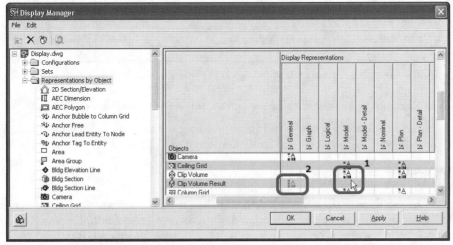

Figure 5–25 *A properties icon indicates editable properties for that object/rep combination*

2. Double-click the property icon in the **Plan** column next to the **Door** object.

This will open the Entity Properties dialog box and reveal each of the sub-components for that Display Rep/Object combination.

 Note: These are the System Default settings. It is not possible to access style or objectlevel overrides from the Display Manager.

You can also display the available Display Reps organized by object type and set.

3. On the left-hand tree beneath the Representation by Object folder, select **Door**.

Another matrix appears on the right. This one shows all of the Display Reps available in this drawing for Door objects in a vertical column and the Sets in which they appear along the top. Changes can be made here to the Set by checking or clearing any of the check boxes and to the Entity Properties by double-clicking on any Display Rep in the vertical column (see Figure 5–26).

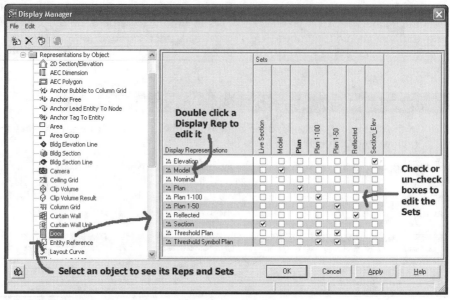

Figure 5–26 *Viewing the Reps organized by object type and set*

> 4. Continue to explore each of the entries in the Representation by Object folder.

Take the time to click on each object and view the settings of each of their Display Reps. Much can be learned about the behavior and configuration of the Display System this way.

Add Custom Display Reps

All of the Display Reps in the *Display.dwg* file are built into the software. Their properties can be freely edited but they can be neither deleted nor renamed. In addition, Display Representations cannot be overwritten the way styles can. This further underscores the need for a well-conceived template file. The fact that we cannot rename or delete these Display Reps can be a bit frustrating if you have reviewed Table B5–2 and are left wondering when you will ever use a Display Rep like Diagram, for instance. Nonetheless, these Display Reps are with us for better or for worse. Much more interesting than this fact however, is that we can create (or rather copy) our own custom Display Reps. These can be named any way we choose and, provided they are not being used at some future date, they can be deleted as well. This is a very powerful feature, full of potential but also ripe for abuse. Be leery of creating excessive custom Display Reps. When used sparingly, they can provide huge benefits to your drawings and office standards, but if too many are made and if they are not thought out and named carefully, they can become a huge management burden and confusing to users.

Some examples of custom Display Reps can be found in each of the two AEC templates we have been exploring in this chapter.

1. In the **D A CH** file, open the Display Manager and click on the **Representation by Object** folder.

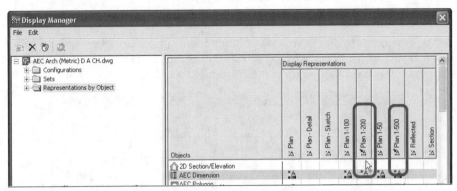

Figure 5–27 *Identifying user defined (custom) Display Reps*

A custom (or user defined) Display Rep has a property icon with a small black arrow pointing to it (see Figure 5–27). In the **D A CH** file, there are four of these: 2D Section 1-100, 2D Section 1-50, Plan 1-200 and Plan 1-500. The names of these make their purpose self explanatory. With the addition of the Plan 1-200 and 1-500 Display Reps, drawings created from this template can be displayed at four distinct levels of detail by simply switching the Display Configuration (see Figure 5–16).

The approach to custom Display Reps in the Imperial file goes a bit further than scale-dependent display.

2. **Close** the Display Manager and switch to the **Imperial** file.

3. In the Imperial file, open the Display Manager and click on the **Representations by Object** folder.

Here there are several custom Display Reps, perhaps a few too many. This collection of custom Display Reps, if not carefully maintained and managed, could become problematic as noted above. There are several "Screened" Display Reps, a few "Design and Poché" ones and some customized versions of the Model Display Rep. It should be noted that the way to create a custom Display Rep is to copy an existing one. Therefore, Model ByBlock and Model Monochrome are copies of the built-in Model Display Rep. This fact should help explain the choice of names used in this template.

4. Expand the **Representations by Object** folder and select **Wall**.

5. In the matrix, right-click the **Plan** Display Rep row and choose **Duplicate** (see Figure 5–28).

Figure 5–28 *Duplicate an existing Display Rep*

6. Type **Test** for the name and then press ENTER.

Test will appear with a row of check boxes next to it.

7. In the tree at left, expand **Sets** and click on **Work_Plan** (see Figure 5–29).

Notice that **Test** now appears in its own column and a single check box appears next to the Wall entry. If you need a "Test" Display Rep for other objects as well, you will need to repeat these steps on each object that requires the custom Display Rep.

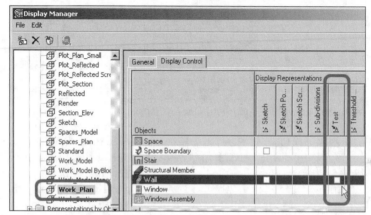

Figure 5–29 *Confirm the addition of the new Display Rep in the Set matrix*

8. Click **Cancel** in the Display Manager to dismiss the Display Manager without saving the changes.

One of the reasons there are so many custom Display Reps in this file is the naming strategy used. The original name is retained, with a descriptive suffix appended to it. Therefore, Model Monochrome is a copy of Model, where all the sub-components use the same color. Plan Screened is a copy of Plan where all the components are set to a color (or Plot Style) that prints screened 50%. The complete collection of custom Display Reps included in both the Imperial and D A CH templates is summarized in Table B5–3 in Appendix B.

DETERMINING WHICH CUSTOM DISPLAY REPS TO USE

As you can see from Table B5–3 in Appendix B, there are many possibilities when it comes to custom Display Reps. Let's try to determine which ones we truly need and eliminate those we don't. Our goal is to establish a list of the custom Display Reps that we will want to include in our main template file.

Understanding Scale-dependent Display Reps

1. **Open** the file named *Scale Dependent Display.dwg*.

A variety of objects is shown in a layout tab named Scales. Each object is displayed in each of the four scale-dependent Display Reps included in the D A CH template file.

2. Double-click in the 1-200 Viewport and then click on the Door object.

3. Right-click and choose **Entity Display**, and then choose the Display Props tab.

All Display Props are assigned at the System Default level in this and all of the template files. Notice that at the top of the dialog box, **Plan 1-200*** appears in the Display Representation drop-down list. (This indicates that Plan 1-200 is the active Display Rep.)

4. Click the **Edit Display Props** button.

Unlike the Plan Display Rep for Doors, the Plan 1-200 (which is a copy of Plan 1-100) has only three sub-components: Panel, Swing and Direction. (Direction is usually expressed as an arrow in door types like folding or revolving doors.)

5. Click the Other tab.

Notice that the **Straight Swing** box is checked.

6. **Cancel** all dialog boxes and click in the 1-100 Viewport.
7. Repeat the same steps.

Notice that everything here is the same except for the straight swing.

8. Repeat the same steps in 1-500.

Upon editing the Display Props here, we find that the small vertical line used to represent the Door at this very small scale is actually a custom display block. To make only this custom block display in place of the default graphics, the default components have been turned off. Perform the same steps in each of the other viewports to understand how the graphic display at each scale is achieved.

9. **Close** the file without saving.

Understanding the Screened and Poché Display Reps

1. **Open** the file named *Poché and Screening.dwg* and then click the Design Development Poché tab.
2. Double-click in the Wall viewport and then click on the Wall object.
3. Right-click and choose **Entity Display**, and then click the Display Props tab.

All Display Props are assigned at the System Default level in this and all of the template files. Notice that at the top of the dialog box, **Plan Poche*** appears in the Display Representation drop-down list. (This indicates that *Plan Poche* is active.)

4. Click the **Edit Display Props** button.

Here you will notice that in the Layer/Color/Linetype tab, all components have been turned off except the Shrinkwrap and the Shrinkwrap Hatch. On the Hatching tab, the hatch pattern is Solid fill. This means that regardless of Wall style, when this Display Rep is used, Walls will display with a single outline and a solid fill hatch.

5. **Cancel** all dialog boxes and repeat these steps on the Slab.

With Slabs and Roof Slabs, the settings are not quite as obvious. If you recall from Table 2–1, "Architectural Objects" in Chapter 2, Slabs and Roof Slabs are actually 3D objects. However, they do have a cut plane and therefore a hatch pattern can be assigned. In this Display Rep, the hatch is turned on and set to solid fill like the Walls. If you perform the same steps on the Structural Members, you will see that hatching is set to solid fill for them as well. Interestingly enough, only these four objects have a Plan Poché Display Rep. If you decide to use this Display Rep in your templates, you may consider adding this Display Rep to other objects that could benefit from solid fill hatching such as Spaces, Areas, Mass Elements and Mass Groups.

6. Click the Screened tab.

Here the entire collection of ADT objects shows in a single viewport assigned to the **Plot Screened** Display Configuration. Notice that the screened effect is reserved for primarily the Architectural objects.

7. Right-click any object in the Architectural Objects group, and choose **Entity Display**.

Notice that the active Display Rep is **Plan Screened**.

8. Right-click the Mass Element and choose **Entity Display**.

Notice that its active Display Rep is **Sketch Screened**. The 2D Plan Display Rep of Mass Elements is actually named Sketch. Therefore, this being a copy of that Display Rep, it was named Sketch Screened so that you would understand from which Display Rep it was copied. The name of this Display Rep could be anything. The downside of this naming scheme is that we end up with a new column in the Display Manager for each custom Display Rep that we add. In the case of the Plot Screened Display Configuration, we have five different Screened Display Reps. (Refer to Table B5–3 in Appendix B.) Add one more, Reflected Screened, which is used for the Plot Reflected Screened Display Configuration. This ought to be simplified.

9. **Cancel** any dialog boxes and open the Display Manager.

10. In the Representations by Object folder, click on **Anchor Lead Entity To Node**.

If you use a Leader Anchor to attach one ADT object to another, a leader will appear connecting the two together. This leader is controlled separately from the objects it anchors together. If you envision needing to screen the anchor, you can follow the lead of this template and add a custom Display Rep. However, you ought to at least rename it to a more common name. Remember that above we used the name "Test." The name is not important, consistency and simplicity are. Let's rename this Display Rep to Plan Screened like the majority of the others.

11. On the right, within the matrix, right-click **General Screened** and choose **Rename** (see Figure 5–30).

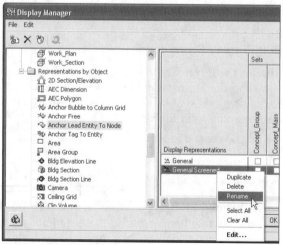

Figure 5–30 *Rename a Display Rep*

12. Type **Plan Screened** and then press ENTER.
13. Select **Mass Element**.
14. Repeat the same steps for **Sketch Screened:** rename it to **Plan Screened**.
15. Rename **Sketch Poché** to **Plan Poché**.
16. Back under Sets, click **Plot_Plan Screened**.

Notice that since the only object that had **General Screened** was **Anchor Lead Entity To Node**, this column is no longer present. Instead, there is a new check box in the **Plan Screened** column next to **Anchor Lead Entity To Node**. Notice that the same is true for the two Display Reps we renamed in Mass Element. In this way, we have simplified the Display Manager by three fewer columns. The purpose of this exercise was to illustrate the process of renaming a custom Display Rep. However; the chances of our actually needing to screen the Anchor Lead Entity To Node object are remote. Therefore, we may decide to simply eliminate this Display Rep altogether.

17. In the Representations by Object folder, click on **Anchor Lead Entity To Node**.
18. Right-click **Plan Screened** and choose **Delete**.

A message will appear stating that it cannot be deleted. This is because it is still being referenced by two sets. To delete it, we must first remove the check boxes in all sets. Remember, a custom Display Rep may be deleted as long as it is not being used in the drawing.

19. Right-click **Plan Screened** and choose **Clear All.**
20. Right-click **Plan Screened** again and choose **Delete.**

It will now be deleted. Unfortunately, we will not be able to simplify the Plot_Plan Screened set any further. Threshold Plan Screened and Sill Plan Screened are both necessary because Doors and Windows show both Plan and Sill/Threshold graphics at the same time. Therefore, they must have separate names.

SIMPLIFYING THE SCREENED DISPLAY SET MAY BE A MOOT POINT

We use a screened display typically to display certain information as an underlay to the focal work drawn on top. This is very common in MEP drawings and useful when checking stacking and other coordination issues between two or more floors. There are a few problems with trying to achieve this with Display Control. First, as has been already stated, Display Control applies only to ADT objects. Therefore, Layer colors will need to be manipulated to achieve the same effect with AutoCAD entities. Second, many of your consultants may still be on older versions of AutoCAD and probably do not have ADT. Although they can use an Object Enabler, many of them choose not to. (The topic of file compatibility is covered in detail in Chapter 13.) In this situation, it is likely that layers will again be the solution to screening your drawings. Finally, even when stacking files internally to check coordination between two or more floors, you are not able to assign two different Display Configurations on screen at the same time. In other words, it is not possible using Display Configurations to display the XREF screened while the work on top remains un-screened. (It is not possible in model space at least.) Here again, it is likely that screening just the XREF layers (by color or with Plot Style Tables) will be the easiest solution.

Model ByBlock, Monochrome, Design Poché and Small Scale

There is much to be learned from the remaining custom Display Reps in this file. The default Model Display Rep has been configured with colors assigned directly to the sub-components for most Architectural objects by default in ADT. The intention of these colors is to represent the material of the object. When the drawing is viewed shaded in 3D in ADT, these colors give some indication of material. Rendering packages such as Autodesk VIZ also use these colors to map materials to linked ADT models. (Refer to Chapter 12 for more information.) Table B5–4 in Appendix B lists many of the typical colors used throughout the content and styles provided in the box with ADT.

There may be occasions when it is desirable to display the model in Layer colors instead of the pre-assigned material colors. This is the purpose of the Model ByBlock Display Rep. Model ByBlock is also used to generate 2D Section/Elevation objects by default. Since 2D Section/Elevation objects use colors to set design rules, this is an important consideration. However, since there are only thirteen unique colors listed in Table B5–4 and perhaps a dozen or so more in layers, the likelihood of encountering conflicts with colors between the needs of rendering and generating sections and elevations is relatively slim. Therefore, you may consider consolidating the two and using the default Model Display Rep for generating sections and assigning rendering materials. Refer to Chapter 11 for more information on 2D Section/Elevation objects.

Similar to the Screened Display Reps covered above, Model Monochrome sets all Model components to Color 254. This Display Rep is referenced by the Design Development Display Configurations. The idea behind this is that materials would not be known at early stages of design, and using the gray color makes it look like a study model.

XREFS AND DISPLAY REPS

Display Control works across all XREFs. When a new Display Configuration is loaded, all ADT objects native to the current file or contained in the XREFs will be affected by the change. This is true regardless of reference type, and level of nesting. There are a few important issues:

- Objects do not always update immediately. You will often need to perform an OBJRELUPDATE to get the drawing to update.
- If the same Display Rep exists in both the XREF file and the host drawing, the settings in the XREF drawing will apply. There is no way to override this behavior.
- If the host drawing contains a custom Display Rep not present in the XREF, the host drawing settings will apply. However, this limits you to the System Default settings.

Understanding Display Control Behavior in XREFs

1. In the *Chapter05* folder, **Open** the file *Display XREF Test.dwg*.

A small floor plan containing a single XREF appears. Most objects are part of the XREF. The small out-building is part of the host file.

2. Load the **Design Development Poché** configuration.

Notice that the out-building has updated immediately, but the XREF has not.

3. At the command line, type **objrelupdate** and then press ENTER.

Tip: If you built the custom icon in Chapter 1, simply click it.

4. Select one of the Walls of the out-building, right-click and choose **Entity Display**.

5. Edit the **Display Props** and make a change to the hatch pattern used by the **Shrinkwrap Hatch**.

6. Click **OK** to return to the drawing and perform another OBJRELUPDATE.

Notice that the hatching of the Walls in the XREF has *not* changed. The only way to change it is to open the XREF and change it there. There is no equivalent to the VISRETAIN setting of layers for Display Control. The XREF settings always apply. However, if Plan Poché did not exist in the XREF, then we would be able to edit it freely from the host file.

7. Open the Display Manager and click on the **Representations by Object** folder.

8. Follow the steps outlined in the "Add Custom Display Reps" step sequence above and add a **Test** Display Rep to the Wall object (see Figure 5–31).

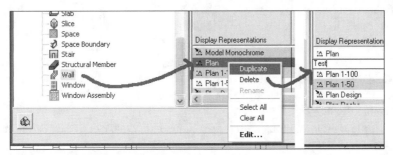

Figure 5–31 *Create a copy of Plan called Test*

9. Double-click **Test** to edit it.

10. On the Layer/Color/Linetype tab, select all sub-components and set the Color to **Red** (see Figure 5–32).

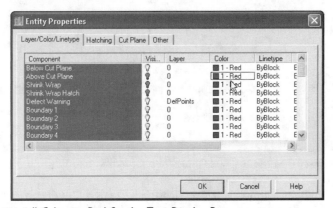

Figure 5–32 *Setting all Colors to Red for the Test Display Rep*

11. Under the Sets folder, right-click **Plot_Plan Poche** and choose **New** (see Figure 5–33).

 Caution: Be sure to right-click the existing set and not the Sets folder. Right-clicking an existing set copies it. Right-clicking the Sets folder makes an empty set.

Figure 5–33 *Right-click an existing set to copy it*

12. Name the new Set **Test Set.**

13. On the right in the Object/Rep matrix, next to Wall, clear **Plan Poche** check box and check **Test**.

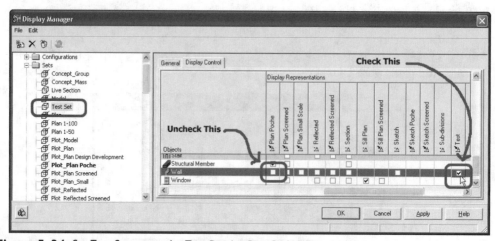

Figure 5–34 *Set Test Set to use the Test Display Rep for Walls*

14. Under the Configurations folder, right-click **Design Development Poche** and choose **New**.

Caution: Again, be sure to right-click the existing one to copy.

15. Name the new Configuration **Test Configuration**.

16. On the right side, on the Configuration tab, next to Top, click on **Plot_Plan Poche**.

17. In the drop-down menu that appears, choose **Test Set** (see Figure 5–35).

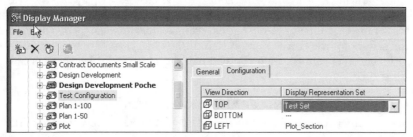

Figure 5–35 *Assign the Test configuration to the Top view*

18. Click **OK** to close the Display Manager.

19. Load the new **Test** configuration and then perform an OBJRELUPDATE.

All of the Walls should now all be red, including the XREF Walls. You could further customize the Test configuration by adding a Test Display Rep to other objects as well. Experiment further if you wish.

XREF DISPLAY CONTROL CHECKLIST

1. In the Display Manager, duplicate a Display Rep and give it a name. (Be sure the name is not in use within any XREFs if you wish to control the XREF's display.)

2. Repeat this step for each Object class that requires the custom Display Rep.

3. Duplicate the set most similar to the one you wish to create. (Right-click the one you wish to copy and choose **New**.)

4. Configure the new set to include the newly created Display Reps.

5. Repeat this step for any additional sets you wish to create.

6. Duplicate the configuration most similar to the one you wish to create. (Right-click the one you wish to copy and choose **New**.)

7. Assign the new set(s) to the configuration and close the Display Manager.

8. Load the new configuration.

If used properly, this very powerful technique can give unlimited potential to your customization of the Display System without overloading your default template files with many Custom Display Reps. For instance, to keep your default *Plan-Sect-Elev.dwt* template file (that we have been developing throughout this book) lean, include only the default hard-coded Display Reps and perhaps a few others that you plan to use frequently. Create additional template files for loading, displaying and printing the drawings in custom Display Configurations. You may wish to create one that includes the screening and poché displays covered here.

There is one potentially significant limitation to the XREF approach covered here. The host file is only able to display the XREF through System Default settings. If a particular object uses style or object based display overrides, these will not be displayed when viewed through the custom Display Configuration through XREF. This is because in order to attach a style or object based override, you must be able to select the object in question directly. This limitation precludes the use of custom display blocks in those object types that support them. Custom display blocks are used by objects to represent some or all of their graphical display in a particular view. The following objects have the ability to use custom display blocks: Curtain Wall, Curtain Wall Unit, Door, Opening, Railing, Structural Member, Window and Window Assembly. Please note that Openings do not have this feature in Plan 1-100 or Plan 1-50, and none has the feature for the (Live) Section Rep. If the objects you are trying to display require this style or object level control, then you will have to add the display settings directly to the XREF file instead.

CUSTOM DISPLAY REPS AND SPANNING ELEMENTS

In Chapter 2, we discussed the dimensionality of ADT Objects. (Refer to Table B2–1, Architectural Objects" in Appendix B.) Specifically we made a point of identifying Curtain Walls, Curtain Wall Units, Stairs and Railings as "Spanning Elements." What that meant for project setup in that chapter was that these items would be kept in separate files wherever possible and XREFed back to the Composite Building Model. When these "vertically" oriented files are XREFed to predominantly horizontal files, they will often need to be referenced at a Z coordinate above or below zero. For instance, consider a three-story building with a single "Spanning" Curtain Wall file and a 12'-0" floor-to-floor height. The Curtain Wall within its file is created at Z=0 for that file. However, when it is XREFed to the other files in the project set, it must be inserted at −12'-0" in the second floor file and −24'-0" in the third floor. The reason for this is that each file when opened is built at its own Z=0 elevation. Therefore, to see the correct portion of the Curtain Wall in the second floor plan (the middle third), it must be inserted below the current working plane.

The only problem encountered with this approach to spanning elements is controlling the Display System relative to these objects. The Curtain Wall in this example will likely vary in plan from floor to floor. Therefore, we must be able to properly display a "slice" of the Curtain Wall at these various heights. We will achieve this when we XREF this Curtain Wall file to the others by using custom Display Reps. Remember: the trick to making this work properly with XREFs is to limit ourselves to System Default settings and be sure to *not* include the custom Display Rep in the Spanning Curtain Wall file.

Setting Up a Spanning Curtain Wall for Correct Display

The following exercise will utilize a collection of files located in the folder named *Spanning* in the *Chapter05* folder. If you did not install the CD files for this chapter, refer to the beginning of the chapter and install them now.

The project included in the *Spanning* folder is a small four-story office structure with a Curtain Wall that spans the top three floors of the building. The Curtain Wall grid is divided horizontally into three bands, one for each floor of its span. Vertically, the Curtain Wall is subdivided into major bays that are expressed with a round pier on the lowest level, a square pier on the middle level and a diamond shape at the top. The bottom bays are subdivided into four smaller bays, while the next two tiers subdivide into three and two respectively, moving up the span. It is obvious that the plan of such a design would vary quite considerably per floor. Figure 5–36 is an illustration of the major features of the design.

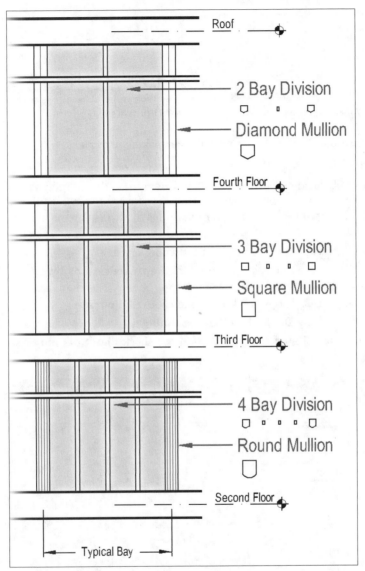

Figure 5–36 *The tier and bay structure of the Spanning Curtain Wall design*

1. In the *Chapter05\Spanning* folder, **Open** the *A-SH02-C.dwg* drawing file.
2. Compare the Curtain Wall in this file to the illustration in Figure 5-34.
3. Open the **Display Manager** and click on the **Representations by Object** folder (see Figure 5–37).

Note that there are no custom Display Reps in this file.

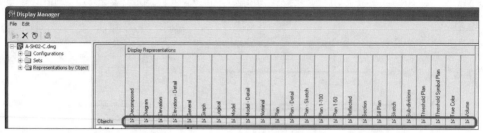

Figure 5–37 *A-SH02-C.dwg contains only the default Display Reps*

4. **Close** the Display Manager and the current file without saving.
5. **Open** the file named *A-104-C.dwg.*

This is the sheet file for the Fourth Floor Plan. **Zoom** in on the Curtain Wall and note that it is not displaying properly. (The graphics used here reflect the Second Floor Plan cut plane.) We need to build a new Display Configuration in this file to display the Curtain Wall correctly at the Fourth Floor.

6. Double-click in the viewport to float model space.
7. Open the Display Manager, expand the **Representations by Object** folder and select the **Curtain Wall** object listing.
8. In the right pane, right-click **Plan** and choose **Duplicate**.
9. Name the new Display Rep **Plan Spanning**.
10. In the **Plot_Plan** column, clear **Plan** and check **Plan Spanning** instead (see Figure 5–38).

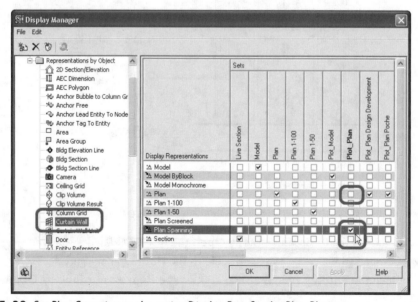

Figure 5–38 *Set Plan Spanning as the active Display Rep for the Plot_Plan set*

11. Double-click **Plan Spanning** and then click the Cut Plane tab.

12. Type in **27'-6"** for the Cut Plane and click **OK** twice to return to the drawing.

13. Perform an OBJRELUPDATE.

Notice that the Curtain Wall now appears correctly cutting through the fourth floor.

14. Open the Display Manager again and right-click the **Sets** folder.

15. Choose **Import/Export** from the right-click menu.

16. Click the **Open** button in the middle of the dialog box.

17. Select the *A-103-C.dwg* file in the same folder and click **Open**.

A complete list of sets from the current file is listed on the left, with the remote file (*A-103-C.dwg*) sets on the right.

18. On the Left (Current Drawing side) select **Plot_Plan** and then click the **Export** button (see Figure 5–39).

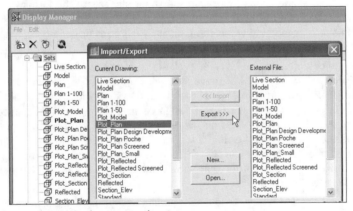

Figure 5–39 *Export the set to the remote drawing*

The Duplicate Names Found dialog box will alert you that the name Plot_Plan already exists in the *A-103-C.dwg* drawing. Always consider your choices carefully in dialog box like this. You could potentially overwrite the wrong item. We want to replace the version of this set in the remote file with the one we just modified in the current file.

19. Choose **Overwrite Existing** and click **OK** twice.

20. When asked to save the *A-103-C.dwg* drawing, click **Yes**.

21. **Close** the Display Manager and return to paper space.

22. **Open** *A-103-C.dwg*.

23. Double-click into the viewport and then open the Display Manager.

24. Under Representations by Object, select the **Curtain Wall** object.

Notice that the Plan Spanning Display Rep is already here. It was imported along with the undated Plot_Plan set.

25. Double-click the **Plan Spanning** Display Rep, change the Cut Plane to **15'-6"** and click **OK** twice to return to the drawing.

26. Perform an OBJRELUPDATE.

Notice that the Curtain Wall now appears correctly cutting through the third floor. The two files we changed were the sheet files. You will also need to change the two model files for the third and fourth floor in the same way to make them display properly as well. The process is nearly the same, except that *A-FP03-C.dwg* and *A-FP04-C.dwg* use the Work configuration, which references the Work_Plan set.

Note: For more complex Curtain Wall designs (those with nested infills and custom display blocks), this technique may not produce entirely satisfactory results. To some degree, this is a limitation of the software. However, there are some workarounds. Nested infills can prove problematic in XREFs because the software does not distinguish between those that occur above and below the cut plane. It simply displays them all.

To overcome this, try any of the following:

- Add custom Display Reps directly in the Curtain Wall XREF, one for each cut plane condition. In the example used here, you would have three: Plan 2nd Floor, Plan 3rd Floor and Plan 4th Floor. In the sheet file, create the Display Rep for the appropriate floor (using exactly the same name) instead of Plan Spanned. This technique is acceptable when the Curtain Wall does not span too many floors. It is not very practical on very tall buildings.

- Use a Slice object in the Curtain Wall file to slice the Curtain Wall at each of the desired heights. Put each of these live Slices on their own layer and display them through Layer control in each of the sheet files instead of the Curtain Wall itself. Slices update automatically as the design changes. However, using layers to control them is less than ideal. Slices do not work on XREFs, so they must be in the same file as the Curtain Wall itself. To facilitate the process, build two Display Configurations: one that shows Curtain Walls only without Slices for working on the Curtain Wall, and the other that shows only Slices without the Curtain Wall for XREFing to the sheet files.

- Use a 3D Section object. Add a section line as if you were cutting a typical section. Set the bottom and top heights of the section line to "cut" though the portion of the Curtain Wall where you want the cut plane to be. Generate a 3D Section object and place it in the same location as the Curtain Wall object. Again, you will need some Display Configurations to help manage this process, one for working on the Curtain Wall and another for generating the sections. Generate one section for each sheet file you need. The Building Section object type (3D Section) has two Display Reps: Sub-divisions and Model. Set it to display in Model so that the properties will match the original Curtain Wall object. The advantage to this technique is that it reads directly from the 3D Model, so even if you have complex designs and custom display blocks, the 3D Section will properly represent them. However, this technique can be cumbersome and difficult to manage. In addition, Section objects do not automatically update, so you will need to periodically update them. More information on this technique can be found in Chapter 11.

Setting Up a Spanning Stair Well for Correct Display

A similar situation to the previous exercise exists with Stairs. At the bottom floor of a flight of stairs, they should only display going up. At the top of the flight, they should only display going down. On any intermediate floor, they should show both ways. Using a similar technique as for the Curtain Wall, we can display the same Stair in all three ways.

1. In the *Chapter05\Spanning* folder, **Open** the *A-FP01-C.dwg* drawing file.

Notice that the Stair displays going up only. A custom Display Rep named **Single Flight** has been added to this file.

2. Open the Display Manager and under the **Representations by Object** folder, click on the **Stair** object.
3. On the right, double-click **Single Flight**.

Notice that all of the "down" components are turned off. In addition, the Outline Down is made dashed to show the Stair continuing overhead (see Figure 5–40).

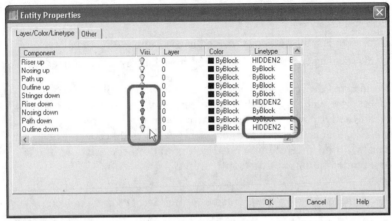

Figure 5–40 *All down components are turned off*

4. **Close** this dialog box and while still in the Display Manager, right-click the **Sets** folder.
5. Choose **Import/Export**.
6. The active set is Work_Plan. Using the same technique as before, **Export** this set to the file *A-101-C.dwg*. Save when prompted.
7. **Close** the Display Manager and **Open** the *A-101-C.dwg* file.

This is the sheet file for the First Floor. If you **Zoom** into the Stair, it will appear as if the **Export** did not work. This is because *A-FP01-C.dwg* had Work_Plan active, while this drawing has Plot_Plan active.

8. Double-click in the viewport and then open the Display Manager.
9. Expand Sets, select **Plot_Plan** and then select **Stair** in the matrix at right.
10. Clear **Plan** and instead check **Single Flight**.
11. Click **OK** and then OBJRELUPDATE.

The Stair should now be properly displaying as an up only Stair. You can repeat the same steps in *A-FP05-C.dwg* and *A-105-C.dwg*. Simply reverse the on and off components; instead of turning off all down components, turn off all up components. Naturally, this is a very simple example, but it illustrates the potential and the benefits of keeping the Stairs separate in their own XREF. With this technique, you will not need to duplicate the geometry of the Stair in two or more files; simply change the way they display in each file to achieve the same effect without the need for extra coordination. Even on very complex Stair designs, you should keep all of the Stairs separated and use this technique to manage their display.

ADDING DISPLAY CONTROL TO THE TEMPLATE

At the end of the last chapter, we added layers to our *Plan-Sect-Elev.dwt* template file. We will now begin adding Display Control settings to this file. In this chapter, we have discussed the following issues related to template files:

- Understanding what is included in a "scratch" drawing
- Recognizing the merits of keeping the Display System simple with a minimum of custom Display Reps
- Understanding the settings provided in the sample AEC templates
- Understanding Spanning Elements and their Display Control needs

Table 5–2 summarizes the recommended Display Configurations to include in your template based on this list. We will begin with our existing *Plan-Sect-Elev.dwt* template as a starting point and import the Display Configurations from both the Imperial and **D A CH** templates. The *Plan-Sect-Elev.dwt* is intended for generating Models, Sheets and Spanning Models. In Table 5–2, you will find the recommended configuration, the AEC template it is imported from and the intended purpose of the configuration.

Table 5–2 *Recommended Display Configurations for Plan-Sect-Elev.dwt Template*

Configurations	Rename To	Imperial or D A CH	Purpose
Standard	N/A	N/A	Default Configuration. Use for Model Construction. Not to scale. (Replaces "Work" in Imperial templates.)
Work_Reflected	Standard Reflected	Imperial	Use for RCP Construction. Not to scale.
Render Below)	N/A	Imperial	Special 3D Model Configuration (See
Plan 1-50	Plot 50 or Plot 48	D A CH	Plotting Plans at 1:50 (1/4"=1'-0")
Plan 1-100	Plot 100 or Plot 96	D A CH	Plotting Plans at 1:100 (1/8"=1'-0")
Plan 1-200	Plot 200 or Plot 192	D A CH	Plotting Plans at 1:200 (1/16"=1'-0" or 1"=20')
Plan 1-500	Plot 500 or Plot 480	D A CH	Plotting Plans at 1:500 (1/32"=1'-0" or 1"=40')
Plot_Reflected	Plot Reflected ##*	Imperial	Plotting RCP at 1:100 or 1:50 (1/8"=1'-0" or 1/4"=1'-0")
Design Development Poche	Plot Presentation	Imperial	Plotting Presentation Style Plans

Note: Substitute the number representing the plot scale for "##" in the Plot Reflected Display Configuration.

Other possibilities include the following:

- Structural – Import from the D A CH template.
- Areas – Import from the D A CH template.
- Mass Modeling – Import from the Imperial template.

Importing Configurations to Your Template

You will need three files to complete the work on your template's Display System: your version of *Plan-Sect-Elev.dwt* and clean copies of *AEC Arch (Imperial - Intl).dwg* and *AEC Arch (Metric) D A CH.dwg*. To save you some effort, these two files have been provided in a folder named *Templates* in the *Chapter05* folder.

1. Open the *AEC Arch (Imperial - Intl).dwg* file.

 Tip: Be sure to use the copy from the CD in the *Chapter05/Templates* folder.

2. Open the Display Manager and under Sets, click on **Plot_Section**.
3. **Right-click the** Model ByBlock **column and choose** Clear All **(see Figure 5–41).**

Figure 5–41 *Clear all checkboxes for Model ByBlock*

We do not want this Display Rep to be imported into our template, so first we must dereference it here.

4. Right-click the **Model** column and choose **Select All**.
5. Click the **Plot_Model** set next.
6. Transfer all check boxes from the **Model ByBlock** column to the **Model** column.

 Caution: Between these two sets, be sure that ALL check boxes in the Model ByBlock column have been cleared.

7. Close the Display Manager and **Close** and **Save** the file.

In an effort to consolidate, you may also wish to rename the Sketch Poche custom Display Rep as suggested above. We will now open your *Plan-Sect-Elev.dwt* file and import the Display Configurations we need. Be sure to close the Imperial file first. You will be unable to use the **Import/Export** command on the AEC templates if the drawings are opening ADT.

8. Open your *Plan-Sect-Elev.dwt* file.

Tip: Be sure to open the actual DWT file, not a DWG copy of it. You must change **Files of Type** to **Drawing Template File** in the **Open** dialog box to do this.

9. Open the Display Manager and beneath Sets, click on **Plan**.
10. Configure the **Plan** Set as shown in Figure 5-42.

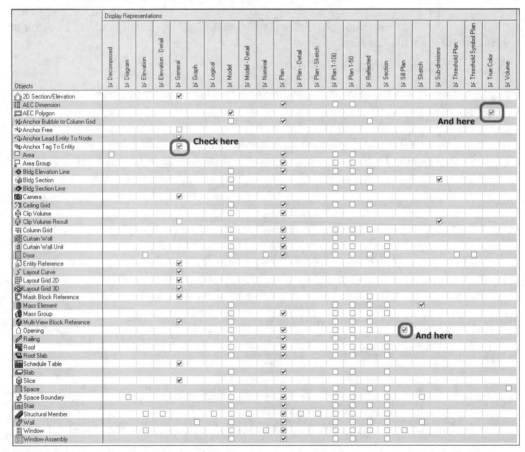

Figure 5–42 *Add or remove any checkmarks to update the Plan Set*

What we have added to the default Plan set are Sills for Opening objects, True Color fill for AEC Polygons and Leaders attached to all object Tags. Remember, as per Table 5–2, we are using this set in place of the default Working set. This is not intended to be a plotting mode (although test plots could certainly be done from this mode). The true color fill and Sills will show only if those individual object types turn them on in the drawing. Here we are simply allowing for that possibility. With the Tag Anchors on, we are making it much easier in a Working mode to *see* which object a particular Tag belongs to. If you wish, you can also make the Tag Anchor non-plotting by placing the sub-component on a no-plot layer such as Defpoints. You may also want to consider turning on the Threshold Symbol Plan for Doors and using Plan 1-100 or 1-50 for Mass Elements. Check the settings of these Display Reps to see what they offer. You may also wish to substitute the Plan 1-100 or 1-50 Display Reps for all of the Plan Display Reps. If you do not envision using both the default Plan Display Rep and the scale dependent version, using only the scale dependent versions will save you some work in configuration of the individual Display Reps.

11. Using the technique covered earlier, right-click the **Configurations** folder and Choose **Import/Export**.

12. Open the Imperial template on the right side of the Import/Export dialog box.

13. Using the CTRL key, select each of the configurations listed in Table 5–2 from the Imperial template and **Import** them. (There are four in total.)

Work always seems to come along for the ride. Simply right-click and delete it.

14. **Rename** the Configurations as shown in Table 5–2.

Go through each set and fine-tune them to your preferences.

15. Return to the **Import/Export** dialog box when you are finished and import each of the scale dependent Plan configurations from the *D A CH* file.

Your completed list of Display Configurations should look something like Figure 5–43.

Figure 5–43 *Results of importing and renaming all configurations*

CUSTOMIZING TEMPLATE DISPLAY REPS

The final step needed to completing the display settings of our template file is perhaps the most tedious. You will now need to edit each Display Rep of each object to fine-tune all of the settings to suit your office standards. This can be done best by selecting the **Representations by Object** folder of the Display Manager. (Recall Figure 5–25 for an illustration of this.) This will show you every Display Rep in one screen. Simply double-click the small Property icon to edit each one. Take the time to carefully go through each Display Rep and consider the proper settings. Read the next passage on Lineweights before performing this task. Following are some tips for recommended settings. There are many other settings to change, but Table 5–3 will get you started.

Table 5–3 *Suggested Display Rep Settings for Template*

Object	Rep	Settings to Change
2D Section/Elevation	General	Use the Lineweight property to assign receding Pen weights to each of the first four or five sub-divisions. Turn on the Erased component and make it Color 1 (Red).
AEC Dimension	All	Assign Dimension styles.
Anchor Tag To Entity	General	Assign the layer to Defpoints and use a primary color like Magenta.
Bldg Elevation Line & Bldg Section Line	All	Assign the Sub-divisions and Boundary components to Defpoints layer; assign primary colors like Magenta and Blue to flag this.
Bldg Section	Sub-Divisions	Set up similar to 2D Section/Elevation.
Door & Window	All Plans	Assign a fine Lineweight to the Swing and Glass components.
Opening	All	Assign components to Defpoints layer.
Schedule Table	General	Turn on the "Out-of-Date" component and make it Color 1 (Red).
Wall	All Plans	Assign layers to each sub-component and set Color, Linetype and Lineweight to ByLayer.

Figure 5–44 shows a sub-set of the available objects and each of their Display Reps. The more familiar you become with each Display Rep and its sub-components, the easier this task will be. The drawing file used to generate Figure 5–44 is in a folder named *Laboratory* in the *Chapter05* folder. You can open this file and experiment with the settings before making them part of your template.

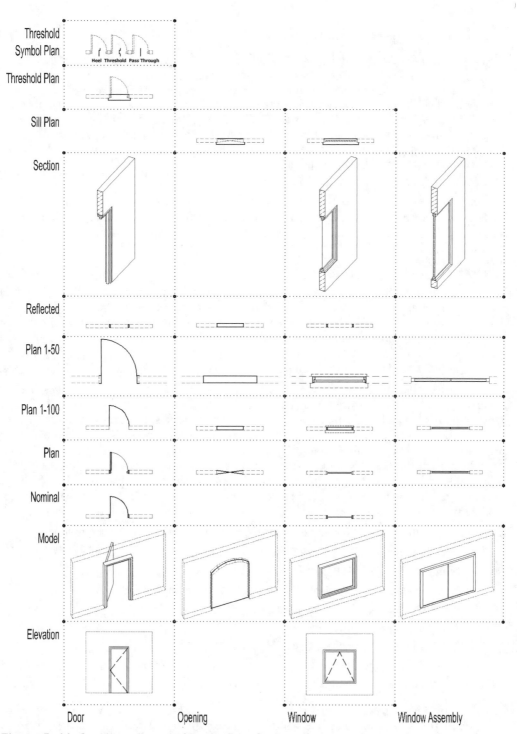

Figure 5–44 *Complete collection of Display Reps for several common objects*

> **CAD MANAGERS NOTE:**
>
> Remember, as you work with all the ADT objects, to make sure a comprehensive graphic standard has been thoroughly discussed and implemented prior to going through this exercise for your office templates. Your firm may not have thought of every method described here and created a consistent standard for all project types. Start with the basics, like doors, walls, openings and windows. Work your way through each potential display set for each sheet file on just those objects. Then move to other common objects like columns, stairs and grids. Before you know it, you have a complete Display Rep set standard, template standard and graphic standard.

A completed version of *Plan-Sect-Elev.dwt* is included in the *Chapter05* folder along with an Excel spreadsheet detailing the System Default display settings of every ADT object and sub-component. Be sure to open these files and explore the settings within them as you build your own template file.

REVISITING LINEWEIGHTS

Just when you thought you were making some progress on your implementation project, a curve ball comes at you. By now, you are starting to see how extensive the reach of the Display System can be. It is possible to control every aspect of the drawing's display, from the type of drawing we want to show, all the way to the minute details of how we would like the individual sub-components of the objects to behave. In previous chapters, we discussed the details of choosing a color scheme, deciding which type of Plot Style Table to use, layering and lineweights.

You are perhaps beginning to appreciate how difficult it is to discuss any one of these tightly integrated topics separately as we have. Therefore, a step back to revisit previously discussed issues is unfortunately inevitable. As previously stated, several graphical characteristics must be controlled in our final plots. Of all of them, lineweight is perhaps the most important. Ultimately the large majority of architectural drawings are printed in black and white. Therefore, apart from the symbology used, the lineweight is the most distinct characteristic we have for producing the desired effect and "read" in our printed drawings. This is not to say that the other graphical characteristics like linetype, screening, join style and end style are not important, just that lineweight is primary. lineweight also requires the most amount of effort and time to set up properly.

If you have been following the chapters in order, you currently have done the following:

- Chosen a type of Plot Style Table: Named or Color.
- Built at least one STB or CTB.
- Decided where lineweights will be assigned: layers or Plot Style Tables.
- Chosen a set of standard lineweights.
- Developed a color scheme.
- Created an office default Layer Key Style.
- Assigned colors, linetypes and lineweights to your standard Layer Keys.
- Added non-keyed layers to your *Plan-Sect-Elev.dwt* template file.

It is an impressive list. The good news is that most of these resources will not need to be modified at this time. Naturally, we will be updating the template file to include our new Display Control settings, and we will want to consider the way in which our lineweights are being assigned again. Consider the following situation: You have a floor plan at 1/8"=1'-0" [1:100] and need a portion of that floor plan as an enlarged plan at 1/4"=1'-0" [1:50]. The enlarged plan can be generated by viewports or XREFs with an XCLIP. In either case, there will potentially be an issue with lineweight scale when the drawing is plotted at the larger scale. This is because lineweights assigned to the layers, either directly or through the plot style, can only plot well for a single scale.

Consider a very common problem in architectural graphics: the desire to show door swings in a lighter lineweight than the door panel. To achieve this, and work in the traditional ByLayer fashion, requires the addition of one or more layers to the drawing. Figure 5–45 illustrates this situation and the two most common ways to achieve it in ADT. For the purpose of this example, the default Door layer is assigned a 0.35mm lineweight. It is desired to have the swing be lighter than this and the door panel slightly heavier. This means that the frame component will be the only component using the lineweight assigned directly to the Door layer.

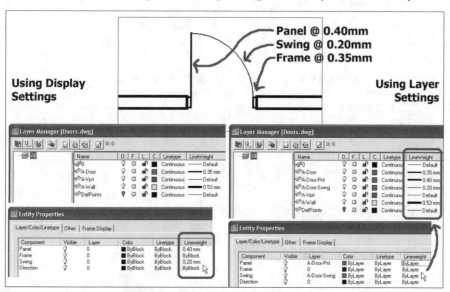

Figure 5–45 *Assigning lineweights to Door sub-components*

If we adhere exclusively to "ByLayer" thinking, we must create two new layers with the desired lineweights and then assign each Door sub-component to the appropriate layer. We must then be certain that these sub-components are assigned ByLayer for their lineweight. This approach is illustrated on the right of Figure 5–45. If we use Display Properties instead, we do not need to create additional layers, rather we simply change the lineweight assigned to each sub-component to desired value. This means that the Door's sub-components will no longer be ByLayer; however, since the Display System will manage when these properties are applied to the object, this will not present a problem. In addition, the Door itself is still ByLayer, we have only editing select sub-components.

This brings us back to the issue of lineweights and scale. Using this same technique, we can easily assign different lineweights to each scale Display Rep without the need to modify Layer Keys, layers or XREFs. Simply switching from a large-scale configuration to a small-scale one will swap not only the graphics used, but the also lineweights assigned to those graphics.

THE RENDER DISPLAY CONFIGURATION

Rendered images can be produced from your 3D ADT Model directly in ADT using the native AutoCAD rendering engine, or through a link to Autodesk VIZ. The assignment of colors to the sub-components of your Model Display Rep to facilitate this process has already been discussed. If you use the native AutoCAD rendering engine built into ADT, you will need to include the Render Display Configuration from the Imperial template file in your own template files. If you were to analyze the Render configuration in the same manner as we have throughout this chapter, you would not notice anything unique about its configuration. It is a Fixed View configuration, which references a set, also named Render. The Render set is virtually identical to the Work_Model set. It has nearly all objects turned on in their respective Model Display Reps. However, there is one significant difference between the Render configuration and any other 3D model configuration. Render is the only configuration that will allow you to use the Mapping coordinates command to adjust the size and application of bitmapped materials. This setting is not exposed in the Display Manager—it is hard-coded into the software. Therefore, if you wish to use native AutoCAD/ADT rendering, be certain to use the Render Display Configuration in those drawings (see Figure 5–46). If you do not like the name "Render," it can be renamed. To test this functionality, do the following:

1. Be sure to set your **Current 3D Graphics Display** settings on the System tab in the Options dialog box as recommended in Chapter 1. (See Figures and 1–33 and 1–34). This will display textures as you work.

2. In the *Chapter05* folder **Open** the file named *Render Test.dwg* to experiment. This simple file has bitmapped materials already assigned by Color to the objects. The brick applied to the Wall and the wood on the Door need to be scaled.

3. Run the Material Mapping command (SETUV at the command line) and select the Wall. The command will fail.

4. Change the current configuration to Render and repeat SETUV. The command will now work correctly.

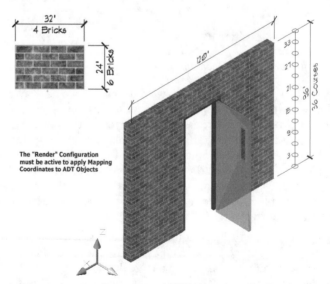

Figure 5–46 *To apply mapping coordinates to an ADT object, use the Render Display Configuration*

THE DISPLAY SYSTEM CHECKLIST

1. Start a new template from scratch and import only those configurations that you intend to use regularly from the "in-the-box" templates.
2. Determine which custom Display Reps to use.
3. For XREFs, decide what approach to take.
   a. Add custom Display Reps for each condition (1st Floor, 2nd Floor, and so on) to the XREF file.
   b. Use custom Display Reps in the host drawing to differentiate Spanning Elements.
4. Customize template Display Reps. Systematically configure all Color, Layer, Linetype, Lineweight and Plot Style settings for each sub-component of each Display Rep of each object.
5. If you intend to use AutoCAD Rendering, import the Render Display Configuration.

A sample template file has been provided on the CD that includes setting for all objects and Display Reps. The complete list of the settings used for this template is included in an Excel file named *Template Settings.xls* in the *Chapter05* folder.

SUMMARY

The Display System is the most powerful tool in your ADT arsenal.

Begin to understand the Display System from the "top down." Start with configurations and work your way down to Display Reps.

The ADT defaults provided in the "in-the-box" templates give great insight into the potential of the Display System.

The Display Manager is a hierarchical, Windows Explorer–style interface to managing the Display System.

Be careful not to get carried away in building custom Display Reps. Managing them can become tedious and confusing.

The display settings of a Display Rep within an XREF cannot be overridden by the host file.

In situations where custom display blocks, nested elements and style level overrides are not present, a custom Display Rep can be used to control visibility of Spanning Elements.

Adding Display Control settings to your template is a time consuming but worthwhile process.

Applying settings like Lineweight and Color directly within the Entity Display Props can often be justified over creating additional layers.

The Render Display Configuration has special "hard-coded" features that make it the only Display Configuration capable of assigning Render Mapping Coordinates.

Content

This section is devoted to the subject of AEC Content. It is divided into three separate chapters, reflecting the variety available in AEC Content. The section begins with an exploration of advanced Schedule Table topics, including information on customizing Property Set Definitions. The next two chapters focus on a variety of possibilities available with AEC Content. There is a large focus on Multi-View Block customization and the AEC Create Content Wizard.

Section III is organized as follows:

> ### WHO SHOULD READ THIS SECTION?
> This section is directed toward the **CAD Software User**, in particular advanced users of ADT, the **CAD Manager** and the **Project Data Coordinator**.

CHAPTER 6

Schedule Tables and Property Set Definitions

INTRODUCTION

Property Set Definitions are the core of the Architectural Desktop Schedule Table functionality. Like the Display System, they can be challenging to master and implement. However, also like the Display System, if they are properly configured initially, you and your users will rarely need to interact with Property Sets at all. This chapter assumes that you already know how to use Schedule Tables in ADT. (See Chapter 11 in *Mastering Autodesk Architectural Desktop* for a complete look at Schedules.) Except for a brief overview of the workflow, the focus here will be entirely on Property Set Definitions, Schedule Data Formats and Schedule Tags.

OBJECTIVES

The goal of this chapter is to equip you for the task of configuring Property Sets and Schedule Tables for your firm. Using Property Sets and Schedule Data can assist you to access and report on data that is already contained in your model. This can help you in meeting certain project-specific demands or, in some cases, reduce the amount of manual coordination and re-drafting required in a traditional CAD process. Using Schedules requires no need to draw in 3D and can automate tasks that were previously more labor intensive. In this chapter, we will explore the following:

- Understand Schedule tool set workflow
- Understand what properties are available automatically
- Understand Property Sets
- Understand Schedule Data Formats
- Work with Schedule Tags
- Customize your template file to include appropriate Schedule tools.

UNDERSTANDING SCHEDULE TABLE FLOW

A Schedule Table is a report of the data/properties contained within a collection of objects. Users create objects within their drawings and then add Schedule Tables to report on them. In Figure 6–1, you can see that this requires a process of object selection, a link to Property Sets and interaction with the Schedule Table Style. (Like most objects in ADT, Schedule Tables receive their formatting and graphical properties from a Schedule Table Style.) You can also see from Figure 6–1 that omitting the Property Set does not prevent a Schedule Table from being created. However, nothing more than a row of question marks will appear in the Schedule if the Property Set linkage is omitted.

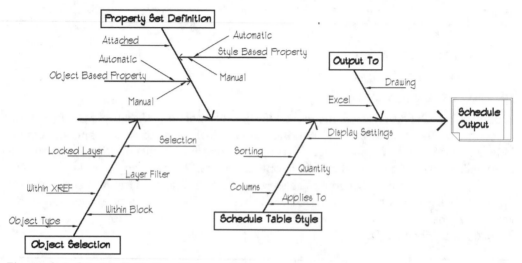

Figure 6–1 *Schedule Table Flow Diagram*

The illustration depicted in Figure 6–1 portrays the scheduling process in its simplest form. Whether you are the CAD Manager in charge of hundreds of seats of ADT or a sole practitioner interested in maximizing your own performance, your goal is to achieve this level of simplicity in your use of the Schedule Table tool set. In most projects, you likely have the same core set of Schedules that you always include in your document sets. One of our primary goals in this chapter is to build all of the underlying resources required to generate this core set of Schedules reliably every time. As you can see from Figure 6–2, there are four major categories of contributing factors: Object Selection, Property Set Definitions, Schedule Table Style and Output destination.

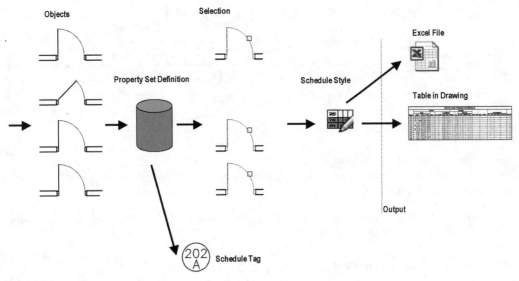

Figure 6–2 *Factors contributing to Schedule Table output*

When reading this diagram, begin from the left and work to the right for the major categories. On this basis, you can see that Property Sets and Object Selection comprise the first consideration in determining what will be included in the Schedule Table. Under Object Selection, there are such considerations as type of object and whether or not objects are contained in XREFs. The Property Set Definition is also a major category, because it determines which parameters and properties of a selected object will be exposed to the Schedule. Next comes the Schedule Table Style. In addition to determining the graphical display properties of the table itself, the style determines which columns of data will be presented and how they should be formatted and sorted. The final consideration is the type of output desired. ADT can create Schedule Tables directly on the drawing or output to an Excel spreadsheet.

The sum total of all of these factors can make configuration of the Schedule Table tool set seem a bit daunting. Therefore, we will take a systematic approach to the process, by beginning with the desired final output and working our way backwards to the pieces required to attain it.

PLANNING

To make Schedule resources for your firm, begin with the Schedules you already use today or refer to the NCS for standard schedules. If they already exist in AutoCAD, simply open a drawing with that Schedule and measure each of its components. For example, Figure 6–3 shows a sample commercial architectural firm Door Schedule. If you have only a printed version of the Schedule you wish to create, you can measure it manually directly on the printout.

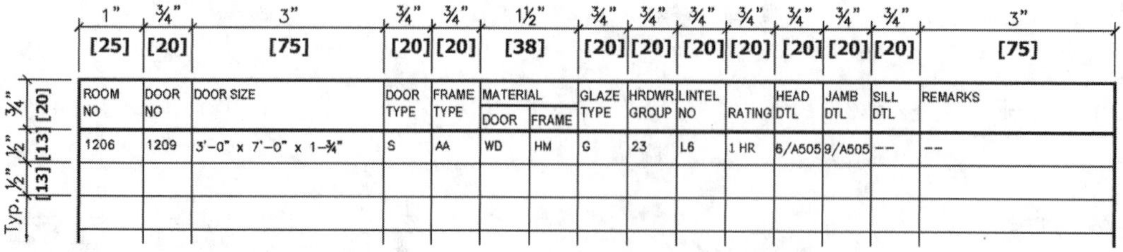

1" ¾" 3" ¾" ¾" 1½" ¾" ¾" ¾" ¾" ¾" ¾" ¾" 3"
[25] [20] [75] [20] [20] [38] [20] [20] [20] [20] [20] [20] [20] [75]
¾" [20]
½" [13]
Typ. ½" [13]
ROOM NO
DOOR NO
DOOR SIZE
DOOR TYPE
FRAME TYPE
MATERIAL
DOOR
FRAME
GLAZE TYPE
HRDWR GROUP
LINTEL NO
RATING
HEAD DTL
JAMB DTL
SILL DTL
REMARKS
1206
1209
3'-0" x 7'-0" x 1-¾"
S
AA
WD
HM
G
23
L6
1 HR
6/A505
9/A505
--
--

Figure 6–3 *Sample Door Schedule with components dimensioned*

This diagram gives us two important pieces of information for our Schedule. First, we have a list of all of the columns needed, and second, we know how large each column ought to be. We can also determine text heights and formatting from this exercise as well.

Note: Throughout the coming tutorial, feel free to work through this example or work on an actual Schedule of your own.

EXPLORING THE SAMPLE SCHEDULE CONTENT

There are several styles included with ADT in the box. Some of these samples are included in the template files provided (see Figure 6–4). However, the best place to look for the styles provided is the Content Library files. There are three such files included with the software. Look for each of these in the *Content* folder of your Management Install.

Schedule Tables (Imperial).dwg is located in *Content\Imperial\Schedules.*

Schedule Tables (Metric).dwg is located in *Content\Metric\Schedules.*

Schedule Tables Styles.dwg is located in the *Content\Metric D A CH\Styles* folder.

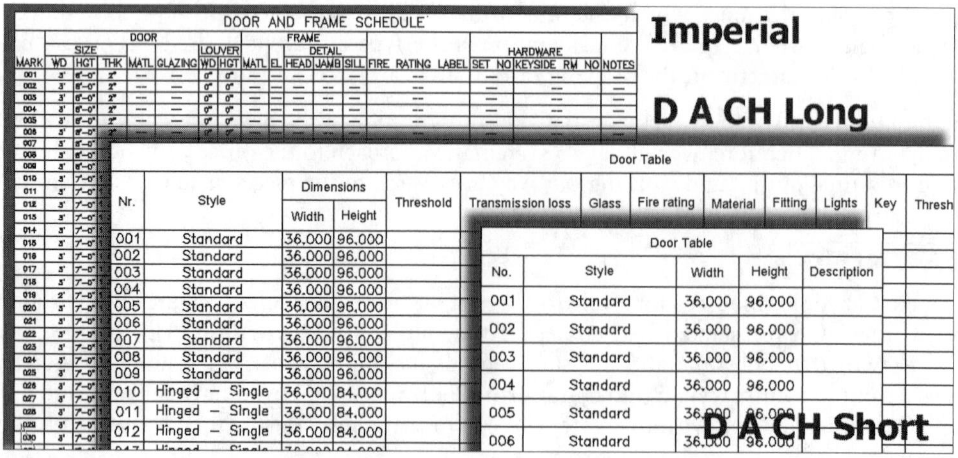

Imperial
D A CH Long
D A CH Short
DOOR AND FRAME SCHEDULE
Door Table
Door Table

Figure 6–4 *Various Door Schedule samples provided in the AEC templates*

You should take some time to explore the styles contained in each of these files before you begin customizing or building your own. You may find that one of these table styles meets your needs as is or simply requires some minor modifications. It may be easier to test these styles out within an actual drawing. A file has been provided on the CD to assist with this. It is called *SimpleModel.dwg* and is located in the *Chapter06* folder (see Figure 6–5). If you have not yet installed the files for Chapter 6, perform the following steps:

Install the CD Files

A folder named *Chapter06* was created containing a collection of sample files that we will use throughout this chapter.

1. Install the files for Chapter 6 located on the Autodesk Architectural Desktop: Advanced Implementation Guide CD ROM.

Refer to the "How to Use the CD" section in the Preface for instructions on installing the sample files included on the CD.

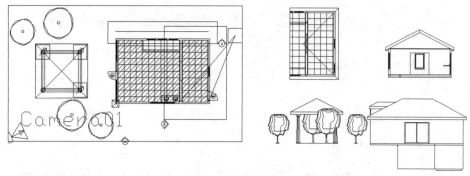

Figure 6–5 *A simple model makes understanding sample Schedule Tables easier*

2. **Open** one of the Schedule Table files noted above and **Insert** the *SimpleModel.dwg* drawing into it.

Be sure to choose **Explode** in the Insert dialog box so that the objects within the file are inserted rather than a Block containing them. Using the Schedule Table Styles within these Content Library files, generate Schedules from the sample objects inserted. In this way, you will be able to study the configurations of the sample styles first hand.

3. **Save** this drawing as *SimpleModel01.dwg* in the *Chapter06* folder.

PROPERTY SET DEFINITIONS

Now that we have the basic dimensions of our table and a list of the columns needed, we need to create a Property Set Definition that includes each of these columns. One of the reasons that people find Property Sets so difficult to understand is that they are often approached from the point of view of the objects to which they are attached. This can often lead to having more Property Sets than are necessarily required. If instead you approach Property Sets the way we are doing here, from the point of view of the columns required in the Schedule Table, they may be easier to grasp. In addition, you will end up with fewer Property Sets, with the purpose of each more readily discernable.

Each column of the Schedule must be associated with a single property from one or more Property Set Definitions. The schedule pictured in Figure 6–3 contains fifteen columns. Therefore, it needs to reference fifteen separate properties. If possible, you should try to include all fifteen of these properties in a single Property Set Definition. If a single Property Set Definition is not possible, then limit it to two. If your schedule will reference both Style-based and Object-based properties, you will require two Property Set Definitions. If all of the properties are of one type or another (all Style or all Object), you will be able to have a single Property Set Definition.

▶ **Style-based Properties** – Properties of this type are attached to ADT object styles (see Figure 6–6). They become a part of the style and thus apply to all objects belonging to the style. As with any style-based parameter, a change affects all objects belonging to that style.

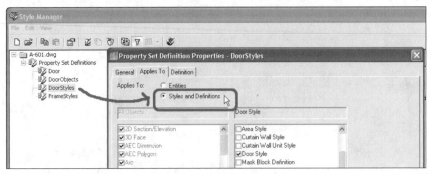

Figure 6–6 *Assigning a Property Set Definition to be Style-based*

▶ **Object-based Properties** – Properties of this type are attached directly to the individual object (see Figure 6–7). Each object therefore contains its own value, and changing one does not affect any other object in the drawing.

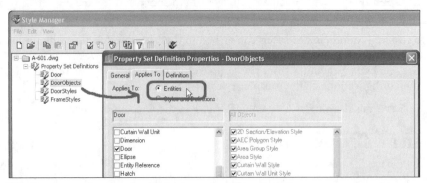

Figure 6–7 *Assigning a Property Set Definition to be Object-based*

We now need to evaluate the sketch shown in Figure 6–3 and determine whether each column should be Style-based or Object-based. The best way to do this is to equate "styles" in ADT with "types" in your architectural document sets. This correlation can be made directly in this example, since Doors are always expressed as "types" in document sets. Simply return to the same set of documents from which you attained the Schedule, and locate the "Door Types" drawing (see Figure 6–8).

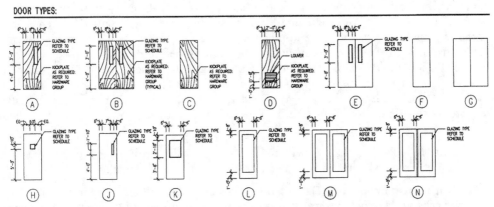

Figure 6–8 *Whatever characteristics separate type "A" from type "B" in the project should also distinguish style "A" from style "B" in ADT drawings*

Using this and the sketch of your Schedule, determine which columns will be Style-based and which will be Object-based. For instance, it is obvious that Door Number will be Object-based, since all Doors must have their own unique Door Number. However, Door Type will be Style-based. After all, we did equate type and style a moment ago. Analyze each column of your Schedule and work up a list like Table 6–1.

Table 6–1 *Object-based and Style-based Properties*

Property (Schedule Column)	Object/Style-based
Room Number	Object-based
Door Number	Object-based
Door Size	Style-based
Door Type	Style-based
Frame Type	Object-based
Door Material	Style-based
Frame Material	Object-based
Glazing Type	Style-based
Hardware Group	Object-based
Lintel Number	Object-based
Fire Rating	Style-based
Head Detail	Object-based
Jamb Detail	Object-based
Sill Detail	Object-based
Remarks	Object-based

You may arrive at different conclusions. This is okay, since virtually every property would work equally well as either a Style-based or an Object-based property. The essential difference lies in how they are applied to the objects and how they are edited. Object-based properties must be attached manually to all of the objects within the drawing before their data will appear in the Schedule. To edit an Object-based property, you must use the **Edit Schedule Data** command. Style-based properties are added and edited within the Style Properties dialog box, on the General tab. Once added to the style, they will apply to all objects that reference that style. Therefore, fewer "undefined" cells (the ones with question marks) will appear in a Schedule that references Style-based properties.

The final bit of preparation that we need to do before actually creating our Property Set Definition is to determine whether each property is "Automatic" or "Manual."

▶ **Automatic Property** – Directly references one of the physical parameters of the object and therefore automatically reflects the current value of that parameter at all times (see Figure 6–9).

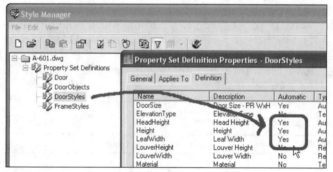

Figure 6–9 *Automatic Properties will have the word "Yes" in the Automatic column*

 Note: In the Edit Schedule Data dialog boxes, Automatic properties are indicated by a small yellow lightning bolt icon.

▶ **Manual Property** – Not linked directly to any object parameter. A Manual property is simply a piece of data attached to an object for purposes of appearing within the Schedule, a Tag or both. It can be freely edited to include any type of value (see Figure 6–10).

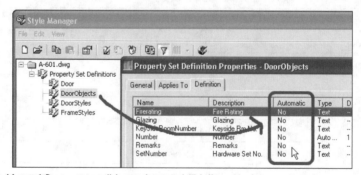

Figure 6–10 *Manual Properties will have the word "No" in the Automatic column*

There are dozens of Automatic properties available to Property Set Definitions. Including both AutoCAD entities and ADT objects, the complete list is quite extensive. Several lists in table form have been provided in an Excel file in the *Chapter06* folder. Open *Automatic Schedule Properties.xls* in Microsoft Excel to view Tables XL6–1 through XL6–5.

Table XL6–1 – Objects with Automatic Properties

Table XL6–2 – Automatic Style-based and Object-based Properties (All Objects - List)

Table XL6–3 – Object-based Automatic Properties – (AutoCAD Objects - Matrix)

Table XL6–4 – Object-based Automatic Properties – (ADT Objects - Matrix)

Table XL6–5 – Object-based Automatic Properties – (Railings and Stairs - Matrix)

For convenience in this exercise, a portion of Table XL6–2 has been reproduced here as Table 6–2 and annotated with descriptions. This table will help us determine which of our fifteen columns can come from Automatic properties. Unlike Style-based versus Object-based, a single Property Set can contain both Automatic and Manual properties. The same list of Automatic properties is available to both Style-based and Object-based Property Sets.

Table 6–2 *Automatic Properties for Door Objects*

Property	Description
Color	Color Number of Door Object (0 = ByBlock, 256 = ByLayer)
Color - Text	Name of the Color for Red through White, ByLayer and ByBlock
Description	Description in the Door Properties dialog
Documents	Name and path of attached Documents (Object level)
Door Size - 2@ WxH	Double Doors shown as a pair combining Width and Height
Door Size - 2@ WxHxT	Double Doors shown as a pair combining Width, Height & Thickness
Door Size - WxH	Combines overall Width and Height in a single Cell
Door Size - WxHxT	Combines overall Width, Height & Thickness in a single Cell
Frame Depth	Frame Depth from Dimensions tab
Frame Width	Frame Width from Dimensions tab
Handle	AutoCAD Object Handle
Head Height	Actual location of the Door Head relative to zero elevation
Height	The Height of the Door itself independent from its location
Hyperlink	URL attached to Object
Layer	Layer of Door Object
Leaf Width	Size of Leaf for Uneven Door Types (as set on the "Design Rules" tab)
Linetype	Linetype of the Door Object (Usually ByLayer)
Notes	Notes in the Door Properties dialog

Table 6–2 *Automatic Properties for Door Objects (continued)*

Property	Description
Rise	Rise from spring of arch to top of Door (for arched Door shapes in Design Rules)
Rough Height	Height of Door Opening in the Wall, includes the Door Height plus Frame size
Rough Width	Width of Door Opening in the Wall, includes the Door Height plus Frame size
Sill Height	Location of the bottom edge of the Door in the Wall
Style	Name of the Door Object's Style
Swing Direction	Swing Direction of Door Object expressed as Right or Left
Swing Direction Reverse	Reverse Swing Direction of Door Object expressed as Right or Left
Thickness	Thickness of Door Leaf
Width	Width of Door Leaf
Width - 2@	Width of Door Leaf for Double Doors expressed as 2 x Leaf Size

If we compile all of the information gleaned from the sketches and tables, we arrive at the following conclusions:

We will need to build two Property Set Definitions, one Style-based containing Door Size, Door Type, Door Material, Glazing Type and Fire Rating, the other Object-based containing the properties for the remaining columns.

The only Automatic property that we need is the Door Size column. For this we will use the **Door Size - 2@ WxHxT** Automatic property format.

PROPERTY SET DEFINITION PREPARATION CHECKLIST

The following summarizes the previous procedure:

1. Procure a sketch of the Schedule you wish to create.
2. Assign dimensions to all major columns and components.
3. Determine which columns ought to be Style-based and which should be Object-based.
4. For each column, decide which properties is Automatic and which is Manual.
5. Determine the Schedule Data Formats and other special formatting required.
6. Using this compiled information, build the Property Set Definition.

Included in the *Chapter06* folder with the files installed from the CD is an Excel file named *Property Set Def Work Sheet.xls*. You can use this to help you plan any new Property Set Definitions you need to create (see Figure 6–11). There are three sections in the form: the top is general information regarding text formatting, the next is for Style-based properties and the last is for Object-based properties. In each of the Property Set sections are five columns: Property, Automatic or Manual, Schedule Data Format, Column Size and Misc. Following the same process as we have here, list each property (Schedule column) that you will need in the first column. Next to it, determine if it is Automatic, and if so, input the Automatic property that it will reference. In the third column, list the way you would like the data formatted in that column of the Schedule using a variety of pre-defined Schedule Data Formats. In the next column, input the width of each column in the Schedule. If you wish to use variable-width columns, input a zero here. The final column is used to note any special needs or features for that entry.

	A	B	C	D	E
	Schedule Table and Property Set Definition Worksheet				
1	**Autodesk Architectural Desktop ~ An Advanced Implementation Guide** by Paul F. Aubin				
2	Schedule Table Name: Door Schedule				
3					
4	**Text Formating**				
5	Style: Schedule Text			Alignment: Left	
6	Height: 3/32"			Rotaton: Horizontal	
7	Gap: 1/16				
8					
9	**Property Set Definitions:**				
10	**Style-based Properties**		Name: DoorStyles		
11					
12	**Property (Schedule Column)**	**Automatic or Manual**	**Shedule Data Format**	**Column Size**	**Misc.**
13	DoorSize	2@ WxHxT	Length - Long	3"	
14	DoorMaterial	Manual	Text	3/4"	

Figure 6–11 *Use the Property Set Definition Worksheet to help you build new Schedule Tables and Property Set Definitions*

BUILDING PROPERTY SETS

Now that the preparation is finished and we have a completed worksheet, it is time to build the Property Set Definition. For now, we will work in a sample file. Later, we can export the Property Set Definition to the appropriate template or Content Library file.

Create the Object-based Property Set

    1. In the *Chapter06* folder, **Open** the file named *A-601.dwg*.

This file is a sheet file for Schedules. It contains an XREF file of a floor plan and a grid representing the space available on the Title Block. (This can be referred to as usable or "live" area.) Three different tools are required to build Schedule Tables: Property Set Definitions, Schedule Table Styles and Schedule Data Formats. Schedule Data Formats are used to present the data within the Schedule Table in an appropriate format, such as text in all uppercase and dimensions in Feet and Inches. To save effort, the Schedule Data Formats from both the Imperial and D A CH files have been included in this file already.

    2. From the **Documentation** menu, choose **Schedule Data>Property Set Definitions.**

There are no Property Sets currently in this file.

3. Right-click **Property Set Definitions** and choose **New**, (or click the **New Style** icon) to create a new Property Set Definition.

4. Name it **DoorObjects.**

PROPERTY SET DEFINITION NAMING CONVENTIONS

The standard naming convention for Property Set Definitions in the Imperial file is *<ObjectType><Association>* where "*ObjectType*" is the name of the drawing object, such as Door or Polyline. The "*Association*" is either Styles or Objects, indicating whether the Property Set references Styles and Definitions (Style-based) or Entities (Object-based). The naming convention used in the D A CH file is a bit more generic, referencing object type only for Object-based and using the word "Styles" as a prefix for the Style-based. When the Imperial file naming convention is used, both the Object-based and Style-based Property Sets for an object type will sort together in a long list. This makes them easier to work with than the D A CH names that do not sort together. The Imperial file naming convention is logical and easy to understand. It is recommended that you adopt it.

It is *critically* important that spaces not be included in Property Set Definition names. The same is true for the names of the properties themselves. This is because Schedule Tags use AutoCAD Attributes to reference the Property Set Data. AutoCAD Attributes do not support spaces in their Tag names; therefore, if you include spaces in the name of a Property Set Definition, you will be unable to link that Property Set or its properties to the text within Schedule Tags. For instance, "**DoorObjects**" is a valid name for use in Schedule Tags, "**Door Objects**" is not.

---

5. Double-click **DoorObjects** (or right-click it and choose **Edit**).

6. Click the Applies To tab and be sure that **Entities** is chosen at the top (at "1" in Figure 6–12).

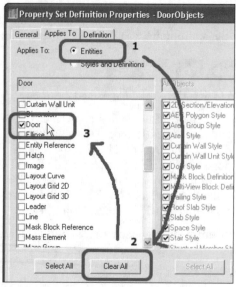

Figure 6–12 *Set up this Property Set to apply to Door entities*

7. Click the **Clear All** button (at "2" in Figure 6-12) and then place a check mark in the **Door** check box (at "3" in Figure 6-12).

8. Click the Definition tab, and at the lower right corner, click the **Add** button (see Figure 6–13).

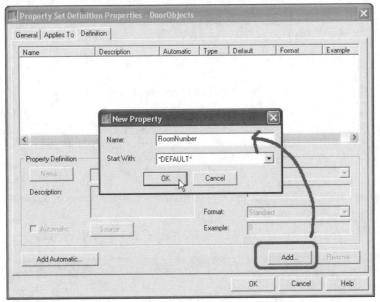

Figure 6–13 *Add the first property to the DoorObjects Property Set*

9. Name the new Property **RoomNumber** and then click **OK**.

 Note: Refer to the "Property Set Definition Naming Conventions" sidebar for tips on this naming convention.

High on the wish list for future enhancements to ADT is the ability to link Schedule Data from two or more objects to the same row in the Schedule. This is not currently possible in this release, so the Room **Name** field here must be a manual text field added to the DoorObjects Property Set.

10. From the **Type** list, choose **Text**.

11. From the **Format** list, choose **Case – Upper**.

These two settings will make this a text-based field that is automatically formatted in uppercase letters. Even though these are room "*numbers,*" choosing one of the numeric formats will make it impossible to include letters in this field. Therefore, room numbers like "1001A" would not be possible.

 Tip: You should always assign a default value to all of your Manual properties. However, since you don't want to assign a default value that might be mistaken as an actual value, it is common to input a double dash (--) as the default. Having a default value makes it easy to see when Property Sets have been added to objects and it makes it possible to use the **Edit Table Cell** command.

12. Type **--** in the **Default** field.

13. Click the **Add** button again and type **DoorNumber** for the **Name** (see Figure 6–14).

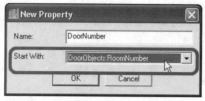

Figure 6–14 *Use the Start With list to copy an existing property*

14. From the **Start With** list, choose **DoorObjects:RoomName**.

This will make a copy of the first property as the starting point for the new one. Unfortunately, it copies everything about the original, including the description. Therefore, you will want to edit the description to match the new name. The description can be any text you want. It can be very helpful to explain the purpose of the property when the property name is abbreviated in some way.

15. Continue to **Add** each of the Object-based properties listed in Table 6–1 (or use your worksheet if you filled one out).

When you are finished, double-check your work before clicking OK (see Figure 6–15).

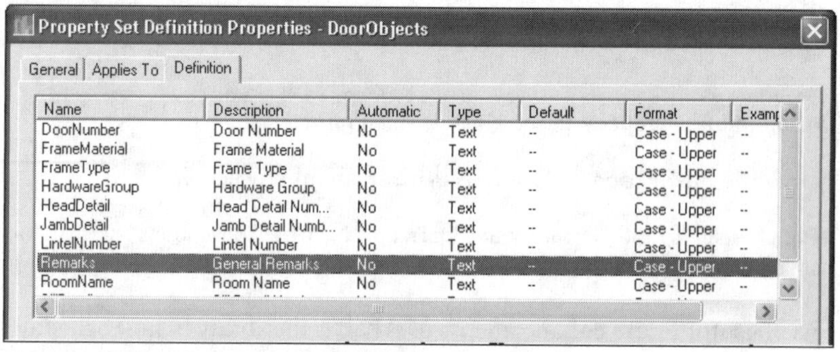

Name	Description	Automatic	Type	Default	Format	Examp
DoorNumber	Door Number	No	Text	--	Case - Upper	--
FrameMaterial	Frame Material	No	Text	--	Case - Upper	--
FrameType	Frame Type	No	Text	--	Case - Upper	--
HardwareGroup	Hardware Group	No	Text	--	Case - Upper	--
HeadDetail	Head Detail Num...	No	Text	--	Case - Upper	--
JambDetail	Jamb Detail Numb...	No	Text	--	Case - Upper	--
LintelNumber	Lintel Number	No	Text	--	Case - Upper	--
Remarks	General Remarks	No	Text	--	Case - Upper	--
RoomName	Room Name	No	Text	--	Case - Upper	--

Figure 6–15 *Your completed DoorObjects Property Set will look something like this*

16. Click **OK** when satisfied.

Build the Style-based Property Set

Earlier in this chapter, we suggested that it might be easier to begin with an existing Property Set Definition and edit it rather than build it from scratch. Let's try that approach for our Style-based Property Set. To do this, we must first import a Property Set to use as a starting point.

1. In the Style Manager, click the **Open Drawing** icon.

2. Navigate to the *Content\Imperial\Schedules* folder of your Management Install, select the file named *Schedule Tables (Imperial).dwg* and then click **Open**.

We are using the Imperial version because it is closer to the Property Set Definition we are trying to build than the D A CH file. However, this process would work equally well on the D A CH (or any) file.

3. Right-click the **DoorStyles** Property Set Definition and choose **Copy**.

4. Also in the Style Manager, right-click the *A-601.dwg* folder and choose **Paste**.

5. Right-click the *Schedule Tables (Imperial).dwg* folder and choose **Close**.

If asked to Save, click **No**.

6. Double-click the new copy to edit it.

There are many more properties in this Property Set than the version that we are creating will require.

7. Select the **HeadHeight** property and click the **Remove** button (see Figure 6–16).

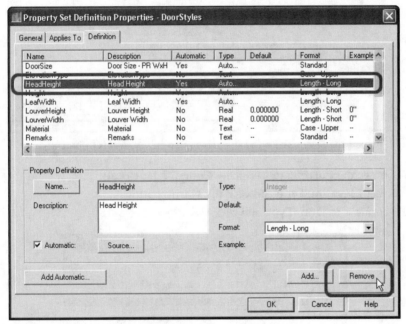

Figure 6–16 *Remove the HeadHeight property*

 Caution: There will be no confirmation warning, so be certain you select the correct one.

8. Following the same process, remove all but **DoorSize**, **ElevationType** and **Material**.

9. Select **DoorSize**.

Notice that it is currently using the Door Size - PR WxHxT Automatic property.

10. Click the **Source** button.

11. Check the check box next to **Door Size - 2 @ WxHxT** and then click **OK** (see Figure 6–17).

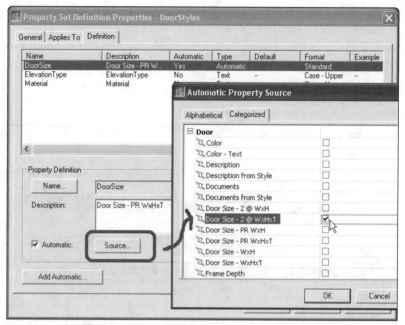

Figure 6–17 *Remap the DoorSize field to the "Door Size - 2 @ WxHxT" property*

12. Change the **Format** to **Length – Long** ["**3 Decimals.**"]
13. Select **ElevationType** and click the **Name** button.
14. Rename it to **DoorType** and click **OK** (see Figure 6–18).

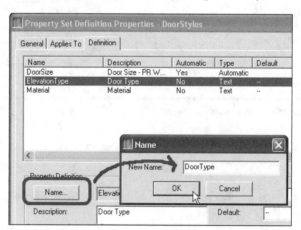

Figure 6–18 *Rename the ElevationType property with the Name button*

15. Repeat the same steps to rename **Material** to **DoorMaterial**.

16. Following the process in the previous step sequence, **Add** the remaining two properties: **GlazingType** and **FireRating**.

17. After double-checking all values, click **OK** to complete the Style-based Property Set.

Build a Custom Schedule Data Format

Perhaps it might be nice to have the Fire Rating column automatically include the letters "HR" after the rating value. We can accomplish this by building a custom Schedule Data Format.

1. While still in the Style Manager, click the **Filter Style Type** icon (revealing all Style types).

2. Expand **Schedule Data Formats**, right-click **Case – Upper** and choose **Copy**.

3. Right-click again and choose **Paste**. Rename the copy **Fire Rating**.

Note: Spaces are allowed in Schedule Data Formats.

4. Double-click **Fire Rating** to edit it and then click the Formatting tab.

Notice that in the **Text** area, this Schedule Data Format is set to **Upper**.

5. In the Suffix field, type " **HR**" (being sure to include a leading space) as shown in Figure 6–19.

Figure 6–19 *Add a suffix to the data in the Fire Rating column*

6. Click **OK** to exit the dialog box.

7. Return to the Property Set Definition dialog box, and for the **FireRating** Property, choose **Fire Rating** from the Format list.

8. Click **OK** when finished.

Updating the Property Set Definitions Master File

A copy of every Property Set Definition in use in your system is stored in a single master library file. There are actually three of these by default provided with the software, one for each type of units:

PropertySetDefs.dwg is located in *Content\Imperial\Schedules*.

PropertySetDefs.dwg is located in *Content\Metric\Schedules*.

PropertySetDefsDACH.dwg is located in the *Content\Metric D A CH\Styles* folder.

These files are primarily used by the Schedule Tags when they are dragged from the DesignCenter. The topic of Schedule Tags is covered extensively below. If you are satisfied with the work completed thus far on your Property Set Definition, you should import it into the appropriate version of the *PropertySetDefs* file. You can do this simply be opening the correct version of the *PropertySetDefs* file within the Style Manager and copying and pasting the two Property Sets built here into that file. Do this each time you build a new Property Set Definition.

Note: Be sure to copy it to the files on your library server (L Drive) and not just your Management Install.

That completes our Door Schedule Property Set Definitions. Property Set Definitions occur behind the scenes in ADT. They must be available and properly configured for the Schedule tool set to function as expected. If configured properly and completely, Property Set Definitions will simply do their job, and users will rarely be confronted with them head on. Getting all of the Property Sets built that may eventually be required will take some effort. Nevertheless, once they are complete, that investment in time will pay handsome dividends in productivity and potentially in downstream data extraction benefits as well.

SCHEDULE TABLE STYLES

There are two primary purposes to the Schedule Table Style. The first is to establish the contents of the Schedule Table, including its columns, the data those columns contain and how they are sorted. The second purpose of the Schedule Table Style is to establish the graphical formatting of the Schedule Table, including fonts, text sizes, color and lineweights. In this passage, we will focus on the content of the Schedule Table and assume that you are already familiar with how to set up the graphical formatting and display properties.

Import a Sample Schedule Table Style

To save some time, we will import a Door Schedule from the D A CH library file and use it as the basis for the graphical formatting of our table.

1. If you closed the Style Manager, reopen it now.
2. Using the steps covered above, within the Style Manager, **Open** the *Schedule Tables Styles.dwg* file from the *Content\Metric D A CH\Styles* folder of your Management Install.
3. Using the **Filter Style Type** icon, isolate **Schedule Table Styles** and expand its list.
4. **Copy** the **Door (1)** style from the D A CH file and **Paste** it into the current file.
5. Rename the style to **AIG-Door Schedule** and then double-click to edit.

Note: "AIG" stands for Advanced Implementation Guide. Feel free to substitute your firm's initials instead.

6. Add a Description on the General tab, and adjust the text formatting on the Default Format tab (see Figure 6–20).

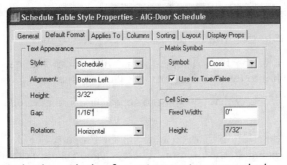

Figure 6–20 *Adjust text heights and other formatting to suit your standards*

 Note: Figure 6–20 shows Imperial units. Feel free to substitute Metric units if you wish.

7. Adjust the formatting of **Headers** and the **Title** on the Layout tab.

In order to use the properties that we built in the previous exercise as columns in this Schedule, the items checked on the Applies To tab of the Schedule Table Style must match those checked on the Applies To tab of the Property Set Definition. This is the case here, but keep this in mind as you build more tables in the future. The easiest way to avoid frustration related to this point is to always check the same objects in both the Property Set Definition and the Schedule Table Style.

8. Click the **Columns** tab.

There are already several columns in this Schedule. Since none of them refers to the custom Property Set Definition that we built, we will delete them all.

9. Click the first column, hold down SHIFT and then click each of the remaining columns (see Figure 6–21).

10. Click the **Delete** button.

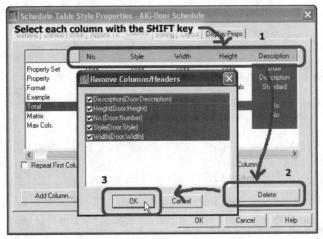

Figure 6–21 *Select all columns and then click Delete*

11. When prompted to verify the deletion, click **OK**.

Add Columns to the Schedule Table Style

1. Click the **Add Column** button (at "1" in Figure 6–22).

In a list at the left, all of the Door Property Sets will appear. We need to add each of the properties that we built into our two Property Set Definitions.

2. Select the property that you want to appear first in the Schedule, **RoomNumber** in this case.

On the right side, the description that you added in the Property Set Definition dialog box will be used here for the Heading of the Column. There is a good chance that this Heading text will be too long to fit in the average Schedule Column heading. Therefore, you may wish to abbreviate the Heading.

3. Change the **Heading** if necessary (at "2" in Figure 6–22).

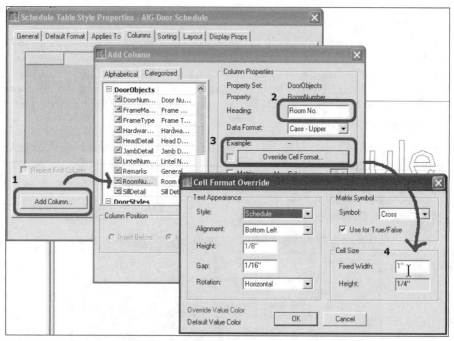

Figure 6–22 *Add a column and set up its formatting*

We now need to use the measurements that we took back in the "Planning" section (refer to Figure 6–3). At that time, we measured each column in the schedule. Here, we can add a fixed dimension to this column with the **Override Cell Format** button.

4. Click the **Override Cell Format** button (at "3" in Figure 6–22).

5. In the **Fixed Width** field, type in the dimension for the first column, in this case **1" [25]** (see "4" in Figure 6–22).

6. Click **OK** twice to add the Column and return to the Schedule Table Style Properties dialog box.

7. Using the dimensions in Figure 6–3 (or your completed Property Set worksheet), **Add** each of the remaining Columns and set their respective Headings and override formatting.

Remember, you will be adding columns from both the Style-based (DoorStyles) and the Object-based (DoorObjects) Property Sets. If you add a column in the wrong order, do not worry; you can simply drag it in the Columns tab to the correct location. When you have finished adding columns, your screen should look something like Figure 6–23.

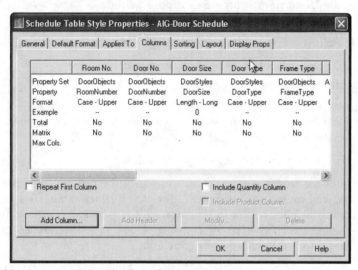

Figure 6–23 *All columns added to the Door Schedule*

You may want to group some of the columns under headers. In this case, the Door and Frame Material columns are meant to be grouped into a single "Material" header.

8. Select the **Door Material** column and, with SHIFT held down, click the **Frame Material** column as well.

9. Click the **Add Header** button, type **Material** and then press ENTER (see Figure 5–24).

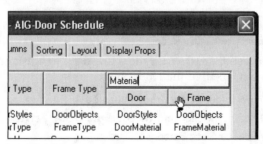

Figure 6–24 *Add a Material header*

10. Click the Sorting tab and then click the **Add** button.

11. Choose the **DoorObjects:DoorNumber** column as the sorting criteria and then click **OK**.

Test the Schedule Table Style

1. Click **OK** to return to the drawing.

2. **Save** the file.

The best way to see if the style is configured satisfactorily is to add a table to the drawing using that style.

3. From the **Documentation** menu, choose **Schedule Tables>Add Schedule Table**.

Choose the style we just created and make sure to check the **Scan XREFs** option.

4. Click **OK** and select the XREF when prompted.

5. Place the Schedule to the side of the plan.

No data (or incorrect data) will appear in the Schedule at this point, since we have not yet added the new Property Set Definitions to the XREF file. However, we are concerned only with the correctness of the Schedule Table Style composition itself.

6. **Zoom** in on the top Heading of the Schedule.

Check each column for accuracy, typos and cases where the text overruns the border of the Schedule Table. When you set the column width to fixed values as we have, you may need to tweak the Header text heights, and/or the Column widths to make everything fit properly. You can also edit the Columns and shorten the Heading names. When you have finished tweaking, your Schedule should look something like Figure 6–25.

Door Schedule				Material									
Room No.	Door No	Door Size	Door Type	Frame Type	Door	Frame	Glazing Type	HrdWR Group	Lintel No.	Fire Rating	Head Dtl	Jamb Dtl	Sill Dtl	Remarks
?	?	?	?	?	?	?	?	?	?	?	?	?	?	?
?	?	?	?	?	?	?	?	?	?	?	?	?	?	?
?	?	?	?	?	?	?	?	?	?	?	?	?	?	?
?	?	?	?	?	?	?	?	?	?	?	?	?	?	?
?	?	?	?	?	?	?	?	?	?	?	?	?	?	?
?	?	?	?	?	?	?	?	?	?	?	?	?	?	?
?	?	?	?	?	?	?	?	?	?	?	?	?	?	?

Figure 6–25 *The result of the completed Door Schedule Style*

Attaching Property Sets to the Model

In order to replace the question marks with the actual data, we must open the plan model file and attach the Property Sets to all of the Door objects.

1. In the *Chapter06* folder, **Open** the file named *A-FP01.dwg*.

2. Select any **Door**, right-click and choose **Edit Door Style**.

3. Click the **Property Sets** button.

Notice that the DoorStyles Property Set is already attached to the style. This is because we used the naming convention that is the default in the ADT Imperial Content. However, the version of the DoorStyles Property Set that exists in this drawing is not the same as the one we just built. Therefore, we must import our new version of the Property Set and overwrite the one resident in this file. The Object-based Property Set will be dealt with later in the chapter.

4. **Cancel** the Door Style Properties dialog box.

5. Open the Style Manager and be sure that the *A-601.dwg* file is showing in the tree list. (If it is not, use the **Open** icon to load it.)

6. Beneath the *A-FP01.dwg* folder in the tree view, right-click **Property Set Definitions** and choose **Purge**.

7. Click **OK** to purge all Property Set Definitions from the Floor Plan XREF.

Three Property Set Definitions will remain: DoorStyles, FrameStyles and WindowStyles. These three are being used by the Object styles in the drawing and cannot be purged.

8. Beneath the *A-601.dwg* folder in the tree view, right-click the **DoorStyles** Property Set Definition and choose **Copy**.

9. Right-click the *A-FP01.dwg* folder and choose **Paste**.

The Import/Export - Duplicate Names Found dialog box will alert you that the DoorStyles Property Set Definition already exists in the floor plan drawing.

10. Choose **Overwrite Existing** and click **OK**.

You have now purged the unwanted Property Sets from the plan file and updated its version of DoorStyles to reflect our changes.

11. Right-click any Door object in the *A-FP01.dwg* file and choose **Edit Door Style**.

12. On the General tab, click the **Property Sets** button (see Figure 6–26).

Notice that DoorStyles now includes only the five properties that we defined earlier.

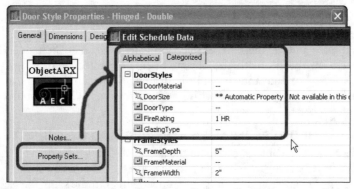

Figure 6–26 *The updated version of the DoorStyles Property Set is already applied to the Door styles in this drawing*

13. **Cancel** the Door Style Properties dialog box.

14. **Save** the *A-FP01.dwg* drawing.

15. Return to the *A-601.dwg* drawing file, **Reload** the XREF and **Update** the Schedule Table. (Right-click the Schedule and then choose **Update**.)

You should now see each of the Style-based properties displaying correctly in the Schedule. Remember, only the five columns that reference the Style-based Property Sets will update. Columns such as "Door Number" will still show question marks.

THE VALUE OF A GOOD NAMING CONVENTION

In the "Property Set Definition Naming Conventions" sidebar earlier in this chapter, the default naming convention was stressed due to its logical and easy-to-understand approach. You now have another reason to stick with the default naming convention. If you change the naming convention, you will have to open every Door style in all of the template and Content Library files and reassign the newly named Property Set to those styles. It is probably not worth the effort in most cases. However, this approach does not completely eliminate the need to open and edit existing Content Library styles. The FrameStyles Property Set is no longer required for the Door Schedule that we have created; however, it is still attached to all of the Content Library styles. If you wish, you could comb through all of the styles, remove this reference from them, and eventually purge the FrameStyles Property Set altogether. In an effort to simplify the content provided, this may be a worthwhile endeavor. Assess carefully the time required for this task against the fact that having the "extra" Property Set will not functionally impede any of the tools. It will however, likely pose a point of curiosity (and possibly confusion) for those users not familiar the history of the problem.

You will encounter similar issues if you use content from both the Imperial and D A CH sources. If you do, consider copying them to an intermediate file first. There you can rename them to the naming convention used by the destination file and then import/overwrite. For example, the Object-based Property Set for Doors in the D A CH file is named Door. In the Imperial file, it is named DoorObjects. So, if you wanted to replace DoorObjects with Door, first copy to an intermediate file and rename **Door** to **DoorObjects**. Then you can import it and have it overwrite the existing, since it will now have the same name.

BUILDING MODEL ISSUES

One of the most important issues you will decide regarding the use of Schedule Tables is in which file the table itself ought to be located. This is because your choice will directly affect workflow of your team and the methods of selection you use to create Schedules. The two basic choices are directly in the file with the objects being scheduled or in a separate file linked through XREFs. There are advantages and disadvantages to each method. In either method, final printing will occur in a sheet file, as with all other drawings. The essential difference lies in where the Schedule object is generated.

GENERATING SCHEDULES IN THE SAME FILE AS THE OBJECTS

If you generate a Schedule Table in the same file as the objects it reports, you can benefit from two-way interaction between the Schedule and the objects. For instance, the **Edit Table Cell** tool works only if the Schedule and the objects reside together in the same file (see Figure 6–27). To print schedules created this way, you must XREF each floor plan model file into the Schedule sheet file and use XCLIP to crop away the floor plan. This will leave you with several Schedule objects that must be arranged manually for presentation on the sheet.

Figure 6–27 *Using Edit Table Cell to directly manipulate data in the Schedule*

Advantages

- Can directly manipulate individual cells in the Schedule Table using the **Edit Table Cell** command.
- Each floor plan will have its own title in the composite Schedule on the sheet file, such as First Floor Door Schedule, Second Floor Door Schedule.
- Can use any AutoCAD object selection method to determine which objects are included within the Schedules.

Disadvantages

- Team workflow can be restricted because one individual becomes responsible for both the Plan and the Schedule.
- Sheet files must be laboriously XCLIPed and arranged to form the final sheet layout.
- If new objects are added to the model, the Schedule Tables must be manually shifted in the sheet file to avoid overlapping rows of data.
- Automatic update severely degrades performance.

Recommendations

- Turn *off* Automatic Update and be sure that the "Out-of-Date" component of the Schedule Table is turned on in Entity Display.
- Generating a Schedule Table in the same file can be appropriate for very small projects and teams.

GENERATING SCHEDULES IN THE COMPOSITE MODEL FILE

If you generate your Schedules through XREF, you gain the flexibility implicit in the separation of data and you maintain consistency with the other Building Model recommendations made throughout this book. The Composite Model file is the ideal location for generating Schedules through XREF (refer to Chapter 2 for more information). Simply add the Schedule directly in model space along with the Composite Building Model XREFs. Be careful to use the Layer Filter mechanism provided in the Add Schedule Table dialog box or on the Settings tab of the Table Properties dialog box (see Figure 6–28). This is the only way you can control which objects contained in the XREF will be included in the Schedule Table.

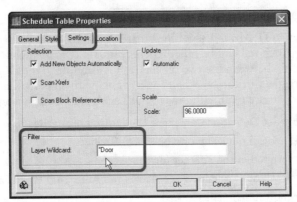

Figure 6–28 *Adding a Layer Filter to limit the selection of objects included in the Schedule*

Advantages

- Team workflow is not restricted, because one individual can work in the plan while another generates Schedule Tables.
- Sheet file setup is simplified, requiring only a one-time setup of viewports.
- If new objects are added to the model, the Schedule Table simply lengthens. As long as the viewport is large enough, no cropping or overlap will occur.
- Automatic update occurs as soon as XREFs load. The schedule is always up to date, without degradation in performance in the model reference files.

Disadvantages

- Cannot directly manipulate individual cells in the Schedule Table using the **Edit Table Cell** command.
- Each floor plan cannot have its own title in the Schedule on the sheet file. There will be a single title for the entire table.
- Schedule becomes a one-way report; no direct editing of cells is possible.
- Can only use layer filtering to limit selection of objects included in the Schedule.

Recommendations

- Turn *on* Automatic Update to fully benefit from automatic XREF loading.
- Be sure, when using Layer Filters, to include a wildcard (*) in front of the layer name.

LAYER FILTERING TIP

Use wildcards to advantage in filtering layers. For instance, using the Layer Filter "A-Door" would select only Doors in the current file that also occurred on the A-Door layer. It would not find any Doors in the XREF file, even though the **Scan XREFs** option is chosen. To capture Doors on the XREF Door Layer, the Filter needs to be "*A-Door" or simply "*Door." By including the wildcard, you are allowing for the XREF name, which is always prefixed to the layer name. See Figure 6–28.

UNDERSTANDING SCHEDULE TAGS

A Schedule Tag is an ADT Multi-View Block that is anchored to another object. The Anchor used is called a "Tag Anchor." Unlike other Anchors in ADT, this Anchor does not determine the physical relationship between the two objects (as for example in the Door to Wall Anchor), but rather it links the data in the object's Property Sets with the Attributes contained in the Tag. Several components are required to make the Tag Anchor relationship function properly:

- A Property Set Definition containing the desired properties
- AutoCAD Block containing one or more Attributes
- Attributes named in the correct format
- A Multi-View Block containing the Attributed AutoCAD Blocks as View Blocks
- A Tag Anchor attachment

Building a custom Schedule Tag requires a systematic approach and a bit of patience. As with most ADT Content, it will be instructional to dissect some of the sample tags provided. To assist you in this task, two files have been provided in the *Chapter06* folder, *Imperial Tags.dwg* and *D A CH Tags.dwg*. The contents of these files are shown in Figures 6–29 and 6–30.

 Note: Tag anchors are shown in these images as small dashed arcs. They would not normally print. They are shown here for information only.

Imperial Schedule Tags

Door & Window Tags

Door Tag   Window Tag

Object Tags

W18x130   W14x30   W18x130

Beam Tag   Brace Tag   Column Tag

Equipment Tag   Furniture Tag
Equipment Tag (Leader)   Furniture Tag (Leader)

Room & Finish Tags

ROOM
001

STANDARD

RM # 001
ROOM

FLOOR: –
BASE: –
N WALL: –
E WALL: –
S WALL: –
W WALL: –
CEILING: –
REMARKS: –

Room Tag   000
Space Tag   L x W   Room Finish Tag

Wall Tags

Wall Tag
Wall Tag (Leader)

Figure 6–29 *Tags provided in the Imperial\Schedules folder of the DesignCenter*

D A CH Schedule Tags

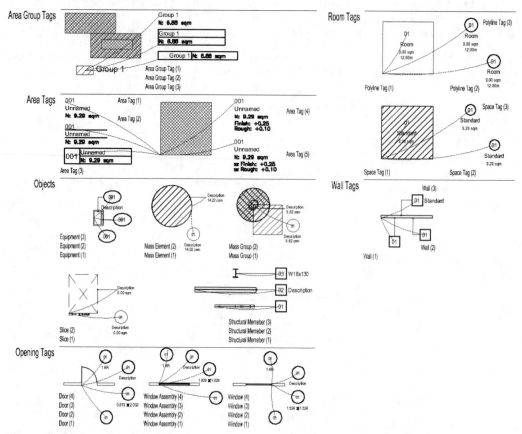

Figure 6–30 *Tags provided in the Metric D A CH\Schedules folder of the DesignCenter*

You will be able to get to only one of these folder structures through the pull-down menus. It depends on what you chose for your default when performing your Architectural Desktop installation. However, once the AutoCAD DesignCenter is open, you can easily navigate to the other folders in the tree and use the tags located there. There are actually three sets of Tags provided with ADT. Only the Imperial and D A CH have been showcased here. The generic Metric Tags are graphically similar to the Imperial. They simply scale to Metric units.

Pay particular attention to some of the unique Tags available in the D A CH Content tree. There you will find Tags for Areas, Area Groups, Mass Elements, Mass Groups and Slices. None of these is found in either the Imperial or Metric folder. There are also some examples of Tags that link to AutoCAD polylines. All Scheduling tools work on AutoCAD entities as well as ADT objects. This can be a great way to bridge the gap between legacy drawings and new ADT practices. Keep in mind that, although many of these Tags are associated with corresponding Schedule Tables, it is not a requirement to have a Schedule Table in order to build an ADT Tag. The reverse is also true; it is not necessary to build Tags to go with each of your Schedules. What both Tags and Schedules have in common is their linkage to Property Set Definitions. Property Set Definitions *are* required by both Tags and Schedule Tables.

CUSTOMIZING SCHEDULE TAGS

Regardless of all of the sample choices available, you will likely wish to create or customize your own Tags. Let's look at that process now. First, we must carry out a few preparatory steps. In Chapter 1, when you built your Client Deployment, it was recommended that you copy all content to the server on the L Drive. You were also instructed to keep a copy of all Content on your C Drive to use as a working copy until you were ready to publish your customized versions to the office library. We will be working in your local "C" copy throughout this exercise. However, if you are using your *OfficeStandard* profile, you will need to temporarily edit the Content Path statement (in the Options dialog box) to point to the C Drive while we work. The best way to do this is to make a temporary "Working" profile.

Create a Working Profile

1. From the **Tools** menu, choose **Options**.
2. On the Profiles tab, choose **Add to List**.
3. Type in the name **Working** and then click **OK**.
4. Double-click **Working** to make it active.
5. Click the AEC Content tab and set the **Content Path** to: *C:\Program Files\Autodesk Architectural Desktop 3\Content.*

 Tip: If your Management Install is located through a different path, use that path instead.

6. Click the Apply button, and then click **OK**.

"Working" is now your active profile and the Content folder is mapped to the C Drive of your Management Install.

Open Content from the DesignCenter

In this exercise, we will build a custom Door Tag. Because there are several components required, as listed at the beginning of this section, it is always easier to edit an existing Tag rather than build one from scratch.

1. From the **Documentation** menu, choose **Schedule Tags>Door & Window Tags**.

If you are using Imperial content, your screen will look like the left side of Figure 6–31; if you are using D A CH content it will look like the right.

 Note: The Tags in the Metric folder are nearly identical to the ones in the Imperial folder and have been omitted here to save space. However, you can use them if you wish—simply follow the steps for Imperial.

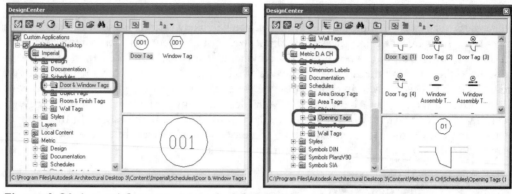

Figure 6–31 *Imperial Content tree is on the left, D A CH is on the right*

2. Right-click the **Door Tag [Door Tag (1)]** and choose **Open**.

3. Perform a **Saveas** and name the new file *AIG-Door Tag.dwg*.

 Note: Substitute "AIG" with the initials of your firm.

This will open the drawing directly into the AutoCAD drawing editor. Each of the bulleted items listed just below the "Understanding Schedule Tags" heading is illustrated hierarchically in Figure 6–32. The Property Set Definition is the central point of the diagram, attaching to the drawing object on the left and the Schedule Tag on the right. The Schedule Tag is composed of three distinct pieces, a **Multi-View Block**, which contains one or more **AutoCAD Blocks** (also referred to as View Blocks), which contains one or more **Attributes**.

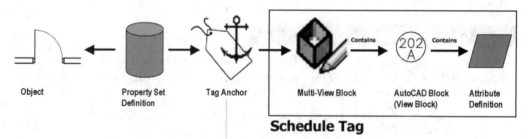

| Object | Property Set Definition | Tag Anchor | Multi-View Block | AutoCAD Block (View Block) | Attribute Definition |

Schedule Tag

Figure 6–32 *Relationship between the various components of a Schedule Tag*

Make Attributes Link to Property Sets

To edit the existing Attribute Definitions in a Block, you can explode the Block, edit the Attribute Definitions and then redefine the Block. However, the easiest way to edit Attribute Definitions is with the Block Attribute Manager (introduced as a new feature in AutoCAD 2002).

1. From the **Modify** menu, choose **Object>Attribute>Block Attribute Manager** (see Figure 6–33).

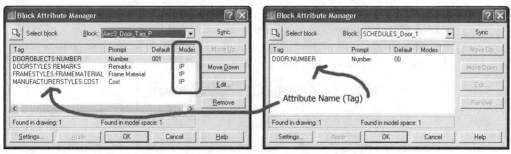

Figure 6–33 *The Block Attribute Manager showing the two Door Tags*

The purpose of the Attributes within the View Block is two-fold: first, the Attribute determines which Property Set(s) to import into the drawing that the Tag is being used in, and second, it serves as the receptacle for the Property Set value. These functions will occur provided the Attribute is named properly.

The required format for the Attribute Definition *Name** is:

<Property Set>:<Property>

Note: Just to confuse things, the Attribute "Name" is technically called the Attribute "Tag."

For example, an Attribute named **DoorObjects:Number** means that the value of the *Number* Property in the *DoorObjects* Property Set will be input as the value of that Attribute. This also means that if this Tag is dragged from the DesignCenter, the **DoorObjects** Property Set will be imported into the drawing and attached to whatever object is selected during the tagging routine.

WHY ARE THERE SEVERAL INVISIBLE ATTRIBUTES IN THIS TAG?

As you can see from Figure 6–33, the Imperial version has four Attributes, while the D A CH has only one. Therefore, the Attributes in the Imperial file are responsible for importing four separate Property Sets into the drawing and the D A CH version only imports one. However, in both cases, there is only one *visible* Attribute (Number in both cases) within the View Block. Looking in the **Modes** column, you can see that the bottom three Attributes in the Imperial file show the letters "IP." The letter "I" stands for invisible and "P" for preset. The fact that they are invisible tells us that they are included in this Block simply to facilitate the importation of the three additional Property Sets. Remember, if you drag this Tag from the DesignCenter, it will prompt for an object selection, at which point, those Property Sets will be imported into the drawing and if they are Object-based, they will also be attached to the selected object.

You may recall that earlier we used a different naming convention within our custom Property Set Definition. Specifically, the Property Set is named DoorObjects and the Door Number Property is named DoorNumber. Therefore, the Door Number Attribute in our View Block must be renamed to DoorObjects:DoorNumber.

2. In the Block Attribute Manager dialog box, select **DoorObjects:Number [Door:Number]** and click the **Edit** button.

3. On the Attribute tab, in the **Data** area, type **DoorObjects:DoorNumber** in the **Tag** field and then click **OK** twice (see Figure 6–34).

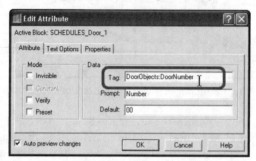

Figure 6–34 *Change the Door Number Attribute Name (Tag) to match our custom Property Set*

Add additional Attributes

Unfortunately, we cannot complete all of the changes that we need to make in the Block Attribute Manager. Therefore, we will use the REFEDIT command to complete our work on this Door Tag.

1. From the **Modify** menu, choose **In-place Xref and Block Edit>Edit Reference.**
2. When prompted, select the Door Tag Block on screen.
3. In the Reference Edit dialog box, place a check mark in the **Display attribute definitions for editing** check box.

Important: If you do not check this box, you will not be able to make any changes to the Attributes for the block.

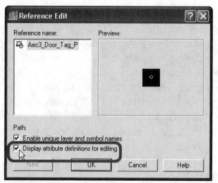

Figure 6–35 *Choose to edit Attribute Definitions within the Block during Reference editing*

4. Using a crossing window, select the entire Block and then press ENTER.
5. If you are working in the Imperial file, **Delete** the three Attributes beneath the Tag.

We will add another Attribute to the Tag that shows the Room Number. We will also add an invisible Attribute below the Tag to import the DoorStyles Property Set.

6. **Copy** the existing **DoorObjects:DoorNumber** Attribute and position both copies to fit vertically within the circle (see Figure 6–36).

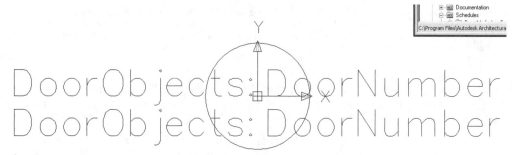

Figure 6–36 *Make a copy of the Attribute for the Room Number*

7. Double-click the top Attribute.
8. In the dialog box, change the **Tag** to read **DoorObjects:RoomNumber** and then click **OK**.

You can also edit the Prompt and Default if you like.

9. Copy one more Attribute below the circle of the Tag. Rename it **DoorStyles:DoorSize**.

This will import the DoorStyle Property Set into a drawing when this tag is dragged from the DesignCenter. In most cases, this Property Set will already be present in the drawing. However, in rare cases it will not be, so adding this extra attribute is simply a precaution to ensure that all required Property Sets are resident in the drawing when we begin tagging.

10. Select this Attribute, right-click and choose **Properties**. In the **Misc** area, set **Invisible** to **Yes** (see Figure 6–37).

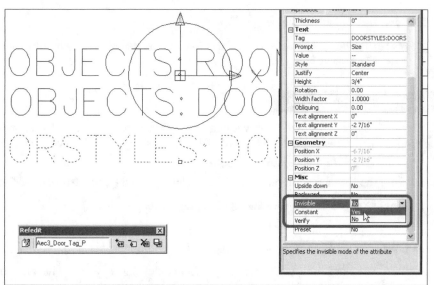

Figure 6–37 *Change the Attribute to Invisible*

At this point, if required, you could make any changes to the geometry of the Tag. In this exercise, we are going to assume that the graphics are acceptable.

11. On the Refedit toolbar, click the **Save** icon and then click **OK** to confirm the Save.

Tip: If you closed the Refedit toolbar, type REFCLOSE at the command line and choose the Save option.

When you complete the REFEDIT session, the Block on screen will appear unchanged. To fix this, return to the BATTMAN command and click the **Synch** button. This will replace the existing Block reference on screen with the newly updated definition.

There is one more step to completing our customized Schedule Tag: We must re-save its AEC Content parameters. All items in the AEC Content tree are saved with a special collection of parameters, which enables them to have ADT functionality when dragged from the DesignCenter. This and many other topics related to creating custom AEC Content are covered in Chapter 8. For this exercise, we will use the AEC Create Content Wizard at a basic level.

12. From the **Desktop** menu, choose **Create AEC Content**.
13. Click the **Next** button twice.
14. On the Display Options page, place a check mark in the **Current Drawing** check box and then click **Finish**.

Figure 6–38 *Save the Content to the Current Drawing file*

15. **Save** and **Close** the file.

Note: The details of the two screens that we just bypassed by clicking **Next** will be discussed in detail in Chapter 8.

Test the New Schedule Tag

To view newly added items to the DesignCenter, it is necessary to collapse all folders of the tree and then reopen them. This forces the DesignCenter to refresh itself and show the new Content.

1. In the *Chapter06* folder, **Open** the *A-FP01.dwg* file.
2. From the Documentation menu, choose Schedule Tags>Door & Window Tags.

Note: You should now have the AIG DoorTag listed.

3. Drag your new Tag into the drawing and follow the prompts.

Congratulations, you have built your first custom Schedule Tag. If you receive any errors, go back, review the steps above, and edit the Tag as necessary. Follow similar steps to edit other tags. If you wish to create one from scratch, you will need to create Property Sets in the default

Property Set Definitions file (as named above at the start of the "Updating the Property Set Definitions Master file" section), draw the graphics using any AutoCAD/ADT objects, and manually create the Attribute Definitions (using the **Draw>Block>Define Attributes** command). Then you must make an AutoCAD Block (**Draw>Block>Make**) and use it as a View Block in a new Multi-View Block (**Desktop>Multi-View Blocks>Multi-View Block Definitions**). Finally, you will need to run the AEC Create Content Wizard to set the new tag's DesignCenter parameters. The type of Content is a "Custom Command." The command to use is AECANNOSCHEDULETAGADD. Again, start with an existing one wherever possible—it is much easier to do so. There are other issues related to scaling and using custom commands; refer to Chapter 8 for more information.

CD RESOURCES

The custom AIG Door Tag built here is included in the *Chapter06* folder. In addition, Chapter 7 will discuss building a custom Room Tag, which is also included with the CD files and located in the *Chapter07* folder. In Chapter 11, we will discuss Sections and Elevations. Included with the CD files in the *Chapter11* folder is a special routine that transforms the standard Section and Elevation bubbles into Schedule Tags that are linked to the Bldg Section and Bldg Elevation Line objects. This is done to take advantage of Property Set Data, in order to keep the Section/Elevation bubble coordinated with the Title Bar information associated with the Sections and Elevations themselves. For instance, when this routine is used, if the Section number changes from 3 to 4, both the Section bubble and the Title Bar will update simultaneously to reflect the change.

The components required to utilize this routine are the custom AEC Content files, a custom Property Set Definition file and a file containing the code named *acaddoc.lsp*. The files include the following:

> The Section mark: *AIG Sect Mark.dwg*
>
> The Elevation mark: *AIG Elev Mark.dwg*
>
> The Property Set Definition file: *AIG-PropertySets.dwg*
>
> The code: *acaddoc.lsp*

If you are already using an *acaddoc.lsp* file, open this one in Notepad and copy and paste the code to your current file. If you are not already using an *acaddoc.lsp* file, simply move this file to the *Support* folder on your I Drive and restart ADT. Copy the other three DWG files to a folder in the Content tree (*Office Standards* is a good choice). To use the routine, simply browse within the DesignCenter to the *AIG Sect Mark.dwg* or the *AIG Elev Mark.dwg* file and drag and drop. Follow the command prompts.

Some of the code within these files comes courtesy of Stardsign cad solutions, inc, © 2002, and is used with permission.

PROPERTY SET AND SCHEDULE TABLE CHECKLIST

I Property Set Definitions

1. Perform the steps in the "Property Set Definition Preparation Checklist" earlier in this chapter.
2. Make sure you have all of the Schedule Data Formats you need within the drawing.
   a. If you do not, build or import them.

3. Decide on a Property Set Definition naming convention.

4. Work in a temporary file to build Property Set Definitions and later import them into the default *PropertySetDefs.dwg* file.

5. Create at most two Property Set Definitions per Schedule Table Style that you need, one Property Set Definition for Object-based, the other for Style-based.

   a. Create Object-based Property Sets by choosing **Entities** on the Applies To tab.

   b. Create Style-based Property Sets by choosing **Styles and Definitions** on the Applies To tab.

6. Add the required properties, both Automatic and Manual.

7. Update your master *PropertySetDefs.dwg* file with your new Property Sets.

8. Open any style libraries that require Style-based Property Sets and attach them on the General tab of the Properties dialog box.

II Schedule Table Styles

9. Import a Schedule Table Style that is close to the one you wish to create.

10. Configure all formatting and be sure that the Applies To tab information matches the Property Set Definition(s) you plan to use.

11. Add Columns and Headers.

12. Determine procedures for adding Schedule Tables to Building Models.

    a. Add Schedule Tables to the same file as data.

    b. Add Schedule Tables to an XREF.

    c. Do both depending on project and team size.

III Schedule Tags

14. Work in a "Working" profile with the path specified to the *Content* folder of your Management Install.

15. Copy a Tag similar to the one you wish to create.

    a. Edit it and save it as AEC Content.

16. Build a Tag from scratch.

    a. Create one or more AutoCAD Blocks containing one or more Attributes named in the correct format.

    b. Create a Multi-View Block containing the Attributed AutoCAD Blocks as View Blocks.

    c. Include Invisible Attributes for Property Sets that you wish to import automatically upon Tag insertion.

    d. Save it as AEC Content.

IV Test Components

17. Add and test all components.

SUMMARY

Property Sets provide the data link to both Schedule Tables and Schedule Tags.

Schedule Tables and Schedule Tags work independently; one does not require the other to function, but both require Property Sets.

Property Set Definitions can be either Object-based or Style-based, but not both at the same time.

Property Set Definitions can include Automatic properties, Manual properties or both.

You should try to limit the quantity of Property Sets referenced by a Schedule Table to two or fewer.

Pre-planning makes the creation of Schedule Table components much easier.

There are advantages and disadvantages to adding Schedule Tables in the same file or remotely through XREF.

Schedule Tags are Multi-View Blocks that use Tag Anchors to link data to their internal Attributes.

Try to build custom tags from existing ones wherever possible.

In order for Property Set Data to link to the Attributes in Tags, the name format of <PropertySet:Property> must be used.

CHAPTER 7

Dimensioning and Annotation

INTRODUCTION

Many tools are available to us for the purposes of dimensioning and annotation. We have the standard AutoCAD tools and a variety of ADT tools as well. This collection of tools is a bit eclectic, with some pieces being more mature and automated than the others. It is important to keep in mind in this chapter that, since annotating drawings does rely heavily on standard AutoCAD tools, layering and other traditional techniques will be more critical to properly managing this type of data.

There are two basic types of annotation in AutoCAD/ADT. Traditional AutoCAD Annotation includes Text, Hatching, Linetypes, Blocks used as drafting symbols and Dimensions. All of these items are typically configured and layered separately. There is no single command or dialog box to establish parameters common to all of these. Many topics related to these issues have already been covered in Chapter 3. ADT Annotation includes Multi-View Blocks used as drafting symbols, AEC Dimensions, Dimension Labels, Chases, Areas and Area Groups. (Areas and Area Groups are covered separately in Chapter 10.) Three tools exist to help establish a set of baseline parameters common to all of these tools in ADT: the **Drawing Setup** command, AEC Content and layer keying. All three of these topics have been at least introduced in the preceding chapters. Here we will explore the specifics as they pertain to ADT annotation.

 Note: Many of the topics in this chapter require the installation of the ADT International Extensions. See the note in the Preface regarding installation of International Extensions.

OBJECTIVES

This chapter is not a comprehensive look at all of the features of the text and dimensioning tools of AutoCAD. It is assumed that the reader is already familiar with the AutoCAD offerings. Rather, the goal of this chapter is to discuss how to effectively incorporate ADT technologies with these AutoCAD tools and to introduce you to some features of the ADT annotation tools with which you may not already be familiar. In this chapter, we will explore the following:

- Understand what annotation tools are available
- Create scale-dependent annotation

- Learn the differences between AEC and AutoCAD dimensions
- Use Dimension Labels
- Work with AEC Chases

DRAWING SETUP

The first place to begin our exploration of ADT Annotation is the **Drawing Setup** command, found on the **Desktop** menu. We have already explored the Layering and Display tabs of this dialog box in detail. The Units and Scale tabs were covered briefly as well. Nearly all ADT Annotation objects respond to the settings on the Units and Scale tabs of Drawing Setup. Here we will discuss their effects in more detail.

Always set the Units first, since the Units setting controls the choices available on the Scale tab (see Figure 7–1).

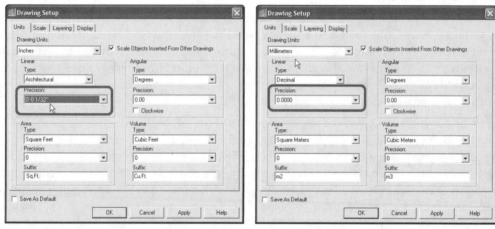

Figure 7–1 *Setting Precision on the Units tab*

You will want to do this in your template file. If you work in Imperial units, set the base unit to **Inches** and the precision to **0'-0 1/32"**. The reason for this is that many office standards and the US National CAD Standard call for text heights of 3/32". You will need the precision to be set to 1/32" to be able to display these values on screen correctly. If you work in Metric units, choose Millimeter, Centimeter or Meter from the **Drawing Units** drop-down list. Precision defaults to four decimal places, which is acceptable in most applications. If you work in Millimeters, you may want to reduce the precision to two decimal places, but this is not terribly important. Bear in mind that the precision set here is "display" precision only, meaning the way it displays on screen. Regardless of the level of precision set here, AutoCAD/ADT saves all files in double-byte precision internally.

All other values on the Units tab adjust automatically to the setting for **Drawing Units**. You can configure them appropriately for your firm and then check the **Save as Default** check box. This will save the settings in the registry and use them when drawings are started from scratch. You will want to save all preferred settings to your office standard template files as well.

You should rarely have reason to change the **Drawing Units** setting once it has been established for a particular drawing, or in many cases for all projects within the firm. However, you will frequently need to adjust the drawing scale for different annotation tasks. For this reason, there is also the **Set Drawing Scale** command available on the **Documentation** menu. This command takes you directly to the Scale tab of the Drawing Setup dialog box (see Figure 7–2).

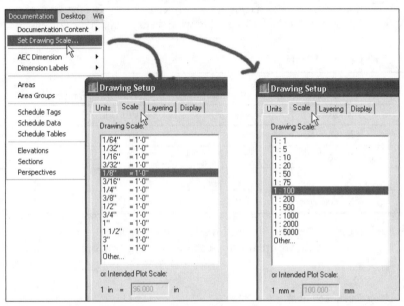

Figure 7–2 *The Set Drawing Scale command loads the Scale tab of the Drawing Setup dialog box*

There are two settings on the Scale tab, **Drawing Scale** and **Annotation Plot Size**. **Drawing Scale** is the plotted scale that we intend for this drawing. **Annotation Plot Size** is the desired annotation text height on paper, expressed in inches for an Imperial drawing or in millimeters for a Metric drawing. There are a few caveats, however: First, the **Drawing Scale** does not actually set the scale used for plotting; rather it "anticipates" the scale that will be used for plotting. In order for this to work, you must follow through and set the corresponding scale in your paper space viewports of your sheet files. Second, **Annotation Plot Size** *does* influence the physical size of AEC Annotation and can therefore produce undesirable results when considered from an office standards point of view. This is because it allows users to change the size of the text to suit personal preferences, rather than forcing them to adhere to an office standard.

The way that **Drawing Scale** and **Annotation Plot Size** function internally is as follows:

Drawing Scale – Establishes a scale factor for the drawing that is used by all ADT annotation from that point on. (It does *not* re-scale previously inserted annotation.) For example, if the drawing uses Imperial units and is set to a drawing scale of 1/4"=1'-0", the scale factor used for symbols and dimensions is **48**. (A complete list of scale factors is included in Table 3–2, "Common Scale Factors," in Chapter 3.) A piece of ADT Content must be *specifically designated* (within the AEC Content Wizard at the time the content is created) to use **Drawing Scale** or it will not use it. For instance, Furniture content does not use **Drawing Scale**.

Annotation Plot Size – Multiplied by the **Drawing Scale** for those Content items that are configured to use it. In other words, if the Annotation Plot Size is 2.5 mm in a 1:100 scale drawing, the amount of scaling applied to a piece of Annotation content is **250**. Only content *specifically designated* (within the AEC Content Wizard at the time the content is created) to use Annotation scaling will take advantage of this setting. All other content will simply ignore it.

Tables 7–1 and 7–2 summarize the behavior of the two scale settings in relation to the options built into AEC Content. Content may be set to use Drawing scaling, Annotation scaling or No scaling.

Table 7–1 *Using Drawing Scaling*

Desired Plotted Text Height	X Drawing ScaleFactor	= Text Height for Model Space	÷ Paper Space Viewport Scale	=Text Height to use in Content
PH	X SF	= MH	÷ VS	=CH
1/8	X 96	= 12	÷ 96	= 1/8 in Inches
3/32	X 96	= 9	÷ 96	= 3/32 in Inches
3.0	X 100	= 300	÷ 100	= 3.0 in Millimeters
2.5	X 100	= 250	÷ 100	= 2.5in Millimeters

Conclusion: **CH = PH** – Use the Text Plot Height (PH) for the Height of Text within Content (CH) when building Content using the Drawing scaling option (in the AEC Create Content Wizard).

Table 7–2 *Using Annotation Scaling*

Desired Plotted Text Height	X Paper Space Viewport Scale	= Text Height for Model Space	÷ Annotation Scale*	=Text Height to use in Content
PH	X VS	= MH	÷ AS	=CH
1/8	X 96	= 12	÷ 12	= 1 in Inches
3/32	X 96	= 9	÷ 9	= 1 in Inches
3	X 100	= 300	÷ 300	= 1 in Millimeters
2.5	X 100	= 250	÷ 250	= 1 in Millimeters

Drawing Scale Factor	X Annotation Plot Size (Height)	= Annotation Scale
SF	X AH	= AS
96	X 1/8	= 12
96	X 3/32	= 9
100	X 3	= 300
100	X 2.5	= 250

Conclusion: **CH = 1** – Use a static height of **1** unit for the Content Text Height (CH) when building Content using the Annotation scaling option (in the AEC Create Content Wizard).

ANNOTATION SCALING AND OFFICE STANDARDS

Most of the AEC Content provided is configured to use Annotation scaling. This means that the size of the text and the symbols dragged from the DesignCenter will adjust in size based upon the value set for **Annotation Plot Size** in the Drawing Setup dialog box. Since most office and industry standards impose a fixed size for text and symbols, it is recommended that you choose Drawing scaling for any AEC Content that you create or edit. This means that you will need to build View Blocks based on the conclusions drawn from Table 7–1. If you intend to use the content provided with ADT, much of it will require rework to make it use Drawing scaling.

However, with nearly 110 Content items in each of the Imperial and Metric folders and nearly 150 in the Metric D A CH, the effort required to change them may not be worth it. Therefore, if maintaining a consistent text height and symbol size is important to your office standards, and you cannot justify the time to edit the existing offerings, be certain that all users know and *use* the correct value for **Annotation Plot Size**. Furthermore, if you work in an environment where multiple standards are required (such as your internal office standard and another standard for a government client), you may also wish to use Annotation scaling.

If you do decide to change the defaults, it is best to do it early before ADT is widely deployed. It can be very frustrating trying to track down an old version of a symbol that keeps popping up unexpectedly while people are trying to get production drawings out the door. This happens frequently as people "borrow" information from one drawing to use in another and unwittingly propagate the use of old "retired" symbols and styles.

TEXT ANNOTATION AND DRAWING NOTES

ADT specific tools to create Text and notes are limited. Mostly you will continue to rely upon tried and true AutoCAD methods. A few points are worthy of mention here.

General Notes

If yours is like most firms, you have a series of standard General Notes that are reused from project to project. You can place these blocks of text directly into the sheet files in paper space. The reason is they are not likely to change that often, so there is no harm in having them with the title block. This will make them easier to manage as well, since they will only show on the one sheet into which they are inserted. If you frequently choose from a collection of possible General Notes for any given sheet type, you can simply include them in a Block library for easy retrieval. In fact, you can even turn them into AEC Content if you wish. This saves your having to devise some other strategy and interface for importing them. The only downside to this technique is that blocks of text do not preview very well in the DesignCenter. However, you can easily overcome this by including a polyline border surrounding the note block. If your General Notes column on your title block does not include a border, you can place this polyline on the DefPoints layer, or other non-plotting layer such as A-Anno-Nplt. Another advantage to creating content from standard notes is that you can control their insertion layers with Layer Keys. You can even create custom Layer Keys for Content items if you wish. That process is covered in the following tutorial.

Install the CD Files and Create a Custom Layer Key

1. Install the files for Chapter 7 located on the Autodesk Architectural Desktop: Advanced Implementation Guide CD ROM.

Refer to the "How to Use the CD" section in the Preface for instructions on installing the sample files included on the CD.

A folder named *Chapter07* was created containing a collection of sample files that we will use throughout this chapter.

2. From the *L:\Layers* folder, **Open** your *AecLayerStd.dwg* Layer Key Standards file.

Note: If you used a different name for this file in Chapters I and 4, choose it instead.

3. From the **Desktop** menu, choose **Layer Management>Layer Key Styles**.
4. Double-click to edit **Office Standard** and then click the Keys tab.

Note: Again, if you used a different name in Chapters I and 4, choose it instead.

5. Click the **Add** button.
6. In the Name dialog box that appears, type **GenNote** and then click **OK**.
7. Click in the **Description** column, and then type **General Notes**.
8. Type in a layer name in the **Layer** column, or click the small browse (...) icon to choose the **Layer** name field from predefined Description lists that match your Layer Standard.
9. Configure the **Color**, **Linetype**, **Lineweight** and **Plot Style** (if using Named plot styles). See Figure 7–3.

| Layer Key | Description | Layer Settings | | | | | | Allow Overrides | |
		Layer	Color	Linetype	Lineweight	Plotstyle	Plot	Discipline	Major
FINFLOOR	Finish tags	A-Flor-Iden	■11	Continuous	—0.25 mm	Style 1		☑	☑
FIRE	Fire system equip.	A-Prot-Eqpm	■90	Continuous	■0.35 mm	Style 1		☑	☑
FURN	Furniture	A-Furn	■10	Continuous	■0.35 mm	Style 1		☑	☑
FURNNO	Furniture tags	A-Furn-Iden	■11	Continuous	—0.25 mm	Style 1		☑	☑
GenNotes	Genearl Notes	A-Anno-GenN	■111	Continuous	—0.35 mm	Style 1			
GRIDBUB	Column grid tags	A-Grid-Iden	■171	Center2	—0.25 mm	Style 1		☑	☑
GRIDLINE	Column grids	A-Grid	■173	Center2	—0.18 mm	Style 1		☑	☑

Figure 7–3 *Add a custom Layer Key and configure its settings*

10. Click **OK** twice to return to the drawing.
11. **Save** and **Close** the *AecLayerStd.dwg* Layer Key Standards file.

Create a Typical General Notes Content Item

1. In the *Chapter07* folder, open the *General Notes Imperial.dwg* [*General Notes Metric.dwg*] drawing file.

This is a simple file with a rectangle representing the General Notes column and a dummy paragraph of General Notes (it may not match your title block exactly, so feel free to make adjustments). Naturally, you will need to substitute the "dummy" text with an actual collection of typical General Notes used in your firm.

2. From the **Desktop** menu, choose **Create AEC Content**.
3. In the **Content Type** area, choose **Drawing** and then click **Next**.

This option means that the entire current drawing will be copied to the Content item and later inserted when it is dragged and dropped from the DesignCenter.

4. In the **Insert Options** area, place a check mark in the **Explode On Insert** check box.
5. Click the **Select Layer Key** button and select the **GenNote** Layer Key from the list, and then click **Next** (see Figure 7–4).

Tip: If you prefer, you can simply type **GenNote** in the **Layer Key** field instead.

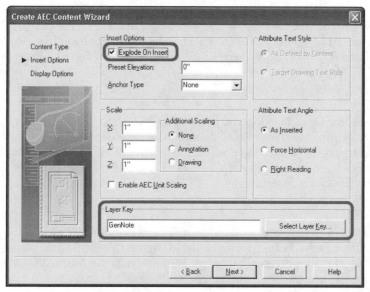

Figure 7–4 *Setting the Content Insert Options*

6. On the Display Options page of the wizard, click the **Browse** button (in the **File Name** area) to locate a folder within your Content path to save the custom Content.

Later in this chapter, we will discuss the creation of an *Office Standards* folder. For this example, you could choose any folder such as the *Local Content* folder, or even place it at the root of your *Content* folder. We can always move the file later. As long as it stays within the *Content* folder, it will function properly in any sub-folder.

7. In the **File Name** field of the Save Content File dialog box, type a name such as **Typical General Notes.dwg** and then click the **Save** button to return to the wizard.

8. Add a description if you wish, such as **Office Standard Typical General Notes - Plan Sheets** and then click **Finish**.

9. **Close** the *General Notes Imperial.dwg* [*General Notes Metric.dwg*] drawing file without saving it.

 Note: It is not necessary to save the original file, since a new drawing was created for the Content item.

10. Open the DesignCenter, navigate to the folder where you saved the Content and test it by dragging it into a new drawing.

This was a very simple example, where the choices were not explained fully. We will build several other Content items in coming exercises and chapters and in them explore many more options in greater detail. In this example, we began by creating a custom Layer Key. This step is not required if the Layer Key you wish to use is already included in the Layer Key Style. (With the other Content items that we create, this will typically be so.) It is important to note that if you type a Layer Key that does not exist into the Layer Key field, the Content will simply insert it to the current layer when it is dragged into a drawing. This can largely negate the purpose of building a Content item such as this, so be careful that you have properly followed through on all decisions you make when building Content items and build custom Layer Keys as required.

Key Notes

Several Text Leaders are included in the default ADT Content library. They are found in the *Content\Documentation\Leaders* folder of all three root-folders, *Imperial*, *Metric* and *Metric D A CH*. These are nothing more that automated versions of the AutoCAD **Leader** command that call a variety of Text and/or Symbol combinations. Like all AEC Content dragged from the DesignCenter, all of these commands layer automatically.

Drawing Notes

The Text Leaders in the DesignCenter are simply automated routines available for drawing notes. None of them is the **Text** command only. However, this would be a very simple piece of AEC Content to add. Simply create a Content item that runs the AutoCAD MTEXT command. This will give you a standard block of text that automatically layers.

Create a Simple Piece of Annotation Content

To see an example of this kind simple Content, look at the *Dimensions* folder in the *Documentation\Miscellaneous* folder in the DesignCenter.

1. Right-click any one of these commands and choose **Edit** (see Figure 7–5).

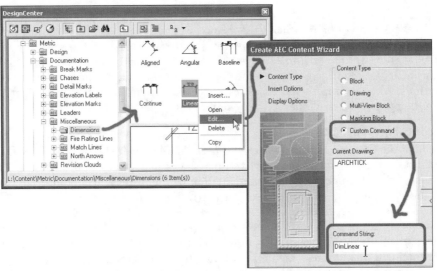

Figure 7–5 *The Dimension Content items are simply AutoCAD commands assigned to Layer Keys*

You will see that these are merely Custom Command AEC Content items that run standard AutoCAD Dimension commands and then assign them to a Layer Key (on the Insert Options page of the Wizard).

2. **Cancel** the AEC Create Content Wizard.

3. Create a **New** Drawing from Scratch.

If you want an icon to appear in the DesignCenter, create a small box on screen and put some text in it. The Content Wizard will create an icon from the extents of whatever objects you have on screen; however, it ignores the extents of text, so the box will help it zoom properly. You can also "draw" some text with lines and fill it in with hatching. (You will see this technique used in many of the standard "out-of-the-box" offerings.)

4. From the **Desktop** menu, choose **Create AEC Content**.

5. In the **Content Type** area, choose **Custom Command**.

6. In the **Command String** field, type **MTEXT**.

7. Click **Next** and at the bottom of the screen, type **ANNOBJ** in the **Layer Key** field (or click the **Layer Key** button and choose one from the list).

8. In the **File Name** area, click **Browse** to locate a folder within your Content path to save the custom Content.

9. Type a name for the file such as **Drawing Notes Text** and then click **Save**.

10. Add a description if you wish, and then click **Finish**.

11. In the DesignCenter, navigate to the folder where you saved the Content and test it by clicking once on the Content item.

You should see a preview of the "dummy" text that you added and its enclosing rectangle.

12. Drag the **Drawing Notes Text** Content item into a drawing to test it out.

The AutoCAD MTEXT command should run with all of the normal prompts. When you are finished in the Mtext editor and click **OK**, the text will automatically move to the layer designated for the "ANNOBJ" Layer Key. As you can see, the only real benefit to making a piece of content such as this is to force it to automatically layer. This can be a very valuable benefit, but weigh this benefit carefully against the time and effort spent making content from these types of commands. Also factor in the potential resistance and/or confusion that you might encounter from other users who may not see the benefit of this type of Content.

DOCUMENTATION CONTENT

Each node of the Content Library (*Imperial*, *Metric* and *Metric D A CH*) contains a *Documentation* folder. Depending on which units you chose as default when you installed ADT, the commands on the **Documentation>Documentation Content** menu will point to one of these folders automatically. You can, of course, use the DesignCenter to manually access Content located in any of the three folders. The Content provided in the *Imperial* and *Metric* folders are nearly identical, except for the difference in units. The *D A CH* folder exhibits a bit more variation, particularly in the area of Elevation Labels. To see what is available, simply open the DesignCenter and click through each folder. You can also load the Documentation – Imperial and Documentation – Metric toolbars to access these DesignCenter folders. If you wish, drag and drop each symbol into a test drawing to explore these items in more detail. This will of course take some time, but it is the only way to truly see what each Content item does. To assist you as you work through the Content items provided, two images in PDF format have been installed along with the other files in the *Chapter07* folder. A portion of one of these is illustrated in Figure 7–6.

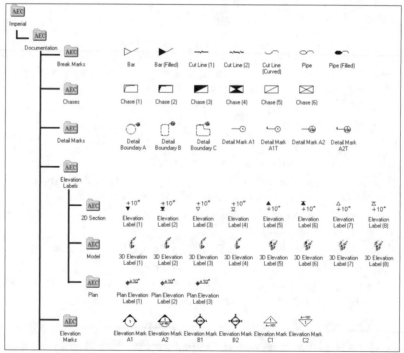

Figure 7–6 *A portion of the Documentation Content Library provided with ADT*

While we evaluate the Content Library, there are three possible outcomes for each item: The item is acceptable "as-is," the item must be edited in some way to be acceptable or the item should be deleted. Open the two PDFs mentioned in the previous paragraph and print them out. They are named *Documentation Content Imperial.pdf* and *Documentation Content D A CH.pdf*. As you go through the Content items, you can cross off the items you do not need, place a check mark next to those that are acceptable as is, and circle those that need to be edited in some way. See Figure 7–7.

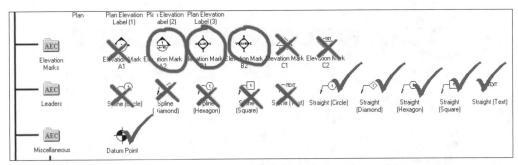

Figure 7–7 *Evaluate each content item for usefulness and readiness in production*

For instance, it is unlikely that you will require twelve different North Arrows, or the same match line with two different fonts. Therefore, in your first pass, it may be very easy to decide what can be eliminated right away. The second pass will require a bit more effort, because you will need to actually drag the Content item into a drawing to understand how it behaves. Another item to consider while you are evaluating the Documentation Content is the folder structure itself. If you do eliminate eleven of the twelve North Arrows, it will not be necessary to have a separate North Arrows folder any longer. You can move the North Arrow Content item to the Title Marks folder for instance.

All of these operations (moving and deleting content, renaming content and creating folders) can be done directly in Windows Explorer, as long as you keep everything within the Root *Content* folder. There is one problem with this type of consolidation of Content resources. The items on the **Documentation>Documentation Content** menu point the default folder structure. If you change the folder names or locations without also editing the ADT menu file, those commands will fail when executed. This will not harm ADT or any drawing; the command will simply fail to open a sub-folder and will instead open the DesignCenter to the root of the *Content* folder.

If you are familiar with the process of editing the menu file, then go ahead and make whatever changes you wish to the folder structure and update the menu accordingly. Be sure to include these customized menu files in the Common Support folder on the I Drive. This way all users will have access to the customized resources. If you are not familiar with editing AutoCAD/ADT menu files but are interested in learning more, pick up a copy of *Customizing AutoCAD 2002* by Sham Tickoo.

Another approach you could take is to simply build a custom root folder within the global Content folder on the L Drive specifically for your office standard resources. You can then add an icon to the MyFavorites toolbar that we built in Chapter 1 to open this folder within the DesignCenter.

Build a Custom DesignCenter Icon

1. If the DesignCenter is currently open, close it.
2. In Windows Explorer, add a folder named *Office Standards* to your *L:\Content* folder (see Figure 7–8).

 Note: If you created a different location for common content, substitute that path instead.

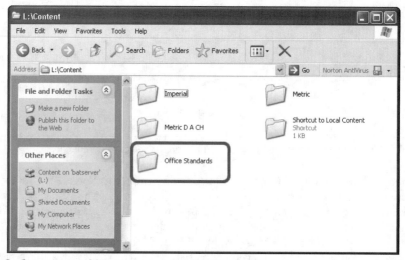

Figure 7–8 *Create a new folder in the network Content folder*

3. Right-click anywhere on the MyFavorites toolbar (the one created in Chapter 1) and choose **Customize**.
4. On the Commands tab on the left side, select **User defined**.
5. From the right side, drag the **User Defined Button** item to a location on the MyFavorites toolbar (see Figure 7–9).

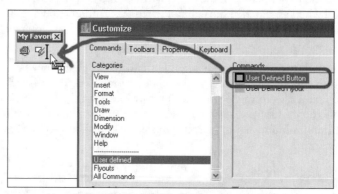

Figure 7–9 *Drag a new User Defined Button to the MyFavorites toolbar*

6. Right-click the new button and choose **Properties** (see Figure 7–10).

Figure 7–10 *Edit the properties of the new button*

7. Type a name such as **Office Standards** and a description such as **Open Office Standards Content Folder**.

8. In the **Macro associated with this button** text field (see Figure 7–10), type the following:

```
^C^C(defun c:AIGADC () (command "adcCustomNavigate" "Architectural Desktop/Office
Standards")) AIGADC
```

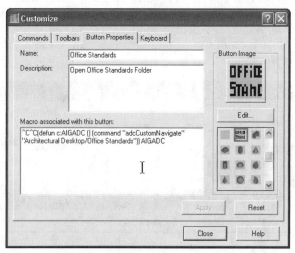

Figure 7–11 *Add the Macro code to the button*

This macro is a snippet of AutoLISP code. Parentheses are very important when writing LISP code. Each piece of the macro is explained as follows:

^C^C – Used to cancel any commands that are currently running.

(defun c:AIGADC () – *defun* stands for "define function." The function we are defining is named AIGADC. This stands for "Advanced Implementation Guide AutoCAD DesignCenter." Feel free to substitute another function name if you wish. Just be certain to preserve all of the parentheses, the "c:" and any spaces.

(command – Simply runs an AutoCAD command. You could put any defined AutoCAD command here. In this case, the command is ADCCUSTOMNAVIGATE. When typed at the command line, this command expects a path statement to be input. Therefore, the path "Architectural Desktop/Office Standards" comes next. Notice the SPACE following ADCCUSTOMNAVIGATE and the quotes surrounding the path statement.

AIGADC – Now that this has been defined as a function, the last step is to run the command. If you substitute another function name, be sure to substitute here as well. Also, be sure to include a SPACE after the function name. (Recall that a SPACE is the same as an ENTER in AutoCAD.)

9. **Edit** the Bitmap image if you wish.

Tip: Remember to click **Saveas** and save the Bitmap to the Local Content folder, so that it will found next time you launch ADT. Otherwise, you will get the little "smiley faces."

10. Click the **Apply** button and then click **Close**.

Your button is now ready to test.

11. Click the new icon.

The DesignCenter should open and go directly to the newly added folder. If the button does not work, review the steps and correct any syntax errors in your macro code. The snippet of code shown above is included in the *Chapter07* folder in the text file named *ButtonMacro.txt*. You can open this file in Windows Notepad and copy and paste the code to the button macro instead. Feel free to move both of the custom Content items created above to this folder for easy access to all users (see Figure 7–12).

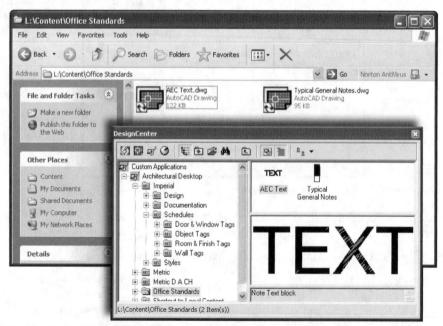

Figure 7–12 *Move both the custom General Notes command and the custom Text command to the Office Standard folder*

CREATING SCALE-DEPENDENT ANNOTATION

It is sometimes desirable to show and print the same content at two different scales. To achieve this, it has long been the norm in AutoCAD to insert the same information twice into the drawing on two different layers. For instance, to show a room tag in both the 1/8"=1'-0" [1:100] scale overall floor plan and 1/4"=1'-0" [1:50] enlarged plan detail sheet, you would need layers A-Anno-Iden-96 [A-Anno-Iden-100] and A-Anno-Iden-48 [A-Anno-Iden-50]. You would then need to be sure to show only one of these layers at a time in any given sheet. Besides the obvious extra steps involved in duplicating the same blocks and annotation to two separate layers, there is the issue of coordination. If the information changes, we must remember to change both versions of the annotation. Moreover, there is the additional frustration encountered when we accidentally forget to freeze one of those layers at deadline time, and we have several plotted sheets showing two superimposed tags. Clearly, it would be ideal if this issue could somehow magically take care of itself. It does not require any magic spells to achieve this goal, only a good understanding of the Display System and a bit of front-end effort in building the Content.

Explore a Sample File

I. In the *Chapter07* folder, **Open** the *Small Room Imperial.dwg* [*Small Room Metric.dwg*] drawing file (see Figure 7–13).

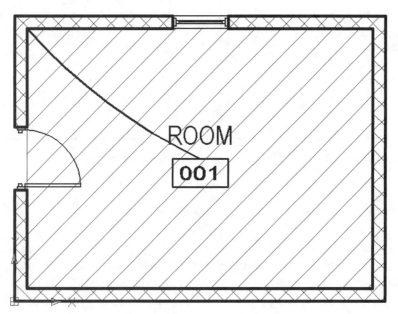

Figure 7–13 *A small room with masonry walls*

This file has been created with the template file that we have been working on throughout this book. Included in the Imperial version of this file are the four scale-dependent Display Configurations: Plot Plans 48, 96, 192 and 480. (In the Metric file, they are Plot Plans 50, 100, 200 and 500.) As we discussed in Chapter 5, the purpose of these configurations is to dynamically adjust the level of detail displayed by each object as they correspond to the various plotted scales. For instance, the configuration designed to work at 1/4"=1'-0" [1:50] shows material poché in Walls, Doors use double lines to draw the door panel and Windows use double lines to draw the glazing. At 1/8"=1'-0" [1:100], Wall poché disappears, and Door and Window components simplify to single line display. In all of these configurations, the Space objects disappear. This is because these configurations are intended for plotting construction documents, and Space objects are not typically plotted on CDs.

Currently Work is the active Display Configuration, which shows detail comparable to the Plot Plan 48 [Plot Plan 50] configuration, but also shows the Spaces and other "working" objects like the Tag Anchors.

2. From the **Desktop** Menu, choose **Select Display**.

3. Load the **Plot Plan 48 [Plot Plan 50]** Display Configuration.

Take note of the display changes.

4. Load the **Plot Plan 96 [Plot Plan 100]** Display Configuration.

Take note of the display changes (see Figure 7–14).

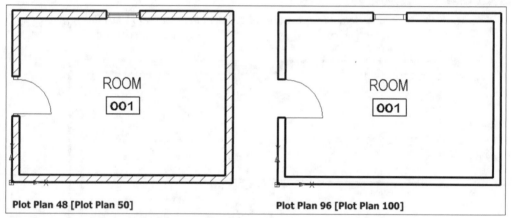

Plot Plan 48 [Plot Plan 50]   Plot Plan 96 [Plot Plan 100]

Figure 7–14 *Variation in detail between small and large scale Display Configurations*

5. Re-load the **Work** Display Configuration.

As you can see, the only object in this example that did not change was the Room Tag. This is not because it isn't possible, but simply because the default ADT Room Tag has not been configured to respond to scale changes in the "out-of-the-box" content. The Room Tag used here is a Multi-View Block named Aec3_Room_Tag [M_Aec3_Room_Tag].

6. Select the Tag, right-click and choose **Edit Multi-View Block Definition**.

7. Click the View Blocks tab.

Notice that an AutoCAD Block named Aec3_Room_Tag_P [M_ Aec3_Room_Tag_P] is loaded into the General Display Representation when viewed from the Top view direction. The way a Multi-View Block (MVB) works is that one or more AutoCAD Blocks are created to represent the object in each of its possible display conditions (Display Representations). The General Display Representation is typically used to represent any two-dimensional views of the MVB (see Figure 7–15). If required, the Reflected and Model Display Reps are used for reflected ceiling plan views and three-dimensional views respectively. If different Blocks are loaded into alternate Display Representations/View Direction combinations, the MVB will dynamically adjust when a new Display Configuration is loaded, just as the Walls and other objects do.

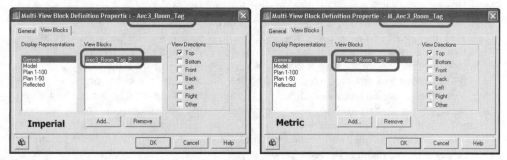

Figure 7–15 *View Block assigned to a Display Rep/View Direction combination*

There are two additional Display Representations available to Multi-View Blocks by default: Plan 1-100 and Plan 1-50. These two are used in the same way as they are for Walls and other objects: to build "scale-dependent" Multi-View Blocks. Since our goal is to create a room tag that adjusts size according to the scale of the drawing, we will want to use these two Display Representations to accomplish this.

Plan 1-100 is intended for Blocks that we wish to display at 1/8"=1'-0" [1:100] scale and Plan 1-50 is intended use at 1/4"=1'-0" [1:50] scale. We will need to create a version of our Room Tag Block sized correctly for plotted 1/8" [1-100] drawings, into the Plan 1:100 Rep and another sized for 1/4" [1-50] drawings in the Plan 1:50. This will make our MVB dynamically resize with a change of Display Configuration, thus making it "dependent" on the scale of the drawing (see Figure 7–16).

Figure 7–16 *The scale-dependent Display Reps of Plan 1-100 and Plan 1-50 are available for an MVB just like Walls and other objects*

Conceptually, this is no different from the traditional AutoCAD procedure outlined at the beginning of this exercise, nor is it different from the method we used to adjust the scale of the other ADT components. The major difference here is that all of the parameters needed to "intelligently" swap scale are built right into the MVB, rather than being a function of data duplication, manual scaling and layer swapping on the part of the user. Building all of those parameters into the MVB will take a little "front-end" effort, but once you have built this MVB, it functions like any other ADT style and can be reused in any drawing from now on.

324

MULTI-VIEW BLOCKS ARE NOT THE SAME AS OTHER ADT OBJECTS

Multi-View Blocks differ from other ADT objects in a few critical ways. First, they do not use Display Control in the same way as other ADT objects, and second, they do not represent any object "type" in particular. Most ADT objects have one or more Display Reps that can be manipulated at three different levels: System Default, Style and Object. Although an MVB does have several Display Reps, they do not use the three levels of Display Properties like other objects. A Multi-View Block's display settings occur within the Multi-View Block Definition (Style) only. This is simply because an MVB does not represent any specific object by default, but rather can represent *any* type of object. For this reason, neither System Default nor object level Display Control settings would make sense for a Multi-View Block. All display variations are handled exclusively by the View Blocks loaded in each representation at the Style level only.

Understanding the Existing Tag

As noted earlier, the View Block used by this MVB is Aec3_Room_Tag_P [M_Aec3_Room_Tag_P]. We need to make a duplicate version of this block sized half as big. We will then remove the View Block referenced in the Plan Display Rep, and instead load the larger one into the Plan 1-100 and the smaller one into Plan 1-50. We will need to do this in the original Content file from the DesignCenter library.

1. From the **Documentation** menu, choose **Schedule Tags>Room & Finish Tags** (see Figure 7–17).

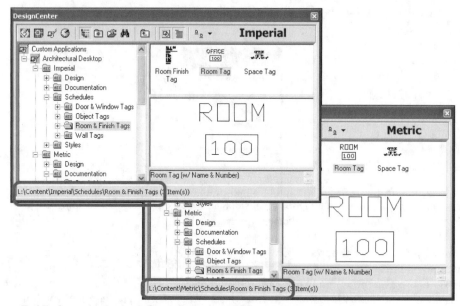

Figure 7–17 *Open the Room & Finish Tags node of the ADT Content Library*

Tip: You may wish to use your Working profile to access the folder structure in your Management Install, rather than work directly in the L Drive Content Library.

2. Right-click the Room Tag icon and choose **Open**.
3. From the **File** menu choose **Saveas** and save a copy of the file in the *Office Standards* folder (the one we built above) with the name **AIG-Room Tag.dwg**.

 Note: Substitute "AIG" with the initials for your firm.

An important thing to understand about working with ADT Content items is that the objects on screen within the Content file are strictly for the purposes of generating the icon preview in the DesignCenter and may have little or nothing to do with the function of the Content item. For instance, if you list the item on screen here, you will find that it is an AutoCAD Block and not a Multi-View Block at all. This Block is actually the View Block used in the MVB in this case, but that will not always be so for every piece of content. Therefore, the first step to understanding the structure of a Content item is to run the AEC Content Wizard and look at the parameters.

4. From the **Desktop** menu, choose **Create AEC Content** (see Figure 7–18).

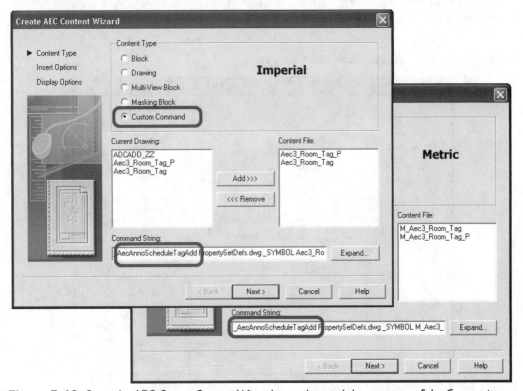

Figure 7–18 *Open the AEC Create Content Wizard to understand the structure of the Content item*

As you can see from the first page of the wizard, this Content item is actually a Custom Command. This command requires two resources, a Multi-View Block named **Aec3_Room_Tag** **[M_Aec3_Room_Tag]** and a Block named **Aec3_Room_Tag_P [M_Aec3_Room_Tag_P]**, which is the View Block for the MVB. The command used is the AECANNOSCHEDULETAGADD command, which is the ADT command for adding Tags.

WHY NOT LISP AND VBA INSTEAD OF THE CONTENT WIZARD?

You will find that many of the Content items in the ADT DesignCenter are actually Custom Commands. We saw this above when we created the AEC Text command. Virtually anything can be inserted through the DesignCenter. You can run any standard AutoCAD or ADT command and then assign a Layer Key to the routine.

There are of course many other ways to achieve the same level of automation in AutoCAD/ADT if you are comfortable with custom programming. AutoLISP and VBA offer very robust environments in which to customize and automate your ADT environment. However, the DesignCenter holds a few advantages over the custom programming methods. First, it supports Layer Keys. Second, it is easy to use it to create new Content quickly (it does not require custom programming). Finally, being part of the core ADT product, it will continue to be supported, enhanced or evolved into whatever mechanisms may appear in future ADT releases, meaning you will not be forced to rebuild all of your code when the new release of ADT comes out.

5. Click the **Next** button (see Figure 7–19).

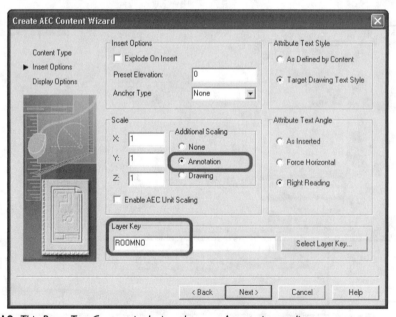

Figure 7–19 *This RoomTag Content is designed to use Annotation scaling*

The **Aec3_Room_Tag_P** [**M_Aec3_Room_Tag_P**] View Block is already inserted in this drawing at a 0,0 insertion point. If you were to measure the height of the Room Number within the box, you would note that it is one unit tall (**1 Inch** in imperial, **1 Millimeter** in Metric). This was done because this Block is designed to use Annotation scaling (as defined in the "Drawing Setup" topic above.) On this page of the AEC Create Content Wizard, we also find the **Layer Key** reference of this Tag is ROOMNO.

6. **Cancel** the AEC Create Content Wizard.

Since our goal is to make a Tag that changes scale with the change of a Display Configuration, we will want the scale factor built *directly into* the View Blocks. This means that we must decide on a fixed plotted height for the text and the symbol size when plotted. We then need to multiply that value by the two drawing scales at which we intend this Tag to operate. In other words, instead of having text that is one unit tall, which later scales to the values in Table 7–2, we will want the text within the View Blocks to be the sizes shown in the Text Height for Model Space (MH) column of Table 7–1. Therefore, assuming you wanted a plotted height of 3/32" [2.5 mm], you will need to build the Plan 1-100 View Block with 9" [250 mm] tall text. For Plan 1-50, the size will be **4.5" [125 mm]**.

Build the View Blocks

1. Select the Aec3_Room_Tag_P [M_Aec3_Room_Tag_P] Block on screen.
2. Right-click and choose **Properties** (see Figure 7–20).

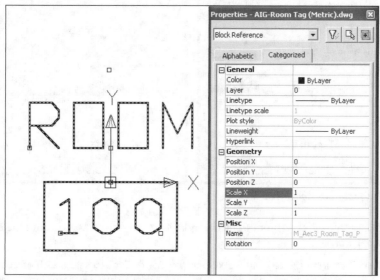

Figure 7–20 *The AutoCAD Properties Manager can be used to scale the View Block*

3. In each of the **Scale X**, **Scale Y** and **Scale Z** fields, type **9 [250]** (see Figure 7–21).

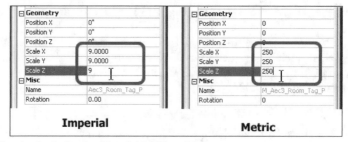

Geometry		Geometry	
Position X	0"	Position X	0
Position Y	0"	Position Y	0
Position Z	0"	Position Z	0
Scale X	9.0000	Scale X	250
Scale Y	9.0000	Scale Y	250
Scale Z	9	Scale Z	250
Misc		Misc	
Name	Aec3_Room_Tag_P	Name	M_Aec3_Room_Tag_P
Rotation	0.00	Rotation	0

Imperial　　　　　　**Metric**

Figure 7–21 *Set the Scale for the Plan 1-100 View Block*

4. **Explode** the Block and then **Zoom Extents**.

5. From the **Draw** menu, choose **Block>Make**.

6. For the name, type **Aec3_Room_Tag_96 [M_Aec3_Room_Tag_100]** (see Figure 7–22).

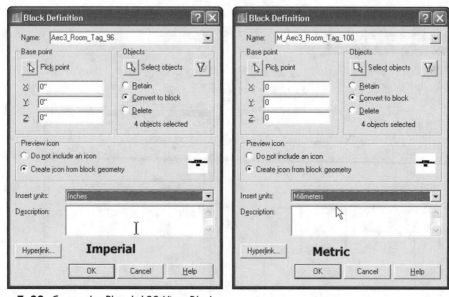

Figure 7–22 *Create the Plan 1-100 View Block*

7. Click the **Select Objects** button and click each object (3 Attribute Definitions and 1 Polyline).

 Tip: Start from the top and work your way down. Click to select each object; do not use a window or crossing selection. With Attributes, the prompts will appear in the dialog box in the order that you select them. You can use the Block Attribute Manager later to reorganize them, but this extra step can easily be avoided using this tip.

8. Be sure that the **Insert Units** is set correctly for your units, leave the insertion point set at 0,0,0 and then click **OK**.

9. In the Enter Attributes dialog box, click **OK**.

The screen will look exactly as it did when we started. However, if you measured the height of the Room Number now, it is 9" [250 mm] tall instead, even though the scale of this new Block is **1**.

10. Select the Block, in the AutoCAD Properties Manager set each of the scale factors to **.5** (see Figure 7–23) and then **Explode** the Block.

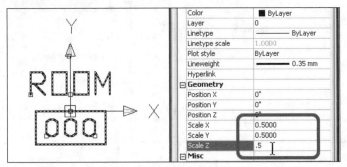

Figure 7–23 *Scale the Block to half its size and then Explode it to create the Plan 1-50 View Block*

11. Repeat the Block steps above, using the name **Aec3_Room_Tag_48 [M_Aec3_Room_Tag_50]** this time.

We now have two View Blocks scaled correctly for their respective Display Reps.

Update the Multi-View Block

1. From the **Desktop** menu, choose **Multi-View Blocks>Multi-View Block Definitions**.

2. Double-click the **Aec3_Room_Tag_P [M_Aec3_Room_Tag_P]** MVB to edit it.

3. On the View Blocks tab, **Remove** the Block referenced by the **General** Display Rep (see Figure 7–24).

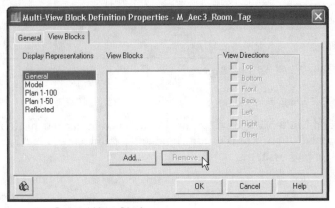

Figure 7–24 *Remove the General View Block*

4. Select the **Plan 1-100** Display Rep on the left, and then click the **Add** button.

5. From the **Select a Block** list, choose **Aec3_Room_Tag_96** **[M_Aec3_Room_Tag_100]** and then click **OK** (see Figure 7–25).

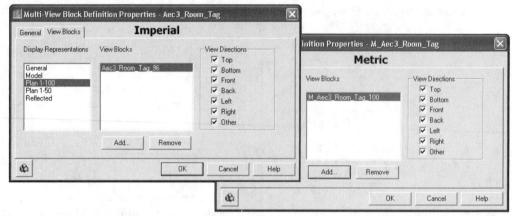

Figure 7–25 *Add the Plan 1-100 View Block*

6. Repeat the same steps for **Plan 1-50** using **Aec3_Room_Tag_48** **[M_Aec3_Room_Tag_50]** this time (see Figure 7–26).

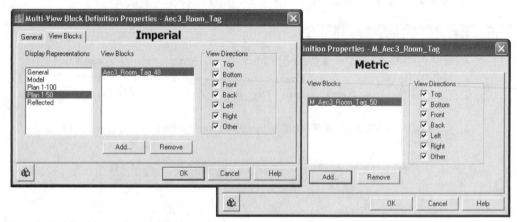

Figure 7–26 *Add the Plan 1-50 View Block*

7. Click **OK** when finished.

All of the Block and Multi-View Block resources for our scale-dependent Room Tag are complete. The only thing that remains is to edit command parameters in the AEC Create Content Wizard for this Content item.

WHAT ABOUT OTHER SCALES?

We are using only the two "out-of-the-box" default Display Reps in this example. However, the template that we are working on has provision for four different scales. If you wish your room tags to show at those scales as well, you will need to add two custom Display Reps to Multi-View Blocks in your office standards. A discussion of custom Display Reps and the advantages and disadvantages of creating them appears in Chapter 5.

If you choose to add Display Reps for 1:200 and 1:500, you will need to repeat the steps shown here for those Display Reps and add properly scaled View Blocks to your MVB for use at those scales.

The Schedule Tag Add Command

We saw above that the Room Tag Content item uses the AECANNOSCHEDULETAGADD command. When this command is launched, several operations are performed. The command first accesses the Property Sets from an external drawing file. It then prompts you to select an object to tag. This prompt is accompanied by other options such as the choice of symbol. Next, a Tag Anchor is established between the object you are tagging and the Tag itself. If you have the option to display the Edit Schedule Data dialog box turned on in the Options dialog box, that dialog box will appear. Finally, the Layer Key kicks in and places the Tag on the correct layer. The command will then simply repeat. Without this routine, each of these steps would need to be performed manually for each object that you wanted to tag. This is why this Content item is a Custom Command and not simply a Multi-View Block.

AECANNOSCHEDULETAGADD has the following options:

- **Property Set Definitions drawing** – This prompt requests the location of the Property Set Definition drawing file that you wish to use. The command will import Property Sets from this external file into the current drawing.
- **Symbol** – With this option, you input the name of the Multi-View Block that will be used for the Tag Symbol.
- **Leader** – With this option, you can add a leader connecting the Tag to the object it is tagging. This is an AutoCAD leader.
- **Dimstyle** – This option works with the Leader option by allowing you to designate which Dimension Style to use for formatting the Leader arrow. If you do not choose the Leader option, this option has no effect. The Dimension Style must exist in the drawing where the Tag will ultimately be used. Unfortunately, it will not be imported from the Content item file. Therefore, be certain to include all required Dimstyles in your office standard template files.

The **Property Set Definitions drawing** file is named *PropertySetDefs.Dwg* and is located in the *Content\Imperial\Schedules* [*Content\Metric\Schedules*] folder by default. If you are using D A CH Content, the name of the file is *PropertySetDefsDACH.dwg*, and it is located in the *Content\Metric D A CH\Schedules* folder. Contained in this file is the complete collection of Property Set Definitions for all ADT Content items, including all Schedule Table Styles, Tags and Object Styles. As we saw in Chapter 6, a Schedule Tag contains Attribute Definitions that map to the properties contained in the Property Sets of this file.

1. From the **Documentation** menu, choose **Schedule Data>Property Set Definitions** (see Figure 7–27).

Figure 7–27 *Content items should not contain Property Set Definitions; they are accessed remotely by the custom command string*

 Important: The Tag's Content file itself contains no Property Set Definitions. All Property Sets are accessed from the remote *PropertySetDefs.dwg* file.

2. **Close** the Style Manager.

 Tip: Whenever you add a Property Set Definition in your firm, you should add it to the *PropertSetDefs.dwg* file first. (Refer to Chapter 6 for more information on Property Set Definitions.)

Create the Room Tag Content Item

1. From the **Desktop** menu, choose **Create AEC Content**.
2. On the right side, select **Aec3_Room_Tag_P [M_Aec3_Room_Tag_P]** and then click **Remove**.
3. Using the SHIFT key, select both **Aec3_Room_Tag_48** and **Aec3_Room_Tag_96 [M_Aec3_Room_Tag_50** and **M_Aec3_Room_Tag_100]** and then click the **Add** button (see Figure 7–28).

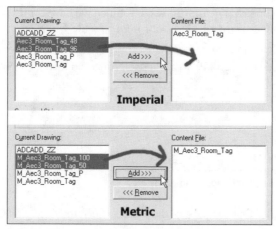

Figure 7–28 *Add the two new View Blocks to the Content file*

4. At the bottom corner of the dialog box, click the **Expand** button.

This will expand the command line area so that you can see it better. The command string is shown in Figure 7–29.

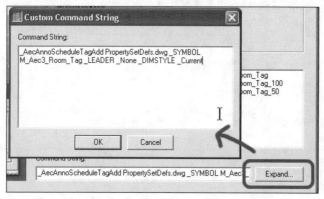

Figure 7–29 *The command string for the Room Tag Content item*

Notice the SPACE after each entry. Remember that a SPACE at the command line in AutoCAD is the same as an ENTER. Therefore, each entry is equivalent to something you could type at the command line. As was outlined above, this command string uses the AECANNOSCHEDULETAGADD command. The next entry is the *PropertySetDefs.dwg* drawing file name, which is in answer to the prompt for which Property Set Definition file to use. The Symbol being used here is **Aec3_Room_Tag [M_Aec3_Room_Tag]**. Finally, the Leader option is set to **None**, and therefore, even though the Dimstyle option is set to **Current**, it is not relevant since the Dimstyle controls only the settings of the optional Leader. Since we did not change the name of the MVB used for the Symbol, there is no need to change anything in this command string.

5. Click **OK**, and then **Next**.

We only need to make one change on this screen. Since we have built the scale factors into our View Blocks, we no longer want this Content item to scale when inserted. Therefore, we will set the scaling to **None**.

6. In the **Additional Scaling** area, choose **None** and then click **Next** (see Figure 7–30).

Figure 7–30 *Change the scaling option to None*

7. On the Display Options page of the wizard, check the **Current Drawing** check box.

This will save the Content parameters to the current drawing rather than create a new Content file. Since we already did a **Saveas** at the beginning, we want this option.

8. Edit the description to include that this Tag is "scale-dependent" and then click **Finish** (see Figure 7–31).

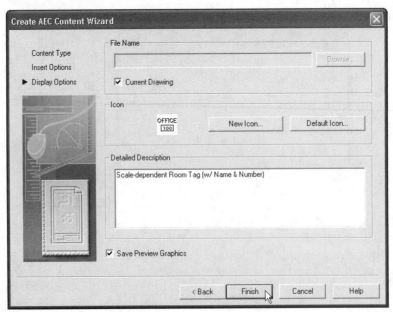

Figure 7–31 *Finalize settings in the AEC Content Wizard and click Finish*

9. **Save** and **Close** the file.

The Scale-dependent Room Tag Content item is now complete and ready to test.

Test the New Scale-dependent Room Tag

Let's test out the results. You should still have the *Small Room Imperial.dwg* [*Small Room Metric.dwg*] drawing file open. If you don't, reopen it now. We must delete the existing Tag and purge its definition from the file before we try to use the new Tag. If we don't, the old Tag will be used when you try to drag the new one in from the DesignCenter. This is because the MVB in the scale-dependent Content item has the same name as the original one.

1. **Delete** the Tag in the drawing.

2. From the **Desktop** menu, choose **Multi-View Blocks>Multi-View Block Definitions**.

3. Right-click the **Aec3_Room_Tag_P [M_Aec3_Room_Tag_P]** MVB and choose **Purge** (see Figure 7–32).

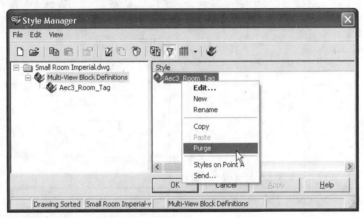

Figure 7–32 *Purge the existing Multi-View Block*

4. In the dialog box that appears, click **OK** to confirm the Purge.

5. Click **OK** again to exit the Style Manager.

If you wish, you can also purge the View Block with the AutoCAD **Purge** command. (Enter PURGE at the command line.) Normally this would be a good idea, since the new MVB may continue to reference a View Block with the original name. In this case, we used different names for our new View Blocks, so this step is not necessary. However, it is good practice to purge Blocks and other items that are no longer needed.

6. Click your custom Office Standards toolbar icon (see the "Build a Custom DesignCenter Icon" step sequence above).

The DesignCenter should open and reveal your new AIG-Room Tag.

7. Drag the **AIG-Room Tag** icon from the DesignCenter into the drawing (see Figure 7–33) and when prompted, select the Space object.

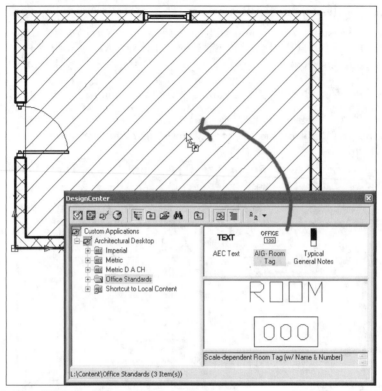

Figure 7–33 *Drag AIG-Room Tag into the drawing*

8. Press ENTER to accept the default position of Centered and then press ENTER again to end the command (see Figure 7–34).

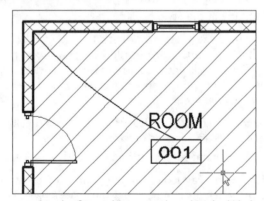

Figure 7–34 *The Tag centered in the Space (shown anchored in the Work configuration)*

9. From the **Desktop** menu, choose **Select Display**.
10. Load the **Plot Plan 48 [Plot Plan 50]** Display Configuration.

The Tag should adjust to half its size.

 11. Load the **Plot Plan 96 [Plot Plan 100]** Display Configuration.

The Tag should switch back to its prior size. To see the results even better, a paper space layout with two viewports, one set to 1/4"=1'-0" [1:50] and another set to 1/8"=1'-0" [1:100] has been provided in this file.

 12. Click the **Layout1** paper space layout tab at the bottom of the drawing.

The tags should appear the same relative size when viewed from paper space, while the 1/4" floor plan image is twice the size of the 1/8" version (see Figure 7–35). This is how they will appear when plotted.

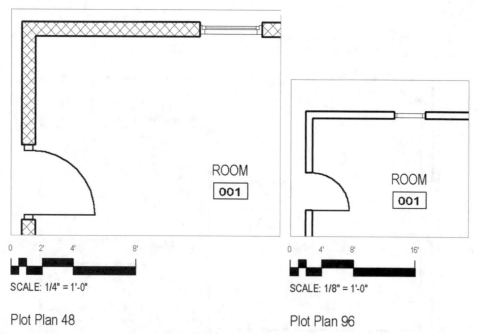

Figure 7–35 *Your Scale-dependent Room Tag is now ready for use*

DON'T BOTHER ME FOR A MONTH; I'LL BE RE-BUILDING TAGS...

In certain applications, the preceding tutorial will be very pertinent. However, before you get carried away and make all of your Tags scale-dependent, think very carefully about which ones *need* to be scale-dependent. This exercise used a Room Tag as an example. This is a good use of a scale-dependent Tag, since rooms need to be identified on nearly all drawings. Column Grid Bubbles are another good use of scale-dependent Tags. If you think through the needs of each Tag you use in a typical document set, you are likely to find that only a few need to be scale-dependent. This will help quite a bit in reducing the amount of effort necessary to build them.

Additional Adjustments

We can make this Tag even more powerful with a few additional steps. First, if we made a small adjustment to the Property Set Definition file (*PropertySetDefs.dwg*) referenced above in the command string, we could link the text of the **Room Name** field with the **Description** field of the Space object. This is potentially useful, because once users get used to the idea of adding the Room Name to the Space object's Description, the Room Names will automatically appear within the Tag as soon as they are added without the need to edit the Schedule Data. In addition, the **Description** field is readily available in the AutoCAD Properties Manager.

1. Select the Space object in the drawing, right-click and choose **Properties**.

Note: You may have to reset you display back to "Work."

Notice that **Description** is available right below Hyperlink in the General portion of the Properties palette (see Figure 7–36).

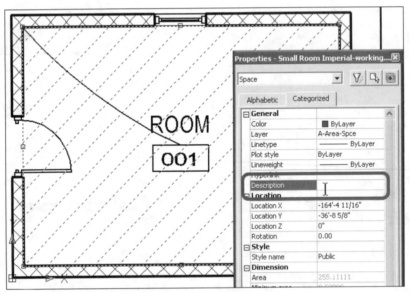

Figure 7–36 *The Description field is one of the General Properties*

2. Type **Office** into the **Description** field and then press ENTER.

To test it out before actually editing the default *PropertySetDefs.dwg* file, we can make this change here in this file.

3. From the **Documentation** menu, choose **Schedule Data>Property Set Definitions**.

4. Edit **RoomObjects** and then click the Definition tab.

5. Highlight **Name** and then place a check mark in the **Automatic** check box (see Figure 7–37).

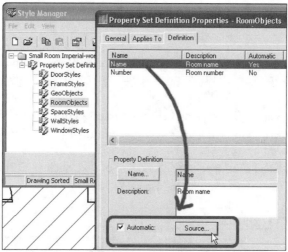

Figure 7–37 *Make Name an automatic property*

6. Next to the **Automatic** check box, click the **Source** button.
7. Choose **Description** and then click **OK** back to the drawing (see Figure 7–38).

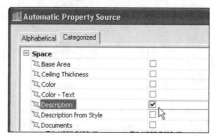

Figure 7–38 *Link the source of the Name property to the Description of Space objects*

Notice how "Office" has automatically appeared as the Room Name (see Figure 7–39). This is because we previously edited the Description of the Space object.

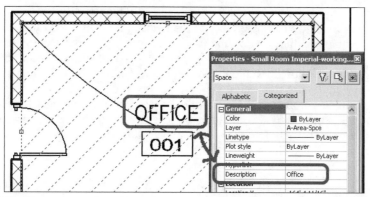

Figure 7–39 *The Description of the Space now reports directly to the Room Name Tag*

340

 Important: If you decide to set up your Property Set this way, be sure to perform these steps in the *PropertySetDefs.dwg* instead.

There is another bonus to setting up the Description link to Room Name this way. You can get the Room Name to display on multiple lines. There are two ways to do it. The easy way is to open the Space Properties dialog box.

8. Select the Space, right-click and choose **Space Properties**.
9. In the **Description** field, type the first line of the Room Name, such as **Executive**, followed by a SPACE.
10. Press ENTER and then type the second line of text, such as **Office**.
11. Click **OK** to view the result (see Figure 7–40).

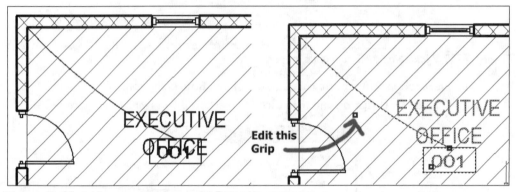

Figure 7–40 *Adding a second line to the Room Name*

Obviously, there is a slight problem with the Room Name's overlapping the rest of the tag. This can be fixed easily with grips. Just select the Tag, click the grip for the Room Name text and move it up a bit.

 Note: You will need to do the grip edit for both View Blocks.

This technique does negate the benefit we gained by linking the Description to the Room Name, since you can't add a carriage return to the fields in the AutoCAD Properties Manager. However, if you do not anticipate needing two lines for many of your Room Names, then this technique will likely still be your best bet. Alternatively, you could use the second technique, which is more complex but does allow editing in any dialog box, including the AutoCAD Properties Manager. To use this technique, you will need to type Mtext control codes in your text. For instance, \mt can be typed into any field to force it to display as Mtext and \P can be used for a carriage return. These codes are case sensitive, so they are not compatible with all Schedule Data Formats, such as Case – Upper. See the "Mtext Control Codes" sidebar for more information on this technique.

MTEXT CONTROL CODES

Several control codes can be input into text fields and output as Mtext within Schedule tags. A complete list can be found in the online Help. Search for "multiline text, format codes." In the list of topics that appears, look for the item titled: "Format Multiline Text in an Alternate Text Editor." This help item shows a table listing all of the Mtext control codes with examples of each. If you decide to input these codes, you should be aware of some rules:

1. Mtext formatting codes work only for attributes in Multi-View Blocks, not attributes in regular AutoCAD Blocks.

2. Mtext formatting codes cannot be included in Style names, because the "\" character cannot be used in Style names. The consequence of this is that Space Style names cannot be put on multiple lines in Space tags.

3. Mtext formatting codes are case sensitive, so they should only be used in properties where the Schedule Data Format does not change the case of the text. Many properties in the Property Set Definitions shipped with Architectural Desktop use the "Case-Upper" Schedule Data Format, so you will need to change the format to "Standard" or something similar.

4. For automatic properties such as Description, Description from Style, Notes, and Notes from Style, there are two ways to get multiple line data and Mtext formatting. One is simply to separate lines in the data with a return (this will only be possible in some edit controls). The other is to put "\mt" in the data. Remember not to change case with the Schedule Data Format!

   Automatic property example: "\mtArchitectural\PDesktop" will display as:

   Architectural

   Desktop

5. For manual text type properties, the only way to get multiple line data and Mtext formatting is to put "\mt" in the data (again, remember not to change case with the Schedule Data Format). In addition, format codes must be preceded with *two* backslashes (\\). When you see the data displayed in the Edit Schedule Data dialog box, you will not see the \\; you will see them when you click on the data to edit it.

   Manual property example: "\\mtArchitectural\\PDesktop" will display as

   Architectural

   Desktop

CHASES

Chases are Multi-View Blocks with interference that can be used to represent all manner of vertical penetrations in Walls, Spaces and Slabs. Some sample Chases have been provided in the DesignCenter. In this section, we will look at the default offerings and build a custom Chase Content item. A Chase is very similar to the Content item that we built above for the

Room Tag. It is a Multi-View Block, and it is brought into the drawing through a Custom Command in the DesignCenter.

The command is AECANNOMVBLOCKINTERFERENCEADD. Try to say that one five times fast! This command establishes an interference condition between the Chase and another object in the drawing. This allows the Chase to "cut" away from the object. This is why they are called Chases.

Add a Chase to a Drawing

1. In the *Chapter07* folder, open the file named *Chases.dwg*.
2. From the **Documentation** menu, choose **Documentation Content>Chases**.

Let's add a cavity to the thick Wall between the two restrooms.

3. Drag **Chase (6)** into the drawing and track it up about **3"** from the bottom left corner of the chase Wall. Click to set the point (see Figure 7–41).

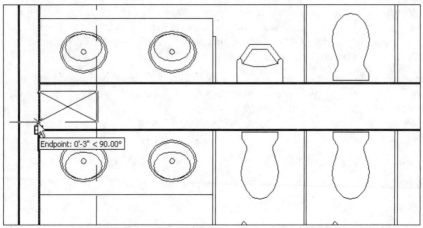

Endpoint: 0'-3" < 90.00°

Figure 7–41 *Using Object Snap Tracking, track from the bottom left corner up 3"*

4. In the dialog box that appears, type a description such as **Toilet Rooms Plumbing Wall**.
5. In the **Scale** fields, type **16'-10"** for **X**, **1'-0"** for **Y** and **12'-0"** for **Z** (see Figure 7–42).

Tip: As an alternative, you can use the buttons in the **Scale** area to set the scale with the mouse pointer in the drawing.

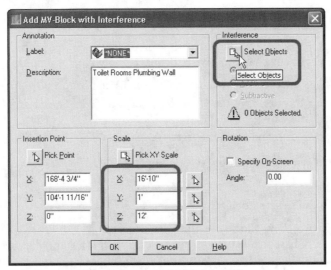

Figure 7–42 *Set the Scales, add a Description and pick Select objects*

6. Click the **Select Object** button in the top right corner.

The objects that you select will receive the Interference. You can select Wall, Space, Slab and Roof Slab objects. In this case, we want to select the Wall separating the two restrooms.

7. Click the Wall between the restrooms and then press ENTER.

8. When the dialog box reappears, verify all of the settings and then click **OK**.

The Chase now creates an interference with the Wall. If you view it three-dimensionally in the Object Viewer, you will see that it appears much like a shaft (see Figure 7–43).

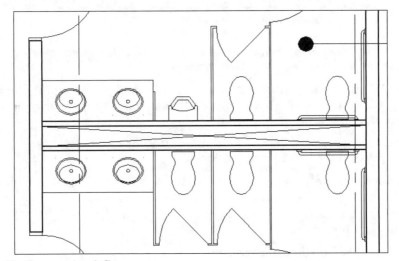

Figure 7–43 *The completed Chase*

Build a Custom Pipe Chase

All of the sample Chases are rectangular with symbols representing shafts, or ducts. You can create a Chase object in any shape you wish, displaying any type of graphic symbol.

1. In the DesignCenter, right-click on **Chase (6)** and choose **Open**.

2. Perform a **Saveas**, save the file to the *L:\Content\Office Standards* folder, and name it **Pipe Chase.dwg**.

3. Hold down CTRL and press the letter **A**. (Select All.)

 Note: If CTRL+A does not work in your version of ADT, simply select all objects with a window.

4. **Delete** the selected objects.

We need to create two View Blocks for the Chase object's MVB. One is a two-dimensional representation of the Chase that is used for plans. The other is a three-dimensional representation that will generate the interference with the other objects.

5. From the **Concept** menu, choose **Mass Element>Add Mass Element**.

6. Choose **Cylinder** from the **Shape** list, set the Height to 1" and the Radius to 1/2" (see Figure 7–44).

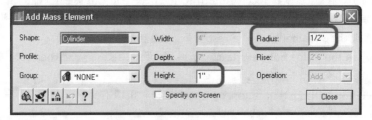

Figure 7–44 *Create a cylindrical Mass Element at 0,0*

7. At the "Insert Point" command line prompt, type **0,0** and then press ENTER. Press ENTER again to accept the default rotation.

This gives us a small cylinder that fits in a one-unit cube. By creating the View Block to fit in a one-unit cube, we can later easily scale it to any size without requiring mathematical calculations.

8. Change the layer of the cylinder to **Layer 0** and set the Color, Linetype, Lineweight and Plot Style to **ByBlock**.

These settings will allow the settings of the Multi-View Block to control the properties of the cylinder.

9. Create a Block from this Cylinder with the insertion point at **0,0** and name it **Chase_Pipe_3D**.

10. **Erase** the block from the screen.

11. **Draw** a circle on **Layer 0** with the center at **0,0** and a diameter of 1"; be sure to use **ByBlock** for all properties again.

12. Create a **Block** from the circle with the insertion point at **0,0** and name it **Chase_Pipe_2D**.

Do not erase this Block; it will be used to generate the preview icon for the Chase Content item.

13. From the **Desktop** menu, choose **Multi-View Blocks>Multi-View Block Definitions**.

14. Rename Chase-6_1 to **Chase_Pipe**.

15. Double-click **Chase_Pipe** to edit it and click the View Blocks tab.

16. **Remove** all View Blocks from all Display Reps.

17. In the **General** Display Rep, **Add** the **Chase_Pipe_2D** to the **Top** view direction and the **Chase_Pipe_3D** to all other view directions (see Figure 7–45).

 Note: Remember, General is used when the MVB is not scale-dependent.

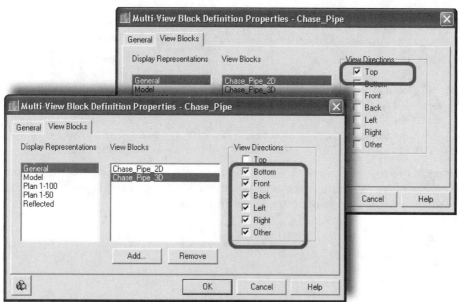

Figure 7–45 *Add the two View Blocks to the appropriate Display Reps*

18. In the **Model** Display Rep, **Add** the **Chase_Pipe_3D** to **all** view directions.

19. In the Reflected, Plan 1-100 and Plan 1-50 Display Reps, **Add** the **Chase_Pipe_2D** to all view directions.

Only the two-dimensional view is required for Reflected Ceilings, as they are 2D drawings.

20. Click **OK** twice to return to the drawing.

21. From the **Desktop** menu, choose **Create AEC Content**.

22. Verify that the Content type is **Custom Command**, select the **Chase_Pipe** MVB on the left and then click **Add**.

23. In the Command string field, replace Chase-6_I with **Chase_Pipe** and then click **Next** (see Figure 7–46).

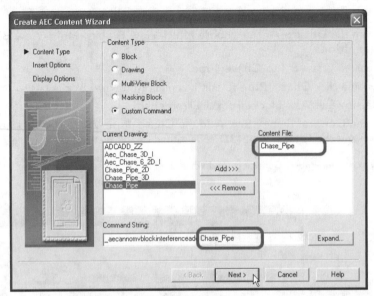

Figure 7–46 *Set up the first page of the Content Wizard*

24. In the **Scale** fields, type **6"** for both **X** and **Y**, leave **Z** set to **8'**, make sure **Additional Scaling** is set to **None** and then click **Next** (see Figure 7–47).

Remember above that we built the View Blocks to fit in a 1x1x1 cube. Therefore, we can "pre-size" the Chase here by inputting the actual dimensions that we want it to be. In this case, the Chase will be inserted at 6" in diameter and 8'-0" in height.

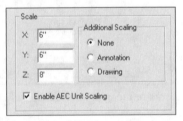

Figure 7–47 *Set the fixed scales for X, Y and Z*

The **Scale** fields are used by the content to preset the scaling values when the Content is inserted into the drawing. These values can be modified at insertion time if needed. The values used here will simply become the "defaults" of the Content. In addition, we have chosen **None** for **Additional Scaling** because this item represents a true chase size in the model, not a piece of annotation. Therefore, we do not want drawing scale to influence its size.

25. On the Display Options page of the wizard, check the **Current Drawing** check box to save these parameters to the current drawing.

26. Click the **Default Icon** button.

This will create an icon for the DesignCenter of whatever objects appear on screen. (This is why we left the 2D Block in the drawing.)

27. Type a description such as **Round Pipe Chase** and then click **Finish**.

Insert the New Chase

The new Chase Content item is complete. Let's test it out in the *Chases.dwg* drawing that should still be open.

1. Click your custom **Office Standards** icon to open the DesignCenter to the *Office Standards* folder.

You should see a new Content item there named Chase_Pipe with a circle icon.

2. **Drag** this item into the drawing.

3. At the "Specify insertion point" prompt, click inside the small bumped-out wall in the Elevator Lobby (see Figure 7–48).

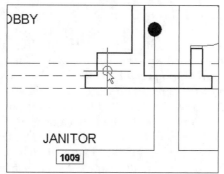

Figure 7–48 *Specify the insertion point in the Elevator Lobby Wall*

4. In the Add MV-Block with Interference dialog box, type a description, set the **Z** Scale to **12'** and then click the **Select Objects** button.

5. In the drawing, click the Wall bump-out and then press ENTER (see Figure 7–49).

 Note: The "bumped-out" portion of the Wall is a Wall Modifier.

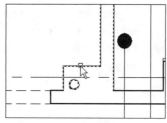

Figure 7–49 *Select the Wall to receive the interference*

6. When the Add MV-Block with Interference dialog box returns, click **OK**.

The Pipe Chase has now been added within the Wall pilaster. Chases can easily be moved. Just be sure to select them carefully so that you don't inadvertently select the Wall instead. Try using CTRL to cycle through the overlapping objects when selecting.

There are many other potential uses for Chases. Use them for Chimney flues, cable shafts, Fire Extinguisher cabinets and floor openings. Although a manual Interference can be established between a Mass Element and Walls, Spaces and Slabs, Chases are valuable because they are Multi-View Blocks and can therefore have custom (and accurate) graphical plan displays in addition to the 3D volume used for the interference. With a Mass Element alone, you don't have the same level of control over the 2D plan graphics.

Chases can be labeled as well. To do so, you need a Schedule Tag that is designed to link to Multi-View Blocks and a Property Set for them as well. Some samples can be found in the Metric D A CH *Chases* folder. Refer to the previous chapter for information on creating custom Schedule Tags.

ELEVATION LABELS

Elevation Labels are Multi-View Block symbols that are used to annotate the elevation of drawing components. Elevation Labels are primarily drafting symbology. They do not reference the heights of the actual ADT objects. Rather they are linked to either the World Coordinate System (WCS) or a User Coordinate System (UCS) within the drawing. Several different styles of labels have been included with the default library. You will find them located in the *Elevation Labels* folder in the *Documentation* Folder in the ADT DesignCenter. As with most of the other Content, each of the three main nodes, Imperial, Metric and Metric D A CH contains an *Elevation Labels* folder. There is also a command on the **Documentation>Documentation Content** menu that opens your DesignCenter directly to one of these folders, based on your default units setting. Each folder contains sub-folders with symbols that can be used for marking elevations in Plans, Models and 2D Sections. In addition, the D A CH library contains another folder named *Standard Blocks*. Within this folder are simple AutoCAD Blocks with attributes. These can be used as an alternative to the Elevation Labels, but they will not remain linked to the UCS.

Using Plan Elevation Labels

1. From the **Documentation** menu, choose **Documentation Content>Elevation Labels**.

2. In the DesignCenter, expand the *Elevation Labels* folder and then select the *Plans* folder.

3. Drag any symbol into the drawing and drop it on the Stair landing.

4. In the Add Elevation Label dialog box, type **TOL** for the **Prefix** and then type **6'-3"** in the **Elevation** field (see Figure 7–50).

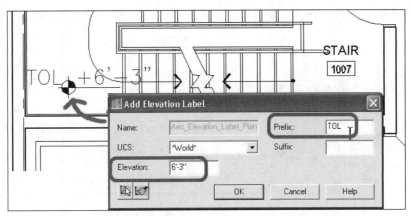

Figure 7–50 *Adding a Plan Elevation Label*

Using 2D Section Elevation Labels

1. In the *Chapter07* folder, open the file name *Elevation Labels.dwg*.
2. In the DesignCenter, expand the *Elevation Labels* folder and then select the *2D Section* folder.
3. Drag one of the symbols into the drawing and, using Object Snap Tracking, line it up with the dashed first floor line as shown in Figure 7–51.

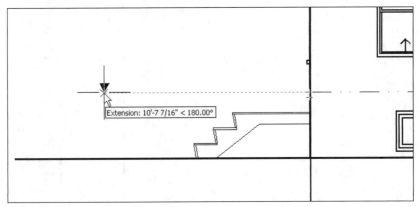

Figure 7–51 *Line up an Elevation symbol with the First Floor line*

4. Click a point to place the symbol.
5. In the Add Elevation Label dialog box, type **First Floor** in the **Prefix** field, and **FF** in the **Suffix** field.
6. In the bottom left corner of the Add Elevation Label dialog box, click the **Define UCS** button.
7. When prompted, click the endpoint of the ground line in the Elevation as the Base point of the UCS.

8. Pull the cursor straight up and click to set the Z Direction (see Figure 7–52).

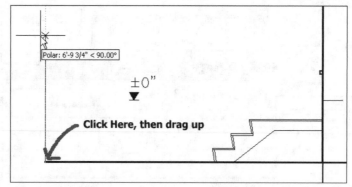

Figure 7–52 *Using Polar Tracking set the Z Direction straight up*

9. When prompted at the command line, type **North Elevation** as the name of the UCS and then press ENTER.

10. Click **OK** in the Add Elevation Label dialog box to complete the routine.

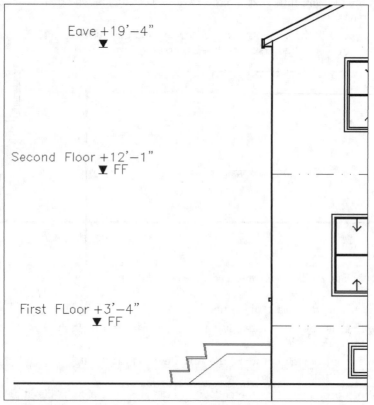

Figure 7–53 *Adding several labels referenced to the same UCS*

Repeat the steps to add additional labels. This time the UCS will already read "North Eleva-tion," so you will not need to define an additional UCS. Simply type in prefixes such as "Second Floor" and "Eave" and place the labels (see Figure 7–53). The nice thing about these routines is that although they reference a UCS, they do not actually make it active. Therefore, your current UCS in the drawing remains the **World Coordinate System** for all other drawing operations. Try moving one of the labels. You will notice that the elevation changes. This is because it is linked to that UCS. Add an Elevation Label to the basement level; notice that it automatically registers with a negative number.

If you later decide to reference another benchmark rather than the ground line we picked here, simply redefine the UCS, and all of the labels that reference it will update automatically.

DIMENSIONS

In many ways, Dimensions were the first "intelligent" object added to AutoCAD many releases ago. Every few releases, they receive enhancements and improvements. AutoCAD 2000 com-pletely revamped the Dimension Style dialog box, and AutoCAD 2002 added the long awaited Associative Dimension feature. Unfortunately, even though all of this functionality is available within ADT, not all of it may be used on ADT objects. However, ADT includes AEC Dimen-sions, which share many features in common with the Associative Dimensions of AutoCAD. These Dimension objects are ADT objects, and they use their own AEC Dimension Styles and support Display Control. The purpose of this section is to help you sort out the difference between these two technologies, not to give a comprehensive tutorial on using dimensioning.

WHEN TO USE ASSOCIATIVE DIMENSIONS

Associative Dimensions link to the points of the objects being dimensioned. They even work "trans-spatially," meaning that an Associative Dimension in paper space remains linked to objects in model space. Since AutoCAD Associative Dimensions do not work with ADT ob-jects, use them only for drawings created with AutoCAD entities such as detail sheets. One thing is important to note: by default, if you wish to use Associative Dimensions on AutoCAD objects, you must have the setting of the DIMASSOC system variable set to **2**. You can type this at the command line or change it in the Options dialog box. If set to **1**, traditional AutoCAD dimensions will be created, which are editable with grips but are non-associative. This value should never be set to **0**. With a setting of **0**, dimensions are exploded as they are created. It is recommended that you set the value to **2** in your template files.

WHEN TO USE STANDARD AUTOCAD DIMENSIONS

There are three separate "semi-automated" Dimensioning routines included with ADT. Each of these uses standard AutoCAD Dimensions (non-associative) to dimension logical points on ADT objects, and they auto-layer.

Column Grids – The first is the Dimensioning routine for Column Grids. AEC Dimen-sions (see next topic) do not work on Column Grid objects; therefore, the Column Grid Dimensioning routine is the fastest way to generate the required dimensions for an ADT Column Grid object. These Dimensions will properly layer based on Layer Keys but must be manually adjusted with grips if the grid spacing changes.

Wall Dimensions – Select any Wall, right-click and choose **Plan Tools>Dimension**, and you will get a few strings of standard AutoCAD (non-associative) dimensions automatically generated to logical points on the Wall or Walls selected. You can control the way in which this command will treat Window and Door openings on the AEC DWGDefaults tab of the Options dialog box. These Dimensions will properly layer based on Layer Keys.

DesignCenter Dimensions – These are located in the *Documentation\Misc\Dimensions* folder of the ADT DesignCenter. These are simply Custom Command Content items that run the standard AutoCAD Dimension commands and then properly layer them when the command is complete, based on a Layer Key assigned to the Content item.

All of these routines use the current AutoCAD Dimension Style set in the drawing.

WHEN TO USE AEC DIMENSIONS

AEC Dimensions are conceptually similar to Associative Dimensions; they also update automatically as the object to which they are linked is changed. However, these true ADT objects use Display Control settings. This potentially makes them even more powerful than traditional dimensions. There are, however, some limitations. AEC Dimensions work only with ADT objects. Specifically, they can be attached to nearly all of the Architectural objects, (excluding Curtain Wall Units, Column Grids and Ceiling Grids). In addition, they may be attached to a few Multi-purpose objects as well, such as Mass Element, Mass Group and Area objects. There is also a special manual AEC Dimension command that attaches AEC Dimensions to selected points on objects rather than directly to the actual AEC object. AEC Dimensions are not as flexible as AutoCAD Dimensions regarding graphic configuration, editing and variety of dimension types (they are linear only).

Consider using AEC Dimensions while designing. If for example, you are working on the layout of a floor plan, you can add some AEC Dimensions at critical points in the plan. As you move Walls and Doors to fine-tune the plan, the dimensions will update dynamically, helping you keep track of critical "hold" dimensions and code requirements. For instance, you could place an AEC Dimension at a Door Jamb, and thereby always know during the design phase if the ADA clearance requirements were being met.

VARIABLE SCALED DIMENSIONS

In addition to being able to create tags that dynamically adjust scale as we did above, we can display the same dimensions at two different scales with AEC Dimensions. The way this works is that each Display Rep of the AEC Dimension refers to an AutoCAD Dimension Style. Since DIMSCALE is a variable assigned by dimension style, we have the ability to assign a 1/4" [50] Dimension Style to the Plan 1:50 Display Rep and an 1/8" [100] one to the Plan 1:100 Display Rep. In this way, the Dimension scale will dynamically adjust with the change from one Display Configuration to another. Furthermore, parameters within the AEC Dimension Style itself allow us to vary the composition of the dimensions at two different scales. For instance, at 1/4" [1:50], which uses Plan 1:50, there may be a detailed string of dimensions, while the 1/8" [1:100], which uses Plan 1:100, might show only the overall dimension.

1. In the *Chapter07* folder, **Open** the file named *AEC Dimension.dwg*.

This file opens with Layout1 active and two viewports at 1/4" and 1/8" scale.

2. Double-click in the left viewport (1/4") to make it active.

3. Select the two upper horizontal Walls, right-click and choose **AEC Dimension**.

4. At the "Pick side to dimension" prompt, pick a point **about 2'** below the window, right-click and choose **Perp**.

Notice that even though the two viewports are displayed at different scales, the size of the dimension components is the same relative to paper space (see Figure 7–54). To achieve this result with AutoCAD Dimensions would require two sets of dimensions on two different layers. With AEC Dimensions, all that is required is two AutoCAD Dimension Styles. Let's have a look.

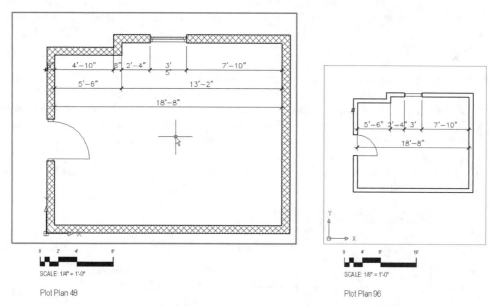

Figure 7–54 *AEC Dimensions show properly at more than one scale*

5. Right-click the AEC Dimension and choose **Edit AEC Dimension Style**.

There are only three tabs in the AEC Dimension Style Properties dialog box. The General tab is the typical name and description tab. The Chains tab has a single parameter that sets the number of strings or "chains" that the AEC Dimension objects of this style will contain. In this case, the AEC Dimension Style is configured for 3 Chains.

Most of the settings that control the AEC Dimension objects occur on the Display Props tab. This file was created from the *Plan-Sect-Elev.dwt* template file. The "Standard" AEC Dimension Style here is provided as an example of recommended settings. Since we added the potential for up to four different scales in our template back in Chapter 5, we also need to think through how we want the AEC Dimensions to display at each of those scales. The AEC Dimensions in this file have five Display Reps. Plan is used in the Work configuration. The four scale-dependent ones apply to each of their named scales. Naturally you would show the most detail at 1/4" [1:50] scale and the least at 1"=40' [1:500].

6. Click the Display Props tab.

Notice that **Plan 1-50** is the active Display Rep. Also notice that the controlling property source is System Default. All settings for all Display Reps in this template are configured at the System Default level. Style and object overrides are reserved for project-specific needs.

7. Click the Edit Display Props button.

There are three more tabs in this dialog box. There is the standard Layer/Color/Linetype tab. Here the AEC Dimension Group Marker is shown, on Defpoints layer in color Magenta. The same has been done for the Removed Points Marker, except that it is assigned Color Red. As we have done in several other places throughout this book, sub-components that are used for informational purposes are set to the non-plotting Defpoints layer. (You could also use a layer such as A-Anno-Nplt that is set to the NoPlot Layer State.) Colors like Red and Magenta have been used consistently to flag this condition. You are reminded that this practice is highly recommended.

8. Click the Contents tab.

Here we assign the behavior of the AEC Dimensions for each of the three chains and for each class of object. Walls, Curtain Walls and Openings are among the special classifications of objects that can be configured separately. For instance, at 1/4" [1:50] scale (the current Display Rep), Chain 1, the one closest to the selected objects, is set to dimension either the Wall length or the individual Wall components. This gives a very detailed dimension string. Chain 2 is set to dimension only the outer boundaries or the Wall width (less detail), and finally Chain 3 is set to dimension just the overall length (see Figure 7–55). Feel free to click through the other Object/Chain combinations.

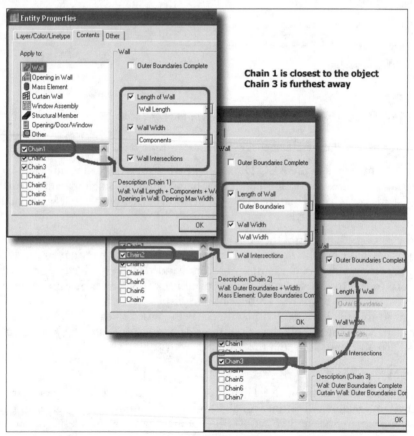

Figure 7–55 The Contents tab includes detailed settings per object type and chain

9. Click the Other tab.

The primary setting on this tab is the Dimension Style. As was stated above, AEC Dimensions reference an AutoCAD Dimension Style. This is good news, since you likely have an office standard Dimension Style already established in your firm. Simply choose it here, and click the **Edit** button to tweak it and set the scale. In this case, the Dimscale is set to **48** as you would expect (see Figure 7–48).

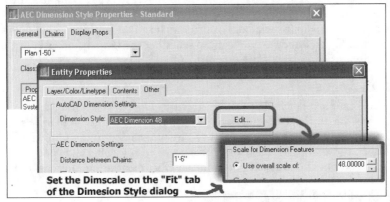

Figure 7–56 *A Dimscale of 48 sets the Plan 1-50 Display Rep scale*

10. Click **Cancel** once, change the Display Rep to **Plan 1-100**, and then click **Edit Display Props** again.

Notice that at this scale (1/8"), the Dimension Style is now **AEC Dimension 96**, which has a Dimscale of **96** (see Figure 7–57).

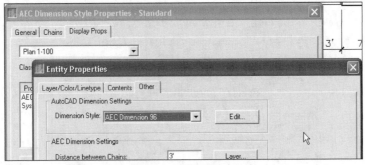

Figure 7–57 *A Dimscale of 96 sets the Plan 1-100 Display Rep scale*

11. Click the Contents tab again.

Here, Walls, Chains 2 and 3 are configured the same as they were in the Plan 1-50 Rep. However, Chain 1 is turned off. This was the very detailed Chain at 1/4" scale, which is not required at 1/8". Repeat these steps for the other two scale-dependent Display Reps as well. You will notice similar strategies being used to limit the level of detail in the Dimension strings as the scale diminishes. Feel free to use these settings in your own template files and tweak them as necessary to suit your firm's needs.

As you can see, with a little up-front effort, you can save yourself a lot of data duplication and layer swapping. Perhaps the best benefit of these techniques is the elimination of the need to coordinate two sets of dimensions. Since you now have a single intelligent dimension string, you can be confident that it is always up-to-date!

If you are not getting the results you wish from AEC Dimensions by letting ADT choose the points automatically, you can either add or remove points from the automated strings, or you can use the **Add Manual AEC Dimension** or the **Convert to AEC Dimension** command. Be careful—these commands actually generate a series of AutoCAD Point (nodes) objects to which the dimension points remain associated. Unfortunately, a flaw in this routine puts these points on the current layer. Therefore, you will want to set either the Dimension layer or the Defpoints layer current before using either of these commands.

DIMENSION LABELS

Dimension Labels are text labels that can be added to Doors, Windows, Openings and Stairs. These labels show dimensional information about the object. Dimension Labels are created on the fly by the software and therefore do not need to be included in the template files. To add a Dimension Label, you simply right-click the object you wish to label and choose the appropriate label command. You can also access them from the **Documentation>Dimension Labels** menu.

Working with Dimension Labels

1. In the *Chapter07* folder, **Open** the file named *Dimension Labels.dwg*.

This very simple file contains a Door, a Window, an Opening and a Stair. The file has opened to a paper space layout where there are three viewports, two at 1/8"=1'-0" scale and the other at 1/4"=1'-0" scale. The 1/4" viewport is set to the Plot Plan 48 Display Configuration. The other two use Plot Plan 96 and Work. We will add a Dimension Label to each object type in this drawing to see the way they behave.

2. Double-click in the Plot Plan 48 viewport to make it active.

3. Select the Door, right-click and choose **Insert Door Label**.

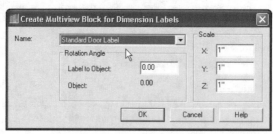

Figure 7–58 *The Door Label is generated on the fly when the command is executed*

The Create Multiview Block for Dimension Labels dialog box gives us a small clue by its name that the Multi-View Blocks used for Dimension Labels are generated on the fly by the software when the command is executed (see Figure 7–58).

4. Click **OK** to add the Label in the default location.

5. Select the Window, right-click and choose **Insert Window Label**.
6. This time in the dialog box input **90°** in the **Label to Object** rotation angle and click **OK** (see Figure 7–59).

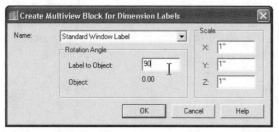

Figure 7–59 *Rotate a label as it is inserted*

7. Following the same steps, add an **Opening Label** and a **Stair Label**.

Once the labels have been added, you can move the labels around on screen and position them wherever it is convenient (see Figure 7–60). They remain linked to the properties in the object. Therefore, if you edit the size or location of one of the objects, the label will update. In addition, you can edit the prefixes used to designate each label. To do this, open the Options dialog box.

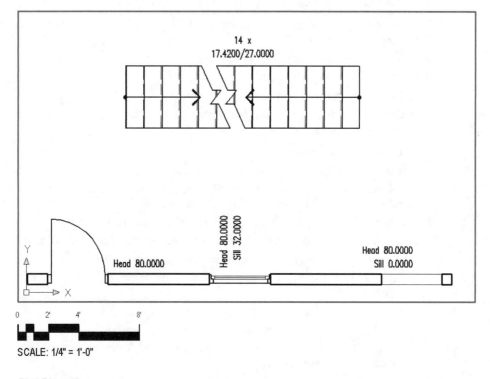

Plot Plan 48

Figure 7–60 *Labels added to each object can be moved around after placement*

8. From the **Tools** menu, choose **Options** and then select the AEC Dimensions tab.

In the **Dimension Label Prefix** area, type your preference for each prefix. Existing labels will update when you return to the drawing. Unfortunately, if you work in Imperial units, you are only able to express your Dimension Labels in Inches or Feet, but not both together. Therefore, you will likely want to keep the setting set at Inches. If you work in Metric units, there is more flexibility in these settings. For instance, Millimeters can be expressed in superscript to make your labels easier to read.

 CHECKLIST

I Text Annotation

1. Configure Drawing Setup settings as the System Default and in template file(s).
2. Create Custom Layer Keys as required for Content items.
3. Create, import or edit all office standard Text Styles for all conceivable uses. Save these to the office standard template file(s).
4. Create Content items for standard Notes, General Notes and other Text.

   a. Create Content items that insert standard blocks of text.

   b. Create Content items that simply run the **Mtext** command.

5. Explore all Documentation Content items in Imperial, Metric and Metric D A CH folders.

   a. Decide which items to keep as is.

   b. Decide which items to edit.

   c. Decide which items to delete.

   d. Restructure the Content sub-folder structure and accompanying ADT menu structure as required.

II Scale-dependent Annotation

1. Decide which tags ought to be scale-dependent.
2. Determine if Custom Display Reps will be required and create them in the office standard template file(s) as soon as possible.
3. Open an existing Tag that is close to the one you wish to create.
4. Save the Tag to a new name and/or location in.the office standard Content tree.
5. Edit/create the required View Blocks for each scale-dependent Display Rep.

   a. Keep in mind the "rules" associated with using Attributes in Tags from Chapter 6.

   b. Use Layer 0 for all Block components.

   c. Use ByBlock for all internal Block component properties.

   d. Create scale-dependent Blocks "pre-scaled" at the Text Height for Model Space (MH) shown in Table 7-1.

   e. Name each View Block with the MVB name plus a suffix indicating scale, such as Room_Tag_100.

5. Create or Edit the MVB to reference each View Block in the appropriate Display Rep.

6. Save the Content file using the AEC Content Wizard.

   a. Use the AECANNOSCHEDULETAGADD command.

   b. Reference the default *PropertySetDefs.dwg* file discussed in Chapter 6.

   c. Be sure **Additional Scaling** is set to **None**.

   d. Remember to purge the existing definition of the Tag from the current drawing before testing.

III Chases, Elevation and Dimension Labels

1. Create Chases similar to any other "full scale" MVB Content.

   a. Optionally, create additional MVBs to use as labels for Chases.

2. Customization potential for Elevation and Dimension Labels is limited. Critically analyze the out-of-the-box offerings and document a procedure for their use.

IV Dimensions

1. Create, import or edit your office standard Dimension Styles for each scale that you require.

   a. Set the parameters for each scale identically save for the "Overall Scale" (Dimscale).

   b. Save these Dimension Styles to the office standard template file(s).

2. Determine and document the extent of your firm's use of AutoCAD Dimensions, AEC Dimensions or both.

3. If using AEC Dimensions, create office standard AEC Dimension Style(s) and configure the settings.

   a. Set chains to 3.

   b. Set up Layer/Color/Linetype settings in Display Props for all Reps.

   c. On the Content tab, configure each object type/Chain combination for each Display Rep.

   d. Allow the level of detail to diminish as the Chain number increases (moves further away from the dimensioned object).

V Save and Document All Configurations

1. Where possible, save items to the template file(s).

2. Items not in the template should be saved to the library.

3. Document all settings and resources for firm-wide dissemination.

SUMMARY

The Drawing Setup command synchronizes the Units and Scale settings and can control the size of drawing annotation.

Using Drawing Scale makes enforcing office standards easier.

Using Annotation Plot Size makes annotation objects insertion more flexible.

ADT uses standard AutoCAD commands for Text.

You can create custom AEC Content to create Text objects and automatically layer them.

The default Documentation library includes many items that you will likely choose to modify or eliminate.

Multi-View Blocks used for annotation can be created to automatically adjust to the scale of the drawing.

Scale-dependent MVBs take effort to build, but once in the library, they work just like all others.

To create interference with objects in 3D, while having complete control over the plan graphics used to represent the condition, use Chase objects.

Elevation and Dimension Labels are "hard-coded" annotation symbols that link directly to certain live information in the drawing.

Use AutoCAD Dimension objects when dimensioning AutoCAD entities.

Use either AutoCAD non-associative dimensions or AEC Dimensions to dimension ADT objects.

CHAPTER 8

Content Library

INTRODUCTION

Although a well-conceived template file(s) is essential to achieving a successful ADT implementation, a well-stocked Content library is just as important. In Architectural Desktop, all styles and display settings are part of the drawing file. Therefore, it can be tempting to include all required symbols and styles directly in the template files. However, each such item that is included in the template increases the size of every drawing file based on that template. This can add up quickly to bloated and unwieldy file sizes. For this reason, it is important to consider carefully those items that are better included in a readily accessible office standard library instead of the template files.

OBJECTIVES

Your Content Library is nothing more than a series of folders on the local network. ADT ships with a large library of ready-to-use Content items. Over the past two chapters, we have been working quite a bit with the Documentation node of this library. In this chapter, we will focus more on the Design node and the features that its content has to offer. In this chapter, we will explore many issues related to evaluating, creating and customizing your Design Content Library, including the following:

- Explore the Design Content Library
- Understand Content naming and orientation
- Understand scaling issues
- Edit existing content
- Build custom content

UNDERSTANDING WHAT IS PROVIDED BY DEFAULT

By now, we have been to the ADT DesignCenter on a variety of occasions throughout the previous chapters. On those occasions, you have been encouraged to click on each folder and explore the various offerings. This is a much bigger task than in the preceding chapter, due to the much larger quantity of Content items contained in the *Design* folders. For instance, the Imperial *Design* folder contains 658 Content items located in eight major categories. The Metric *Design* folder contains 384 Content items in seven categories, many of which do not appear within the Imperial folders. The Metric D A CH *Design* folder contains only three categories and 256 Content items.

Since there is not much duplication between the three *Design* folders, you will want to spend some time exploring all three. The total collection from all three folders includes 1298 Content items. It will certainly take some time to acquaint yourself with all of these resources, so be sure to allow an appropriate amount of time.

UNIT SCALING ISSUES

Naturally, if you intend to use Content items from both Metric and Imperial folders, you will run into issues related to scale. Metric sizes do not translate exactly to Imperial and vice versa. You may even encounter this within the Metric folders as well, due to the different building standards holding jurisdiction in each country. For instance, a standard brick is sized slightly differently in nearly every major country. For this reason, you will need to have these issues in mind while exploring the Content offerings.

Let's try an experiment:

Install the CD Files and Understanding AEC Unit Scaling

1. Install the files for Chapter 8 located on the Autodesk Architectural Desktop: Advanced Implementation Guide CD ROM.

Refer to the "How to Use the CD" section in the Preface for instructions on installing the sample files included on the CD.

A folder named *Chapter08* was created containing a collection of sample files that we will use throughout this chapter.

2. In the *Chapter08* folder, **Open** the file named *Units.dwg*.

This file is in Imperial units. We are going to explore what happens when you insert a piece of Metric Content into an Imperial drawing or vice versa.

3. At the command line type **undo**, press ENTER, then type **m**, and press ENTER again (see Figure 8–1).

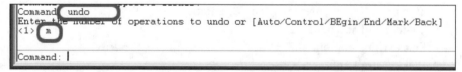

```
Command: undo
Enter the number of operations to undo or [Auto/Control/BEgin/End/Mark/Back]
<1> m

Command: |
```

Figure 8–1 *Setting a "Mark" in the Undo command allows you to later undo several operations in one step*

We will be performing a few steps that we will want to reverse shortly. By entering the UNDO command and "marking" the current spot, we can jump back to here in one step later.

4. Open the DesignCenter and navigate to the **Metric>Design>Bathroom> Accessories** folder.

5. Drag the **Towel Rail** symbol into the drawing and place it somewhere in one of the restrooms (see Figure 8–2).

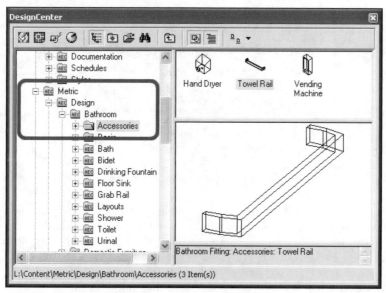

Figure 8–2 *Accessing Content items in one of the Metric folders*

Visual inspection reveals that it came into the drawing at a size appropriate to Imperial units. This can be confirmed by measuring actual sizes. Even though the towel bar is drawn 515 millimeters long in its own file, it measures as 1'-8 9/32" here.

6. From the **Desktop** menu, choose **Drawing Setup** and click the Units tab.

Notice the check mark in the **Scale Objects Inserted From Other Drawings** check box. This setting determines whether objects built in different units will scale or not when they are inserted (see Figure 8–3).

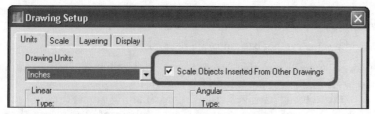

Figure 8–3 *Scale Objects Inserted From Other Drawings automatically resolves differences in scales between the library and the current drawing*

7. Click **Cancel** to dismiss Drawing Setup.
8. From the **Desktop** menu, choose **Multi-View Blocks>Add Multi-View Block**.

9. From the **Name** list, choose **M_BATH_ACC_Towel Rail** and set all three scales (**X**, **Y** and **Z**) to **1** (see Figure 8–4).

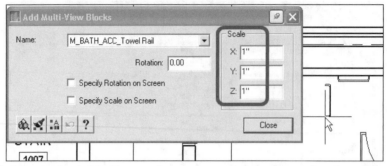

Figure 8–4 *Once the Symbol has been scaled, it remains the new size in that drawing*

Notice that the Symbol definition itself has been scaled, not the reference. (This is true of each of the View Blocks as well.) This is important, as it will control the behavior of this particular symbol within this particular drawing from now on, regardless of whether it is inserted manually or through the DesignCenter, even if we clear the **Scale Objects Inserted From Other Drawings** check box in Drawing Setup, (unless we first purge this symbol from the current drawing). We can confirm this by turning it off and dragging it again.

10. Return to the Drawing Setup dialog box and clear the **Scale Objects Inserted From Other Drawings** check box and then click **OK**.

11. Drag the same symbol back into the drawing.

Notice that there is no change in its behavior. It is still scaled, because the local definition in this drawing has been modified.

12. At the command line type **undo**, press ENTER, and then type **b**, and press ENTER again.

The drawing will be returned to the state that it was in above when we set the mark.

13. Return to Drawing Setup and clear the **Scale Objects Inserted From Other Drawings** check box and then click **OK**.

14. Drag the same symbol back into the drawing.

Notice that it is now huge! From this exercise, you can see the value of the automatic scaling feature and can begin to appreciate some of it ramifications. For instance, 1'-8 9/32" is very close to 20"—a standard towel bar size in the United States. At most drawing scales, this 9/32" disparity would go unnoticed. It is only if an actual dimension was placed that the disparity would be noticed. In most projects, the exact size of the towel bar could be off by even more than 9/32" without causing any issues down stream. However, many other building component sizes are much more critical. Therefore deciding exactly which pieces of content to use from the other libraries is an important decision, as is deciding whether to let ADT scale them automatically (as admirable a job as it does) or to open the Content items and scale them manually and with control that is more exact.

COMMON FEATURES OF DESIGN CONTENT

One of the features of the ADT Content path is that it makes that node of the DesignCenter's interface interactive. In other words, you can select any Content item in the DesignCenter icon window and three-dimensionally rotate it in the preview pane. You can also choose any of the six orthographic views in the DesignCenter Viewer pane as well (see Figure 8–5).

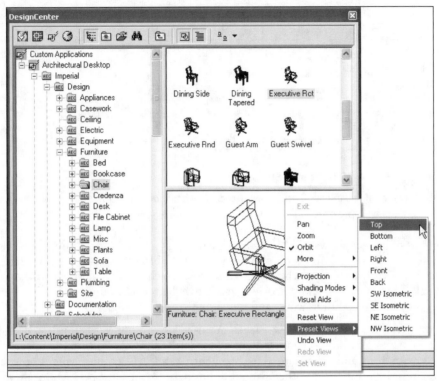

Figure 8–5 *Right-click in the Preview Pane to access additional viewing options*

You can even shade the preview window to view 3D views better. As nice as these features are, the process can be a bit tedious if you are trying to view several symbols this way and from many different views.

You may recall, when using the AEC Create Content Wizard in the previous chapter, that AEC Content can be entire Drawings, Blocks, Multi-View Blocks, Masking Blocks or Custom Commands. Most of the Documentation content was Custom Commands. Most of the items in the Design nodes of the Content Library are Multi-View Blocks. A few, in particular many of the ceiling fixtures, are Masking Blocks. Even fewer are simply AutoCAD Blocks.

Multi-View Blocks (MVB) are used as content for two major reasons. The first is implied by its name, "Multi-View." An MVB can contain several AutoCAD View Blocks, which represent the object from different viewing angles. The benefits of this are evident when comparing the top of Figure 8–6 with the bottom.

366

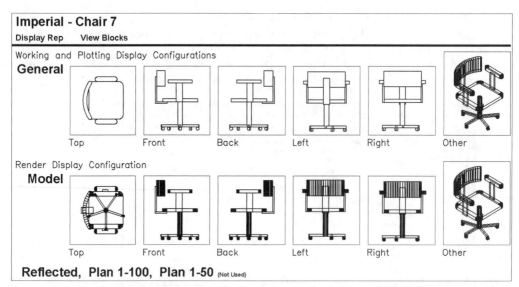

Figure 8–6 *A typical Multi-View Block shown from all orthogonal views in both General and Model Display Representations*

The top row of images shows the MVB from each orthogonal view direction when the active Display Rep is General. The bottom row shows what the object looks like if the Model Display Rep is active. Clearly for most architectural drawings, which require a level of abstraction, the View Blocks used in the General Display Rep will give better results. The Model Display Rep is usually only used for 3D Drawings and renderings.

The second reason that Multi-View Blocks are used as content is demonstrated when the object must be an ADT object rather than a standard AutoCAD entity. This occurs most frequently with Tags and other "anchored" content. AutoCAD entities cannot be anchored; therefore, in order for a Tag or other symbol to be anchored, it must be a Multi-View Block. However, this does not mean that they must contain several View Blocks. You have likely noticed in the previous chapters that most of the Multi-View Blocks that we worked with had only a single View Block.

DO WE NEED THE 3D?

Many of the Design Content Library items are hybrid 2D/3D Multi-View Blocks. This means that they have several View Blocks like the chair shown in Figure 8–6 that intelligently respond to the drawing's current view direction. Many ADT users question the necessity of including 3D View Blocks in their Content library, contending that they only serve to bloat file sizes and slow down drawings. It is important to understand the precise behavior of an MVB thoroughly before answering the question of 3D necessity.

First, how do we identify when a Multi-View Block contains a 3D View Block without inserting it? There is not a "guaranteed to work" way to do this, but if we follow the lead of the creators of the default library, a 3D MVB will be very easy to spot. Every MVB that contains 3D geometry in the default ADT library was saved in a 3D view direction. This means that when these files were saved as AEC Content, the 3D view was used to generate their icon preview in the DesignCenter. Since there is no way to force users to remember to save their 3D files in 3D, this tip is not "fool-proof." However, it is a safe bet that saving 2D files in Plan view and 3D files in 3D would be obvious to most people. It is recommended that you make this a documented policy in your firm (see Figure 8–7).

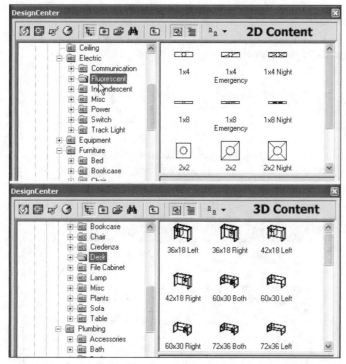

Figure 8–7 *Identifying 2D Content and 3D Content in the default DesignCenter*

So now that we know how to identify when a Content item contains 3D, the question of whether it is necessary still remains. For instance, many an ADT user has questioned the wisdom (and necessity) of including 3D blocks within toilets and other plumbing fixtures. The simple answer to the question is:

> *Yes, the 3D should be included, and it is necessary.*

The reason why is that even though a Multi-View Block can contain several View Blocks, it can only contain four elevation View Blocks. This means that if you do not happen to be viewing the MVB from exactly 0°, 90°, 180° or 270°, it is considered "Other." These View Directions are relative to the MVB, not the drawing. So if you insert a chair at 45° to go with a corner desk, the chair will display its "Other" Block even though you may be viewing the drawing from the "Front" (or 270°). Let's have a look.

Understanding View Blocks

  1. In the *Chapter08* folder, **Open** the file named *3D View Blocks.dwg*.

If you examine the Section, you can see that every Multi-View Block in this file is currently displaying its 3D View Block. Without those 3D Blocks, those items would not display at all, even though this section is viewing the drawing from the standard orthogonal view of Front. You can test this out by removing the 3D Blocks from the General Display Rep.

2. In the plan, select one of the urinals, right-click and choose **Edit Multi-View Block Definition**.

3. On the View Blocks tab, for the General Display Rep, select the **I_Plumb_Urinal_Wall-Hung_M** View Block.

Notice that this View Block is used for the Other view direction. Many people get confused by this. If General is intended for 2D views, why does it contain a 3D Block in "Other?" Again, the answer is simple. Although the directions of Front, Back, Left and Right are the most common, they are not the only views that we will need to view this object as an elevation. Other is used when none of the six fixed view directions is active. By including a 3D Model in the General Display Rep, you make it possible to create an elevation of your model from *any* view direction and be confident that certain objects will not simply disappear in that view.

4. With the 3D Bock still selected, click the **Remove** button and then click **OK** (see Figure 8–8).

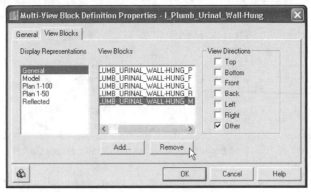

Figure 8–8 *Remove the 3D View Block from the Urinal MVB*

5. Select the Section, right-click and choose **Update**.

6. In the Generate Section/Elevation dialog box, click **OK**.

Notice that the Urinals have disappeared from the section. This is because the Display Set that is used to generate sections and elevations references the General Display Rep for Multi-View Blocks. The conclusion regarding 3D Blocks is that, even though it is unlikely you will create too many 3D renderings of toilet rooms, the 3D Block is still a necessary component of the typical piece of Design Content. It is recommended that you not strip these out of your library.

DECIDING WHAT TO KEEP AND WHAT TO REMOVE

As you analyze each of the library items for its usefulness for your firm, many of the Content items will likely be rejected in this process. However, many will prove to be quite useful. Still others will prove useful after they have been modified in some fashion. This same process was suggested in Chapter 6 when we were evaluating the Schedule Tags and again in Chapter 7 when analyzing the Documentation Content. Take the time to perform this analysis of the included content before committing any time and resources to customizing existing items and/or building new items from scratch. In the end, daunting as it may seem to explore 1300 Content items, it will likely take less time to do this than to build everything from scratch. The other benefit to this process is that you may discover ways of performing certain tasks or representing certain items that you had not considered before. Reverse engineering existing content can prove a tremendous learning experience.

WHERE TO STORE THE CONTENT LIBRARY

The default ADT content is contained in a collection of folders within the *Content* folder installed with ADT. As you recall, we have moved this folder to the server on the L Drive and mapped our OfficeStandard profile to this location. A copy of the complete structure has also been kept in our Management Install to use as a working copy until we are ready to post our changes to the office's public library on the L Drive. When you installed ADT, you chose between Metric and Imperial and if you chose Imperial, you chose between Architectural Desktop and CSI Master Spec. This choice is responsible for determining which of the three main nodes (Imperial, Metric or Metric D A CH) is accessed from the drop-down menus. (It has been noted on many occasions that regardless of units choice, you can always access all Content items located in the root *Content* folder from the DesignCenter itself.)

BUILDING A CUSTOM FOLDER STRUCTURE

As you begin to consolidate, expand and customize the existing ADT Design Content Library, you may find yourself wanting to add, delete or rename folders within the hierarchy. If you can avoid this, your implementation of Design Content will be simpler. However, the chances are good that the existing folder structure will fail to suit your firm's needs at some point. You will have a few options for dealing with this inevitability. You can add a completely separate folder and pull-down menu for your firm's customized library, you can access the separate *Office Standard* folder only from a customized toolbar icon (as covered in Chapter 7), or you can customize the default ADT folder and menu structure to better suit your firm's needs.

If you decide to customize the folder structure, then the menu pointers on the **Design>Design Content** and **Documentation>Documentation Content** menus may no longer work. This will be the case if you rename or remove a folder. If a folder referenced by a menu command no longer exists where it expects to find it, the command will simply fail. It will not cause any other detriment to the system. However, this can certainly be frustrating to the user, who will receive no message explaining the reason for the failure. Therefore, if you customize the default folder structure of the ADT Content Library, you will want to consider modifying the menu structure to match it. There are a few basic steps to perform to do this.

1. Two files must be copied and edited, *AecArchx.mnu* and *AecArchx.mnl*.
2. Also, copy the *AecArchx.dll* file, which contains the toolbar bitmaps.

Never edit a menu file without first making a backup copy for safekeeping. In Chapter 1, we discussed the possibility of editing the default menu files and recommended locating the customized versions on the server in the I Drive *Support* folder. The two files listed here are the menu template file for the ADT menus (.MNU) and the menu lisp source file (.MNL). These two files must have the same name (but do not change the extensions), so if you rename your copy, be sure to rename both to the same root name.

3. Add one or more functions to the MNL file for any custom routines.

In the MNL file, copy the function that we defined in Chapter 7 for the *Office Standards* folder icon. Copy just the function definition and remove the command execution at the end. In other words, in the code we used in Chapter 7, we defined the function "AIGADC" and then immediately executed it. In this case, we only want to define the function, but not execute it (see Figure 8–9).

```
                    Add a comment here for future reference
;; Load Office Standards Design Center Folder
^C^C(defun c:AIGADC () (command "adcCustomNavigate" "Architectural Desktop/Office Standards"))
                    Define the function here without the command line call
```

Figure 8–9 *Adding a function to the MNL file with a comment*

It is good practice whenever working in menus or other code to add comments to your additions. In the MNL file, this is done with a double semi-colon (;;).

4. Add a menu item to the appropriate section of the MNU file.

For the Design menu, the appropriate section is Pop2 for Imperial and Pop5 for Metric (see Figure 8–10).

```
;;; ================================================================
;;; Load Architectural Desktop menus...
;;; ================================================================

;; First, remove all of our pull-down menus
;; to prevent loading both Imperial & Metric versions
(menucmd "GAECARCHX.pop1=-")    ;; [Concept]
(menucmd "GAECARCHX.pop2=-")    ;; [Design]          <- Imperial (ADT or CSI)
(menucmd "GAECARCHX.pop3=-")    ;; [Documentation]   <- Imperial
(menucmd "GAECARCHX.pop4=-")    ;; [Desktop]
(menucmd "GAECARCHX.pop5=-")    ;; [Design]          <- Metric
(menucmd "GAECARCHX.pop6=-")    ;; [Documentation]   <- Metric
(menucmd "GAECARCHX.pop8=-")    ;; [Help]            <- ArchDesktop's
(menucmd "GACAD.pop10=-")       ;; [window]          <- Acad's
(menucmd "GACAD.pop11=-")       ;; [Help]            <- Acad's
(menucmd "GAECARCHX.pop7=-")    ;; [Design]          <- Imperial CSI (renamed as pop2)
```

Figure 8–10 *Understanding the AecArchX POP menu designations*

You can see the complete list of Pop designations for the ADT menus by reading the MNL file. Open it in Notepad and scroll to the "Load Architectural Desktop menus" section (see Figure 8–11).

```
***POP2                             Add your menu item where
**ARCHX2_IMPERIAL_ADT               you want it to appear
AECID_MnDesign              [Desi&gn]
AECID_MnDesignContent       [->Desig&n Content]
AECID_DesignContentAppliances      [&Appliances...        ]^C^C_AecDcSetImpAppliances
//       AECID_DesignContentCasework    [&Casework...       ]^C^C_AecDcSetImpCasework
                                    [office Standards]^C^C_AIGADC
AECID_DesignContentCeiling         [Ceiling Fixtures...    ]^C^C_AecDcSetImpCeiling
AECID_DesignContentElectric        [&Electric Fixtures...  ]^C^C_AecDcSetImpElectric
AECID_DesignContentEquipment       [E&quipment...          ]^C^C_AecDcSetImpEquipment
AECID_DesignContentFurniture       [&Furniture...          ]^C^C_AecDcSetImpFurniture
AECID_DesignContentPlumbing        [&Plumbing Fixtures...  ]^C^C_AecDcSetImpPlumbing
AECID_DesignContentSite            [&Site...               ]^C^C_AecDcSetImpSite
                                    [--]
                    If you want to remove an item, just comment it out
```

Figure 8–11 *Add a menu item or comment out ones you want to "remove"*

Use comments to eliminate menu items you want to eliminate. This is better than actually deleting the item, as it will be very easy to restore it later if required. In the MNU file, the comments use two slashes (//).

5. Create a "Test" profile to test the menu before you load it to your OfficeStandard profile.

6. In the Test profile, unload the AecArchX menu and then load the customized version.

ADT will warn you about overwriting the source menu. This is why we copied the files and kept a backup of the original. This is also why you were instructed to load it to a "Test" profile first. These two measures will ensure that you do not end up with a non-functioning system should something go wrong. If the new menu does not work, repeat the steps to fix the problem or reload the originals and start again. Also, remember to consult a copy of Sham Tickoo's book *Customizing AutoCAD 2002* for additional information on customization of AutoCAD resources, including menus. If you do change the menu structure to the point where a modification to your OfficeStandard profile (created in Chapter 1) becomes necessary, refer to Chapter 1 in the "Argggg… Why Won't the ARG Work?" sidebar for techniques on how to quickly propagate the new version of your OfficeStandard profile.

CONTENT BUILDING CONSIDERATIONS

When the time comes for you to build your own MVB Content, several standard conventions should be followed to keep your Content consistent with those provided by default (and those created by third party providers or manufacturers). There is an entire document describing these guidelines available on the Autodesk Point A web site. The document is named "Style Guide: Creating i-drop Content Specific to Autodesk Architectural Desktop." The URL to the latest version of this document is included in Appendix A, and a copy has been provided in PDF format in the *Chapter08* folder. It is named *1887354_Style_Guide_-_Creating_i-drop_Content_Specific_to_ADT.pdf*. However, please download the latest version from the URL in Appendix A, because the document has likely been updated since press time. This document is included courtesy of Autodesk, Inc. © 2002, used with permission, all right reserved. Highlights of these guidelines are summarized in the following topics.

WHAT TO INCLUDE IN THE CONTENT FILE

There are three distinct pieces to the typical Content item. (Refer below to Tables 8–1 and 8–2.) These are the AutoCAD View Blocks, the Multi-View Block and the Content file itself. In reality however, these designations are not completely correct, when you consider that there are five different types of Content items: Blocks, Drawings, Multi-View Blocks, Masking Blocks and Custom Commands. Perhaps then we ought to refer to the three pieces as "components," "Content items" and "Content files."

▶ **Component** – Any item required by the Content item for graphical display or data integrity. View Blocks are components of Multi-View Blocks, a Multi-View Block could be a component of a Masking Block and Blocks and Multi-View Blocks could be components of a Custom Command.

▶ **Content Item** – The final object or objects that result from insertion through the DesignCenter. It might be a single Block or Multi-View Block, or it could be a collection of objects imported by a Custom Command or the result of an exploded Block.

▶ **Content File** – The DWG drawing file saved in the Content Library folder structure. It is simply the container for storing an individual Content item. Within it are saved all of the components and the instructions for insertion through the DesignCenter (as established by the AEC Create Content Wizard).

You can include any objects in the Content File, even those that are not part of the Content item. Only those items specifically selected in the Content Wizard will be imported into other files when the Content is dragged and dropped from the DesignCenter. However, you will likely want to purge these files anyway to help keep them small. To help prevent accidental deletion of required items, insert all Blocks and required components into the Content file so that they will not be inadvertently purged. Be certain that you include only those items that you want to appear in the preview graphics of the Content item on screen and delete all other extraneous elements. When you click the **Default Icon** button in the AEC Create Content Wizard, an icon is generated from the extents of the Content file.

NAMING

You will need to select three distinct names for every piece of AEC Content that you create: the Content file, the Multi-View Block and the View Blocks. Content should be named in a consistent manner. If you select any one of the Multi-View Blocks in the file from the previous exercise and edit its definition, you will see that names all have a prefix of "I_," which stands for Imperial. This is followed by several fields of descriptive abbreviations. Table 8–1 shows all of the MVBs in the current file.

Table 8–1 *Content File Naming*

Content File Name	Style	Typical View Block Name	Description
Chair 7	I_Furn_Chair_Chair 7	I_Furn_Chair_Chair 7_P	Furniture: Chair: Chair 7 [12500]
Desk 1	I_Furn_Desk_Desk 1	I_Furn_Desk_Desk 1_P	Furniture: Desk: Desk 1 [12500]
36in + plate	I_Plumb_Grab_36in + plate	I_Plumb_Grab_36in + plate_P	Plumbing: Grab Bar: Grab Bar - 36in (w/ plate) [10800]
42in + Plate	I_Plumb_Grab_42in + Plate	I_Plumb_Grab_42in + Plate_P	Plumbing: Grab Bar: Grab Bar - 42in (w/ plate) [10800]
Counter 4	I_Plumb_Lav_Counter 4	I_Plumb_Lav_Counter 4_P	Plumbing: Lavatory: Counter Lavatory 4 (2D) [15400]
Wall-Hung	I_Plumb_Urinal_Wall-Hung	I_Plumb_Urinal_Wall-Hung_P	Plumbing: Urinal: Wall Hung Urinal [15400]
Flush - Wall	I_Plumb_WC_Flush - Wall	I_Plumb_WC_Flush - Wall_P	Plumbing: Toilet: Flush valve, wall hung WC [15400]

As you can see, the View Blocks include the complete name of the Multi-View Block with a suffix indicating the intended view direction. The complete list of recommended suffixes is included in Table 8–2. The first few fields of the name indicate the location of the file in the folder structure of the Content library, which is also a way of categorizing the Content items. Therefore, "I_Plumb_Grab_42in + Plate" is located in the *Imperial>Plumbing>Grab Bars* category/folder. This Content item is a 42-inch grab bar with wall plates. The Name that shows in the DesignCenter is simply "**42in + Plate**."

Table 8–2 *View Block suffixes*

Suffix	Meaning
_P	Plan - used for Top View in General and Reflected
_L	Left View
_R	Right View
_F	Front View
_K	Back View
_B	Bottom View
_M	3D Model used for Other in General and all views in Model Rep
_C	Ceiling - use for Reflected Rep when P is being used in General
_##	Numeric Scale Factor for scale-dependent Block in Plan 1-50 and Plan 1-100

If the item is symmetrical in both directions, the "L" View Block is typically used for both Left and Right directions and the "F" View Block is used for both Front and Back. When the item is Ceiling content that only appears in Reflected Ceiling Plan, the "P" View Block is typically used to represent the Top view for the Reflected Rep. If you have an item that spans from floor to ceiling and therefore shows in both floor and ceiling plans, use the "P" Block for the General Display Rep Top view and the "C" Block for the Reflected Display Rep Top view.

BLOCK ORIENTATION

It is very important that all views be built true to the model in the correct sizes, scales and orientations (see Figure 8–12).

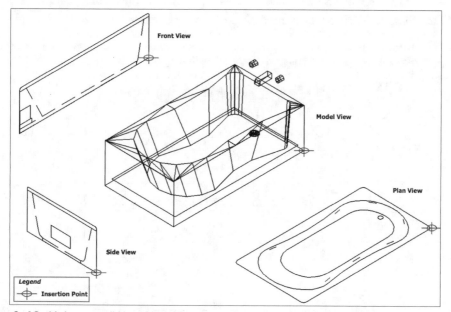

Figure 8–12 *Make sure all View Blocks have consistent insertion points and orientations*

USE OF PROPERTIES

There is no single Multi-View Block Layer Key—there are in fact several. This is because Multi-View Blocks can be used to represent any type of object. You can add your own custom Layer Keys if required for the Content items that you build. (Refer to the previous chapter for an example.) As a general rule of thumb, it is not recommended that you "hard code" layers into the View Block of your Content. You are encouraged to use Layer 0 internally instead. This will make the entire MVB respond to the Layer Key of the Content item when inserted. In this way, your content will remain flexible and able to conform to future changes in layering standards without rework.

This same recommendation is made for the other AutoCAD properties as well. However, it is most important that you avoid hard-coded Colors. All entities in the View Blocks that make up your Multi-View Blocks should be drawn on Layer 0 with the Color, Linetype, Lineweight (and Plot Style if using Named Plot Styles) set to ByBlock. If it is absolutely necessary to convey intent, you can hard code the Linetype and/or Lineweight. Nevertheless, consider this choice carefully before making that decision. Again, by avoiding hard-coded properties, your Blocks will remain very flexible and will migrate easily to any future office standard.

ALTERNATE INSERTION POINTS

Multi-View Block Content items can be created with alternate insertion points. When such points are present, you can cycle through them at the time of insertion to facilitate better placement of the object in the drawing. Continuing to work in the same file as above, let's see an example of this feature. (If you already closed the file, in the *Chapter08* folder, open the file named *3D View Blocks.dwg*.)

Insert an MVB with Alternate Insertion Points

1. From the **Design** menu, choose **Design Content>Furniture**.
2. Click the **Chairs** folder and double-click any chair (see Figure 8–13).

Note: You must double-click to insert the Content item in order to cycle through the alternate insertion points. Cycling will not work with drag and drop.

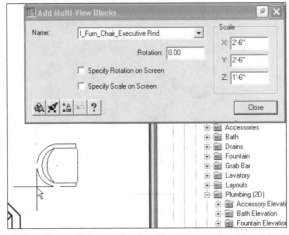

Figure 8–13 *Double-click a Content item to access its alternate insertion points*

3. Simply press CTRL to cycle to the next insertion point. Continue to press CTRL to cycle through all points.

Notice the symbol shift relative to the cursor. This is a very handy way to place objects precisely without the need for an extra step. To add alternate insertion points within a Multi-View Block, simply add one or more AutoCAD Point objects on the Defpoints layer within the plan View Block. If these points exist, the Block will cycle through them with CTRL. The order that the points are cycled through is determined by the order that they are added to the Block definition at the time of creation (see Figure 8–14).

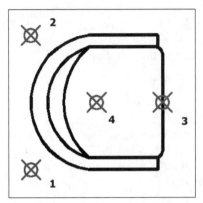

Figure 8–14 *Insertion points cycle in the order in which they were added to the View Block*

USING AEC POLYGONS IN CONTENT

The View Blocks used for the General Display Rep are typically two-dimensional drawings that are standing "upright" in the correct location relative the insertion point of the MVB. They are like cardboard cutouts of the object used for generating the elevation or section. Normally, other objects will exist "behind" the furniture and equipment MVBs that ought to be concealed when you generate Sections and Elevations from your models.

To have your Content items conceal other items in the model automatically, you will need your 2D Blocks to contain a surface. It is recommended that you use AEC Polygons for this purpose. (AEC Polygons use fewer resources than Regions do, so drawings will perform better.) Simply draw a polyline in the shape of the outer profile of the object. (Use the **Draw>Boundary** command to quickly create a polyline within any closed shape.) Then use the **Desktop>AEC Polygon>Convert to AEC Polygon** command to convert this shape to an AEC Polygon. Be sure to place the AEC Polygon on Layer 0 with all other properties set to ByBlock, as noted above. You will need to do this manually after creation, as AEC Polygons will default to their own layer when created. Be certain to include the AEC Polygon in the selection set with the other objects that make your View Block.

CONTENT SCALING OPTIONS

We have already seen that there are several different ways to create and configure AEC Content, particularly when it comes to scale. Throughout the Design and Documentation Content libraries, are examples of each of the following scaling strategies.

Full Size (1 to1) Content

This is perhaps the easiest scale option to discuss. Content items that are built at full size use no scaling at all. When inserted, the X, Y and Z scale factors will all be **1**. They are created at the size they should be used. Usually Blocks that represent real-life objects are built at real-life size. You simply add them to the drawing as you would any other ADT object. Look in the Casework and Furniture folders to find examples of "Full Size" content.

Annotation Scaling – (Drawing Setup)

This type of scaling was discussed in detail in the last chapter. Annotation Content items, like symbols and Tags, usually make use of Annotation scaling. The final scale of the object is determined by the desired height of the text contained within the symbol. This gives users flexibility in determining the size of their drafted symbology, but it makes it more difficult to enforce office standards.

Scale-dependent Content

Also covered in detail in the last chapter, this type of content typically does not use scale factors externally upon insertion, but rather has them "built into" the sizes of the View Blocks. Therefore, a symbol that needs to be 5/8" [16 mm] in diameter on the plotted sheet would be built 60" [1,600] in diameter in the Content file. See the previous chapter for a detailed example involving a scale-dependent Room Tag.

Variable Size Content

This is a very interesting approach to the issue of scaling content. In Variable Size content, the View Blocks are built to fit within a 1x1 square, in the case of two-dimensional blocks, and a 1x1x1 cube in the case of three-dimensional blocks. When the content is inserted, the user inputs the actual dimensions desired for the item directly into the scale fields. Therefore, a variable size table symbol would be created 1x1x1, but when inserted, we could designate it as a 36" x 72" x 30" [1,000 x 2,000 x 780] table by simply typing those dimensions into the **X, Y** and **Z** scale fields. (A similar strategy was used to create the custom Pipe Chase in the last chapter.) This approach works very well for simple objects without complicated profiles or shapes, like square or rectangular tables, cabinets and other similar items. It can also work well for non-rectilinear items such as toilets or chairs, which have a limited range of possible sizes in the real world.

Working with Variable Size Content

1. Continue working in the same file.
2. From the **Design** menu, choose **Design Content>Furniture**.
3. Select the **Chair** folder and then double-click **Executive Rnd**.

In the Add Multi-View Block dialog box, notice that the scale factors are X=2'-6", Y=2'-6" and Z= 1'-6" (see Figure 8–15). These are the actual overall dimensions of the chair once it is inserted. Several examples can be found in the Metric and Metric D A CH folders as well, for example, 3D Executive Chair – Rect in the *Metric\Design\Furniture\Office Furniture\Chairs* folder, and OfficeChair (2) in *Metric D A CH\Design\Furniture\Office\Office Chairs* folder. You will find several others as you explore these folders.

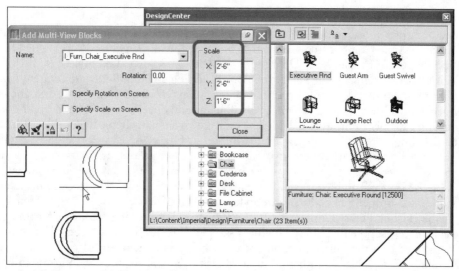

Figure 8–15 *The actual dimensions of the chair are typed in as scale factors*

4. Click a point in the drawing to place the chair.

If you wanted the chair a little bigger or smaller, you can simple change the dimensions in the scale fields.

5. In the **X** and **Y** fields, type **36"** and then click a point in the drawing to place another chair.

Naturally, with content of this type, at some point the values that you type might exceed the limits of "believability." For instance, if we typed **6'** in the Y field, the chair would become too distorted (see Figure 8–16). However, this is not a valid size anyway, so the chances of this becoming an issue are slim.

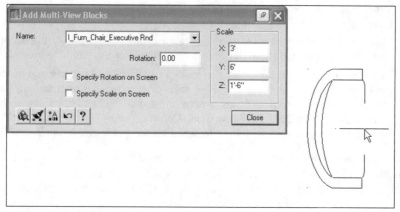

Figure 8–16 *Certain sizes will exceed reasonable limits for this type of object*

6. **Cancel** the Add Multi-View Block dialog box.

PROPERTY SETS

There is one final consideration when building custom AEC Content. If you would like to include any of your Content items in your Schedules, they will require Property Sets. If you recall in Chapter 6, Property Set Definitions may be Object-based or Style-based. Consider this decision very carefully while you are exploring, editing and building your Content Library. For instance, there are dozens of furniture Content items included with ADT, and due to the rather generic quality of many of them, you will likely add dozens more. The logical extension of a Furniture Plan is a Furniture Schedule. You were given many items to consider in Chapter 6 when deciding to build a Property Set Definition as Object-based or Style-based. Without reiterating those consideration here, it is fair to say that Furniture Content would be ideal for Style-based Property Sets. In fact, if you examine the sample Furniture Schedule provided, by default (in the Imperial and Metric folders with ADT), you will find that its columns reference two Style-based Property Sets: **FurnitureStyles** and **ManufacturerStyles**. The problem is that none of the existing library Content has these Property Sets pre-attached. It is recommended that you explore the potential offered by these sample Property Sets and Schedule Table Styles and consider editing the Furniture Content to include the attachment of the required Style-based Property Sets. This is another example of tasks that will require a significant amount of front-end work but will yield enormous ongoing benefits in production. Consider the increase in productivity when users add furniture to their drawings and when ready, simply add a Schedule, and all of the furniture appears with all automatic properties already input.

Furthermore, recalling what we did with the Room Tag in Chapter 7, do not overlook the potential of the Description, Style Name and even the Notes fields. When you drag an item in from the library, it has a very generic name. The name "I_Furn_Chair_Executive Rnd," though very useful for organizing large quantities of library data, provides us very little useful information for use in a particular project. However, if a user were to drag in this generic symbol, and then rename the Definition to something more pertinent to the project, like "Herman Miller Aeron Chair," this field could become very valuable, both for the other project team members and for Schedules and reports generated from data.

Consider the following procedure:

1. When adding Furniture and Equipment Content items to a project in the early design phase, simply drag the generic symbols in from the office standard library.

2. Generate Schedules whenever appropriate for project workflow. At the early stages, these Schedules would show the generic Style Name of the Multi-View Block Definition.

3. When it comes time in the project to specify a specific manufacturer and model, rename the Definition using the Make or Model name or number of the actual item being specified. Users can rename Styles directly in the drawing by right-clicking the actual items and choosing Edit Multi-View Block Definition or by opening the Style Manager. The Schedule will update to show the actual Manufacturer and Model number.

To make this process work, you simply add a Style-based Property Set to your furniture and equipment Multi-View Blocks that references the automatic property Definition Name. You then make the Model column of the Schedule reference this property. That's all there is to it. You may prefer to do this technique using the Description or Notes field instead. It is not important exactly which field you use, only that the process is clearly communicated to all users.

 Note: If your projects are competitively bid, and it is not possible to specify a particular manufacturer and model, you can still use this technique to reference the description or some other identifying aspect of the item.

Working with Property Sets and AEC Content

1. In the *Chapter08\Furniture* folder, **Open** the file named *A-RP03.dwg*.

2. From the **Documentation** menu, choose **Schedule Tables>Add Schedule Table**.

A Furniture Schedule Style has been included in this sample file. It was imported from the defaults provided with ADT.

3. In the **Layer Wildcard** field, type **A-Furn**.

It is necessary to use a Layer Wildcard with any Schedule that applies to Multi-View Blocks. This is because an MVB can represent many types of objects, and without the wildcard, this Schedule would have no way of knowing the difference between furniture, equipment and Schedule Tags.

4. Make sure that both **Add New Objects Automatically** and **Automatic Update** are checked and then click **OK** (see Figure 8–17).

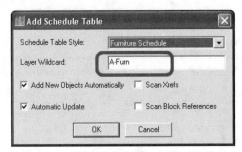

Figure 8–17 *Be sure to include a Layer Wildcard when adding the Schedule*

5. When prompted, select all of the furniture in the file and then press ENTER.

Tip: Since we used the Layer Wildcard, you can use a window or crossing selection.

6. Click a point for the upper left corner of the Schedule and then press ENTER to finish placing it.

Notice that all fields in the Schedule Table are missing their data (they all have question marks). As was pointed out above, this Schedule uses Style-based Property Sets, and they have not been attached by default to the Furniture Content items included in this sample file. This is why the data does not appear.

7. Select any piece of furniture, right-click and choose **Edit Multi-View Block Definition**.

8. On the General tab, click the **Property Sets** button.

9. Click the **Add** button, place check marks in both **FurnitureStyles** and **ManufacturerStyles** and then click **OK** (see Figure 8–18).

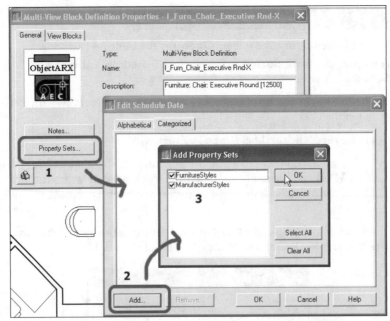

Figure 8–18 *Add both Style-based Property Sets to the Multi-View Block Definition*

Notice that the row (or rows) of the Schedule for that item have now filled in with default values (see Figure 8–19).

FURNITURE SCHEDULE

NO	DESCRIPTION	MANUFACTURER	MODEL	COST	HYPERLINK
?	?	?	?	?	
?	?	?	?	?	
000	— —	— —	— —	0.00	
000	— —	— —	— —	0.00	
000	— —	— —	— —	0.00	

Figure 8–19 *Schedule updates with attached Style-based properties*

10. Repeat the previous steps for any remaining furniture items in the drawing.

The Schedule and Property Set Definitions used in this example are the default versions shipped with the product. Now let's explore the possibilities mentioned above for the use of the Style **Name** and **Description** fields. In this exercise, we are going to map the Style Name to the Model name or number.

11. From the **Documentation** menu, choose **Schedule Data>Property Set Definitions**.

12. Double-click **ManufacturerStyles** to edit it.

    Click on the Definition tab and select the **Model** property.

13. Place a check mark in the **Automatic** check box, and then click the **Source** button.

ManufacturerStyles applies to several Style types. As a result, all of those objects are listed in the Source dialog box. You can select a single check box for each object type (see Figure 8–20).

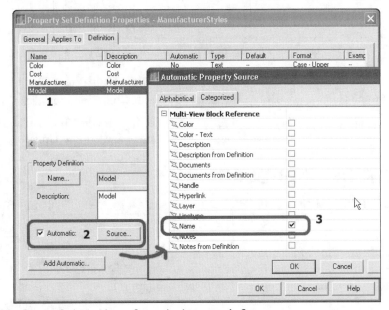

Figure 8–20 *Choose Style or Name for each object type's Source*

14. In the Automatic Property Source dialog box, select **Style** for each object.

15. For Multi-View Block Reference and Mask Block Reference, choose **Name**.

A warning dialog box will state: "Warning. Multiple sources with different names were selected." This is because we chose **Name** for some objects and **Style** for others. In this case, this will not cause any problems, so simply click **OK** to dismiss the dialog box (see Figure 8–21).

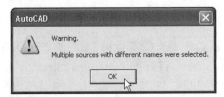

Figure 8–21 *Dismiss the warning dialog box when it appears*

16. Click **OK** back to the drawing.

Notice that the Schedule has updated to reflect the change in the Property Set Definition (see Figure 8–22).

FURNITURE SCHEDULE

NO	DESCRIPTION	MANUFACTURER	MODEL	COST	HYPERLINK
000	— —	— —	I_FURN_DESK_DESK 1	0.00	
000	— —	— —	I_FURN_CHAIR_EXECUTIVE RND	0.00	
000	— —	— —	I_FURN_CHAIR_CHAIR 7	0.00	
000	— —	— —	I_FURN_CHAIR_EXECUTIVE RND	0.00	
000	— —	— —	I_FURN_CHAIR_EXECUTIVE RND	0.00	

Figure 8–22 *The table now reports the Style Name in the Model column*

17. Select one of the chairs, right-click and choose **Edit Multi-View Block Definition**.
18. On the General tab, change the name to **Aeron Chair** and then click **OK**.

Notice the change to the Schedule Table. The Model now reads "Aeron Chair" for those entries (see Figure 8–23). (If necessary, right-click the table and choose **Update**.)

FURNITURE SCHEDULE

NO	DESCRIPTION	MANUFACTURER	MODEL	COST	HYPERLINK
000	— —	— —	I_FURN_DESK_DESK 1	0.00	
000	— —	— —	AERON CHAIR	0.00	
000	— —	— —	I_FURN_CHAIR_CHAIR 7	0.00	
000	— —	— —	AERON CHAIR	0.00	
000	— —	— —	AERON CHAIR	0.00	

Figure 8–23 *The Schedule column remains linked to the Style name as it becomes more specific*

As you can see, this can be a very useful process once the proper pre-configuration has been completed. If users must attach all of the Property Sets to their content, then rename them, and then map the Description to the Model column of the Schedule, they will become frustrated and abandon the process. If however, you have all of the Property Sets attached within the library and the Schedule Table Style already configured to use the **Description** field, the process will be very straightforward and easy to use.

BUILDING A PIECE OF AEC CONTENT FROM SCRATCH

We have covered quite a lot of ground in this chapter. There are several considerations necessary when building custom content. As with the previous chapters, a checklist appears at the end of this chapter that summarizes many of these considerations. In addition, included with the files installed in the *Chapter08* folder is an Excel spreadsheet named *Create AEC Content Checklist.xls* that you can use as a checklist when building new Content. The first tab of the

Excel file gives you a tool to plan the necessary View Blocks that a Multi-View Block will require (see Figure 8–24).

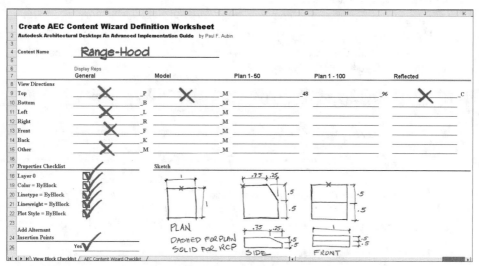

Figure 8–24 *Tool for planning the required View Blocks in a new MVB*

The second tab provides a tool to pre-plan your choices in the AEC Create Content Wizard (see Figure 8–25).

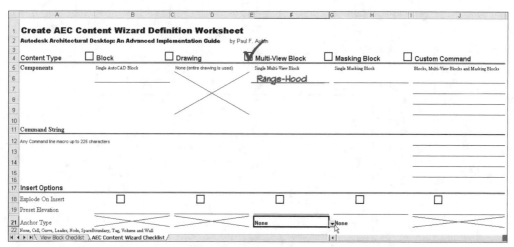

Figure 8–25 *Tool for planning your inputs in the AEC Create Content Wizard*

In the following exercise, we will use these tools to assist us in building a Range Hood Content item from scratch. The Content will be a Variable scale Multi-View Block Content item. We will save it to our Office Standard folder. We are going to use the name "Range-Hood" for the MVB. We will also include some Property Sets in the Content file so that this item will be ready to Schedule once inserted.

Create the Components (View Blocks)

1. Start a **New** drawing from scratch.

2. **Save** the file to the *Office Standards* Content folder and name it **Kitchen Range Hood.dwg**.

3. Set the current Color, Linetype, Lineweight and Plot Style (if using Named Plot Styles) to **ByBlock** (see Figure 8–26).

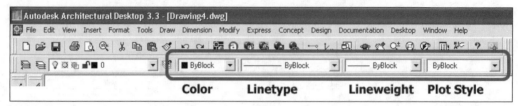

Figure 8–26 *Set all of the current properties to ByBlock*

4. Create a Rectangle **1 unit** square.

 Note: You can work in either Imperial or Metric units in this tutorial.

5. **Move** the rectangle so that the midpoint of the top edge is at **0,0** (see Figure 8–27).

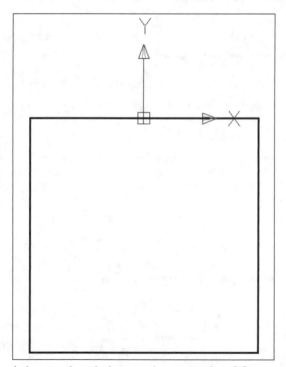

Figure 8–27 *Create a 1x1 rectangle with the top edge centered on 0,0*

6. Select the rectangle in the drawing and then from the **Draw** menu choose **Block**.

7. Type **Range-Hood_C** for the name, leave the **X**, **Y** and **Z** base points set to **0**, and click the **Retain** option in the **Objects** area (see Figure 8–28).

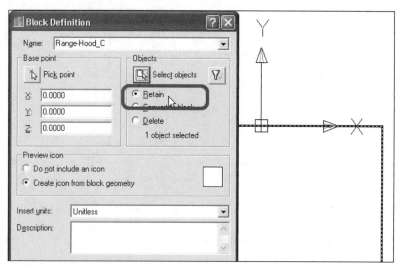

Figure 8–28 *Create the Reflected Ceiling View Block*

8. Click **OK** to create the Block.

With the **Retain** option chosen, we can reuse the rectangle to create the other View Blocks.

Create the Plan View Block with Alternate Insertion Points

Now let's create the Plan Block for the General Display Rep. In this Block, we will include some AutoCAD Point objects on the Defpoints layer. These points will be used as alternate insertion points in the completed AEC Content.

1. Select the rectangle and change its Linetype to **Hidden2**.

Note: You will need to load the linetype into the drawing first.

2. Add the Defpoints layer to the drawing.

Tip: The easiest way to add the Defpoints layer to any drawing is to simply draw a Dimension object and then erase it.

3. From the **Draw** menu, choose **Point>Multiple Point**.

4. Add points to each location where you would like an alternate insertion point (see Figure 8–29).

The center of the object and the two back corners are good choices.

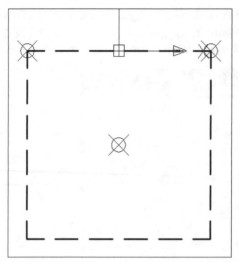

Figure 8–29 *Add Points at each location where you want an alternate insertion point*

5. Press ESC to cancel the command when finished.

6. **Move** all of the points to the Defpoints layer.

Note: Regardless of the current Point Style active in the drawing, points on the Defpoints layer will display as small dots.

7. Select the rectangle and all of the points, and repeat the **Block** command.

8. Type **Range-Hood_P** for the name, leave the **X**, **Y** and **Z** base points set to **0**, leave **Retain** selected and then click **OK** to create the Block (see Figure 8–30).

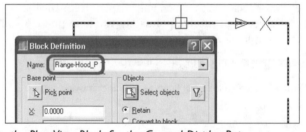

Figure 8–30 *Create the Plan View Block for the General Display Rep*

For the Side View Block, we will draw a new polyline. The shape is based on the sketch in Figure 8–24.

9. Draw a polyline starting at the midpoint of the right side of the existing rectangle.

10. Track the second point **.25** units to the left of the top right corner of the rectangle (see Figure 8–31).

11. Move counterclockwise around the rectangle and snap the next three points to the endpoints at the top left, bottom left and bottom right corners.

12. Right-click and choose **Close** to finish the polyline.

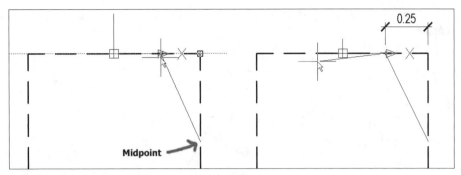

Figure 8–31 *Complete the shape of the Side View Block*

Both the Front View and the Side View Blocks need to be oriented correctly in 3D for them to function properly as View Blocks. (Look back at Figure 8–12 for an example.) Before we rotate the objects into their correct orientation, let's create the necessary geometry for the "front" View Block as well. We will then rotate the objects for both views and create the Blocks.

13. Draw a horizontal line across the original rectangle from one midpoint to the other (see Figure 8–32).

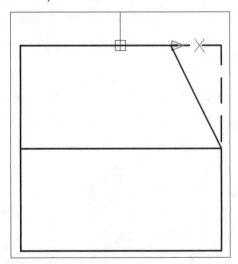

Figure 8–32 *The geometry for both Front and Side views superimposed*

14. From the **Modify** menu, choose **3D Operation>Rotate 3D**.
15. At the "Select Objects" prompt, select all of the objects on screen and then press ENTER.
16. At the "Specify first point on axis or define axis by" prompt, type **X** and then press ENTER.
17. When prompted for "point on the X Axis" press ENTER to accept the default of **0,0,0**.
18. At the "Specify rotation angle" prompt, type **90** and then press ENTER.

After the objects have been rotated in 3D, the drawing will appear like a single line from plan view.

19. From the **View** menu, choose **3D Views>SW Isometric** (see Figure 8–33).

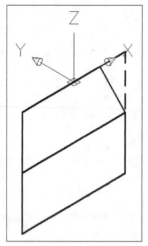

Figure 8–33 *Rotate the objects into 3D, and then rotate the view point to see them*

20. Using the standard **Rotate** command, rotate the Side View polyline from its top left corner, **–90°**.

21. **Move** the Side View polyline so that its top left corner is at **0,0,0** (see Figure 8–34).

22. Change the **Linetype** of the rectangle back to **ByBlock**.

Caution: Be sure to choose ByBlock and not Continuous.

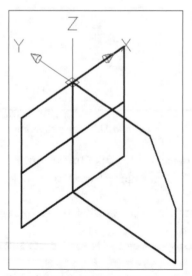

Figure 8–34 *Rotate and Move the side view polyline to the correct orientation and position ·*

We now need to add the AEC Polygons for these two View Blocks. Remember, the AEC Polygons are surfaces that will enable the 2D View Blocks we are creating to conceal any objects occurring behind them when Sections and Elevations are cut.

23. From the **Desktop** menu, choose **AEC Polygon>Convert to AEC Polygon**.

24. At the "Select Polylines" prompt, select the rectangle and the Side View polyline and then press ENTER.

25. At the "Erase layout geometry" prompt, type **N** and the press ENTER.

Since AEC Polygons are ADT objects, they use Layer Keys to insert on the correct layer when added to the drawing. This is the normal and usually preferred behavior. However, since we are creating View Blocks, we want them to be as flexible as possible regarding their properties. Therefore, we will want to change the layer of these AEC Polygons to Layer 0 like the other objects we have drawn so far.

26. Change the layer of the AEC Polygons to **Layer 0**, and verify that all of the other properties are set to **ByBlock**.

27. Select the rectangle, the line and the corresponding AEC Polygon and then from the **Draw** menu choose **Block**.

28. Type **Range-Hood_F** for the name, leave the **X**, **Y** and **Z** base points set to **0**, and click the **Delete** option in the **Objects** area.

29. Click **OK** to create the Block.

30. Repeat the steps for the Side View polyline and its AEC Polygon using **Range-Hood_L** for the name and choose **Retain** this time.

31. Erase the AEC Polygon.

Note: Make sure you keep the Side View polyline, as we will need it in the next step.

Create the Model View Block

The next step will be a little easier if you have the Productivity Enhancement Extension (PBX1) for Architectural Desktop. Included with this Extension is a command that allows you to directly extrude a polyline to a Mass Element. We will use that command to create our last View Block, the 3D Model View Block.

If you do not have the Productivity Enhancement Extension, you will first need to create a profile (**Desktop>Profiles>Profile Definition**) from the polyline and then use it to create an Extrusion Mass Element object.

1. Select the polyline, right-click and choose **Convert to>Free Form Mass Element** (see Figure 8–35).

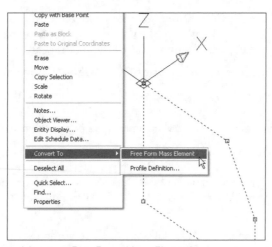

Figure 8–35 *Convert a polyline to a Free Form Mass Element*

2. At the "Erase selected polyline(s)" prompt, type **Y** and press ENTER.

3. At the "Specify extrusion height" prompt, type **I** and press ENTER.

 Caution: If using Imperial units, be sure to type **I"** and not **I'**.

4. **Move** the Mass Element so that the Midpoint of the top back edge is at **0,0,0** (see Figure 8–36).

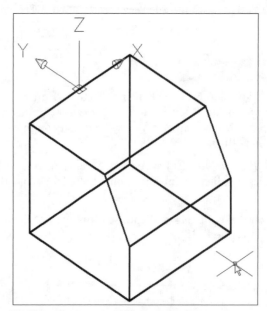

Figure 8–36 *Move the 3D Mass Element into the correct position*

Since Mass Elements are ADT objects, they use Layer Keys to insert on the correct layer when added to the drawing. This is the normal and usually preferred behavior. However, since we are creating View Blocks, we want them to be as flexible as possible regarding their properties. Therefore, we will want to change the layer of this Mass Element to Layer 0 like the other objects we have drawn so far.

5. Change the Layer of the Mass Element to **Layer 0**, and verify that all of the other properties are set to **ByBlock**.

6. Create a **Block** out of this Mass Element named **Range-Hood_M.** Choose **Delete** in the **Objects** area.

Import the Necessary Property Sets

As we mentioned above, it will be valuable to have some Property Sets attached to this MVB automatically. To do this, we must import them into this drawing.

1. From the **Documentation** menu, choose **Schedule Data>Property Set Definitions**.

2. Click the **Open Drawing** icon and navigate to the folder that contains your default *PropertySetDefs.dwg* file.

If you are using Imperial units, this folder is located in the *Content\Imperial\Schedules* folder. For Metric it is in the *Content\Metric\Schedules* folder and for D A CH it is located in the *Content\Metric D A CH\Schedules* folder and is named *PropertySetDefsDACH.dwg.* (Refer to Chapter 6 for more information.)

3. Copy and paste the **ManufacturerStyles** and **EquipmentStyles** Property Set Definitions from this file to the current drawing.

4. In the left-hand tree view of the Style Manager, right-click the *PropertySetDefs.dwg* folder and choose **Close**.

5. Click **OK** to close the Style Manager.

6. **Save** the file.

Create the Multi-View Block Definition

We have finished all of the View Block Definitions and we are now ready to build the Multi-View Block.

1. From the **Desktop** menu, choose **Multi-View Block>Multi-View Block Definitions**.

2. Create a new Style and name it **Range-Hood**.

3. Right-click **Range-Hood** and choose **Edit** (see Figure 8–37).

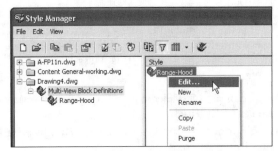

Figure 8–37 *Create a new Multi-View Block Style and Edit it*

4. On the General tab, type a description and then click the **Property Sets** button.

5. Click the **Add** button and add both of the Property Sets previously imported, **ManufacturerStyles** and **EquipmentStyles** (see Figure 8–38).

6. Click **OK** to return to the Multi-View Block Definition Properties dialog box.

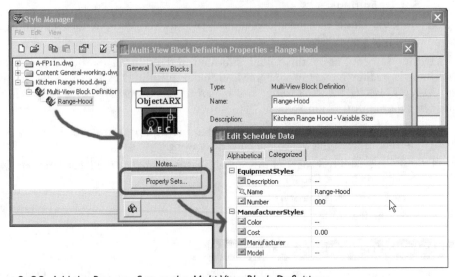

Figure 8–38 *Add the Property Sets to the Multi-View Block Definition*

7. Click the View Blocks tab.

8. With the **General** Display Rep selected, click the **Add** button, choose the **Range-Hood_P** Block and click **OK**.

   Clear all **View Directions** check boxes except for **Top** (see Figure 8–39).

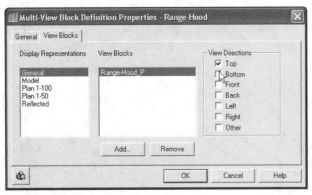

Figure 8–39 *Add the first View Block for the Top View*

9. Add **Range-Hood_F** to Front and Back.
10. Add **Range-Hood_L** to Left and Right.
11. Add **Range-Hood_M** to Other.
12. Select the **Model** Display Rep, and add **Range-Hood_M** to all View Directions.
13. Select the **Reflected** Display Rep, and add **Range-Hood_C** to Top.
14. Click **OK** twice to return to the drawing.

Add the Multi-View Block to the Drawing

Now let's add the Range-Hood Multi-View Block to the drawing and see how we did. You may want to zoom out first, as we will be scaling the MVB up to an appropriate size.

1. Change the current Color, Linetype, Lineweight and Plot Style (if using Named Plot Styles) back to **ByLayer**.
2. From the **Desktop** menu, choose **Multi-View Blocks>Add Multi-View Blocks**.

The Name list already shows "Range-Hood."

3. Set the **X Scale** to **30" [760]**, the **Y Scale** to **18" [450]** and the **Z Scale** to **4" [100]**. Leave the **Rotation** set to **0** (see Figure 8–40).

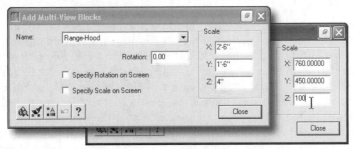

Figure 8–40 *Set the Scale factors to insert the MVB at the correct size*

4. For the insertion point, type **0,0** and then press ENTER.

As you can see, the Scale factors directly correspond to the linear dimensions of the object (see Figure 8–41). This is the benefit of building the View Blocks to fit within the 1x1x1 cube as we did.

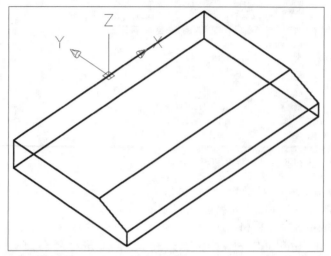

Figure 8–41 *The completed object inserted to scale factors that correspond to "real" dimensions*

Offset the View Blocks

All that remains is to create the custom AEC Content file in the Content Library. We simply need to run the AEC Create Content Wizard and input the values that we already planned for at the start for this exercise.

However, one issue remains to be discussed, which will reveal that we could build the Content file in two different ways. When we created the View Blocks, we built them with their insertion points at the ground plane (0,0,0). This was done because the default elevation for AEC objects should be Z=0. (Recall the dimensionality discussion in Chapter 2.) However, this means that the object itself must somehow be "raised" to the correct mounting height when used in a project. The mounting heights could vary from one project to the next, so it would not be desirable to "hard code" the height into the Content file. (Nor would it have been good to build to the View Blocks that way.)

There are two ways to handle this problem. The first is to insert the Content item at the correct mounting height when used in a project. This is a simple and direct way to solve the problem. However, the downside of this solution is that it requires that the Z elevation of the object itself be moved above zero. Although this solution will generate correct Sections and Elevations, it could cause other problems, since the Plan view will also be inserted above the ground plane. The other solution is to "Offset" just those View Blocks that require it within the Multi-View Block properties of the individual object. This means that the Content item must first be inserted into the drawing at Z=0 (as we have done here) and then, by editing the Multi-View Block Properties, we can shift the Front, Side and Model Views up to the correct Z Elevation, while leaving the Plan view at 0. The advantage of this approach is that it maintains the integrity of the hybrid nature of ADT objects. In other words, the Plan views remain "flat" and at Z=0, while the elevation and 3D views can occur at the correct mounting heights. The disadvantage of this approach is that it requires more manual manipulation after insertion of the object and it is a bit counter-intuitive.

There is, however, one more option that combines the best of both approaches. We will first offset the appropriate views here in this file as just described. Then, rather than create a Multi-View Block Content item, we will choose the **Drawing** option instead. This will insert the MVB as it is in this file, rather than a fresh unaltered copy.

1. Right-click the MVB and choose **Multi-View Block Properties**.
2. Click the Offsets tab.
3. Highlight the **Range-Hood_F** View Block and in the **Z Offset From Insertion** field, type **5'-5" [1,650]** (see Figure 8–42).

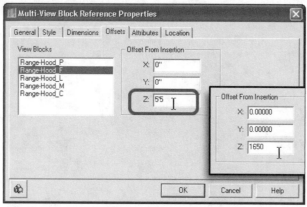

Figure 8–42 *Add a Z Offset to the Elevation and Model View Blocks*

4. Repeat this step for the **Range-Hood_L** and **Range-Hood_M** View Blocks.
5. Click **OK** when finished.

The MVB will appear to shift up, since we are viewing it from the 3D vantage point.

Create the AEC Content

1. **Purge** the drawing to remove all un-referenced objects.
2. From the **Desktop** menu, choose **Create AEC Content**.
3. On the Content Type page, choose **Drawing** and then click **Next** (see Figure 8–43).

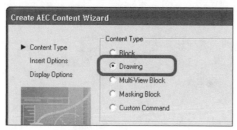

Figure 8–43 *Choose Drawing as the Content Type*

4. On the Insert Options page, place a check mark in the **Explode On Insert** check box and choose the **APPL** Layer Key.

 Note: If you prefer, you can build a custom Layer Key for this content. Refer to the previous chapter for an example.

Since we have chosen the **Drawing** option, it will insert as a "Wblock" would unless we choose **Explode On Insert,** which will insert the contents of the file as separate entities rather than as a single Block (see Figure 8–44).

5. Click **Next** to continue to the last page.

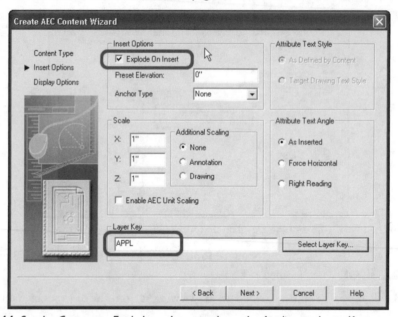

Figure 8–44 *Set the Content to Explode on Insert and use the Appliances Layer Key*

6. Click the **Current Drawing** check box.

This will save the AEC Content instructions to the current file.

7. Click the **Default Icon** button to create an icon from the onscreen objects.
8. Type a detailed description and then click **Finish** (see Figure 8–45).

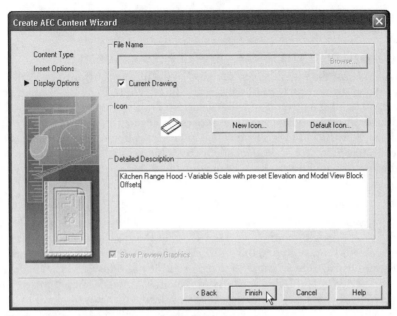

Figure 8–45 *Set the final parameters and click Finish*

9. From the **File** menu, choose **Close**. When prompted, **Save** the file.

Test the Content Item

1. **Open** any other drawing.
2. Click your custom **Office Standards** DesignCenter icon.
3. Drag and drop the new **Range-Hood** Content item into the drawing.

If all is functioning properly, the Range-Hood Content item should appear on the Appliances layer, correctly scaled and with the Left, Front and Model View Blocks already offset to 5'-5" [1,650]. Be sure to set the LTScale to an appropriate value to verify that the Plan View Block is showing as a Hidden Line. Switch to a Reflected Display Configuration. The dashed line should change to solid in the reflected ceiling plan display. If anything is not functioning as expected, go back, review the steps and try again.

To change the size of the hood, simply set the scales to the new dimension, for instance to make it a 36" [900] hood; just type **36 [900]** in the **X Scale** field of the Multi-View Block Modify dialog box. To change the mounting height, open Multi-View Block Properties, and change the Offsets of the Left, Front and Model blocks to the new height. Remember that, based on the way we built the View Blocks, the height measures to the top of the object.

BUILDING CUSTOM SCHEDULE TAGS

Schedule tags do not need to have the Property Set Definition embedded in the file, since the AECANNOSCHEDULETAGADD command automatically imports the required Property Sets from an external file. The external Property Set Definition file is called from the **Schedule Tag Add** command, and the individual Property Set and properties are imported based on the Attribute

Names in the View Blocks of the content. The concepts and procedure are covered in detail in Chapter 6.

BUILDING CONTENT WITH VARYING LEVELS OF DETAIL

In a manner similar to the way in which the level of detail was controlled in Walls, Doors and Windows, using custom Display Reps we can also build Multi-View Blocks that contain high and low levels of detail. You would add a new custom Display Rep to Multi-View Block Definitions within the Display Manager. (The process is covered in Chapter 5.) You must then draw two sets of View Blocks for each view that requires a different level of detail, similar to the steps covered in Chapter 7 for scale-dependent annotation. Finally, you add these View Blocks to each of the appropriate Display Reps in the same way that we have in the previous examples in this chapter.

Although this is within the realm of possibility, you should consider the ramifications very carefully before deciding to add a custom Display Rep to Multi-View Blocks. The reason for this is that it is not enough to simply add the Display Rep to those objects that need it. You must add the custom Display Rep to the entire drawing. Therefore, all Multi-View Blocks will have the custom Display Rep whether they require it or not. However, the bigger issue is the Display Sets and Display Configurations. As we saw in Chapter 5, simply creating the custom Display Rep is only the first step. You must then decide whether it is on or off in every Display Set. This naturally must be done in the template file, or it will need to be re-done in every file that is opened. Therefore, if you feel it is justified to add a custom Display Rep to Multi-View Block Definitions, do so early in your office standard template file, and then open those library items that require the additional level of detail and add the custom Display Rep directly in those Content files. It is not necessary to add the Display Rep to every Content file and item; only those Multi-View Blocks that will actually contain a View block in the custom Display Rep will need to have it.

MILLWORK

Several Casework Multi-View Blocks are included in the Imperial Design Content tree. These include several standard-sized upper and lower cabinets. Also included in the standard Style Library files (refer to the next chapter) are some examples of Wall Styles that are used to generate countertops and other millwork. In addition to these examples, industrious ADT users have built casework from Railing and Curtain Wall Styles. Examples of these types of clever uses of ADT objects can be found at the Point A Download Center or in the Autodesk "Content" Discussion Group. URLs for both sites are located in Appendix A.

MASKING BLOCK CONTENT

A Masking Block is another kind of ADT Block. It differs from an AutoCAD Block or a Multi-View Block in that it contains a two-dimensional Mask object that can be attached to and "cover" up parts of other objects. Masking Blocks are often used for Ceiling content. For instance, most of the fluorescent lighting fixtures in the DesignCenter are Masking Blocks. A Masking Block has two parts: the Mask Boundary and additional graphics. The Mask Boundary is made from one or more closed polylines.

PUTTING THE TEAM TO WORK

As you can see, there are dozens of options to consider when building AEC Content. The creation of content is an ongoing task, and one that should be shared amongst all members of your team. Your colleagues will undoubtedly discover methods and tricks that you did not consider. Try to create a strategy for getting others involved and build a procedure to add their custom content to the office standard library. The more ideas in the mix, the more "good stuff" will be generated, resulting in a better content library. If users are encouraged to share their custom Content creations, you will witness less "hoarding" and everyone can potentially benefit from each other's creations.

CHECKLIST

I **Creating Content**

1. From the offerings provided, decide what to keep, what to edit and what to remove.
2. Create new or edit existing Content items following a consistent process.
   a. Decide the type of Content: Full Size, Annotation, Scale-dependent or Variable Scale.
   b. Create all Block elements on Layer 0 with all other properties set to ByBlock. Try to avoid exceptions to this rule.
   c. Build all View Blocks with the correct orientation and insertion points relative to the 3D View Block.
   d. If desired, include alternate insertion points within the Plan View Block (AutoCAD Point objects on the Defpoints layer).
   e. Add AEC Polygons to Elevation View Blocks to "hide" items behind in Sections and Elevations.
   f. Attach any required Property Sets to the Multi-View Block prior to saving as content.
   g. Adopt a standard naming procedure, such as the practices outlined in the Autodesk Content Style Guide.
   h. 2D files should be saved in Plan view (to create a 2D icon preview) and 3D files should be saved in 3D (to create a 3D icon preview).
3. Publish updated content to the server and document the changes.

SUMMARY

There are over 1200 Content items provided in the combined Imperial, Metric and Metric D A Ch Design Content Nodes of the DesignCenter.

When content from Metric is used in Imperial drawings, or vice versa, the AEC Unit Scaling feature can automatically scale the Content file to match the current units.

Beware of small discrepancies when using AEC Unit Scaling.

Even though it may seem that the 3D View Blocks are unnecessary for a particular piece of Content, remember that an orthogonal view at 45° is considered "Other" by a Multi-View Block.

Most Content items can be thought of in three parts: components, Content items and Content files.

Multi-View Blocks can have several alternate insertion points by the inclusion of nodes on the Defpoints layer within the Plan View Block. Cycle through them with the CTRL key on insertion.

Content can be created in a variety of scaling and sizing options.

Be sure to include Style-based Property Sets within your Multi-View Blocks.

Styles

This section is devoted to the subject of Styles. It is divided into three separate chapters, the first of which covers a variety of ADT Object Style–related issues. Included in its inventory of Styles, Chapter 9 covers important issues related to Wall Styles and their impact of a successful ADT implementation. Area objects and Sections/Elevation objects each have a chapter devoted to their use and configuration.

Section IV is organized as follows:

Chapter 9	AEC Object Styles
Chapter 10	Area Objects and Area Groups
Chapter 11	Sections and Elevations

WHO SHOULD READ THIS SECTION?

This section is directed toward the **CAD Software User**. Most topics are of equal interest to the **CAD Manager** and the **Project Data Coordinator**.

AEC Object Styles

INTRODUCTION

ADT Object Styles are one of the most commonly used and customized pieces of the ADT tool set. Many custom effects and overall global control of objects can be achieved by simply tweaking or swapping a Style or two. What all ADT Styles share in common is that they are a group of settings that control many aspects of an object's parameters, thereby enabling the change of a parameter that ultimately affects an entire collection of objects from a single central location. In addition to this basic functionality, you will find some unique functionality offered by each individual Style type.

OBJECTIVES

The goal of this chapter is to expose you to a variety of tips and tricks when working with ADT Object Styles. This is not intended to be a tutorial-based chapter. Although there will be some tutorial exercises, the main focus of this chapter will be in listing several considerations and issues to keep in mind when you build Object Styles. In this chapter, we will explore the following:

- Understand the Style Manager
- Ways to use Object Styles
- Issues when customizing Object Styles
- Store, publish and re-use Object Styles

GETTING THE MOST OUT OF STYLE MANAGER

The Style Manager is the central interface for working with Styles of all types in ADT. Its interface mirrors Windows Explorer (see Figure 9–1). If you use ADT with any kind of regularity, you visit the Style Manager dozens of times each day. In this topic, we will look at a few features of the Style Manager that often go unnoticed.

Figure 9–1 *The list of Style types in the Style Manager*

FILTER STYLE TYPE

Thirty-three separate menu choices within the four ADT drop-down menus launch the Style Manager dialog box (ADT 3.3 with the International Extension installed). One additional menu choice calls the Style Manager itself, for a total of thirty-four. The convenience of these commands lies in the filtering mechanism. For instance, when you choose the **Door Styles** command from the **Design>Doors** menu, it opens the Style Manager with the Doors node selected and filtered (see Figure 9–2).

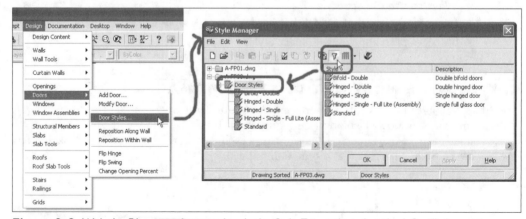

Figure 9–2 *With the Filter icon depressed, only the Style Type selected at the left will show*

The **Filter** icon is very useful when working in the Style Manager. Open any drawing that has some ADT Object Styles within it. Open the Style Manager (use any command) and try it out. Select a Style type on the left, and then click the **Filter Style Type** icon to show only that Style type. Click it again to show the whole list. This can be very helpful when working with related Style types and long Style or File lists. Suppose you are working on a Schedule Table Style and then realize that you need to define a Property Set Definition. Using the **Filter Style Type** icon, you can easily switch between the two Style types (see Figure 9–3).

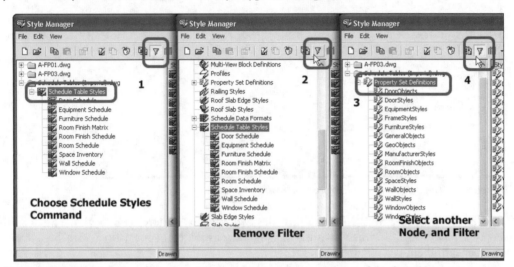

Figure 9–3 *Use the Filter Style Type icon to minimize trips in and out of the Style Manager*

TOGGLE VIEW

The Toggle View feature of the Style Manager allows you to compare the styles within two or more drawings side by side. This setting is available from the **View>Explore** menu, or the **Toggle View** icon on the Style Manager toolbar. Open two drawings in ADT. Be sure to pick two drawings that have some Styles in them. Open the Style Manager. Both files appear as folders on the left, beneath which is the list of Styles that each one contains. Selecting a Style type shows those Styles on the right.

406

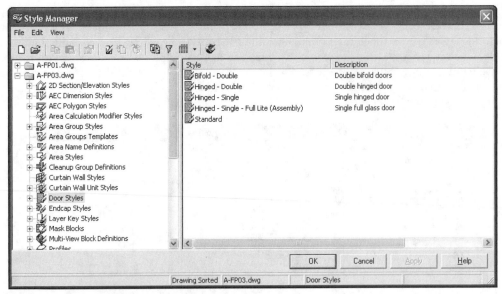

Figure 9–4 *The default view is Explore by Drawing*

The **View** menu of the Style Manager provides two ways to "explore" your Styles. The default view is "By Drawing." In this view, drawings (shown as folders) appear first, with Style types listed beneath them (see Figure 9–4).

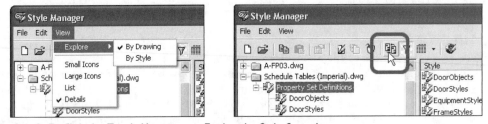

Figure 9–5 *Click the Toggle View icon or Explore by Style from the menu*

Click the **Toggle View** button as shown in Figure 9–5 (or choose **By Style** from the Style Manager **View** menu), and the list will reverse to show Style types first. Beneath each Style type, each open drawing will be shown (again as a folder). On the right will appear a matrix-style view showing the drawings across the top and the Styles along the left side (see Figure 9–6). This view is particularly useful if you are trying to compare two drawings to see if they both contain the same Styles.

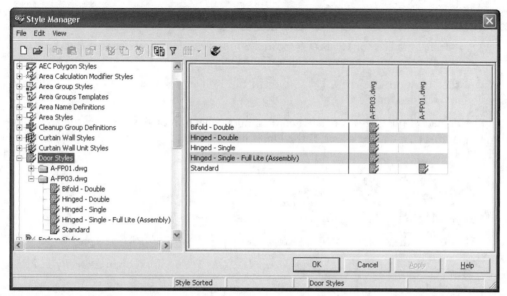

Figure 9–6 *Exploring by Style*

STYLE LIBRARIES

Styles are saved as part of the drawing file. They can be easily imported from and exported to other files, but they are stored directly in DWG files. As you work through the topics in this chapter, you will need to consider where and how you wish to store your office's collection of ADT Object Styles.

There are three basic options: you can store Style Content in dedicated library drawing files on the server, you can include all styles in your template files, or you can do a little bit of both. The third option is the most common. Most of us will have a select set of Styles that are used frequently enough that it makes sense to store them within the template files. If a Style is used 90 percent of the time or more, it makes sense to include it in the template file. However, we have taken great care to build a lean and efficient template file throughout the chapters of this book. The goals of any good template are to make common tasks more efficient, by making all of the required settings and resources readily available, and to standardize workflow. These goals can sometimes seem at odds with the need to keep file sizes in check. Library files help to alleviate this situation. Several are provided with the software by default. As with the other content provided, they are stored in the Imperial, Metric and the Metric D A CH folders. The paths are as follows:

Imperial – *Content\Imperial\Styles*

Metric – *Content\Metric\Styles*

Metric D A CH – *Content\Metric D A CH\Styles*

Install the CD Files and Explore Some Sample Library Files

1. Install the files for Chapter 9 located on the Autodesk Architectural Desktop: Advanced Implementation Guide CD ROM.

Refer to the "How to Use the CD" section in the Preface for instructions on installing the sample files included on the CD.

A folder named *Chapter09* was created containing a collection of sample files that we will use throughout this chapter. Let's look at the Door Style Content Library files provided in the box with ADT.

2. Launch ADT and close all files.
3. **Open** the following three files:

Imperial – In the *Content\Imperial\Styles* folder, open *Door Styles (Imperial).dwg*.

Metric – In the *Content\Metric\Styles* folder, open *Door Styles (Metric).dwg*.

Metric D A CH – In the *Content\Metric D A CH\Styles* folder, open *Door Styles.dwg*.

The Imperial and Metric files open to a title page in paper space, while the D A CH file opens to a 3D view of all the Doors inserted in model space. The same is available in the other two files. To see it, simply click the Model tab to switch to model space (see Figure 9–7). Now we can make use of both the **Filter** icon and the Explore by Style view as we explore these Door Style Library Files.

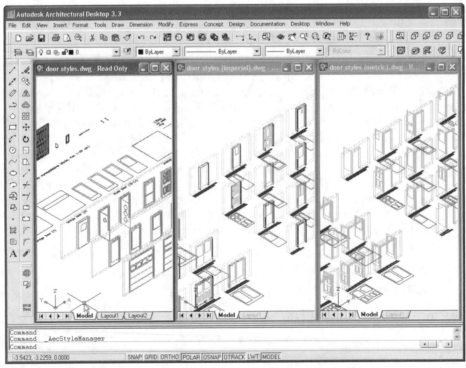

Figure 9–7 *All three drawings have examples of the Styles inserted in model space*

4. In any one of the three drawings, open the Style Manager.

5. Click the **Toggle View** icon (or choose **View>Explore>by Style**).

6. Select the Door Styles node and then click the **Filter Style Type** icon (see Figure 9–8).

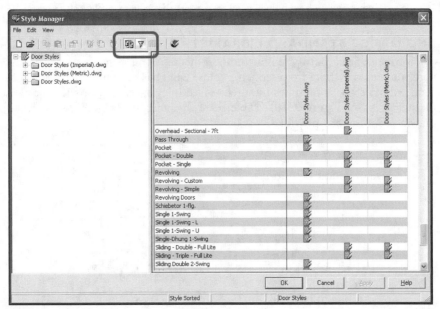

Figure 9–8 *With both Filter and Explore by Style active, we can compare the three files*

While in this view, you can also expand the individual drawing files. This gives you access to each individual Style for editing. Single-click a Style on the left tree view to preview it (in the same way as when the Style Manager is sorted by drawing) or double-click it to edit (see Figure 9–9).

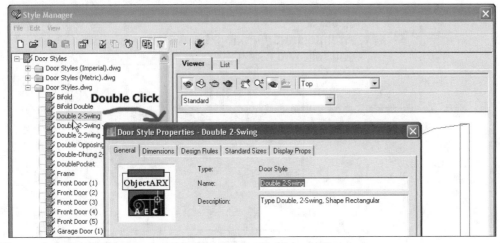

Figure 9–9 *Expand the list, single-click to view, double-click to edit*

Repeat this process on several other Style types as well. This will help you judge the differences and the value of each of the provided samples.

 Note: Unlike the Content items explored in the last chapter, there is no automatic scaling feature for Object Styles. Therefore, if you discover a Style in another folder that uses a different unit system than you do, you will need to convert the Style manually to the preferred units.

BUILDING OFFICE STANDARD LIBRARIES

By now, you have noticed that there is a common theme running through each of these chapters. Before you build anything from scratch, explore what has been provided. Object Styles are no different—dozens and sometimes hundreds of samples have been provided in most of the common categories (see Figure 9–10). Table B9–1, included in Appendix B, summarizes the library drawing files that are provided.

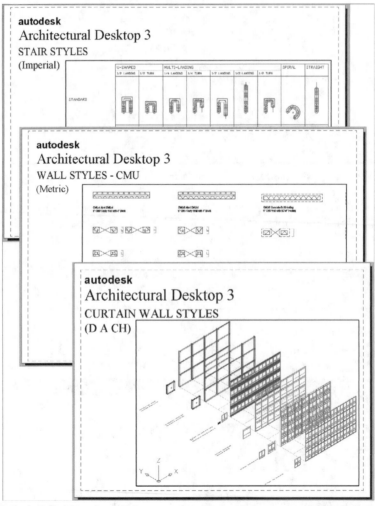

Figure 9–10 *Each Style Library's model space tab is filled with instances of each Style*

DEVELOPING A FOLDER STRATEGY

The first thing that we ought to do is move the *Styles* folders to a more logical and easily accessible location. You may recall that we made the same decision regarding the *Layers* folder back in Chapter 1. There are two reasons for this recommendation. First, with the *Styles* folder being a sub-folder of *Content*, it appears in the ADT DesignCenter folder structure. However, if you click on it there, it will appear empty. This can be a source of confusion among users. Since you will be accessing the Styles contained within the Style Library files through the Style Manager, there is no reason for its remaining in the Content tree and fostering this confusion. Second, by moving it to the root of our L Drive, we will find it easier to quickly locate and load styles, as we will have fewer folders to sift through before arriving at the library files. It is not necessary to evaluate all of the Styles contained in the three different *Styles* folders before moving the folder. Simply consolidate the contents of all three folders (and the two "Schedules" Style Libraries noted at the bottom of Table B9–1) into a single *Office Standard Styles* folder at the root of the L Drive (see Figure 9–11). You will be able to do this without renaming, since the names of the files in each original folder are already unique.

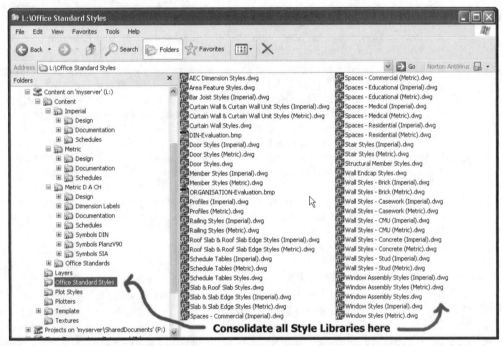

Figure 9–11 *Consolidate all Style Libraries into a single Office Standard Styles folder and delete the separate Styles folders*

As you can see from the samples provided, separate library files have been developed for each major type of Style. Style types that are integrally linked to one another are saved in the same file. For instance, you will find Area Styles, Area Group Styles, Calculation Modifiers, Area Group Templates and Area Name Definitions all within the same file named *Area Feature Styles.dwg*. However, there is only one sample Area Styles file provided, and it is contained originally in the *D A CH* folder. (Refer to Chapter 10 for complete coverage of the Area Features of ADT.) On the other hand, Wall Styles are spread over five separate files in the Imperial and Metric folders. There really is no right or wrong way to organize these files. There are, however, a few things to consider as you develop your system.

- You can load remote drawings into the Style Manager, which will stay loaded until closed. These files will be remembered even after you close and re-launch ADT. This can make it very easy for users to access Library Style Content. However, if you leave several files loaded in the Style Manager at once, it can potentially affect performance. This may suggest keeping the number of Style Libraries loaded at one time to a minimum.

- Use a Project Content folder system for Styles and other Content that is pertinent to only a single project team. This will keep the library files from growing needlessly large and will help prevent unnecessary scrolling through lists of Styles with highly specialized purposes. Simply add a "Style Library" drawing file (or folder if you plan to have several library files for a project) at the root of the *Project* Folder. Team members will then be able to load the file or files in the Style Manager and import the Styles to individual drawings.

- Personal Styles can be saved to the Local Content folder, but be careful of "hoarding" by users who never learned to share. In addition, if users save locally without any backup regimen, it becomes almost impossible to recover that information quickly. Therefore, make sure that a suitable backup procedure is in place.

- If you do not need many Styles, you can consider saving all Styles regardless of type to a single library file named *Office Standard Styles.dwg*. This approach can work well in smaller firms or in cases where the type of work you do does not require large quantities of Styles. The advantages of the single file approach are that file management is much simpler and there is no question as to which file should be loaded. The disadvantage to this approach is that the file can grow to become excessively large over time and potentially inhibit performance. Also, with only one file, you are limited by the Style Manager's filtering mechanism, which although useful, has no way to filter other than by "Style type." In other words, if you examine the sample Wall Style libraries provided, there is one for each major type of construction. This makes it easier to access the precise Style you need without excessive searching. Again, let the quantity of Styles that you anticipate supporting be your guide in developing your file strategy.

LOOK OUT FOR DEPENDENCIES

In many of the chapters so far, we have imported and exported Styles and Definitions from one file to the next. It bears mention here, however, that Styles can often come with a variety of dependencies. For instance, suppose you import a Curtain Wall Style. Curtain Wall Styles can reference other Styles like Doors or Windows. Doors and Window Styles can reference custom Display Blocks. If the custom Display Blocks are not built carefully, they can have other items like text style and layers hard-coded within them. Finally, any Object Style can have custom Display Reps embedded within them. As we discussed in Chapter 5, if not thought through carefully, custom Display Reps can become a management burden.

The aim of this passage is not to discourage you from looking for outside resources of styles or from building or using Styles with dependencies; rather, the point is simply to make you aware of the potential linkages between various Object Styles and to stress the importance of pre-planning and good naming.

WALL STYLES

Walls tend to be the most prevalent objects in any Architectural Desktop Model. This is true not just in the physical quantity of Wall objects typically contained in the Model, but also in the fact that there is almost no drawing type in an architectural document set that does not display Walls. What this means is we must try to consider all of the needs of Walls in all the ways that we present them, both on screen and in print, not just in plans.

PHYSICAL MODEL PARAMETERS

This manual assumes that you already know how to build a basic Wall Style. What we hope to accomplish here is to give you solid guidelines for building Wall Styles that remove some of the potential ambiguities from the process. All recommendations are made based on the following philosophical premise:

 GUIDING PRINCIPLE: A Wall Style should show the least amount of detail possible to convey the most amount of intent.

The physical model parameters (the Wall components) both dictate the physical dimensions of the Wall and provide for its graphical display. However, the component structure of a Wall Style is not accessible through Property Set Data. This means that we should consider adding components to a Wall Style if we wish to physically change the Wall structure and/or display.

There are a few issues to consider when setting up the components of a new Wall Style: the most important is to "**be consistent.**" This is the golden rule. No matter what you decide, actual or nominal, do it consistently. If you like to use negative Edge Offsets and positive Widths, do it consistently. If you change the component priority for brick in a new Wall style, do it consistently. Making good decisions that support the way your firm works is half the battle. Adhering to those decisions with consistency will make your job and those who work with you *much* easier.

Nominal versus Actual Dimensions – The industry seems to be leaning toward actual dimensions. Clearly, the software and new hardware support it, and the down-stream benefits can be significant, when you consider the types of accurate measurements and takeoffs that can be extracted from the Building Model. This is not say that quantities extracted from a Model built with nominal dimensions would not be useful. What is most important is that a strategy be employed *consistently*.

There is a place for both approaches in the typical firm. At the beginning of a project, it is typically easier to work nominally. Most architects have no clue as to the precise size of the Walls at these early stages. They will have an estimate of overall size and component composition. Therefore, you could have a collection of generic Wall Styles with names like "Masonry – Exterior" and "Stud – Interior." As a project progresses, simply modify the Wall and swap the Style to an appropriately detailed Style for the phase of work. At this point, a Wall using actual dimensions could be chosen. Naturally, it will be important to choose a Style that is close in width and composition to the nominal one used for design; otherwise many "issues" are likely to arise.

Variable versus Fixed – There is a place for both types of Wall Style in the typical firm. Variable Walls ought to be used early in the design process when few details are known about the project. In many cases, a simple Standard Wall Style is all that is needed to convey intent in the early SD and DD phases. As a project progresses, simply modify the Wall and swap the Style for an appropriately detailed Style for the phase of work. At this point, a fixed width Wall Style could be chosen. Naturally, it will be important to choose a Style that is close in width to the variable width used for design; otherwise many "issues" are likely to arise.

Baseline Position – The baseline ought to be positioned relative to the structural component of a Wall. For consistency, use the outside face of the structural component where applicable. In cases where no outside edge is inherently implied, use the Left edge of the structural component. In other words, if the Wall is drawn from the bottom of the screen to the top, the left edge of the structural component should be the baseline (zero point) of the Wall (see Figure 9–12).

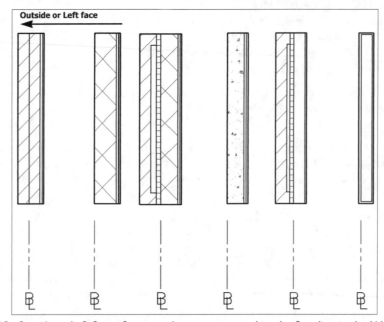

Figure 9–12 *Outside or Left face of structural component used as the Baseline in the Wall Style*

A complete set of guidelines has been developed by Autodesk for the creation of Styles and other content. This document and others related to it can be found on the Autodesk Point A Web Portal. The *Style Guide* in PDF format has been included with the files on the Autodesk Architectural Desktop: Advanced Implementation Guide CD ROM. The document is named *1887354_Style_Guide_-_Creating_i-drop_Content_Specific_to_ADT.pdf* and it is included in the *Chapter09* folder.

File reproduced "Courtesy of Autodesk, Inc. Used with permission; all rights reserved."

BRING THE CASE TO MANAGEMENT

Unfortunately, a common trend seems to be a lack of consistency from one project team to another, on some or all of these points. Try to avoid this type of discrepancy in your firm. Engage all project teams and their leaders, and find a solution that works for the largest number of users. Then make your case to the upper management. If you make the case compelling, you will get their buy-in. This is critical to any successful implementation. Although it is possible to provide, maintain and support two sets of standards and Wall Style libraries, it is not advisable. With the buy-in from upper management, you can settle these types of disputes definitively and in so doing create a process that works for all members of the firm.

LEVEL OF DETAIL

Many of the previous issues have philosophical underpinnings. It can therefore be difficult to establish a consistent set of standards that satisfy all parties sufficiently. Overall, the most important question to answer is this:

 GUIDING PRINCIPLE: How much detail do your Walls really need?

When you have the answer to this question, many of the other answers will fall into place. Start with the document sets from several completed projects in your firm. Look at all the drawings where Walls are a major component. This will naturally include a variety of plans, sections and elevations. Next, decide how much of the detail required by each of these drawings can reasonably be accommodated by an ADT Wall Style.

There are two basic strategies to working with ADT Wall Styles: the "Two-Line" Wall approach and the "Multi-Line" Wall approach (see Figure 9–13).

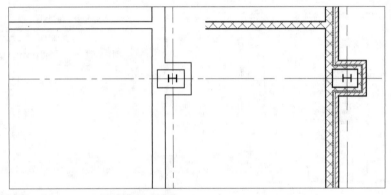

Figure 9–13 *Two-Line versus Multi-Line Wall approach*

In the Two-Line Wall approach, Walls rarely if ever show any detail other than two parallel lines. This is regardless of the scale of the drawing. Wall composition is shown only in the Wall Type Details and never in plan views. In the Multi-Line approach, Wall composition is shown in some or all plan views as well as in Wall Type Details. In some cases, the level of detail steadily increases as the scale of the drawings increases. An ADT Wall Style can contain up to twenty sub-components. These do not always have to represent a plan component; components can also be used to represent section and elevation components as well. Just keep one thing in mind when you consider how many and which components to define in your Styles:

 GUIDING PRINCIPLE: Just because you *can*, doesn't mean that you *should*.

If, upon analysis of your current drawings, you find that most often Walls are shown with only two lines, it is probably because two lines always was sufficient to convey design intent, and not because it was too hard or impossible to add the additional lines. In that case, stick with the strategy if it is working for your firm. If, on the other hand, you do sometimes need to display the internal components, or you typically generate a large quantity of Wall Sections, you may find benefit from the multi-component Wall Styles.

If you do decide to use multi-component Wall Styles, try to keep the number of components to a minimum and be certain to use the standard default Cleanup priorities. Refer to Table B9–2, "Standard Wall Cleanup Priorities" in Appendix B. (The complete list is also included in the Autodesk Style Guide PDF referenced above.)

Most importantly, be sure to check your work in a paper space layout with the Lineweights *turned on* at each of the scales that you intend to plot (see Figure 9–14). When you do this, everything that we discussed in the first several chapters begins to make sense. If you add detail to your Walls only to have it plot out as a "muddy mess," there is little point to adding that detail, at least from a Construction Document point of view. There are sample drawings in the *Chapter09* folder that help to illustrate this point.

- *A-FP01-TwoLine.dwg* – A drawing using only Two-Line Wall Styles. These can be the "Standard" Wall Style or preferably, copies of it with unique names.

- *A-FP01-MultiLine.dwg* – A drawing using Multi-Line (or multi-component) Styles. They are used throughout the plan.

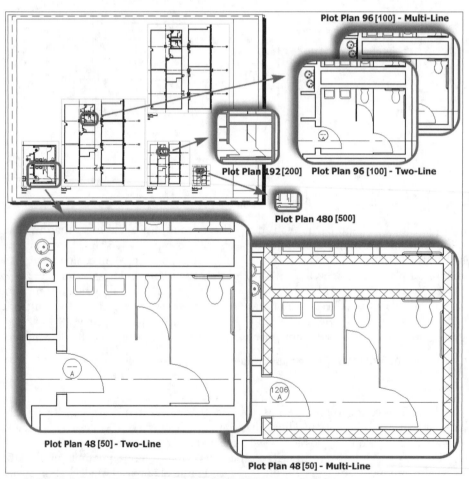

Figure 9–14 *Comparing the Two-Line to the Multi-Line approach*

Both of these drawings have the same section of a floor plan in model space and a Layout tab set up with five viewports in paper space. Each of the viewports is labeled to indicate which Display Configuration it uses. Each of the four scale-dependent Display Configurations (suggested in Chapter 5) is represented here as well as the Work display. The Plot Style Table is displayed on screen in this Layout tab, and **Display Lineweights** is turned on. In particular, the Plot Plan 192 Display Configuration requires a "halftone" Plot Style to display the Walls properly. (This Plot Style is included in the *Office Standard.stb* Plot Style Table, also included in the *Chapter09* folder with the files from the CD.)

Using a Layout tab like this is the best way to preview your drawings to see what they will actually look like when printed. To display Plot Styles, open the Page Setup dialog box and on the Plot Device tab, place a check mark in the **Display Plot Styles** check box (see Figure 9–15).

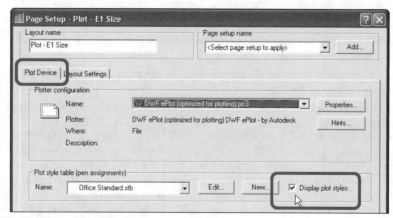

Figure 9–15 *Displaying Plot Styles in paper space*

Lineweights can be displayed properly relative to the sheet size and scale of the paper space viewports. To do so, simply click the LWT toggle button on the AutoCAD Status bar (see Figure 9–16).

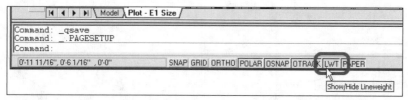

Figure 9–16 Display Lineweight in the drawing window (works best in paper space)

Study what is shown in each viewport carefully and compare the two drawings to one another. You should begin to get a clear idea of how different the final product will be depending on which of the two approaches you favor.

Two-Line Wall Styles (single component) have the following benefits:

- Wall cleanups are extremely simple.
- Regen time when switching Display Configurations is reduced.

- No need for complex Endcaps.
- Fewer Styles to manage.
- Shorter learning curve for existing and new team members to understand and master.
- Success is easily repeatable.
- Ability to reference Style Name as Wall Type in Property Set Definitions.

Multi-Line Wall Styles (multiple component) have the following benefits:

- Graphical differentiation between Walls of different Styles.
- Ability to accurately represent the construction of the Wall.
- Take advantage of scale-dependent Display Configurations.
- No math at the user level (all component sizes are built into the Style).
- Ability to link data to components through third-party applications (this is *not* "in-the-box functionality").
- Can dimension to internal components (for instance run dimension strings accurately to face of stud).
- Ability to reference Style Name as Wall Type in Property Set Definitions.

The two lists are not necessarily mutually exclusive. In fact, the choice of simple or complex Wall Styles is not an "all or nothing" proposition. For instance, suppose you were most interested in the simplicity of Wall Cleanups afforded by Two-Line Walls, but also wanted to reap the benefits of scale-dependent display for your enlarged toilet room plans. In this case, you can build a hybrid strategy whereby simple Two-Line Walls are used throughout the plan, except in areas that will be "blown-up" in enlarged plans. Therefore, your toilet room areas would utilize the more complex Multi-Line Wall Styles and the remainder of the plan would use simple Two-Line Walls. An example of this approach is shown in the file named *A-FP01-Hybrid.dwg* (see Figure 9–17).

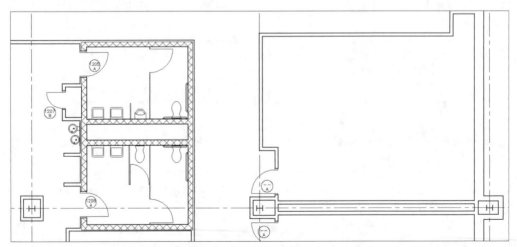

Figure 9–17 *The "Hybrid" approach utilizes Multi-Line Walls only where required for enlarged plans*

As you can see if you open and explore this file, unlike the Multi-Line example, the exterior Wall here uses simple Wall Styles. This is because these Walls will likely only appear in 1/8" [1:100] plans and possibly in some Plan Details. The Plan Details are not plentiful enough to justify the additional time that would be required to use complex Walls throughout the entire drawing set. Therefore, they will be generated using more traditional means, such as XREFing the plan and using XCLIP to crop out the area of the detail, and then tracing the additional information "on top." However, the toilet room plans will be enlarged to 1/4" [1-50] scale and are likely to change frequently. Therefore, it is justified to spend a bit more time building the Walls in these portions of the model, in exchange for the coordination benefits they will bring. The biggest issue you must deal with in this type of approach is managing the "seams" between the two approaches. The important thing to keep track of there is the Wall Cleanup Priorities. If the Component Priorities are coordinated properly based on Table B9–2 in Appendix B, your high detail and low detail plan sections should mesh together just fine.

ENDCAPS

Keep in mind that complex Walls will often require complex Endcaps. Endcaps are mostly Wall Style specific, meaning that they will only function properly on the particular Wall Style for which they were originally designed. (The Standard Endcap Style is a notable exception to this.) Therefore, even though a sample *Wall Endcap Styles.dwg* file is provided among the sample ADT Style Libraries, it is recommended that you store Endcap Styles within the same library file as the Wall Style to which they apply. A Wall Style can reference a Wall Endcap Style directly at the Style level, making it necessary to include them in the Wall Style library file regardless of their inclusion in a separate library file. Therefore, by forgoing the creation of the separate Endcap Library file, you eliminate a possible source of disconnect between two versions of the same Style, should edits be required in the future.

Carry this suggestion forward to all of your Style libraries, regardless of your decisions about the quantity or composition of any particular Style Library file.

GUIDING PRINCIPLE: Avoid including the same Style Definition in two separate Library files.

There is no way to guarantee that both copies will always be synchronized to one another, so you only increase the possibilities of coordination errors in the future if you choose not to follow this suggestion.

NAMING CONVENTIONS

The following naming convention is currently in use by the default ADT Imperial and Metric Styles, and it is the one required for compliance with the Autodesk Style Guidelines document referenced above. The name includes the general name of each of the components and its size. The format is:

<Component name>-<Size> <Component name>-<Size> ...

Note: If your firm does consistent renovation work, a designation determining an existing type may also need to be considered.

Even though there is a 256-character name limit, try to avoid cryptic abbreviations that require a lexicon to decode. However, do be aware that the Add Wall dialog box can display only the first 22 characters of the Style name. It is recommended that you adopt this naming convention for your libraries and for publishing of Styles outside of your firm. Use logical and easily recognized abbreviations for long material names, such as "CMU" for Concrete Masonry Unit, "Block" for Concrete Block and "GWB" for Gypsum Wall Board.

DATA PARAMETERS

Like all ADT objects, Walls can have Property Set Data attached. The automatic properties for Walls includes dimensional information like the length, surface area and volume of the Walls, as well as other types of information like the Wall's Style name. The complete list of automatic properties for Walls and other objects can be found in Tables XL6–1 through XL6-5, contained in an Excel spreadsheet named *Automatic Schedule Properties.xls* in the *Chapter06* folder. Refer to Chapter 6 for more details.

Many issues have been raised in this chapter regarding the use of Wall Styles and their level of detail. Be sure to also consider the potential of Property Sets when devising a Wall Style strategy. For instance, Walls are typically assigned a number in document sets that identifies each unique "type" of Wall. These types are then flagged throughout the drawing so that each Wall can be associated properly with the detail that explains its construction. The Wall Type indicator ought to be included in the Wall's Property Sets. This can be done as an Object-based or Style-based property. Depending on the procedures that you adopt, it can also be either an Automatic or a Manual property.

Putting aside the issues already discussed regarding Wall Styles, consider the potential if Wall Type were set as an *Automatic Style-based* property that linked directly to the Wall Style Name. We began to make the case for this type of procedure back in Chapter 6 at the start of the "Property Set Definitions" topic. There you were encouraged to think of "Door Types" and "Door Styles" as synonymous. Now we are suggesting that Walls ought to be treated in the same way. Making this link does not require you to use highly detailed Wall Styles. The two decisions are related but separate. Let's compare the workflow for tagging Walls with their respective types in a traditional "tag-driven" process versus the proposed "Style-driven" process proposed here.

The traditional Object-based Wall Type tagging process:

1. Draw all Walls.
2. Add Tags to each Wall.
3. Edit the Schedule Data (Attributes) within each Tag to the appropriate Wall Type.

Three steps don't seem very bad, but the data in those tags is not linked to anything about the Wall. If the Wall changes, the Tag must be manually changed. There is no way to verify that a change has been made. There is also no way to globally change "all Walls of Type A" in some way.

The Style-based Property Set driven tagging process:

1. Draw all Walls, assigning a particular Wall Style as you work.
2. Add Tags to each Wall.

There is no need to edit attributes, because the data comes directly from the Wall Style. If a Wall type must change, you edit the Wall Style and as a bonus, if the change affects the Wall type designation, all Wall Tags will update. As you work in your plan and assign Wall Styles to objects, that information is immediately available to the Wall Type tags when they are added. The only thing that is required to make this process work is a separate Wall Style for each Wall type. It does not matter what the graphic properties or component makeup of those styles is. They could all be simple "Two-Line" Styles, if you wish, that simply have different names (and possibly different cleanup parameters). Even if this is the case, this process offers benefits because all of the "Type A" Walls are linked by a common Style. What you do to one will affect all. If a particular Wall requires a change that cannot be made globally, then it is likely time to consider a Wall Type A1 anyhow, and this process helps to make that situation evident.

This type of functionality can be attained with either an Automatic or a Manual Style-based property.

Using a Manual Style-based Property for the Wall Type

If you decide to use a Manual Style-based property for the Wall type, you must edit the WallStyles Property Set Definition to include the Wall Type property and the Wall Type tag to reference this new property.

1. In the WallStyles Property Set Definition, add a new Manual property named WallType. The value of this property would need to be input manually once per Style (not per Wall object).
2. Following the process outlined in Chapters 6 and 7, edit or create a Wall Type Schedule Tag that contains an Attribute linking to this new property (**WallStyles:WallType**).

Using an Automatic Style-based Property for the Wall Type

Likely candidates for use with an Automatic property solution are the **Name**, **Description** and **Notes** fields of the Wall Style. As we recommended for Doors in Chapter 6, consider using the Style Name as your Automatic property source. To do this, you simply reference the Wall Style Name in the WallStyles Property Set to the Wall Type property and then change the Tag to reference this as well.

1. Edit the WallStyles Property Set Definition, make the Wall Type property Automatic and set the Source to Wall Style Name.
2. Following the process outlined in Chapters 6 and 7, edit or create a Wall Type Schedule Tag that contains an Attribute linking to this new property (**WallStyles:WallType**).

With the Automatic property approach, we will follow a procedure similar to one suggested for Door Type in Chapter 6. Continue to use the existing descriptive Wall Style names within the library, but once they are imported and used in a particular project, they would be renamed to the appropriate Wall Type Designation for the project. Therefore, the "CMU-8 Rigid-1 Air-2 Brick-4 Furring-2" Wall Style in the office library might become Wall Type "B1" in the current project. Therefore, you would rename this Style to **B1** in your project files (not in the library.) Leave the original description intact for reference.

Wall Style Property Sets and Linked Tags

1. From the *Chapter09* folder, **Open** the file named *A-FP02n.dwg*.

In this small floor plan, the Wall Styles have all been renamed to their respective Wall Types. (The same has been done for Doors.) To change a Wall's construction type, Modify the Wall and swap the Style. To globally change the group of Walls belonging to the same Style, edit the Style.

2. From the **Design** menu, choose **Walls>Wall Styles**.

Most of the Wall Types in this file are Stud Walls with minor variations. A Type A Wall has drywall on both sides. A Type B Wall is simply furring strips and drywall on the interior face of the existing masonry Walls. The number variations like A1 and A2, reference the trim conditions for each Wall.

3. Click on each Wall Type in the Style Manager to view a preview. Preview them in both Top and 3D views (see Figure 9–18).

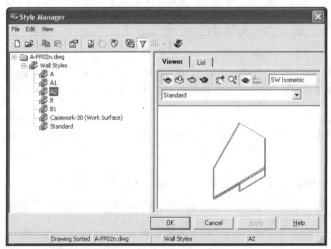

Figure 9–18 *Click on each Wall Type to view the differences*

4. Click on the **Wall Styles** node of the tree to view a list of all Wall Types on the right (see Figure 9–19).

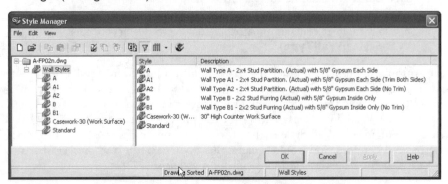

Figure 9–19 *Read the descriptions for another way to understand the differences*

5. Rename Style **A** to **C** and then click **OK** to return to the drawing.

Notice that all of the A Tags have changed to C (see Figure 9–20). If the update did not occur automatically, click your **Object Update** icon (created in Chapter 1) or type OBJRELUPDATE.

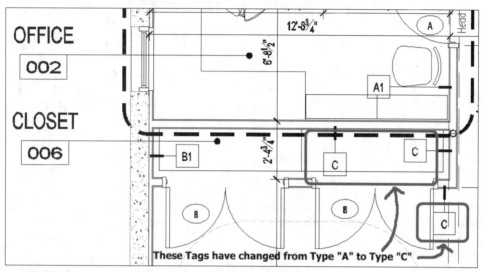

Figure 9–20 *Renaming the Style updates all of the referenced Tags*

6. Select any Wall in the drawing, right-click, choose **Wall Modify** and change it to a different Style.

Notice that the Tag automatically updates to reflect the change. There is no way to make such a change more quickly or easily. Once the Style, Property Sets and Tags are built, one need only follow this simple process to achieve this functionality.

7. From the **Documentation** menu, choose **Schedule Data>Property Set Definitions**.

8. Double-click **WallStyles** and click the Definition tab.

9. Scroll down and locate the Style property.

Notice that the Description reads **Wall Type** (see Figure 9–21).

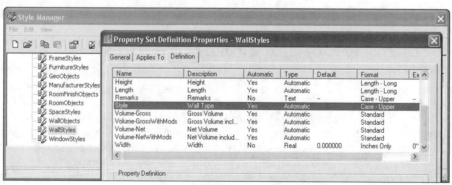

Figure 9–21 *The Description identifies this property as the Wall Type*

10. Click the **Source** button.

Notice that the check mark appears next to **Style** (see Figure 9–22).

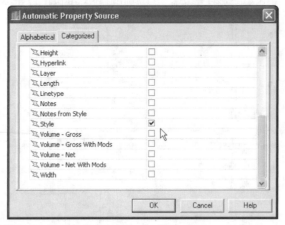

Figure 9–22 *The Wall Type property is linked to the Style Automatic Source*

11. **Cancel** all dialog boxes to return to the drawing.
12. **Insert** the Block named **AIG_Wall_Tag_P** to the side of the drawing and **Explode** it.

Zooming in, you can see that the Attribute within the Block references the WallStyles:Style property (see Figure 9–23). Refer to Chapter 6 for more information on linking Tags to Property Sets.

Figure 9–23 *The attribute within the View Block for the Wall Tag references the Style property*

13. **Close** the drawing without saving.

THE BUILDING MODEL IS LOADED WITH UNTAPPED POTENTIAL

So far we have talked only of traditional construction document requirements and devising Property Sets to meet those traditional needs. There will certainly be a payoff in just achieving a more efficient path to well-coordinated CDs. However, if we broaden our thinking just a bit, it is not too unreasonable to think of a Property Set that assigns a unique ID number to every Wall in the drawing. (This can be done easily with an "Auto-Increment" property.) We can then assign whatever other additional information to that Property Set we see fit. Now, imagine a phone conversation with a colleague, consultant or contractor who has a copy of the drawing open in ADT on his or her own computer screen. All you would need to do to discuss a specific Wall in the drawing is refer to it by its number. A simple Tag could be devised that shows the number on screen only when needed (using a custom Display Configuration).

Now suppose that we began to tap into the nearly 500 Automatic properties available to the 42 ADT objects and 26 AutoCAD objects that can use Property Sets. Imagine the possibilities for scheduling, cost estimating, coordination and design.

Certainly, for most of us, our first priority is to improve the processes that we are currently required (and paid) to complete. But if value can be added to the product with little or no additional work, then we will begin to recoup our investment in technology that much sooner, and perhaps even discover new services that we could reasonably offer and charge to our clients.

CREATING FIRE RATED WALL TYPES

Designating Walls as being fire rated has long been achieved by applying "tape" to the rated Walls. In the days before CAD, there was literally a tape applied along the Walls that needed to have a fire rating. In AutoCAD, this "tape" is often simulated with wide polyline that has a linetype assigned. Custom commands stored in the *Documentation>Miscellaneous>Fire Rating Lines* folder of the DesignCenter (for both Imperial and Metric) automate this traditional CAD simulation of a manual drafting process. These routines simply automate the process of adding the wide polyline tape to the Walls in your drawings.

As an alternative to the "fire tape" approach, you could consider building the fire rating information into the Wall Style. Again, this does not necessarily require that you use complex Wall Styles. All you need is a second version of each of your Wall Styles that will be rated. For instance, the default library contains a Wall Style named Stud-4. All you need to do is copy this style and rename it for as many different ratings as you need such as "Stud-4 - 1 Hour Rated" and "Stud-4 - 2 Hour Rated." There are two ways that you can graphically represent the rating information in the Wall Style. Both involve a Linetype and Lineweight designation.

If it is important that the fire rating "tape" always occur at the center of the Wall, then you can define a Wall component at the geometric center of the Wall Style that has a zero-width (see Figure 9–24). In the Display Props for that Wall Style, attach a Style Override and assign the Boundary lines of the Fire Rating Tape component to an appropriate Linetype and Extra Wide Lineweight.

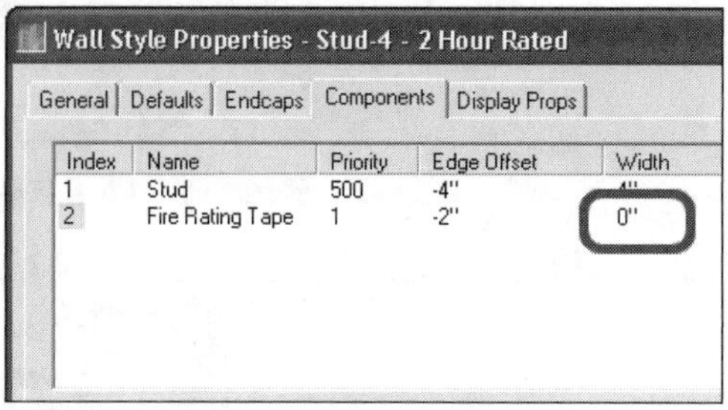

Figure 9–24 *Adding a zero-width Fire Rating component*

Advantages to the zero-width component approach:

- Fire rating graphic stays with the Wall because it is part of the Wall Style.
- Wall Tags can reference the Fire Rating of the Wall (requires the addition of a manual Style-based property).
- Fire Rating display can be controlled with a Display Configuration rather than layers. (Build a "Fire Plan" Display Set and Configuration that shows the Fire Rating component.)
- Down stream and estimating (third party) applications would be able to access Fire Rating information in your Wall Styles.
- The potential of Third party applications raises all sorts of interesting scenarios, such as the checking for Wall Penetrations that violate the rating and other such things.

Using the Sketch Display Rep

As compelling as the zero-width component approach is, if you can live with the Fire Rating symbol's coinciding with the Wall's justification line, rather than true geometric center, an even better approach is to use the Wall's Sketch Display Rep for Fire Rating. The concept is the same as the zero-width component, except that no components are necessary. Simply customize the Sketch Display Rep to shown Fire Rating linetypes on a per Style basis. To make this work, you need to do the following:

1. Create a new Display Set (copy your **Plan 96 [100]** or **Plan 48 [50]** for best results). Turn on the Wall Sketch Display Rep for that Set, in addition to the Plan display already active for that scale (see Figure 9–25 on the right).

2. Create a new Display Configuration (copy your **Plot Plan 96 [100]** or **Plot Plan 48 [50]** for best results) that references the new Set in Top View (see Figure 9–25 on the left).

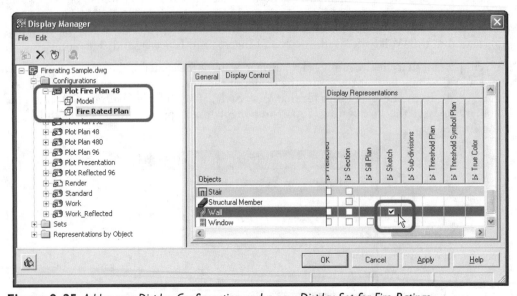

Figure 9–25 *Add a new Display Configuration and a new Display Set for Fire Ratings*

3. Set the **System Default** Display Properties for Sketch to have both the **Baseline** and **Marker** components turned **off**.

4. Set the default **Baseline** parameters to the preferred Layer, Color, Lineweight and Plot Style for the typical Fire Rating symbol. (For instance, set System Default parameters to match a 1-hour rating or whatever rating is most typical. See Figure 9–26)

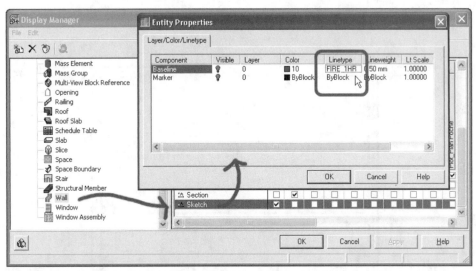

Figure 9–26 *Setup the Sketch Display Rep default settings to be off (not rated)*

5. For any Rated Wall Style, attach a Style Level override to the Sketch Display Rep Display Props and turn **on** the **Baseline** component. Adjust the **Linetype** to the required hour rating. (Follow the steps shown in Figure 9–27 to build the Style override.)

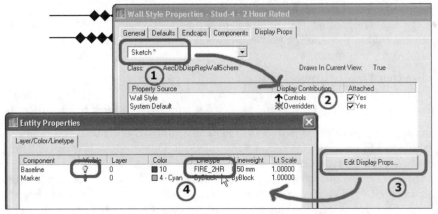

Figure 9–27 *Create a two-hour rated Wall Style*

To see a rated plan, simply load the new Display Configuration. Since it was a copy of the normal Plot Plan Configuration, the only change will be for the rated Styles to show their Sketch lines, which you will have configured to represent the Fire Rating of that Wall Style (see

Figure 9–28). An example of this approach is provided in the *Chapter09* folder, named *Firerating Sample.dwg*. Again, note that the only potential limitation to this approach is that the Fire Rating symbol is applied to the Wall Justification line.

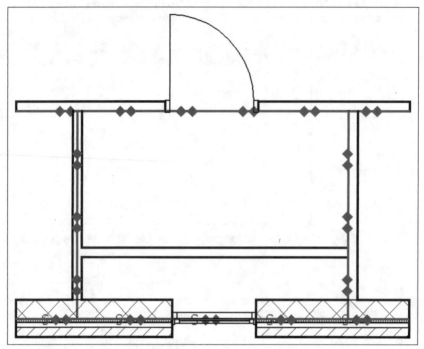

Figure 9–28 *Display the Rated Walls Display Configuration*

GRAPHICAL PARAMETERS

Most of the issues regarding the graphical parameters of Walls and Wall Styles have been sufficiently covered in this and the previous chapters. However, there are a few additional procedural tips worth mentioning.

Standardize your Display Control settings – Develop a material chart that includes standard material names, priorities, Layer, Color, Linetype, Lineweight, Plot Style and Hatching settings. This will make it very easy to add new Styles later that conform to the office standards.

Graph line display override – When Wall Cleanup becomes troublesome, you can use the **Wall Merge** command to manually clean up an intersection. This can be a lifesaver in a tough situation. However, it becomes very difficult later to locate merged Walls if you do not know that they are there. There is no way to flag them automatically; however, you can address this by following a procedure. When you are working on Wall Cleanup, you should always have the Wall Graph displayed (**Design>Wall Tools>Toggle Wall Graph Display**). When you use the **Wall Merge** command, do the following after the merge is completed successfully:

1. Select the Wall that was merged, right-click and choose **Entity Display**.
2. Make sure that **Graph** is selected for the Display Rep, attach a **Wall** level override (see Figure 9–29) and click **Edit Display Props**.

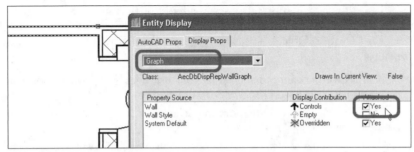

Figure 9–29 *Attach an Entity Level Display Override to the Graph Display Rep*

3. Change the Color of the Graph **Edge** to a color like **Magenta** (one of the primaries that we have been using to flag special conditions). See Figure 9–30.

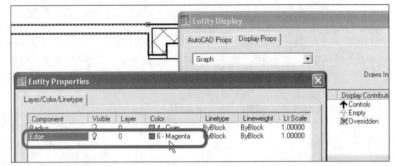

Figure 9–30 *Change the Edge component to Magenta to flag the Merge condition on this Wall*

Remember, the purpose of **Wall Merge** is to "manually" clean up "stubborn" Wall intersections. It will usually work where other techniques fail. There are a variety of recommended steps to perform before resorting to a **Wall Merge**. An entire chapter was devoted to the subject in *Mastering Autodesk Architectural Desktop*. If you have a copy, you may want to review it now. If you don't, a checklist from that book is included in the *Chapter09* folder for your reference. It is named *MAD-AppB.pdf*.

It is good to be able to quickly locate Wall merges later in the production process in case they need to be edited or removed. By following this procedure, you will gain an easy way to locate merged Walls quickly. Again, there is no way to force this to occur automatically, but if you make a procedure that all users are "highly encouraged" to follow, it will make later edits to Wall merges much easier.

WALL DETAILS

If you are using complex Multi-Line Wall Styles for some or all of your Wall Styles, and if you have applied pertinent Property Sets to them, you may also be wondering how feasible it would be to generate Wall Type Details directly from the Model. The answer to this query depends on a great many factors. The simple answer is that it is possible. The more important question, however, is whether it is practical. Included in the sample Schedule Table Styles provided with the software is a Wall Type Schedule Style. With this Schedule, you can get a list of all Wall Styles that are being used in a particular drawing (or several drawings using XREFs). Unfortunately, there is no automated way to generate drawings from this report. However, you can use this Schedule Table for coordination purposes as a tool to help you locate and detail all of your various Wall Types. Then, if sufficient detail exists in your Wall Styles, it may be worthwhile to use a 2D Section/Elevation Object as the basis for the Wall Type Detail. The process would follow these steps:

1. Add Walls with Property Set Data to your drawings. For proper coordination, be sure that all Walls have Property Sets attached.

2. Generate a Wall Schedule (import the Style from the *Schedule Styles (Imperial)* or *(Metric)* library file).

3. Using the Schedule as a guide, locate one instance of each Wall Type and add an AEC Section Line cutting through it (**Documentation>Sections>Add Section Line**).

4. Generate the Section for each Section Line. (Use the custom routine in the *Chapter06* folder, mentioned in the "CD Resources" section of Chapter 6 to keep the details coordinated with their section cuts.)

5. Use these 2D Section/Elevation objects as the basis for your Wall Type Details (see Figure 9–31). Simply trace over the top, add notes, and then turn off the Section object. If you prefer, you can explode and use the linework, but then it will no longer be linked to the Section Line. Tracing gives you the opportunity to update and cross-reference if the Wall Types change.

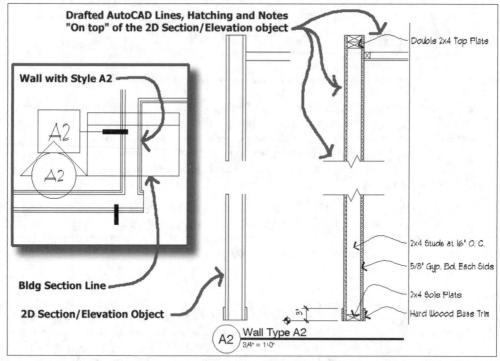

Figure 9–31 *Diagram of the steps required to use a 2D Section/Elevation object as the basis for a Wall Type Detail*

More information on using the 2D Section/Elevation objects can be found in Chapter 11.

PROFILES

AEC Profiles (or simply "Profiles") are shapes made from one or more closed polylines. Profiles are helper objects that provide a means to customize the shape of other ADT objects. For instance, Profiles are used to create custom Door and Window shapes. When "inner rings" are included in a Profile used for a Door, the inner ring becomes a lite. When inner rings occur in a Profile used for a Mass Element, the ring becomes a hole. Sometimes, you are required to use a Profile to customize the shape of an internal component of an object, like the shape of the Railing or Curtain Wall components. Sometimes you can simply draw a polyline and convert it to the object directly. Table 9–1 gives a complete list of the objects that use either polylines or Profiles, together with their use and a suggested prefix.

Table 9–1 *Profiles and Polylines used by ADT Objects*

Object Type	Objects that can be converted from Polylines	Objects that use AEC Profiles	Suggested Name Prefix
Architectural Objects			
Ceiling Grid	Clip		
Column Grid	Clip		
Curtain Wall	Convert, Division Type, Convert Linework	Edge Shape	CW_Edge_
Curtain Wall Unit	Convert, Division Type, Convert Linework	Edge Shape	CW_Edge_
Door		Shape	Door_
Opening		Shape	Open_
Railing	Convert	Component Shape	HR_
Roof	Convert		
Roof Slab	Convert	Edge Shape	Roof_Edge_
Slab	Convert	Edge Shape	Slab_Edge_
Space	Convert		
Stair	Customize Edge		
Structural Member	Convert, Custom Member Shapes		
Wall	Convert	Sweep	Sweep_
Window		Shape	Win_
Window Assembly	Convert, Division Type, Convert Linework	Edge Shape	Win_

Table 9–1 *Profiles and Polylines used by ADT Objects (continued)*

Object Type	Objects that can be converted from Polylines	Objects that use AEC Profiles	Suggested Name Prefix
Other Objects			
Area	Convert		
AEC Polygon	Convert		
Mask Block Reference	Shape of Mask		
Mass Element		Shape	Mass_

Creating AEC Profiles is simple. They appear in the Style Manager like all other Style types and are referred to as Definitions in the drop-down menu (**Desktop>Profiles>Profile Definitions**). The steps to creation are as follows:

1. Draw one or more closed polylines. The polylines must be "lightweight" polylines—no spline or fit curves. If there is more than one polyline, they must not touch or overlap.
2. Choose **Desktop>Profiles>Profile Definitions** and create a New Style.
3. Right-click the New Profile and choose **Set From**.
4. In the drawing pick the outermost polyline first and work your way into the center as prompted. Answer "Yes" to make internal rings "Voids."

You cannot insert a Profile as an object in its own right. It must be used to customize the shape of another ADT object. You can insert the Profile as a polyline for editing purposes. Edit the polylines, and then redefine the Profile by repeating the "Set From" process. If you keep the original name, all existing references to that Profile will update to the new shape.

There are some sample Style Library files included with ADT in the *Style* folder, one for Imperial and one for Metric. The shapes are basic geometrical shapes that can be used for nearly any purpose. Profile definitions of a more specific nature, such as those used to customize the shapes of Door and Window Styles, can be found directly in Style Files for those object types. Since Profiles are generic objects used for a variety of purposes, it is very important to follow a logical and consistent naming convention. For instance, you will find that by default, generic shapes are named "Unit_Square" or "Unit_Hexagon" while specific shapes such as those for handrails are named "HR_Cap," and for Doors, you might see "Door_Single_Lite." This type of naming convention, beginning with a code to designate the intended usage, will make it easier to locate and reuse Profiles in the future. Refer to the rightmost column of Table 9–1 for a complete list of suggested name prefixes.

FENESTRATION STYLES

Three separate Style types exist for fenestration in ADT, Door, Windows and Window Assemblies. Window and Door Styles are extremely similar; Window Assemblies are more complex. Many of the same considerations should be given to fenestration Styles as were applied to Walls. For instance:

- Consult the Autodesk Style Guide referenced above for naming conventions.

- Once a Style has been imported into a project, consider renaming it for the "type" that it will represent.
- Use Style-based Property Sets where possible to minimize the amount of manual data input that is required.

DISPLAY REPS

The "level of detail" issue discussed with Walls above is pertinent to Doors and Windows as well. With these objects, however, the discussion is less about the quantity of components and more about the specific Display Rep that ought to be used to represent the object for a particular scale of drawing. This topic was covered extensively in Chapter 5. Refer back to Figure 5–44 for the complete collection of hard-coded Display Reps for Doors, Windows, Window Assemblies and Openings. Remember to look seriously at the Plan 1-100 and Plan 1-50 Display Reps, particularly the Plan 1-100, which shows only the Panel and Swing for most Door types and shows a circular enclosure for revolving doors. It also shows only two lines for the Window. These present a nice alternative to the default Plan and Nominal Display Reps and are designed specifically to represent varying degrees of detail at different scales.

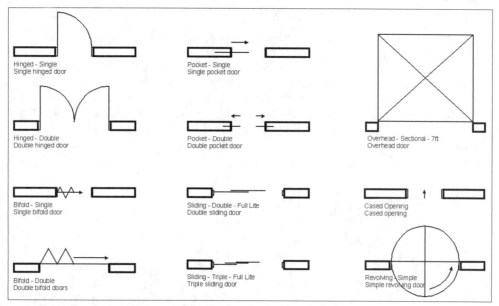

Figure 9–32 *A sampling of Door Styles displayed in the Plan 1-50 Display Rep*

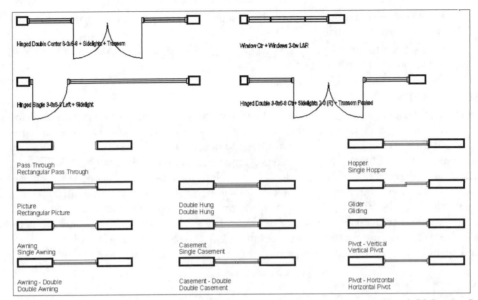

Figure 9–33 *A sampling of Window and Window Assembly Styles displayed in the Plan 1-50 Display Rep*

Figures 9–32 and 9–33 are taken from the file named *Fenestration Display Reps.dwg*, which is included in the *Chapter09* folder. Feel free to open it and explore the various Display Reps further. In addition, be sure to open each of the default Style Libraries (listed in Table B9–1) and click the Model tab (to switch to model space). In virtually every Style Library file, all of the Styles have been inserted as objects into the drawing with text annotations labeling each one (see Figure 9–34). This is an excellent way to become acquainted with the existing styles.

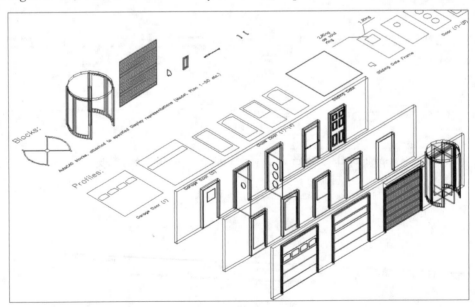

Figure 9–34 *Doors in the Metric D A CH Door Style Library*

COORDINATING PHYSICAL, GRAPHICAL AND DATA PARAMETERS (CREATING A SINGLE HINGED DOOR WITH LOUVER STYLE)

In Chapter 2, the distinction between physical, graphical and data parameters was defined. An example of a Door with louvers was given to illustrate the point. Let's go through the creation of such a Style here to help you cement these concepts and understand their implications to your Style Library.

As was stated in Chapter 2, the presence of louvers affects both the graphical displays of the Door object in Elevation and Model and must appear on the Schedule. However, the louver portion of a Door is not a parametric property. Therefore, we need to "build" the louvers into our Door Style. Once the Style is created, it can become part of our Library and reused in any project.

Create the Door Shape (AEC Profile)

1. Create a New Drawing using your *Plan-Sect-Elev.dwt* template file.
2. Save the file as *Office Standard Doors.dwg* in the *Office Standard Styles* folder on the L Drive.
3. Draw a rectangle in the size of a typical Door leaf (**3'-0"x7'-0" [900 x 2100]**.)
4. Draw another rectangle within it matching the size and position of your most typical louver condition.

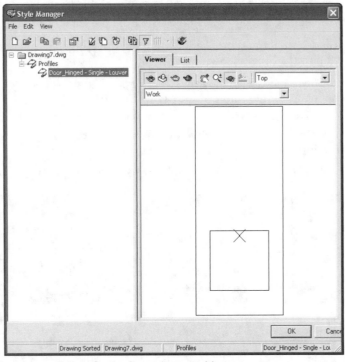

Figure 9–35 *Draw two polylines for the Door shape and louver position*

5. Create an AEC Profile from these two Polylines named **Door_Hinged - Single – Louver** (see Figure 9–35).

Use the large polyline as the main ring and the small (louver) shape as a void ring. Use the Centroid as the base point.

Build the Door Style

The graphics of the louver grill itself cannot be generated parametrically. We will need custom Display Blocks for this. To save time, we will Import the Louver Display Blocks from two files are included in the *Chapter09* folder: *Louver Blocks (Imperial).dwg* and *Louver Blocks (Metric).dwg*.

1. **Insert** the appropriate version (for the units you are using to build your Style) into the current drawing.

The file will insert into the current drawing as a Block and bring with it two additional Block Definitions: **Door_Louver_M** and **Door_Louver_E**. (If you prefer, you can use DesignCenter to import just these two definitions.) You can erase the Block once inserted. We only need to have the two Block Definitions within it available to us when we build our Door Style. If you wish, you can build your own custom Display Blocks instead of those provided here. These are provided for expediency's sake only.

It is always easier to build a new Style from an existing one. Therefore, we will first import from one of the Door Style Library files.

2. Open the Style Manager to Doors and Load one of the default Door Style Libraries.

3. **Copy** an existing single-hinged Door Style from the library to the current drawing.

If you are using the Imperial or Metric Library, import the **Hinged – Single** Style; if you are using D A CH, import the **Single 1-Swing** Style.

4. **Edit** the Style and change the name to **Hinged - Single – Louver** and the Description to **Single hinged door with Louver**.

5. Click the Dimensions tab and verify that the settings meet your approval.

6. Click the Design Rules tab, change the Shape to **Custom** and choose **Door_Hinged - Single – Louver** from the list (see Figure 9–36).

We created this Profile above. Leave the **Door Type** set to **Single**.

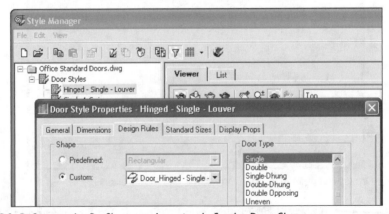

Figure 9–36 *Reference the Profile created previously for the Door Shape*

7. Click the Standard Sizes tab and edit any of the sizes, as you require.

Here you see one of the benefits to beginning with an existing similar Style. It is much less work to edit a few sizes than to re-input the entire list.

Set Up the Display Properties

When you add a Profile to a Door as we did on the Design Rules tab, the "hole" in the Profile defaults to glass within the Door. To change this, we need to change the Display Properties of the Door to show a louver in place of the glazing. This must be done using a custom Display Block for each view where we would like to see the louver. We will use the Blocks that we just inserted into this file.

1. Click the Display Props tab and choose **Elevation** from the Display Representation list.

Elevation is not the active Display Rep, but it is one of the ones where the louver would show graphically.

2. **Attach** a Style Level Override and then click the **Edit Display Props** button (see Figure 9–37).

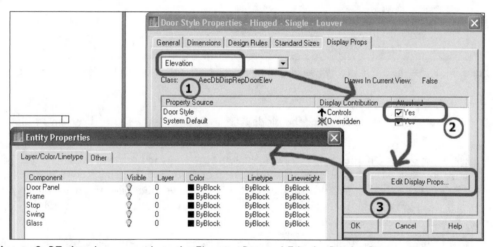

Figure 9–37 *Attach an override to the Elevation Rep and Edit the Display Props*

3. Click the Other tab and then the **Add** button in the **Custom Display Blocks** area.

Since this is the Elevation, we want the 2D Elevation Block. Set the Viewer window to Front view to make it easier to see the effect of the block parameters as you manipulate them.

4. Click the **Select Block** button and choose **Door_Louver_E**.

5. In the **Component** area at the bottom, choose **Glass Component** and **All**.

6. In the **Scale to Fit** area, choose both **Width** and **Height**.

7. In the **Insertion Point** area, choose **Center** for both **X** and **Y** and leave **Z** set to **Bottom** (see Figure 9–38).

The Blocks provided had their insertion points aligned to the bottom center of the louver for convenience when inserting them. Always plan your Base points carefully to save yourself extra effort in this dialog box.

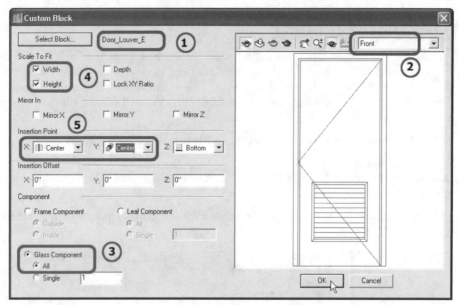

Figure 9–38 *Set the parameters to scale and center the block to the available Glass Component area*

8. Click **OK** to finish.

9. Repeat the same steps for the Model Display Rep using the **Door_Louver_M** Block.

There are actually two Model Display Reps in this template file, if you followed the recommendations set out in Chapter 5. If you want the louver to show in both, you will need to repeat these steps for both Model Display Reps, and any other additional Display Reps you may have added. Be careful, however, because applying a "Style level" override to a particular Display Rep compels it to be imported into any drawing that uses this Style. Furthermore, that Display Rep cannot be purged, because this Style will be *using* the Display Rep.

10. **Edit** any Layer, Color or Linetype settings that you wish and click **OK** to complete the Display Props.

Property Sets

1. Click the General tab and then the **Property Sets** button.

Notice that there are already some Property Sets attached to this Style because we imported it from the default library. If you are satisfied with these, you can leave them as is. If you followed the tutorials in Chapter 6, you likely have a new Property Set Definition for Doors that would be more appropriate here. Import that Property Set Definition into this file, and then remove and add Property Sets as necessary here in this Style. When you are finished, close the Style Manager.

2. **Save** and **Close** the *Office Standard Doors.dwg* file.

This Library file is now ready to be used. A sample is included in the *Chapter09* folder. Simply load it into the Style Manager and copy this Style into the drawings where it will be used. A similar process can be used to build Pivot Doors, Center Mullion Doors, sliders with four or more panels, etc.

STRUCTURAL MEMBERS

Display Control was discussed extensively in Chapter 5, but we keep finding ourselves revisiting the topic. Here again as we look at Structural Member Styles, it is appropriate to begin with a review of the Display Control issues. In the *Chapter09* folder is a file named *S-FP01.dwg*, which contains a small structural bay with columns, beams and joists. Open this file while reading this section. A few nice structural sample files are contained on the ADT 3.3 CD ROM. Look in the *ADT Samples\Commercial* folder of your ADT Installation CD. A portion of the *10001S1.dwg* file from that location was used as the basis for the sample file used here.

DISPLAY REPS UNIQUE TO STRUCTURAL MEMBERS

You may recall in Chapter 5 that the D A CH template file had some Display Configurations designed specifically for working with Structural Members. From an implementation point of view, this is a sound strategy, considering that Structural drawings will usually be files separate from the architectural floor plans.

In the sample *S-FP01.dwg* file included here, four layout tabs have been created, each one displaying the structural model at a different scale and with different level of detail. There are three levels of display detail for Structural Members (see Figure 9–39).

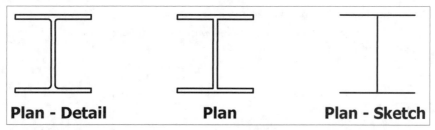

Figure 9–39 Three levels of detail available for Structural Members

If you click through each of the tabs in the sample file, you can see these Display Reps in use. Table 9–2 lists the names for the Display Configurations in this file:

Table 9–2 *Structural Display Rep Matrix*

Display Configuration Name	Plan Rep	Elevation Rep	Model Rep
Structural Member 48	Plan – Detail	Elevation – Detail	Model – Detail
Structural Member 96	Plan	Elevation	Model
Structural Member 192	Plan – Sketch	N/A	N/A

The configuration names can be changed if you wish. The idea is simply to show increasingly more detail at the larger scales.

BUILDING A CUSTOM MEMBER SHAPE

The Structural Member Catalog is very extensive. However, it does not contain every conceivable member shape that we may need. When the catalog fails to provide the specific shape you require, or if you wish to use Structural Members for some "non-structural" purposes, you will be faced with creating your own member shapes. The process is not difficult, but it must be performed at the command line. To build a custom shape, you must draw at least one closed polyline to use for the medium level of detail. However, if you wish, you can use lines, arcs, polylines and circles to represent a simplified "Sketch" display and closed polylines to represent the detailed display. The medium (Design) and high (Detail) levels of detail must be drawn with one or more closed polylines. The potential of Structural Members is far reaching. In the following tutorial, we will only scratch the surface and create a custom Structural Member Style that will be used as a design element. This Style will represent a piece of dentil molding, complete with "teeth." This molding adorns the top of the façade in the Model rendered in VIZ in Chapter 13, available on the Autodesk Architectural Desktop: Advanced Implementation Guide CD ROM. One of the most fascinating uses of Structural Members (that has little to do with Structural engineering) is their ability to display at three levels of detail. This provides enormous flexibility when importing these models into VIZ. We can literally decide at what detail we would like the rendering to be generated.

Create a Custom Shape

To create a custom shape, first draw your geometry and then use the –MEMBERSHAPEDEFINE command at the command line. You must type it exactly, including the leading dash. Then you simply follow the prompts.

1. In the *Chapter09* folder, **Open** the file named *Dentil Molding01.dwg*.
2. The file contains three shapes labeled with their intended uses in the custom shape (see Figure 9–40).

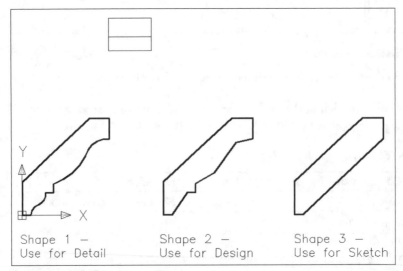

Figure 9–40 *The Dentil Molding.dwg drawing contains three shapes ready to be used in a custom member shape*

3. At the command line, type **–membershapedefine** and then press ENTER.

4. Type **n** (for New) and then press ENTER to create a new shape.

5. When prompted, type **Dentil** and then press ENTER.

6. Type **d** (for Description) and then press ENTER to give the shape a description.

7. When prompted, type **Dentil Molding** and then press ENTER.

8. Type **g** (for Graphics) and then press ENTER to configure the graphics for the Shape (see Figure 9–41).

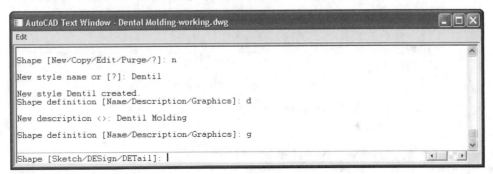

Figure 9–41 *Using the Command Line driven –MEMBERSHAPEDEFINE command*

9. Type **s** (for Sketch) and then press ENTER to configure the Sketch Display Graphics.

10. When prompted for the insertion point, click the lower left endpoint of the Sketch polyline (Shape 3). See Figure 9–42.

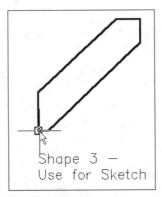

Figure 9–42 *Set the Base point of the Sketch Display*

11. When prompted to select objects, click to select the Sketch polyline (Shape 3).

12. Type **des** (for Design) and then press ENTER to configure the Design Display Graphics.

13. When prompted to "Erase Polyline" type **n** to keep it.

14. When prompted to select a polyline, click to select the Design polyline (Shape 2 in Figure 9–40).

15. When prompted to "Add another ring" type **n** to answer "No."

16. When prompted for the insertion point, click the lower left endpoint of the Design polyline (Shape 2).

This is the same relative point that was used for the Sketch shape.

17. Type **det** (for Detail) and then press ENTER to configure the Detail Display Graphics.

18. When prompted to "Erase Polyline" type **n** to keep it.

19. When prompted to select a polyline, click to select the Detail polyline (Shape 1 in Figure 9–40.)

20. When prompted to "Add another ring" type **n** to answer "No."

21. When prompted for the insertion point, click the lower left endpoint of the Detail polyline (Shape 1).

This is the same relative point that was used for the Sketch and Design shapes.

22. Press ENTER three times to return to the command prompt.

23. **Save** the drawing.

Create a Structural Member Style

We will now create a new Structural Member Style that uses the new shape.

1. From the **Design** menu, choose **Structural Members>Member Styles**.

2. Create a **New** Style named **Dentil Molding** and then **Edit** it.

3. On the General tab of the Structural Member Style Properties dialog box, type in a **Description** (see Figure 9–43).

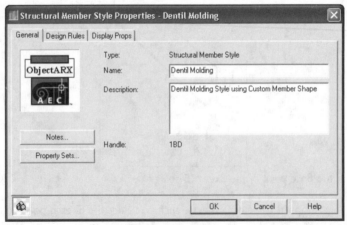

Figure 9–43 *Add a Description to the new Member Style*

4. Click the Design Rules tab, and next to the "unnamed" component, choose **Dentil** from the list (see Figure 9–44).

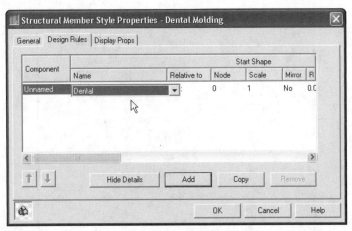

Figure 9–44 *Choose the custom shape from the list of shapes for the unnamed component*

5. Click the Display Props tab, and then choose **Model – Detail** from the list of Display Reps.

6. **Attach** a Style level override and then click the **Edit Display Props** button.

7. Click the Other tab, and then click the **Add** button in the **Custom Block Display** area.

Custom Blocks are used in many Styles to represent items that cannot be represented parametrically. We used them in the last tutorial to represent Door louvers. In this exercise, we will use a display Block (already resident in this file) to represent the "teeth" on the molding.

8. Click the **Select Block** button, choose **DentilTooth** from the Block list and then click **OK**.

9. In the **Position Along (X)** area, choose **Midpoint of Curve**.

This will center the teeth starting from the middle of the total molding length.

10. In the **Insertion Offset** area, type **–3** in the **Y** field and **1** in the **Z** field.

This shifts the Block relative to the Structural Member based on the way the Block was created.

11. In the **Mirror In** area, place a check mark in the **Mirror y** box.

Again, this compensates for the orientation of the Block Definition. To make several teeth appear along the length of the molding, do the following:

12. Place a check mark in the **Repeat Block Display** box.

13. In the **Space Between** field, type **2**.

14. Place a check mark in the **Fit** box, and choose the **Scale to Fit** option.

The complete settings should look like Figure 9–45.

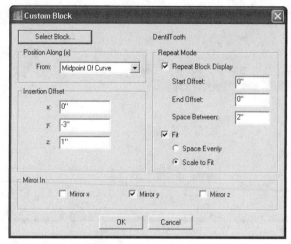

Figure 9–45 *Configure the repeating Block display to show "teeth" along the molding*

15. Click **OK** all the way back to the drawing.
16. **Save** the drawing.

Use the New Style

1. From the **Design** menu, choose **Structural Members>Add Beam**.
2. Choose **Dentil Molding** from the Style list and then draw at least two points on screen.
3. **Close** the Add Beams dialog box.
4. Choose an isometric view from the View menu.

Notice that the shape is displaying correctly, but there are no teeth. This is because we added the teeth to the Model – Detail Display Rep, and not the Model Display Rep. The current Display Configuration is Work, which uses the Model Display Rep to display Structural Members. To see the teeth, we must load a Display Configuration that displays the Model – Detail representation. In this drawing, **Structural Member 48** is configured to do so.

5. From the **Desktop** menu, choose **Select Display**, and then load **Structural Member 48**.

Notice that the teeth now appear. Using grips, stretch the length of the Member and notice that more teeth will appear. The Sketch display that we added really applies only to Plan views. Therefore, for a "molding" example, you will not see the Sketch Rep much. However, if you followed the same steps to create a custom Member for use in a Column, you could make use of the Sketch Display Rep. A single paper space viewport has been provided in this file to help you view the two levels of Model Detail. Continue to experiment with it if you wish.

6. **Save** the drawing (see Figure 9–46).

One other technique that is particularly useful with a Molding Style is to convert a polyline of any shape in plan to a Beam that uses the Molding Style. Draw a multi-segment polyline in plan view and then from the **Desktop** menu, choose **Structural Members>Convert to Beam**. Select the polyline and note the results. This is a very effective and simple way to add complex moldings (that display at varying levels of detail depending on the active Display Configuration) to your 3D Models. You can see the results of this technique in the Model used in Chapter 13 for interoperability with Autodesk VIZ. (Chapter 13 is on the CD ROM in PDF format.)

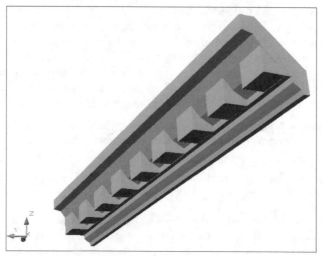

Figure 9–46 *The completed custom Structure Member Style*

THE LOGICAL DISPLAY REP

A very interesting Display Rep is the "Logical" one. This Display Rep is a diagnostic mode. It shows a small circle at a connection between two structural members. For instance, if two structural members' endpoints physically meet, this Display Rep will flag the condition with a circle. Connections that do not physically touch are considered to not "logically" connect (see Figure 9–47).

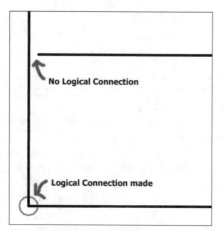

Figure 9–47 *Small circles indicate when logical connections are made*

Set the System Default Display Settings for the Connection component within the Logical Display Rep to a color like cyan to make them easier to see on screen.

GROUPED WINDOWS AND DOORS WITH SIDELIGHTS

ADT offers a Window object and a Window Assembly object. The Window object can be used to represent a variety of typical Window configurations. However, multiple groupings are not part of its capabilities. Window Assemblies are designed specifically to achieve multiple groupings of Windows and Doors in any combination, and since the Window Assembly object is essentially a Curtain Wall, you can model complex Window/Store front configurations directly within the Window Assembly Style interface. This leaves the potential of Window Assemblies wide open and allows for a wide range of Window Assembly configurations, from simple to complex. Let's discuss two very simple situations that arise in many architectural projects: "Grouped" Windows and Doors with Sidelights. Here are several possible solutions:

- **Option 1** – Add several adjacent Windows (for Grouped Windows) or adjacent Door and Window (for sidelights) within the same Wall.
- **Option 2** – Build a Window Assembly that contains nested Doors or Windows.
- **Option 3** – Create custom View Blocks within the Display Props of the Door or Window Styles representing the complete grouping.

Option 1 – All of these solutions offer their pros and cons. In the first scenario, you will literally end up with several separate objects that must be selected, edited and scheduled separately. There may be problems with the graphical display as well. For instance, if you group three Windows and then modify the Sill Plan display properties, the graphic representation will be that of three separate sills, not one continuous sill for the grouping. You will also have issues with "double" frames showing where the separate objects meet (see Figure 9–48).

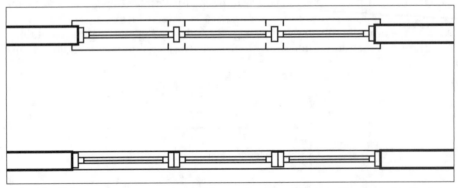

Figure 9–48 *When using separate units, problems arise when Sills are offset and with the potential for "double frames"*

Option 2 – The Window Assembly approach is certainly the most logical of the three. It creates a single object that in turn references the appropriate Window and Door components. The entire grouping can be moved by simply moving the Assembly. Although the interface to edit Window Assemblies (same as the Curtain Wall) is challenging, these types of Assemblies are simple and usually require only a single grid division. Problems can arise when you consider the frames and when including these "ganged" units on Schedules. Both the Window Assembly and the Window/Door have a frame component. You will need to decide which to display. Although you can build it either way, displaying the Window Assembly frame will give the most benefit. First, if the Window Assembly frame is displayed, it makes it easy to select the Window Assembly to edit the Window spacing. Second, there will be only one frame component between Windows when using the Window Assembly frame, while there would be two when using the Window or Door Frame.

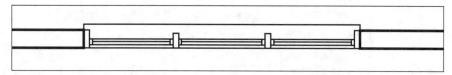

Figure 9–49 *A Window Assembly with the infill Window frames set to zero, and Sills off, offers the best solution*

You will also want to turn off the Sill display for the Window Style referenced by the Assembly. Instead, you can create a "Sill Frame" component in the Window Assembly Style (see Figure 9–49). This will allow a single sill to span across all three Windows. The final challenge is in how these items will be scheduled. You can include Window Assemblies on your Window Schedule by simply including them in the **Applies To** for the Schedule Table Style. The problem is that you (and your teammates) must remember that in the case of ganged Windows, you must select the Window Assembly for the Schedule and not the Windows themselves.

This can cause some coordination issues that can unfortunately only be solved by following good procedure. They cannot be solved automatically. One way to help alleviate this situation is to be sure not to include any Property Sets for the Infill Window Style used by the Window Assembly Style. In this way, if someone inadvertently selects the Windows instead of the Window Assembly, question marks will appear in the Schedule for the contained Windows. This should alert users to the situation so that they can correct the Schedule Table selection. To create a Window Assembly Style as described here, do the following:

1. Create one or more Window Styles (and if necessary Door Styles) whose Frame dimensions are set the proper depth and a zero width. You will need one such Style for each type of Window infill (for instance, a Fixed, a Casement or a Hinged Single Door). For Window Styles, be sure to also turn off the Sill Components in the Sill Plan Display Rep (see Figure 9–50).

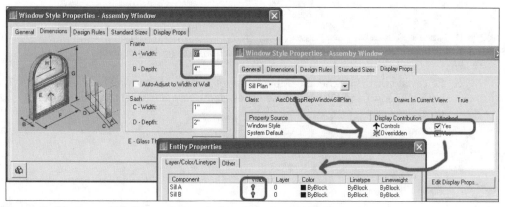

Figure 9–50 *Create Window and Door Styles with a zero-width frame*

2. Create a Window Assembly Style for each condition you need (for instance, a 3-gang Window or Door with Sidelight).

3. For a group of three, set the Default Division to Vertical orientation with a **Fixed Number of Cells** of Cells set to **3** (see Figure 9–51).

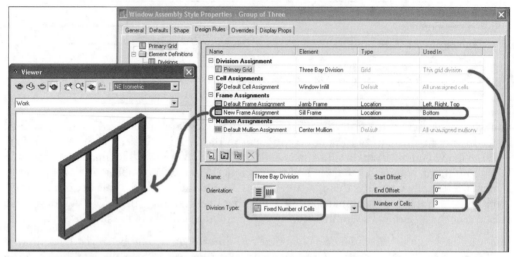

Figure 9–51 *Create a Window Assembly Style divided by three*

4. Create any Mullions and Frames required, including a Sill Frame if you wish. For the Infill(s) reference one of the Window/Door Styles created in step 1 (see Figure 9–52).

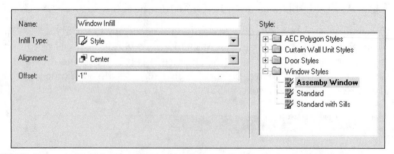

Figure 9–52 *Reference the Window Style with the zero-frame in the Infill definition*

5. Be certain that the Schedule Table Style you use for Window Schedules **Applies To** both Windows and Window Assemblies (see Figure 9–53). (The default ones do already.)

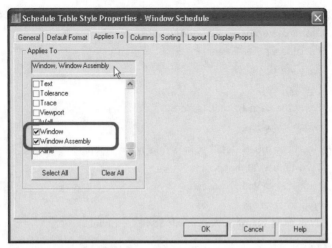

Figure 9–53 *The Window Schedule Style must "Apply To" both Windows and Window Assemblies*

For a Door with Sidelight Assembly, simply select the Door when scheduling, like any other Door. You can add the appropriate information regarding the sidelight to the Frame columns, Remarks or other appropriate column of the Schedule. Only if you reference two Door objects in the same Window Assembly will you run into issues similar to those with the Windows when scheduling.

Option 3 – The third option (custom Display Block), although possible, is the least desirable of the three. The only advantage that it brings is in scheduling. Since there will be only Windows and Doors, a typical Door or Window Schedule will work.

Regardless of which approach you take to this issue, be sure to include plenty of typical conditions in your Office Standard Style Library files so that the process during production is smooth and predictable. Most users will happily follow the designated procedure if all of the resources have been provided and well planned in advance. Frustration arises when production must be interrupted to solve problems with content and Styles that could have been resolved more easily earlier.

CHECKLIST

I Style Libraries

1. From the offerings provided, decide what to keep, what to edit and what to remove.
2. Create new or edit existing Content items following a consistent process.
3. Move the *Style Library* folder to the root of the L Drive (and out of the Content tree).
4. Set up a "Project" Style Library procedure if appropriate for your firm.

II Wall Styles

1. A Wall Style should show the least amount of detail possible to convey the most amount of intent.

2. Determine your firm's standing on the following philosophical points:
   a. Nominal versus Actual dimensions
   b. Variable versus Fixed
   c. Baseline position
   d. Level of detail
   e. Bring the case to management
   f. Be consistent
3. Decide how much detail your Walls really need.
   a. Two-Line Wall approach
   b. Multi-Line Wall approach
   c. Hybrid strategy
   d. Wall style Naming, Autodesk Style Guidelines
4. Property Sets
   a. Change Wall Type to a Style-based property
   b. Update the Wall Type Tag to reference this Style-based property.
5. Build fire rated Wall Styles if using.
   a. Use a "zero" width component at the center of the Wall Style, or
   b. Use the Wall Sketch Display Rep.

III Fenestration Styles

1. Build a library of commonly used AEC Profiles for customizing various Style types. Save them with the associated Style Library.
2. When considering the level of customization required, consider the Plan 1:100 and Plan 1:50 Display Reps.
3. Use the Autodesk Style Guide naming for library components, and rename for project-specific needs.

IV Structural Member Styles

1. Build a Structural Member template that incorporates the unique Display Configurations of the Structural Members.
2. To create custom Member Styles do the following:
   a. Draw the Sketch graphics using lines, circles, arcs and polylines. Keep them schematic and simple.
   b. Draw the Design and Detail graphics using closed polylines.
   c. Use the –MEMBERSHAPEDEFINE command to create the custom Member Shape.
   d. Create a new Structural Member Style. For simple Styles, use a single component referencing the custom shape. For complex members, optionally add additional components on the Design Rules tab referencing catalog shapes, or other custom shapes.
   e. Configure the Display Props and add Custom Blocks if required. Be sure to build the Custom Block before entering the Display Props dialog box.
   f. Add Custom Member Styles to the library as appropriate.

V Grouped Windows and Doors

   1. The recommended approach is to use Window Assembly Styles.

      a. Create custom Window and/or Door Styles that do not contain a Frame (set the Frame width to zero).

      b. Build Window Assembly Styles that reference one or more of these styles as Infills.

      c. Use the Window Assembly Design Rules to configure the Frames and Mullions of the grouping.

      d. Include Window Assemblies in the Window schedules. Do not attach Property Sets to the Infill Window Components. For Doors, just schedule the Door.

VI Document All Procedures

SUMMARY

Your Style Library should be centrally located for all users.

Take the time to explore the default offerings, because they may save you much time in customization.

A Wall Style should show the least amount of detail possible to convey the most amount of intent.

Remember, just because you *can* doesn't mean that you *should*.

Avoid including the same Style Definition in two separate Library files.

Consider using "Two-Line" Wall Styles over "Multi-Line" to keep project files more responsive and simple.

Use Property Sets to add complexity to Wall Styles.

Use the Wall Style Name for the Wall Type Tags. Rename Styles in projects to correct Type designations.

The Wall Sketch Display can be used for fire rating.

Coordination of physical, display and data properties in fenestration Styles is sometimes challenging.

Structural Members can be used for more than just Structural.

Window Assemblies are the best choice for creating "ganged" Windows and Doors with Sidelights.

CHAPTER 10

Area Objects and Area Groups

INTRODUCTION

Most building projects require calculation of critical areas in some form. The extent of such need will vary depending on whether you seek to calculate the building Floor Area Ratio (FAR,) determine leasing rates based on BOMA (Building Owners and Managers Association), or generate a Proof of Areas report for building department officials. There are almost as many ways to extract this information from an AutoCAD/ADT drawing as there are reasons to do it. In this section, we will look specifically at the AEC Area tools included with ADT.

There is often confusion among ADT users regarding when to use Space objects and when to use Area objects. Both objects can report their included area in a schedule. The Space object has the ability to use a Space Inquiry command and export directly to Microsoft Access. The Area object can export information directly to a spreadsheet in Excel format. So which one is best for area calculations of your drawings? If you need only a simple report of areas for each individual space in a building, either tool could be used. However, Area Calculation Modifiers and Area Groups make the Area object more attractive for complex area takeoffs such as those required by building codes and leasing agents.

Note: Many of the topics in this chapter require the installation of the ADT International Extensions. See the note in the Preface regarding installation of International Extensions.

OBJECTIVES

The primary goal of this chapter is to overcome the hurdles that stand in the way of incorporating Areas and Area Groups into your firm's workflow. Many firms using ADT have taken the attitude that Areas are "an interesting tool showing potential and worthy of exploration at some undetermined future date." This chapter is your chance to make the time for that future exploration now. In this chapter, we will explore the following:

- Understand the relationships of Areas to Area Groups
- Learn how to create and manipulate Areas and Groups
- Understand and use Area Group Templates
- Explore the possiblities of Display Control settings for Areas
- Utilize Area and Area Group Property Sets and Schedule Tables
- Generate reports of Area information

BASIC AREA TECHNIQUES

Area objects are two-dimensional shapes designed specifically to report building area information. They graphically represent positive and negative areas and, with a special "Decomposed" Display Rep, can even show an area broken down into a series of simplified triangular sections with dimensions for generating area proof. Area objects can also be grouped together using Area Groups. An Area Group is a simple "wrapper" object, used to report the total area of several Area objects and/or other Area Groups. You are able to create Areas from existing geometry in your Building Model, but to get the exact Area needed, you will often need to manipulate the Area manually. Grip editing works best for this.

Understanding the Relationship of Areas to Groups

If you do not live in the D A CH countries, getting started with Area objects is a bit more challenging, since there are no pre-configured templates or styles in the Imperial or Metric offerings. There are some great samples in the D A CH Content, however. Let's begin with a look at the basic tool set.

1. **Open** the file named *Area Feature Styles.dwg*.

This file is located in the *C:\Program Files\Autodesk Architectural Desktop 3\Content\Metric D A CH\Styles* of your Management Install. A copy of this file has also been provided in the *Chapter10* folder for your convenience.

There is a sample floor plan included in this file, and two Area Group structures beneath it. There is a note explaining that the Area Groups were inserted from Area Group Templates. Let's start by exploring the Groups.

2. Click on the Group marker labeled **Organisation**. You can click either the marker or the label (see Figure 10–1).

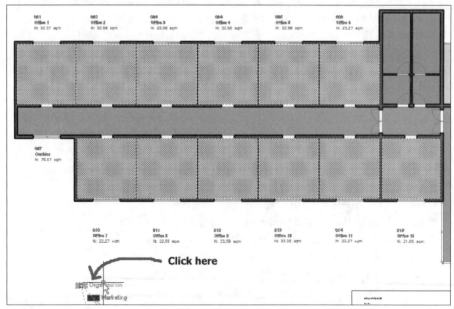

Figure 10–1 *Click a Group marker to highlight the entire Group*

Notice that all of the offices highlight in the plan, both above and below the corridor.

3. Press ESC to deselect the Group and try clicking another such as the **North** sub-group beneath Sales.

Note that this time, only the two offices in the top left corner of the plan highlight.

4. Continue clicking the various Group markers to understand their relationship to the floor plan.

Look at the right side of the plan at the Entrance Hall space. Note that there is a void in the middle of this space. It will be difficult to click on the Area object for this space in the current Display Configuration. You could use CTRL to cycle through the objects until the Area object highlights, but there is a special Display Configuration in this template that displays only Area objects. Making this one active will make it easy to work with the Area objects by themselves.

5. Load the Display Configuration named **Areas**.

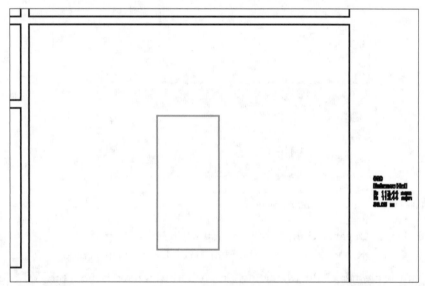

Figure 10–2 *Displaying only Area objects reveals the "subtractive" Area in the center of the Entrance Hall*

Notice the red outline around the void area (see Figure 10–2). This subtractive area is part of the Entrance Hall Area object. Areas can be comprised of several rings, both positive and negative. To make a ringed or "donut" type space, you can use a subtractive ring like this example.

6. Click on the Red Rectangle. Select the midpoint grip on right side of the red rectangle and move it to the right I meter.

Notice the change in Area in the tag to the right. The void was enlarged, which reduced the overall area of the Entrance Hall space.

7. **Pan** to the top right corner of the floor plan and click on either **Rest Room** Area.

Notice that the Rest Room Areas are comprised of two rings each. Both of these rings are "positive" Areas. The Rest Rooms are simply made up of an anteroom and the main rest room.

Even though they are two physical spaces, they are considered a single space for area calculation purposes.

8. Select the **WC Men** Area, right-click and choose **Area Modify**.

There are two drop-down lists in this dialog box. The **Name** list accesses the list of Area Name Definitions referenced by this Area. The other is the **Style** list.

9. In the Area Modify dialog box, click the **Area Group Viewer** icon (see Figure 10–3).

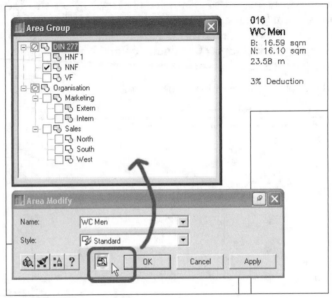

Figure 10–3 *Click the Area Group Viewer*

10. Under **Organisation,** place a check mark in **Marketing**.
11. Click **OK**, and then restore the **Plan 1-100** Display Configuration.

Notice that when you click the Marketing Group marker, it now includes the WC Men space as well. This is just one way that you can manipulate the contents of an area.

Area Creation

Areas can be drawn with the **Add Area** command, or created from existing objects like polylines or Walls.

1. From the **Documentation** menu, choose **Areas>Add Area**.
2. Make sure that the **Composed Areas** check box is not checked and begin adding points in the same way as adding Walls.

Add as many points as you wish.

3. Choose **Polyline** or **Ortho Close** to complete the Area.

When **Composed Areas** is turned off, you are creating simple "one-ring" Areas. If you want to create a more complex Area like the WC Men or Entrance Hall space viewed above, use the Composed Area feature. Let's create some composed Areas.

4. While still in the Add Area dialog box, check **Composed Areas** and click the positive (**+**) button.
5. Draw an Area about **4** meters by **3** meters.
6. Leave some space and draw another positive enclosed Area next to it.
7. Click the negative (**-**) icon and add a **2** meter by **1** meter Area within the first one (see Figure 10–4).

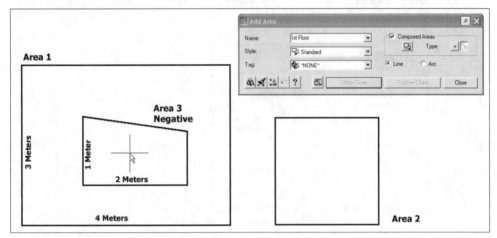

Figure 10–4 *Creating a composed Area*

8. Click **Close** to end the Add Area dialog box.

Notice that the inner ring turns red when the command completes. This indicates that it is a void Area. Be careful when creating composed Areas not to let positive and negative areas overlap. This will cause an error marker to appear as shown in Figure 10–5.

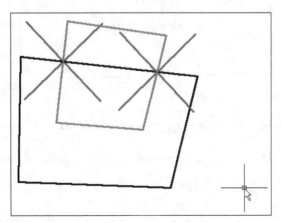

Figure 10–5 *Error markers when positive and negative intersect*

458

Modify Area Objects

The shape of Area Objects can be modified easily with grips and a variety of Area "operation" commands.

1. In the *Chapter10* folder, **Open** the file named *Area Objects.dwg*.

Shape 1 is a compound Area including three rings: two positive and one negative.

2. Select the compound Area, right-click and choose **Operation>Trim** (at "1" in Figure 10–6).
3. When prompted, choose the midpoint of the square ring at right as the first point of the trim (at "2" in Figure 10–6).

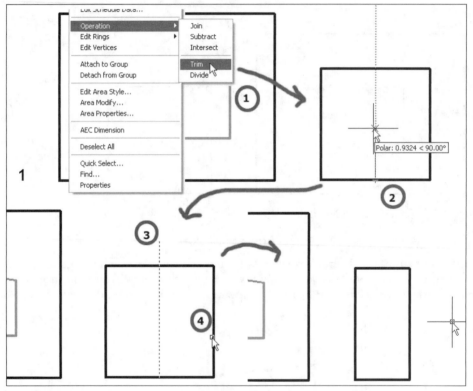

Figure 10–6 *Using the Trim operation on Areas*

4. Click a second point straight up from the first (at "3" in Figure 10–6). It does not need to snap.
5. Select the right side of the Area object to trim away (at "4" in Figure 10–6).

A similar operation is **Divide**, which will break the Area into two pieces and keep both as separate Area objects.

6. Select the same object, right-click and choose **Operation>Divide**.

7. Following the prompts, pick two points across the top of the Area as shown in Figure 10–7.

The top portion will now be a separate Area object.

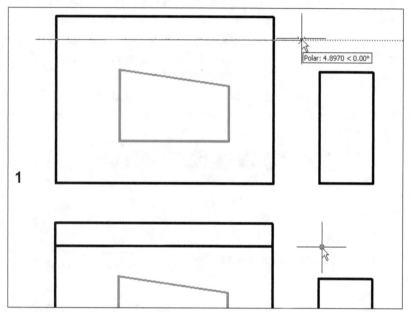

Figure 10–7 *Divide an Area into two objects*

8. Repeat the same steps, but this time cut across the negative portion of the Area object.

Notice that since the negative ring is no longer completely contained by the positive one; the shape of the new Areas simply incorporates the void, and the edge is no longer red (see Figure 10–8).

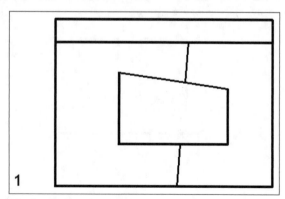

Figure 10–8 *Result of dividing across the void Area*

Area objects can perform Boolean operations. Shape 2 includes four separate Area objects. The two at the top are compound Areas with negative "holes" in the center. The two at the bottom are separate overlapping shapes (see Figure 10–10).

9. Select the large Area at the bottom, right-click and choose **Operation>Subtract**.

10. When prompted, select the small Area in the middle and press ENTER.

Notice that the Area in the middle turns red. This indicates that it is now a negative portion of the larger area. Another way to say this is that the larger Area is now a donut. To see this more clearly, we can manipulate the Display Props of the Area object.

11. Select the same Area, right-click and choose **Entity Display**.

12. Place a check mark in the **Yes** column next to **Area Style** and then click the **Edit Display Props** button (see Figure 10–9).

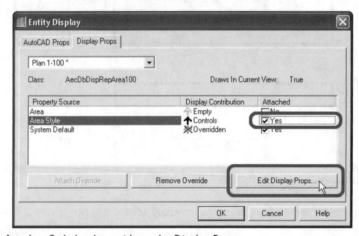

Figure 10–9 *Attach a Style level override to the Display Props*

13. Turn **on** the **Additive Hatch** component (click the dim light bulb) and click **OK** back to the drawing (see Figure 10–10).

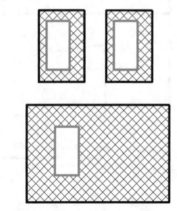

2

Figure 10–10 *Turn on Additive hatching*

We turned on the Additive hatching so that it is much easier to see the "donut" effect of these Area shapes. You can also turn on Subtractive hatching if you wish. If you do, you should pick a hatch pattern that is not used for anything else and make the color the same red as the rings. This will help the Subtractive Hatch stand out against the additive portions of the Area objects.

14. **Move** the Area at the top right down into the larger Area at the bottom.

Make sure that it is completely contained in the larger Area as shown in Figure 10–11.

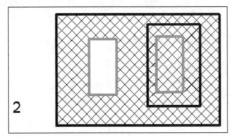

Figure 10–11 *Move the top Area into the larger one*

15. Select the large compound Area, right-click and choose **Operation>Join**.
16. When prompted, select the small overlapping Area (the one we just moved).

The result is not very exciting. Since you joined a "donut" with a solid, it simply became solid.

17. **Undo** the **Join** command.
18. Select the large compound Area again, right-click and choose **Operation>Subtract**.
19. When prompted, select the small overlapping Area (the one we just moved).

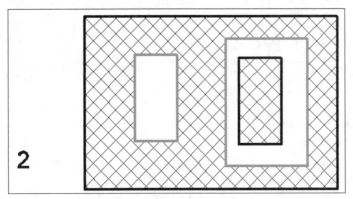

Figure 10–12 *The result of subtracting a "donut" from a solid*

This time the result is much more interesting (see Figure 10–12). Anything that was solid (additive) in the small Area, cut away from the solid in the large Area. However, since there was a hole in the small area, it left some of the material behind when carved away from the large one.

You can also manipulate the additive and subtractive rings after they are created. Therefore, if you would like to turn a void into a solid, you can simply "reverse" the ring.

20. Select the Area, right-click and choose **Edit Rings>Reverse Ring**.

21. When prompted, click on any internal red (subtractive) ring.

The void will hatch to indicate that it is now solid. Do a few more experiments if you wish. If you choose **Reverse Profile**, this will reverse the entire shape; anything that was positive will become negative, and vice versa. Use the **Join** command to join the large bottom shape with the remaining small shape above it. This will yield a compound shape like Shape 1.

Shape 3 contains two overlapping Areas. Use this shape to experiment with the **Intersect** command.

22. Select one of the two Areas in Shape 3.

23. Right-click and choose **Operation>Intersect**.

24. When prompted, select the overlapping Area.

You will be left with a small Area the size of just the overlapping portion.

Create Area Object from Other Objects

1. **Pan** the drawing to the right and locate a small room with four Walls and a Roof.

2. From the **Documentation** menu, choose **Areas>Create Area from Objects**.

3. In the Create Areas from Objects dialog box, click the **Select Objects** button (see Figure 10–13).

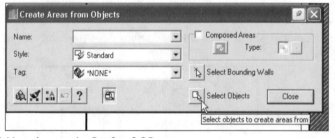

Figure 10–13 *Add an Area in the Roof at 3.25 meters*

4. When prompted, select the Roof object and press ENTER.

5. At the "Enter Cutplane Height" prompt, type **3.25** and press ENTER.

You can make Areas from virtually any ADT object this way. You will always get to assign the Cutplane where the Area is extracted. This can prove a valuable way to generate takeoffs of your Model. You can also generate areas with this command from the Walls that bound the rooms in a plan.

6. Delete the Area just created and then return to the **Areas>Create Area from Objects** command.

7. Click the **Select Bounding Walls** button (see Figure 10–14).

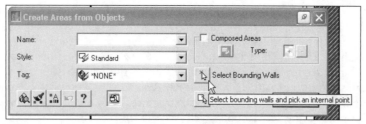

Figure 10–14 *Use the Select Bounding Walls button to create Areas from Walls*

8. When prompted, select all four Walls and then press ENTER.

9. At the "Pick internal point" prompt, click a point inside the room and then press ENTER.

10. Press ENTER again and then **Close** the Create Area from Objects dialog box.

Notice that an Area object has been created along the inside perimeter of the Walls. You can use grips editing to fine-tune the exact shape. For instance, you could grip edit the Area to the centerline or outside edge of the Walls. The **Edit Vertices** command is also very useful. Using this command, you can add or delete Vertex points from the Area's shape. This is a very helpful way to manipulate an Area after creation. You can find the **Edit Vertices** command on the **Documentation>Areas** menu or the right-click menu of an Area object.

WORKING WITH AN AREA OBJECT DRAWING TEMPLATE

Once you have mastered the mechanics of working with Area objects, you must figure out how best to use them in your organization. There are many building code requirements that Area objects and Groups can help to satisfy. In addition, they have potential as tool for generating takeoffs of all sorts of critical building quantity information.

To get started, let's look at the available tools. The Area object tool set consists of five distinct Style types:

▶ **Area Calculation Modifier Styles** – Formulas that are applied to the actual area and perimeter values derived from Area objects.

▶ **Area Group Styles** – Define the name and graphic display properties of an Area Group object.

▶ **Area Groups Templates** – Predefined collections of pre-named and styled Area Groups organized hierarchically.

▶ **Area Name Definitions** – Lists of names that can be associated with Areas and Area Groups to help standardize the names of individual Area objects.

▶ **Area Styles** – Control the name and graphical properties of an Area object.

The Area object is the basic component of the tool set. As we saw above, it behaves much like a polyline and is therefore simple to use and understand. However, the other tools, and more importantly, the relationships between each of the tools, are a bit more complicated. In addition to each of these tools, you will work with Schedules, Property Sets and Area Evaluations to export the data from ADT and into a program such as Microsoft Excel for analysis and reporting.

The online help system for Areas suggest two approaches to their use: "From the Top Down" and "From the Ground Up." The Top Down approach suggests that you first build and configure all of the pieces of the tool set and then put them to use to create an Area Evaluation. The Ground Up approach has you building the components on the fly as needed while you simultaneously create your Area Evaluation. Although both approaches are theoretically viable, the Top Down approach will enable you to quickly grasp the tool set and techniques required to use Areas effectively much more quickly than the Ground Up approach will. We will use the Top Down approach here. Usually, if you do the same type of Area calculations frequently, the best approach is to build a template file that includes all of the basic Styles you will need. Area Group Templates can be a great help for this. An Area Group Template contains a pre-built collection of Area Groups and their hierarchical structure. They are useful for pre-assigning the Area Name Definitions to each Group as well (see Figure 10–15).

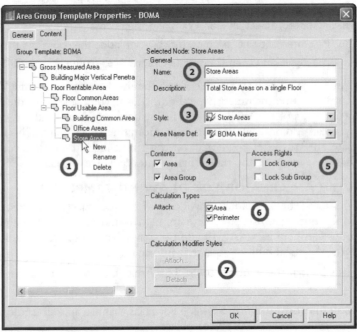

Figure 10–15 The Area Group Template Properties dialog box

Figure 10-15 notes:

1. Right-click a node on the left in the tree to add, delete or rename a sub-group.

2. Add a **Name** and **Description** for this Node. (This will be the name and description of a single Area Group.)

3. Assign the Area Group an Area Group Style and an Area Name Definition. With an Area Name Definition assigned, individual Area objects that are later attached to this Group will be able to reference names from that list. This will help to maintain consistency in naming.

4. By default, Area Groups can include both Areas and other Groups. If you wish, you can clear one of the check boxes and make the current Group able to contain only Areas or Area Groups.

5. **Access Rights** allow you to keep this particular node (in the case of **Lock Group**) or any sub-nodes (in the case of **Lock Sub Group**) from being renamed, moved or deleted. (There is no password for this feature—its intention is primarily to protect against accidental changes.)

6. By default, Area Groups calculate both Area and Perimeter of the objects they contain. You can clear either one of these check boxes if you wish.

7. Calculation Modifiers can be attached to either Areas or Area Groups. They are attached directly to the objects themselves, not through the Styles.

In the following tutorial, we will look at an example of using the Area Objects and Groups to meet the calculation requirements of BOMA (Building Owner and Managers Association), which is prevalent in the United States and Canada for determining rentable areas in buildings. Included in the data-set files for this chapter are some drawings with pre-built Area Group Templates and Area Styles. This will make the process flow much more smoothly and give you a chance to work with the tool set first hand.

Referring back to the file structures recommended in Chapter 1, you should do your work with Areas in a separate XREF file. This enables you to keep these objects from cluttering your main floor plan Models. You can screen the XREF background to make it easier to work with the Area objects, while keeping the XREF floor plan for reference. In addition, this approach allows you to have a different person work on Area calculations while others continue working on the Building Model.

Three of the four files used here have already been created, and all the XREFs have been put in place. The file prefix of "AA" has been used for these files. (AA was chosen since it is the AutoCAD shortcut for the **Area** command and is likely a familiar abbreviation to most users.) In the AEC Layer Manager, a Snapshot has been made in each file that changes the layer color of all XREF layers to a gray color. If you also intend to plot these Area files, you can also assign a "screened" Plot Style Table to these layers to make the backgrounds print halftone.

Tools developed in this chapter are designed to assist the reader in performing area calculations based on the guidelines outlined in the "Standard Method for Measuring Floor Area in Office Buildings" publication, © 1996, Building Owners and Managers Association International. To avoid misinterpretation, information in this chapter should not be used without the complete ANSI/BOMA Z65.1-1996 publication.

Exploring the Dataset Files

1. In the *Chapter10\Leasing* folder, **Open** the file named *A-AA02.dwg*.

This is the second floor of a small office building. At the bottom of the plan, we see the upper level of an atrium space, a large open leasable office space and a building core. There are some Schedule Tables to the right. The top one references Area objects on the **A-Area-Grss** layer. This is used specifically to report the Gross Building Area. A special Area Style has been provided for this that displays with a blue outline. There is only one Area using this Style in both files, and it has been placed on the **A-Area-Grss** layer in order to be isolated on the top schedule. The remaining Area objects are on the default **A-Area** layer. The middle schedule is set to filter this layer and therefore includes all Areas except the Gross Building Area object. The last one lists the Area Groups.

2. Right-click either Schedule and choose **Table Properties**.

On the Settings tab, notice the entry in the **Layer Wildcard** field (see Figure 10–16).

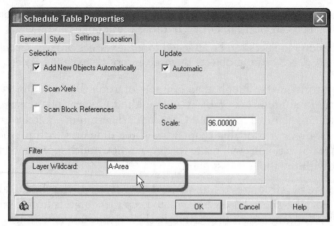

Figure 10–16 *Using a layer wildcard allows only certain Areas to be included in the Schedule*

3. Click **Cancel** to dismiss the Schedule Table Properties dialog box.

This file also includes an Area Group structure. Graphically a small rectangular icon with a text label is used to represent a Group within the drawing. These icons are organized hierarchically in a tree-style organization. Each one also includes a line to show how it is attached to other groups. Those lines are colored green in these sample files. Another line is used to link Area Groups to the Areas they contain. Those lines are colored cyan with a dashed linetype in these files.

Another important tool to help us control, understand and convey our Area and Group structure is the Display Properties. In these files, the Display Properties of each Area Style and Area Group Style have been configured to make it easy to understand at a glance the Style to which a particular object belongs (see Figure 10–17). Table 10–1 summarizes these settings. For more information on how to configure Display Settings, refer to Chapter 5.

Table 10–1 *Display Settings used for Area & Area Group Styles*

No	Area Style	Color	Pattern	Angle
2	Gross Building Area	5 - Blue	User Single	135
4	Major Vertical Penetrations	222	User Single	90
7	Office Area	183	User Single	60
8	Store Area	132	User Single	30
9	Building Common Area	83	User Single	120
11	Floor Common Area	23	User Single	150

Notes on Table 10-1

1. The colors and hatch settings shown in Table 10–1 are used for the Additive Boundaries and Additive Hatch for Area objects in the Plan and Plan 1-50 Display Reps.

2. They are used for Area Groups in the Plan 1-50 and Plan 1-100 Display Reps.

3. In Plan 1-100, additive Hatch has been turned off for Area objects and set to Solid Fill for Area Groups.

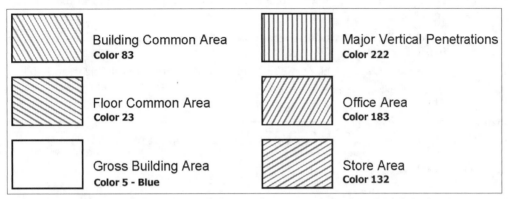

Figure 10–17 *Varying the angles of the hatch patterns helps identify each of the Styles easily on screen*

Three Display Configurations show Areas and Groups in these drawings. Work is the currently active Display Configuration. It is configured to use the Plan Display Rep for both Areas and Groups.

4. **Load** the **Plot Plan 48** Display Configuration (**Desktop>Select Display**).

The display of the Areas has adjusted to reflect the scale, but more importantly, the Groups now show only their icons. This view could be used to plot out Areas without the Group lines.

5. **Load** the **Plot Plan 96** Display Configuration.

The reverse display situation occurs here. The hatching of the Area objects has been turned off and the Groups turned on. Furthermore, the Area Groups now use a solid fill display, which could be used to plot a color presentation key plan of the critical Area divisions in a plan.

6. **Load** the **Work** Display Configuration again.

As the name implies, "Work" is a working display; therefore both Areas and Groups are shown here. The Area objects are each coded by color and hatch, as indicated above. The Groups are shown here with a simple hatch pattern with wide spacing at 0 degrees and green in color. This enables you to easily select either the Area or the Group without the need to use complex selection techniques.

Use an Area Group Template

An Area Group Template is like any other Style in ADT. It is accessed through the Style Manager and can be imported from or exported to other files. The files in this dataset already include a sample Area Group Template. (Feel free to edit it to suit your firm's needs.)

1. In the *Chapter10\Leasing* folder, **Open** the file named *A-AA01.dwg*.

This is the first floor of the office building. It has an atrium entrance, a main lobby and a core in the rear. To either side are open retail spaces. Most of the spaces have already been marked by Area objects. You will notice that there are no Area Groups in this file and that the retail space to the right has not been drawn yet.

2. From the **Documentation** menu, choose **Area Groups>Create Area Groups from Template**.

3. In the **Row Offset** field, type **3'-6"** and in the **Column Offset** field, type **7'-0"** (see Figure 10–18). You can also click the buttons to set the offsets in the drawing by picking.

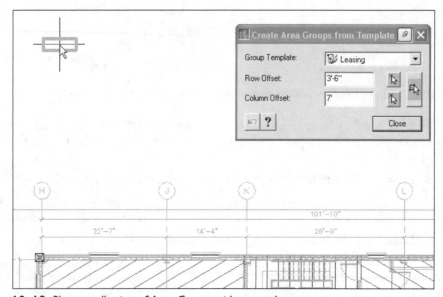

Figure 10–18 *Place a collection of Area Groups with a template*

4. At the "Specify insert point" prompt, click a point above the plan.

5. **Close** the "Create Area Groups from Template dialog box.

You will now have eight Area Groups inserted into the drawing and linked to one another. All we need to do now is attach the Area objects to the appropriate Groups. If you frequently use the same grouping structure, the productivity boost from using an Area Group Template cannot be overstated.

6. **Zoom** in on the bottom Schedule Table.

Notice that all of the Area Groups have already populated the Schedule. This is because the schedule references Style-based Schedule Data and all of the pre-built Area Group Styles in this file already have the appropriate Property Set attached. For more information on Property Sets and Schedule Tables, refer to Chapter 6. All of the values are currently zero (see Figure 10–19). This is because there are no Area objects attached to any of the Groups yet.

Area Group Table		
No.	Name	Calculated Area
3	Gross Measured Area	0.00 SF
4	Building Major Vertical Penetrations	0.00 SF
5	Floor Rentable Area	0.00 SF
7	Office Areas	0.00 SF
8	Store Areas	0.00 SF
9	Building Common Areas	0.00 SF
10	Floor Usable Area	0.00 SF
11	Floor Common Areas	0.00 SF

Figure 10–19 *The Group structure populates the existing Schedule, but the values read zero until Areas are attached to the Groups*

Attaching Areas to Groups

1. Select the **Building Major Vertical Penetrations** icon, right-click and choose **Attach Areas / Area Groups**.
2. Select the two Area objects at the Stair and Elevators and then press ENTER (see Figure 10–20).

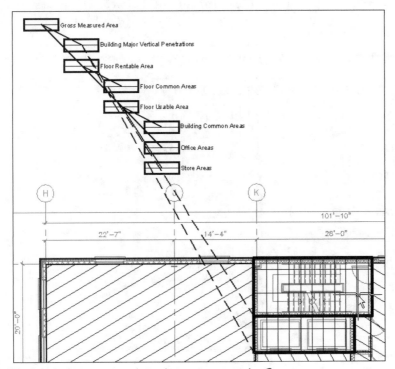

Figure 10–20 *Attach the two vertical circulation spaces to the Group*

3. **Pan** over to the Schedules and check the numbers (see Figure 10–21).

Notice that the Stair is 210.02 SF and the Elevator is 109.83 SF. The sum of these two values is 319.85 SF, which is in fact the value that shows in the Area Group Table for Building Major Vertical Penetrations.

	Area Identification			Area	
No.	Name	Style	Description	Base Area	Ca
002	Stair	Major Vertical Penetrations	First Floor	210.02 SF	
003	Elevator	Major Vertical Penetrations	First Floor	109.83 SF	
004	Atrium	Building Common Area	First Fl...		

	Area Group Table	
No.	Name	Calculated Area
3	Gross Measured Area	319.85 SF
4	Building Major Vertical Penetrations	319.85 SF
5	Floor Rentable Area	0.00 SF

Figure 10–21 *Comparing the values in each Schedule Table*

This value also appears in the Gross Measured Area field as well. This is because the Building Major Vertical Penetrations Group is a member of the Gross Measured Area Group, and currently is the only member.

4. Using Figure 10–22 as a guide, attach the remaining Areas to the appropriate Groups.

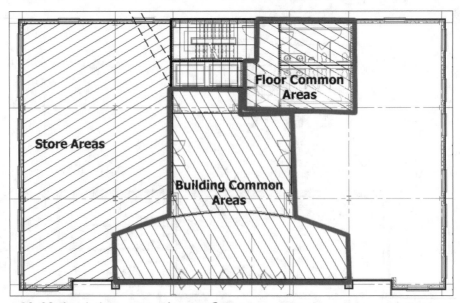

Figure 10–22 *Attach the remaining Areas to Groups*

Compare the total at the bottom of the Area Schedule with the value for the Gross Measured Area Group in the Group Schedule. Both values read 4501.92 SF (see Figure 10–23). This is a good way to check that you haven't missed Areas, since the Gross Measured Area of the building should include all of the Areas in the drawing.

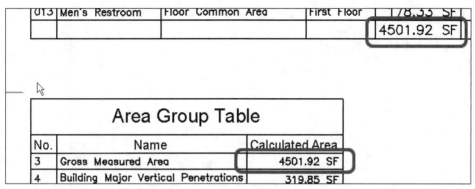

Figure 10–23 *Verifying both totals in the Schedule serves as a good check*

Add a New Area

There is one Area missing for the Retail Space at the right. Let's add it now.

> 1. From the **Documentation** menu, choose **Areas>Add Area**.

Notice that the **Name** list is empty. We can correct this by adding this new Area to a Group before we begin drawing it.

Figure 10–24 *Click the Properties icon to attach the Area to a Group*

> 2. Click the **Properties** icon in the lower left of the Add Area dialog box (see Figure 10–24).
>
> 3. Click the Area Group tab.

Here is listed all of the Area Groups in the current drawing.

> 4. Expand the list completely and place a check mark in the box next to the **Store Areas** Group (see Figure 10–25).

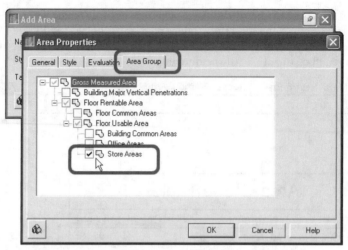

Figure 10–25 *Expand the Group list and check the Store Areas Group*

5. Click **OK** in the Area Properties dialog box.
6. In the Add Area dialog box, open the **Name** list.

Notice that there is now a list of pre-defined names from which to choose. This list comes from an Area Name Definition that is attached to the Area Group object. This is why we attached the Area to the Group first.

7. Choose **Store Area** from the **Name** list (see Figure 10–26).

Figure 10–26 *Choose a pre-defined Name from the list*

8. Choose **Store Area** as the **Style** as well.
9. Begin clicking points around the open retail space on the right.

Use Figure 10–27 as a guide as you work. The Area should measure to the inside face of the right and rear exterior Walls, and the Core Walls. It should measure to the center of the common Walls between the retail space and the lobby. Finally, the Area should project all the way to the outside face of the street frontage. Be sure to switch to Arc on the segment along the atrium.

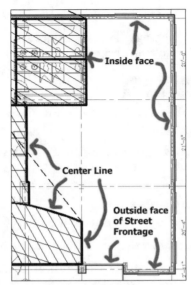

Figure 10–27 *Create a Area for the right side Retail Space*

10. Look at the Schedule again.

Notice that since we added the Area directly to the Group, both Schedules have updated and the totals still match. However, notice that there is a question mark in the number column of the new Area. This is because the number property is an Object-based property.

11. Right-click the new Area and choose **Edit Schedule Data**.

 Tip: Remember to select the Area using the diagonal hatching. The straight green colored hatching is the Area Group.

12. Click the **Add** button, confirm that only the **AecArea** Property Set is chosen, and then click **OK** (see Figure 10–28).

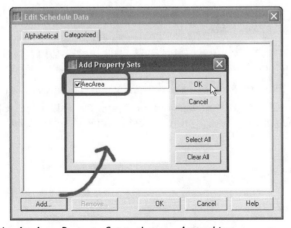

Figure 10–28 *Add the AecArea Property Set to the new Area object*

13. Click **OK** again to return to the drawing.

The Schedule updates to reflect this change. If you wish, you can use the **Documentation>Schedule Data>Renumber Data** command to renumber all of the Area objects.

Edit the Second Floor Area Plan

The Elevator Lobby space on the second floor is incorrectly included as part of the Building Common Areas Group. It ought to be included in the Floor Common Areas instead.

1. If the second floor drawing (*A-AA02.dwg*) is still open, switch to it now; otherwise open it from the *Chapter10\Leasing* folder.

2. Select the Area just beneath the Elevators (Elevator Lobby).

3. Right-click and choose **Area Modify.**

4. Click the small icon located next to the **OK** button (Area Group Viewer).

This opens a small window that shows the Area Group hierarchy.

5. Expand the list, clear the **Building Common Areas** check box and check **Second Floor Common Areas** instead (see Figure 10–29).

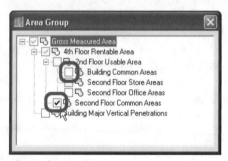

Figure 10–29 *Re-assign the Group for this Area*

6. Click **OK** in the Area Modify dialog box.

Create the Fourth Floor Area Plan

This building has four floors, but only three have been provided in this dataset. However, the fourth floor is substantially similar to the second floor and will be easy to generate.

1. From the **File** menu choose **Saveas** and name the new file *A-AA04.dwg*.

2. Open the XREF Manager, change the name of **A-FP02-C** to **A-FP04-C** and change the path to *C:\ADT_AIG\Chapter10\Leasing\A-FP04-C.dwg* (see Figure 10–30).

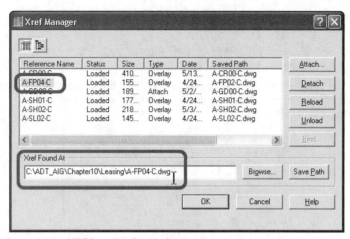

Figure 10–30 *Re-path the XREF to the fourth floor plan*

3. Detach the **A-SL02-C** XREF.

4. **Zoom** in to the cluster of Area Group icons.

5. Right-click the **2nd Floor Rentable Area** Group and choose **Area Group Modify**.

6. Rename it to **4th Floor Rentable Area** and click **OK**.

7. Repeat this procedure for all Group names with reference to the second floor.

The fourth floor does not have an atrium space, so the final step is to merge the atrium space from the second floor with the office space.

8. Select the large Office Area, right-click and choose **Operation>Join**.

9. At the "Select areas to join" prompt, select the atrium space and then press ENTER.

The plan, the Groups and the Schedules should all update to reflect this change.

10. **Save** the file.

CREATING AREAS FROM WALLS

The following process is a bit involved, but the several steps are simple to execute. The goal is to quickly generate a large collection of Area objects from an existing floor plan. There are two factors which make this process complex: first, the Areas need to go to the centerline of the Walls and not the inside face, and second, some of the bounding Walls are in XREF files. These two issues will be common in many medium-scale to large-scale projects. Therefore, the following process has been developed as a way to minimize the total amount of work required to achieve results. As has been mentioned, minimizing the workload does not necessarily mean minimizing the steps.

This tutorial is broken down into four major parts. We must first open the file containing the Walls for the third floor and copy them out to a temporary file. Next, we will build a custom Display Configuration showing only the centerline of Walls. Next we will generate Area objects from this Display Configuration and finally, we will move the Areas back to the appropriate file.

Build a Temporary Working File

1. In the *Chapter10\Leasing* folder, **Open** the *A-FP03-C.dwg* drawing file.

2. From the **Tools** menu, choose **Quick Select**.

3. Choose **Entire Drawing** from **Apply To**, set the **Object Type** to **Wall**, choose **Style Name** from the **Properties** list and type **standard** in the **Value** field (see Figure 10–31).

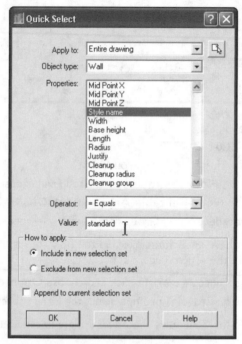

Figure 10–31 *Use Quick Select to select all standard Walls*

4. Click **OK** to retrieve the selection.

5. In addition to the Walls already selected, manually select the two curved Counter Top Walls as well.

6. Right-click and choose **Copy**.

7. In the *Chapter10\Leasing* folder, **Open** the *Third Floor Outline1.dwg* file.

This file already contains a previously drawn polyline of the Building perimeter line. This line represents the rentable area of the entire third floor and has been provided here to save time (see Figure 10–32).

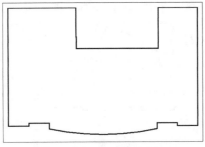

Figure 10–32 *The file contains a polyline of the floor rentable area*

8. Right-click and choose **Paste to Original Coordinates**.

We will need to clean up the plan a bit. There are Walls that are not required to generate proper Areas, such as the column enclosures.

 Note: The file up to this point has been provided as *Third Floor Outline2.dwg*. For convenience, you can open up this file and continue, or you can keep working on your own.

9. **Delete** all column enclosures, including the round ones (see Figure 10–33).

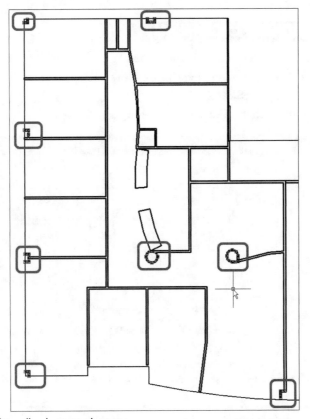

Figure 10–33 *Delete all column enclosures*

10. **Extend** all Walls to the perimeter polyline (see Figure 10–34).

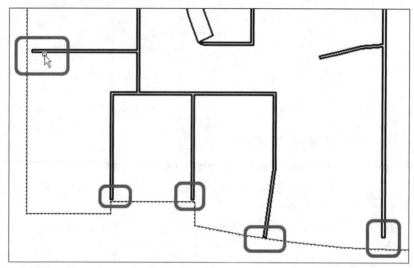

Figure 10–34 *Extend all perimeter Walls to the polyline*

Build a Custom Display Configuration

1. Open Display Manager and expand **Sets**.
2. Right-click the **Sets** folder and choose **New**.
3. Name the new Set **Sketch Walls**.
4. Next to **Wall**, check **Sketch** only (see Figure 10–35).

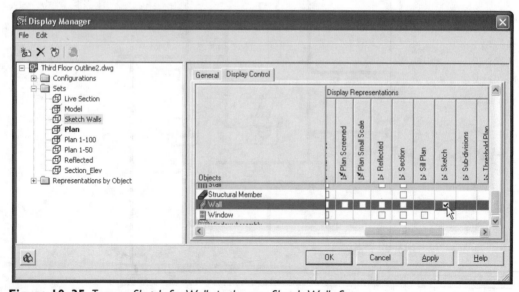

Figure 10–35 *Turn on Sketch for Walls in the new Sketch Walls Set*

5. Next to **Areas**, check **Plan** only.

6. Right-click Configurations and choose **New**, and name it **Generate Areas**.

7. On the Configuration tab, choose **Sketch Walls** from the list of Sets next to the **Top** View Direction (see Figure 10–36).

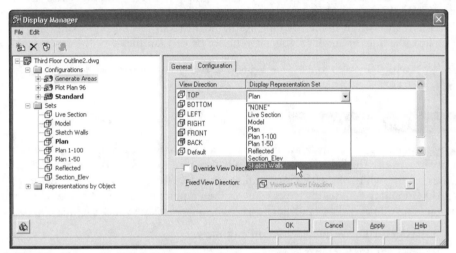

Figure 10–36 *Choose the Sketch Walls Set to display from Top view in this new configuration*

8. Right-click **Generate Areas**, choose **Set to Current Viewport** and then click **OK**.

9. Select any Wall, right-click and choose **Entity Display**.

10. Make sure that **System D**efault is selected and click the **Edit Display Props** button.

11. Turn off the Marker component and click **OK** twice to return to the drawing (see Figure 10–37).

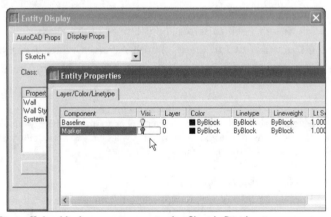

Figure 10–37 *Turn off the Marker component in the Sketch Display*

12. **Regen** the drawing.

The small triangle markers should no longer be visible.

13. Perform any final cleanup necessary.

The goal is to have all "gaps" closed. The command that we will use to generate the Area objects is very similar to the **Hatch** command. Therefore, if the spaces are not completely closed, it will fail. You can use **Extend, Trim** and grip editing as necessary. Also, remember that you can easily manipulate the Area objects later. These Walls are a means to an end, not an end in themselves; so don't get carried away making them perfect. Sometimes it is easiest to just add a small Wall segment to close a gap. Use this technique to "close" the Conference Room and Reception areas off from the corridor. Remember that in this Display Configuration Walls will look just like lines. Therefore, this task should be very simple.

 Note: The file up to this point has been provided as *Third Floor Outline3.dwg*

Generate the Area Objects

Before we can generate the Areas, we must create a polyline in each room. This step is unfortunately necessary since the **Select Bounding Walls** option of the **Create Areas from Objects** command uses the true width of the Walls regardless of the Display Set currently active. However, the AutoCAD **Boundary** command "looks" only at the graphics on screen. Therefore, we will first convert our single-line Walls to polylines and then convert the polylines to Area objects.

1. From the **Draw** menu, choose **Boundary**.
2. Click the **Pick Points** button.
3. Following the "Select Internal Point" prompt in the drawing, click once in each closed space.

 Note: The current layer is 0. However, this will prove unimportant in the coming steps.

Polylines should appear within each space as you click. If you get an error, go back and check to see that the space is closed and that all gaps are filled.

4. Use **Quick Select** again to select and delete all of the Walls.
5. From the **Documentation** menu, choose **Areas>Create Area from Object**.
6. In the Create Areas from Objects dialog box, choose **Store Areas** from the **Style** list.

 Note: Store Areas has been included in this file for convenience in the tutorial. When you use this technique on your own, you need to remember to import this Style into this drawing to do this step. It is not critical that you do this, but it will save you some effort later in not having to change the Style of the Area objects.

7. Click the **Select Objects** icon.
8. Select all polylines in the drawing (except the original overall one) and then press ENTER.
9. At the "Enter Cutplane Height" prompt, simply press ENTER to accept the default.

You should now have 17 Area objects (this includes one for the small shaft in the middle of the plan).

Move the Areas to Correct File

1. Using **Quick Select**, select all Area objects (see Figure 10–38).

Figure 10–38 *Quick Select all Area objects*

2. Right-click and choose **Cut**.
3. From the *Chapter10\Leasing* folder, **Open** the *A-AA03.dwg* drawing.
4. Right-click and choose **Paste to Original Coordinates**.
5. **Attach** all of these new Areas to the **Tenant A Office Areas** Area Group (see Figure 10–39).

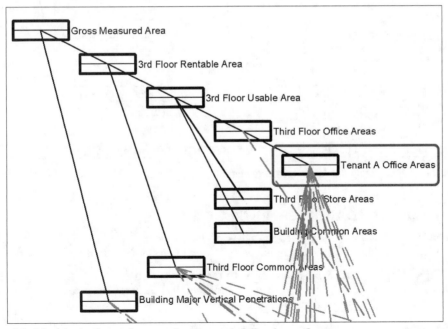

Figure 10–39 *Attach all new Areas to the Tenant A Office Areas Group*

6. Using grips, tweak any of the Area Boundaries as required.

7. Check the totals in the Schedules and **Save** the file.

It is not necessary to save the *Third Floor Outline.dwg* file. Again, it is stressed that although there were several steps involved in the previous process, the result was less labor intensive and therefore faster than the process of manually tracing each room to generate the Area object.

AREA TAGS

The only sample Tags that have been provided with out-of-the-box ADT are the ones contained in the Metric D A CH *Schedules* folder of the DesignCenter. You will need to manually select this folder in the DesignCenter to use them. They are functionally the same as all other Tags in ADT (see Figure 10–40). You can use them as a starting point to get ideas for creating tags of your own. Refer to Chapter 6 for detailed instructions on how to modify an existing Schedule Tag to suit your own needs. Keep the following points in mind as you explore these Tags:

- These are Scale-dependent Tags and rely on either **Plan 1-50** or **Plan 1-100** being turned on for Multi-View Blocks in the current Display Set (see Chapter 7).

- The Attributes built into the View Blocks of these Tags reference the Object-based Property Set Definitions of **AECAREAGROUP** and **AECAREA** (see Chapter 6 and Figure 10–40).

- The Schedule Data Formats used by these Property Sets are based on Metric units. (See Chapter 6 for information on Schedule Data Formats.)

- Default values used in the Attributes and the DesignCenter icons reference German terminology for Area measurement. (See the "DIN 277 Standard Terminology" sidebar.)

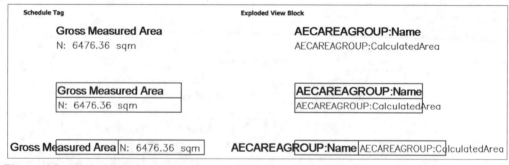

Figure 10–40 *Exploring the Attributes contained in some of the D A CH Area Group tags*

DIN 277 STANDARD TERMINOLOGIES

The default values shown in the icon preview of the DesignCenter (HNF 1) is an abbreviation used in the DIN 277 (Deutsche Industrie Norm) German Standard.

HNF 1 - Hauptnutzfläche, which is similar to the dominant portion of a usable area in a building. There are a variety of such classifications: HNF 1 includes living rooms, HNF 2 - office areas, HNF 3 - manufacturing areas and so on.

Other designations include:

> **NNF** – Nebennutzfläche, which includes Floor common areas like sanitary rooms and wardrobes.
>
> **HNF** + **NNF** = usable area
>
> **VF** – Verkehrsfläche, which is an area for "connection" like corridors, stairs, elevators.
>
> The DIN standard is very detailed. Following it is very important in acquiring a building permit in Germany.

CALCULATION MODIFIERS

Calculation Modifiers are simple mathematical formulas that are attached to either the Area objects or the Area Groups. You can assign these formulas to work on the Area, the Perimeter or both. The values generated are referred to as the "Calculated Area" and the "Calculated Perimeter." You might use Calculation Modifiers to assist with meeting the requirements of a particular building code or with performing quantity takeoffs.

The dataset we will use for this exercise is a small two-story residential addition (see Figure 10–41). The first and second floor plans have been XREFed side by side into an Area Plan file. Included in this file are some Area Styles, Area Group Styles, Property Set Definitions and Schedule Tables. The same strategy has been used in the Display Properties as the previous dataset. An additional column has been added to the Area Group Schedule to report the Perimeter of each Group.

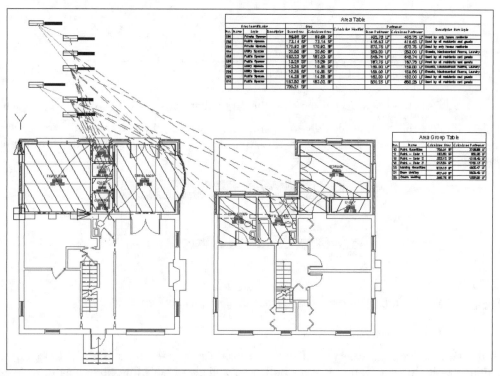

Figure 10–41 *A simple residential two-story addition*

An Area object has both Area and Perimeter values associated with it. Let's assume that we need to know how much paint to order in each color for the residential addition. Therefore, we need a formula that multiplies the Perimeter of the space by the ceiling height (8'-0" in this case).

Create a Calculation Modifier Style

1. In the *Chapter10\Residential* folder, **Open** the file named *A-AA01.dwg*.
2. From the **Documentation** menu, choose **Areas>Calculation Modifier Styles.**
3. Create a new Style and name it **Paint Quantity**.
4. For a Description, type **Perimeter x Ceiling Height** (see Figure 10–42).

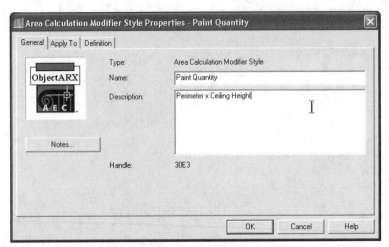

Figure 10–42 *Add a new Calculation Modifier Style*

5. Click the Apply To tab and clear the **Area** check box.

This will make this Style apply only to Perimeters. However, if you wanted to also calculate the volume of the room, you could leave this Area check box checked and use this formula for both purposes.

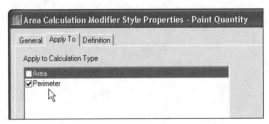

Figure 10–43 *Clear the Area check box to make this formula apply only to the Perimeter value*

6. Click the Definition tab.

In the **Parameters** area, we can add variables to use in our formulas. This makes complex formulas easier to read. In this case, we only need a single variable for ceiling height.

7. Click the **Add** button and type the name **CeilingHeight** for the name.

8. Click in the Value field and change it to **96** (see Figure 10–44).

 Note: We are using inches here because as we will see below, the Imperial measurement system is not fully supported by Calculation Modifier Styles.

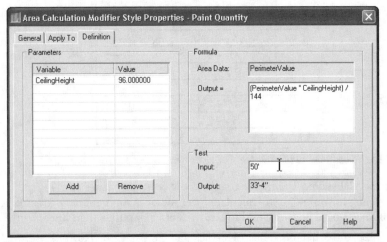

Figure 10–44 *Add a variable and a formula to multiply the Perimeter by the ceiling height*

9. In the Formula **Output =** field, type **(PerimeterValue * CeilingHeight) / 144**.

This formula will multiply the perimeter value from the object (in inches) by the height (also in inches) and then divide the result by the number of square inches in a square foot (144). If you use Metric units, you can avoid some of these mathematical gymnastics. You can use a variety of keywords in the formula. The keyword **PerimeterValue** is a pre-defined variable. It will retrieve the actual value of the Perimeter from the Area object. The pre-defined variable **AreaValue** retrieves the actual value of the area instead. You can use any arithmetic expression (+,-,* and /) and even VBA expressions. (Check the online help for VBA for more information on using these.)

10. To test your formula, type a number in the Test **Input** field and then press ENTER.

11. Click **OK** twice to return to the drawing.

Attach the Formula to Areas

We now need to attach this formula to our Area objects.

1. Using **Quick Select**, select all of the Areas in the drawing.

2. Right-click, choose **Area Properties** and click the Evaluation tab.

3. Click the **Attach** button and select **Paint Quantity.**

4. Click **OK** to return to the drawing.

5. **Pan** over to the Schedules.

	Perimeter	
fier	---	---
	Base Perimeter	Calculated Perimeter
	33'–7"	22'–5"
	34'–9"	23'–2"
	56'–1"	37'–5"
	21'–1"	14'–1"
	53'–10"	35'–10"
	15'–8"	10'–5"
	13'–3"	8'–10"
	13'–3"	8'–10"
	15'–2"	10'–1"
	54'–2"	36'–2"

Figure 10–45 *Values now appear in the Calculated Perimeter column*

Notice that the Calculated Perimeter column now contains values (see Figure 10–45). There are two problems, however. First, Calculated Perimeter is too generic a column heading. Second, this column should be formatted in square feet (SF) rather than lineal feet (LF.)

6. Select the Schedule Table, right-click and choose **Edit Table Style**.

7. Click the Columns tab, select the **Calculated Perimeter** column (at "1" in Figure 10–46) and click the **Modify** button (at "2" in Figure 10–46).

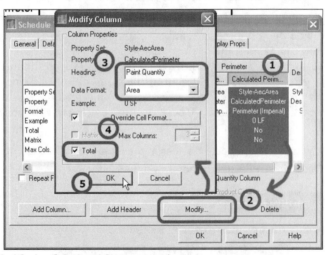

Figure 10–46 *Modify the Calculated Perimeter column*

8. Change the value in the **Heading** field to **Paint Quantity** and choose **Area** from the **Data Format** list (at "3" in Figure 10–46).

9. Put a check mark in the **Total** box (at "4" in Figure 10–46) and click **OK** (at "5" in Figure 10–46).

10. Click **OK** again to return to the drawing.

Modifier	Perimeter	
	Base Perimeter	Paint Quantity
y	33'−7"	268.50 SF
y	34'−9"	277.75 SF
y	56'−1"	448.50 SF
y	21'−1"	168.67 SF
y	53'−10"	430.49 SF
y	15'−8"	125.17 SF
y	13'−3"	106.00 SF
y	13'−3"	106.00 SF
y	15'−2"	121.33 SF
y	54'−2"	433.50 SF
		2485.91 SF

Figure 10–47 *Results of the formula in the Schedule*

The Schedule Table will now appear something like Figure 10–47. If you wish, you can create a Schedule Data Format that shows square feet rounded off to the nearest foot (no decimal places). You could also move the Calculation Modifier column closer to the result and rename it as well. However, considering that we do not purchase paint in square feet, but rather in gallons, it may make more sense to apply another modifier to these values to report the number of gallons of paint rather than the square footage.

Make a More Complex Formula

1. From the **Documentation** menu, choose **Areas>Calculation Modifier Styles.**
2. Double-click **Paint Quantity** to edit it.
3. Add a new variable named **PaintCoverage** and set the value of it to **400**.
4. In the Formula **Output =** field edit the expression to read: **((PerimeterValue * CeilingHeight) / 144) / PaintCoverage** (see Figure 10–48).

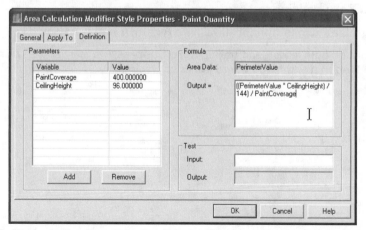

Figure 10–48 *Divide existing formula by the amount of coverage of a single gallon of paint*

What we are doing is taking the existing formula, which determines the total square footage of Walls of the room, and dividing that by 400 (the amount of coverage of a single gallon of paint).

 Tip: If you want to plan for two coats, reduce the variable to 200.

Looking at the Schedule Table, there is one additional change we can make to this column. Paint cans are not measured in square feet. Therefore, we should create a Schedule Data Format that reports the numbers of cans.

5. From the **Desktop** menu, choose **Schedule Data>Data Format Styles**.
6. Create a new Style named **Gallons** and edit it.

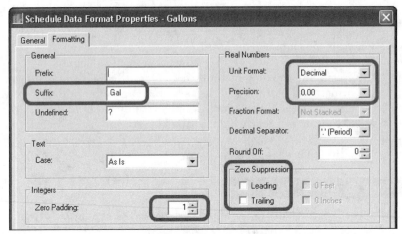

Figure 10–49 *Set up a custom Schedule Data Format for Gallons*

7. On the Formatting tab, add **Gal** in the Suffix field, being sure to include a leading space (see Figure 10–49).
8. Set **Zero Padding** to 1.

Zero Padding adds a zero in front of all values rather than simply leaving them blank.

9. Turn off both leading and trailing **Zero Suppression** and set the **Unit Format** to **Decimal** with two decimal places of precision.

Figure 10–50 shows the result.

er	Perimeter		
	Base Perimeter	Paint Quantity	
	33'–7"	0.67 Gal	Used
	34'–9"	0.69 Gal	Used
	56'–1"	1.12 Gal	Used
	21'–1"	0.42 Gal	Clos
	53'–10"	1.08 Gal	Used
	15'–8"	0.31 Gal	Used
	13'–3"	0.27 Gal	Clos
	13'–3"	0.27 Gal	Clos
	15'–2"	0.30 Gal	Used
	54'–2"	1.08 Gal	Used
		6.21 Gal	

Figure 10–50 *The Schedule now shows the Paint Quantity in gallons*

There are some additional gaps using this method. The formula does not take into account the paint for the ceiling. Unfortunately, we have pushed the limits of our current formula within ADT. We could build additional Area Calculation Modifiers and apply them separately to the Area and/or Groups to achieve the results we need. However, this would require a bit more effort than might be justified. To perform complex calculations, we will get better results if we export the data to an application like Excel. This formula also does not take into account the voids in the Walls for Windows, Doors and Openings. Depending on the number of such openings, this could amount to quite a bit of extra paint. To get a more accurate count in this case, you may want to consider creating a Schedule directly from the Walls instead, which do have an Automatic property for Wall Area minus voids.

GENERATING REPORTS

Now that we have seen nearly the complete Area tool set in action, it is time to learn how we can work with all of this data. If you are able to solve all of your calculation and reporting needs directly using these tools, you will need only to set up a sheet file that organizes and presents the data in a way suitable for printing. However, in most cases you will want to export the data from ADT for further manipulation and presentation.

There are two basic ways to export the data from Areas and Area Groups in ADT: using the Area Evaluation feature and by directly exporting a Schedule Table. Area Evaluations provide a rather interesting report, complete with bitmapped graphic representations of each Area and Group. The problem with them is that the format is a bit inflexible. An Excel Template file (*.XLT) is used to designate the format of the columns in the report, and an ADT interface allows you to select which data to include.

Area Evaluation Reports

1. From the **Documentation** menu, choose **Areas>Area Evaluation.**

Notice the tree structure at the left. This includes all of the Groups and their attached Areas. If you had something selected when you opened this dialog box, it is already selected here in the tree.

2. Close the dialog box, select the Paint Quantities Group in the drawing and then repeat the **Area Evaluation** command.

Notice that now the entire Paint Quantities Group is already selected.

3. In the **Evaluation** area, click the **Options** button.

In the dialog box that appears, you can decide which information to include in the exported report, whether to include images and which Excel template to use for column formatting (see Figure 10–51).

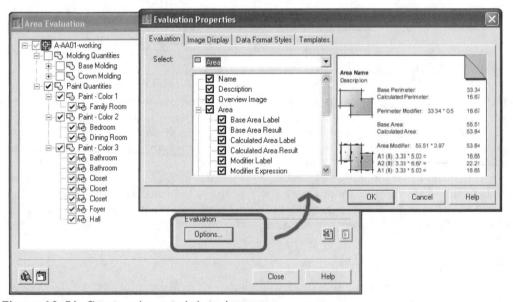

Figure 10–51 *Choosing what to include in the report*

4. On the Evaluation tab, choose **Area** from the **Select** list, and check all boxes except the **Decomposed Image** box.

5. Choose **Area Group** from the **Select** list, and check all boxes.

Most of the defaults on the remaining tabs will give satisfactory results. The Image Display tab allows you to set the size of the bitmap image and set the Display Configuration from which it is generated. If you wish to format any of the data differently in Excel than the way it is in the drawing, click the Data Format Styles tab and make your choices. If you would like to include the "raw" unformatted value of the area or perimeter, you can check the **Additional Exact Value** box (see Figure 10–52).

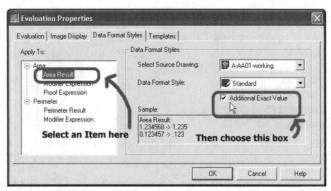

Figure 10–52 *Add the Additional Exact Value to your report*

Finally, you can choose a different Excel Template to determine the organization of the columns in the report. There are several samples provided with ADT in the *Template\Evaluation Templates* folder of your Management Install. Check the online help for information on how to customize these templates.

      6. Click the small **Excel** icon in the lower right corner of the dialog box.

      7. When the export is complete, **Open** the Excel file.

The report will look something like Figure 10–53.

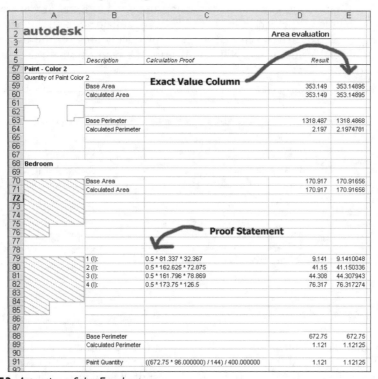

Figure 10–53 *A portion of the Excel report*

8. Repeat the steps, and try clearing several of the items in the **Select** list.

Although you can build your own Excel Template and rearrange the columns, you do not have any control over the order in which the rows appear. The order will be Area - Proof - Perimeter. However, you can choose not to include some of these rows. You need to do any post processing manually in Excel.

Exporting a Schedule Table

In addition to Area Evaluations, we can export Schedule Tables directly to Excel. Here we get a bit more flexibility in organizing the columns, including the ability to add headers.

1. From the **Documentation** menu choose **Schedule Tables>Export Schedule Table**.

2. Click the **Browse** button and choose a name and location to save the Excel file.

If you want to export a table that is already in your drawing, place a check mark in the **Use Existing Table** check box. When you click OK, you will be prompted to select the table. Alternatively, you can clear this check box, choose the Style and other settings, and create a new Schedule Table directly in Excel.

3. Make your choices and then click **OK**.

4. **Open** the Excel file and explore the results (see Figure 10–54).

Architectural or Engineering Units formats are not supported in Excel. If you are using either of these in any column of the Schedule, you will receive a dialog box asking how you wish to format the exported data. If you intend to perform calculations on the data, choose **Use Unformatted Decimal Value**. If you will simply print the report from Excel, you can choose **Convert to Formatted Text**.

	A	B	C	D	E	F	G	H	I	J
1							Area Table			
2	No.	Name	Style	Description	Base Area	Calculated Area	Calculation Modifier	Base Perimeter	Paint Quantity	Description from Style
3	001	Bathroom	Private Spaces		69.66 SF	69.66 SF	Paint Quantity	403.	0.67 Gal	Used by only house residents
4	002	Bathroom	Public Spaces		73.14 SF	73.14 SF	Paint Quantity	417.	0.69 Gal	Used by all residents and guests
5	003	Bedroom	Private Spaces		170.92 SF	170.92 SF	Paint Quantity	673.	1.12 Gal	Used by only house residents
6	004	Closet	Utility Spaces		20.50 SF	20.50 SF	Paint Quantity	253.	0.42 Gal	Closets, Mechanical Rooms, Laundry
7	005	Dining Room	Public Spaces		182.23 SF	182.23 SF	Paint Quantity	646.	1.08 Gal	Used by all residents and guests
8	006	Hall	Public Spaces		15.26 SF	15.26 SF	Paint Quantity	188.	0.31 Gal	Used by all residents and guests
9	007	Closet	Utility Spaces		10.35 SF	10.35 SF	Paint Quantity	159.	0.27 Gal	Closets, Mechanical Rooms, Laundry
10	008	Closet	Utility Spaces		10.35 SF	10.35 SF	Paint Quantity	159.	0.27 Gal	Closets, Mechanical Rooms, Laundry
11	009	Foyer	Public Spaces		14.28 SF	14.28 SF	Paint Quantity	182.	0.30 Gal	Used by all residents and guests
12	010	Family Room	Public Spaces		183.52 SF	183.52 SF	Paint Quantity	650.	1.08 Gal	Used by all residents and guests
13					750.21 SF				6.21 Gal	
14										

Figure 10–54 *Open the exported Schedule Table in Excel*

 CHECKLIST

We have covered quite a bit of ground in this chapter. As you can see from the examples included, it is recommended that you generate Areas in a separate file that XREFs the model files. This helps you keep them separated from the rest of the production drawings. Also, since the display needs are a bit different, having the Areas separated is an easy way to accommodate the potentially conflicting display needs.

If you have decided to begin incorporating Areas and Area Groups into your firm's workflow, create a separate template file for the Area Plan files. This template can begin with the *Plan-Sect-Elev.dwt* file that we have been working on throughout the book. Make a copy and name it something like *Area-Group.dwt*. Add to this template the following:

1. Display Configurations and Sets like those discussed throughout this chapter.

   a. Set up Work to display both Areas and Groups together. Display Areas in varying hatch patterns designed to identify the particular Area Style. Display Groups in the same hatch pattern with wide horizontal spacing for easy selection.

   b. Set up a Display Configuration showing only Areas, and another showing only Groups. Make the Groups one use Solid Fill Hatch for presentation plans.

2. Area Styles for most common uses (following BOMA for example).

3. Area Group Styles for most common uses (following BOMA for example).

4. Area Group Template with pre-built hierarchical Group structure (following BOMA for example).

5. Area Calculation Modifiers for most common uses.

6. Area Name Definitions including the most common Area Names (following BOMA for example).

7. Property Set Definitions, Schedule Data Formats and Schedule Table Styles designed for Areas and Groups.

8. Area and Group Tags (these should be saved in the library rather than directly in the template file).

From the standpoint of productivity and implementation, the Area Group Template is one of the most important components of all those listed. Having a well-thought-out Group Template will make generating a new Area Group Plan very easy. It will also help to eliminate inconsistencies from one project to the next.

SUMMARY

Creating and manipulating Area objects is similar to working with polylines, with some Boolean functionality thrown in.

Composed Areas show both positive and negative rings in a single Area object.

Area Group Templates provide a complete set of Area Groups, predefined and attached to one another and ready to use.

Areas can be created from existing objects such as bounding Walls and other 3D objects like Roofs and Slabs.

By configuring the Display Configurations for Areas to meet different display needs, you can attain much flexibility in display.

Area Tags work like other Tags. Samples are in the D A CH folders. You can create your own following the steps in Chapter 6.

A Calculation Modifier is used to adjust the results of the Area or Perimeter of an Area or Group for reporting purposes.

Reports can be generated directly to Excel from the Areas and Groups themselves or through Schedule Tables.

CHAPTER 11

Sections and Elevations

INTRODUCTION

Often in architectural design and production, much attention is paid to Plans and not enough to Sections and Elevations. However, Sections and Elevations are no less important to communicating design intent. In addition, Sections and Elevations can actually prove more challenging to produce in some circumstances. For instance, while a Plan is actually a horizontal cut taken from 3'-6" [1,000] to 4'-0" [1,200] above the floor, no such "standard" applies to a section; they can be cut from any location in the building. In addition, where we would expect to have a Plan for each unique floor of a building, there is no direct correlation between the building itself and the quantity of Sections required.

Sections and Elevations are not rendered the same way as Plans. For instance, a bold profile line typically outlines the entire edge of an Elevation but not a Plan. Perhaps these distinctions can help to explain why there are actually three ways to create Section and Elevation drawings in Architectural Desktop. We have the 2D Section/Elevation object, the 3D Section/Elevation object and the Live Section Display Configuration. The first two are similar and use the same basic tools and procedures. They both generate a new object that remains linked to the objects in the original model. They are graphical reports of the building model data. The third option, the Live Section, is actually a display option and operates directly on the model itself.

OBJECTIVES

In this chapter, we will take a comprehensive look at all of the Section and Elevation objects available in ADT. The goal of this chapter is a complete understanding of the Section/Elevation tool set. In this chapter, we will explore the following:

- Understand the 2D Section/Elevation tool
- Customize Section/Elevation Display Set
- The "right" way to use Edit Linework and Merge Linework
- Know when to Explode
- Live Sections and other issues

3D SECTION/ELEVATION OBJECTS

If you have used Architectural Desktop since its early releases, you might be familiar with the 3D Section/Elevation as the "first generation" Section/Elevation tool in ADT. The concept and process of this object work like this:

1. In Plan, add a Section (or Elevation) Line object indicating where you want to cut the Model and in which direction you want to "look."

2. Add Sub-divisions to this Section (or Elevation) Line object that will be interpreted by the Section/Elevation object and rendered in lineweights that recede as they get further from the cut line.

3. Generate a Section/Elevation object and place it somewhere in the drawing.

4. View and/or print this "linked" object to represent Section or Elevation cuts of the Model. (This has to be done from the appropriate "elevational" viewpoint such as Front or Right, and benefits from having **Hide Lines** turned on.)

Conceptually, the theory and even the procedure are sound. In practice, several limitations exist both *physically*, related to the Model, and *interpretively*, related to the level of abstraction typically associated with an architectural Section or Elevation drawing. On the physical front, items that are not modeled precisely, or that are not modeled at all, will not display properly in the Section/Elevation object. In many cases, physical limitations can be overcome simply by adjusting the Model. The interpretive aspects of Section/Elevation drawings are the more challenging criteria to satisfy. A great deal of abstraction is assumed and implied by the very definition of a Section or Elevation drawing. These abstractions are not easy to simulate through direct linkage to a Model that most people will build, edit and print mostly in Plan view. For many of these reasons, the 3D Section/Elevation object is seen by many (perhaps even the Autodesk developers) as an obsolete stepping-stone to the 2D Section/Elevation object introduced with Release 3 of Architectural Desktop.

On the plus side, this new object type shares the same procedure and all of the features of its predecessor. It also addresses many of the limitations of its predecessor in the areas of hidden line removal and level of abstraction. (We will discuss the pros and cons of this object in detail next.) As a result, for many, the 3D Section/Elevation object has been rendered obsolete and is no longer used. Although this manual will not propose to reverse that position or decision, it is worthwhile to consider some lesser known qualities of this object.

In Chapter 5, we discussed the issues and some procedural solutions to the issues of "spanning." In particular, the issues of multi-story Curtain Wall objects and Stairs were discussed, and solutions to each of these object's unique display requirements proposed. Let's consider one other option to the problem of a multi-story Curtain Wall with different display needs on each floor. It is not well known that a 3D Section/Elevation object has two Display Reps. This is because the most common Display Rep, Sub-Divisions, is the one used to indicate receding lineweights. However, a 3D Section/Elevation object can also be displayed using its Model Representation. The Model Display Rep for this object is identical to the Model Display Rep of nearly all other ADT objects, with one small exception: it does not have Display Properties of its own. It inherits its Display Properties from the objects to which it is linked. Therefore, what you get is a three-dimensional linked replica of your Model (or whatever portion of it you selected) contained in a single entity. Being a single entity, it cannot be edited directly; however, it does remain linked to the original objects and may be updated as needed.

Install the CD Files and Open the Sample File

1. Install the files for Chapter 11 located on the Autodesk Architectural Desktop: Advanced Implementation Guide CD ROM.

Refer to the "How to Use the CD" section in the Preface for instructions on installing the sample files included on the CD.

A folder named *Chapter11* was created containing a collection of sample files that we will use throughout this chapter.

2. In the *Chapter11* folder, **Open** the file named *3D Section.dwg* (see Figure 11-1).

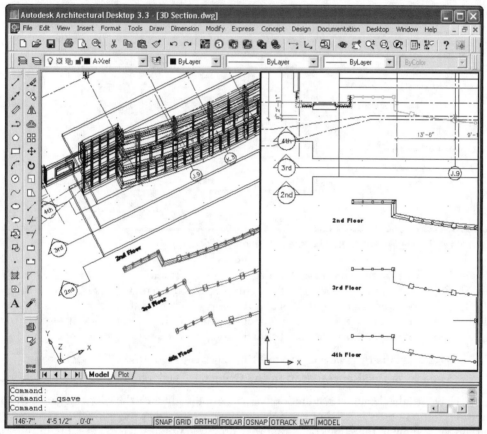

Figure 11-1 *Open the sample file using 3D Sections to show spanned Curtain Wall plan conditions*

The file opens into model space with two viewports. The left viewport shows the file in 3D while the right shows it two-dimensionally. Three Section Line objects have been placed in this file at different Z heights. (This is easiest to do by placing them in plan and then switching to a Front or Side view and moving them vertically.) The trick to this technique is to limit the amount of the Curtain Wall that each Section line slices through vertically. Each of these

Section Line objects is only 2' tall. This gives us a "thin slice" through the Curtain Wall at each height required for each of the three different Floor plan heights across which it spans.

3. With the left viewport active, Open the Display Manager, expand **Representations by Object** and click on the **Bldg Section** object.

Notice that the Bldg Section object type has only two Display Reps, **Model** and **Sub-divisions**. In this viewport, **Sub-divisions** is currently active, as shown in Figure 11–2.

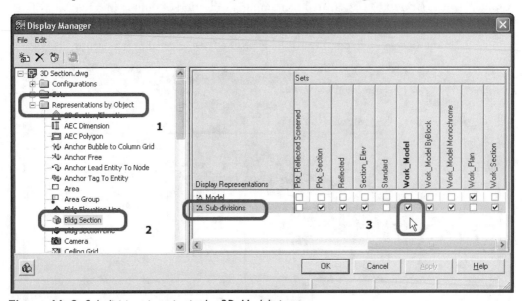

Figure 11–2 *Sub-divisions is active in the 3D Model viewport*

This is the default configuration for the Bldg Section object and is rarely changed by anyone. The Section objects in the 3D viewport are completely red in color. This is because they are reflecting the ByLayer color of the A-Sect layer onto which they are inserted. Had we added Sub-divisions to the Section Line object (open the Section Line Properties dialog box to do this), the color of this object would be "banded" to reflect the sub-division colors.

4. Click in the right viewport to make it active, and re-open the Display Manager.

5. Again, expand **Representations by Object** and click on the **Bldg Section** object.

In this viewport, the default configuration has been modified to use the Model Display Rep for the Bldg Section object instead of the Sub-divisions. Notice that the individual pieces of the Curtain Wall show in the same colors that they do in the physical Model (see Figure 11–3).

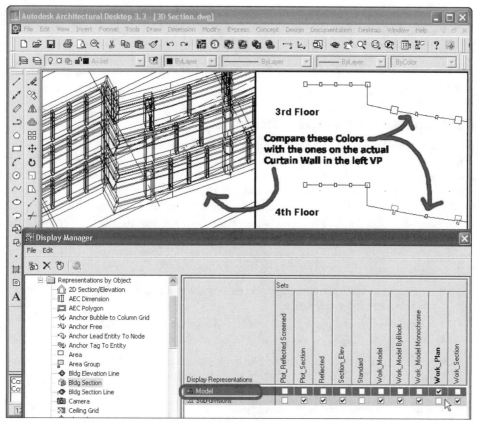

Figure 11–3 *The Bldg Section object is colored to match the original Model*

6. **Close** the file when finished.

As you can see, this technique does offer some promise. However, it is not a necessarily *easy* process, nor is it necessarily superior to the approach taken in Chapter 5. It is merely another alternative, should you discover difficulties with spanning elements in the approach recommended in Chapter 5. For instance, you will note that this technique properly sections through custom Display Blocks used in the Model Display Rep of the fourth floor of the Curtain Wall. This would not be easily achieved in the Display Rep process used in Chapter 5. In addition, you can sometimes encounter difficulties with nested infill components in the Chapter 5 method. This method avoids those issues as well. However, as you can see here, the 3D Section objects are floating precariously in space in this sample file. In a real project, you would need to move these elements correctly into place, thereby adding another level of steps and potential complexity to the process. If your Curtain Wall designs tend to be fairly simple in plan view, stick with the procedures recommended in Chapter 5. If your Curtain Walls routinely become very complex, particularly in Plan view:

- Do they include nested Infill components like Window Assemblies and Curtain Wall Units?
- Do they "step" or "tier"?

- Do they include Custom Display Blocks that you wish to show in Plan?
- Do you have someone in charge of the Curtain Wall that is meticulous and detail oriented?

If you can answer yes to most of these questions, consider this technique over the one covered in Chapter 5.

2D SECTION/ELEVATION OBJECTS

In many ways, the 2D Section/Elevation object is a vast improvement over the 3D version. The 3D version will not be used in production, aside from very specialized uses similar to the one outlined above (using the object as a Plan and not a Section at all). Rather, if you decide to use ADT tools to create Sections and Elevations, the 2D Section/Elevation object is the best choice. Unfortunately, many ADT users have decided against implementing this tool in its current release. One of the primary goals of this chapter is to assist you toward the regular use of the 2D Section/Elevation object in normal production. Although there is plenty of room for this object to mature in future releases, it is still very useful in its current incarnation if you know how and when to use it. Some of the key benefits to the 2D Section/Elevation object:

- Produces a "flat" 2D drawing inserted in the World Coordinate System, not standing upright in 3D space.
- Uses Styles to configure its parameters like other ADT objects.
- Automatically removes the "hidden" lines from the object.
- Uses a sophisticated system of Design Rules to intelligently "re-index" components from the Model to proper display components within the 2D Section/Elevation for plotting.
- Can be edited manually if necessary.

USES OF THE 2D SECTION/ELEVATION OBJECT

The 2D Section/Elevation object can be used as the basis for a variety of drawings (see Figure 11–4) required in a typical set of documents:

- Full Building Exterior Elevations
- Interior Elevations
- Full Building Sections
- Wall Sections
- Detail Sections
- Wall Types
- Roof Plans

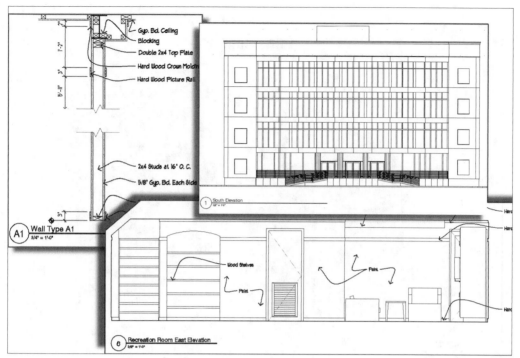

Figure 11–4 *Many different types of drawings can be generated from the 2D Section/Elevation object*

The last one, Roof Plans, may seem a bit unusual, but in some cases the 2D Section/Elevation object provides the best solution for generating a simplified Roof Plan from a complex Model. To use a 2D Section/Elevation this way, we must rotate the UCS before adding the Bldg Section Line.

THE BLDG SECTION LINE AND BLDG ELEVATION LINE

The basic mechanics of creating 2D Section/Elevation objects are nearly the same as that described above for the 3D Section/Elevation. In fact, the same Section or Elevation Line object is used to generate all Section/Elevation types in ADT (including the Live Section). The Bldg Section Line and Bldg Elevation Line objects are actually three-dimensional "boxes." The purpose of this three-dimensional box is to determine what portion of the Building Model will be included in the Section or Elevation.

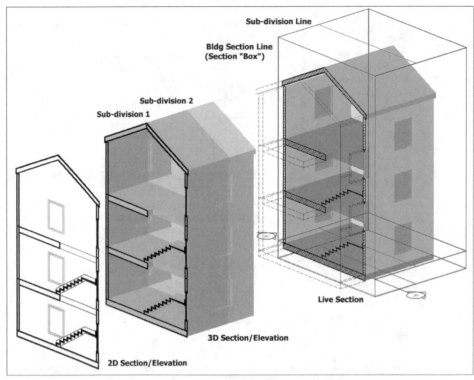

Figure 11–5 *Section Line (Box) at right can generate 2D, 3D or Live Sections*

In the example shown in Figure 11–5, the Bldg Section Line is sized to give a full building cross-section. If the size of the box is adjusted, in both plan and Z heights, we can effectively create an entirely different type of drawing like the interior elevation shown in Figure 11–6.

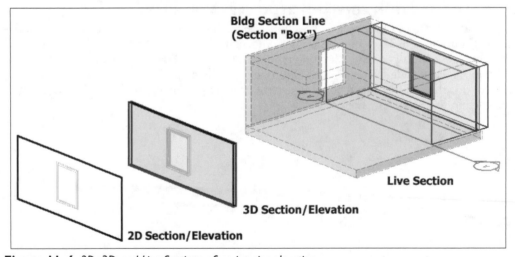

Figure 11–6 *2D, 3D and Live Sections of an interior elevation*

The Bldg Section Line and Bldg Elevation Line objects have five Display Reps, each of which contain three sub-components. The Bldg Section Line and Bldg Elevation Line objects are classified in Chapter 2 as Annotation objects (refer to Figure 2–1). Nevertheless, in many respects they are really more of a Helper object with the potential to also be used as an Annotation object. The reason for this clarification is that you will not want to print two of the three subcomponents mentioned above: the Boundary and the Sub-divisions.

The third component is the Defining Line, which is the only component that would have value printed, at least for Sections (you will likely not want it printed for Elevations). It can be used in conjunction with a Section Bubble in your drawings and is the reason for the Annotation classification. The Defining Line is the "cut line," drawn through the building in plan view. It forms a plane three-dimensionally that determines where the bold cut line in the Section will be. The Boundary *is* the "box" (referred to above). Nothing outside the Boundary is included in the Section or Elevation. The Sub-divisions are drawn as lines parallel to the back edge of the Section box and determine where the Lineweight zones occur (see Figure 11–7).

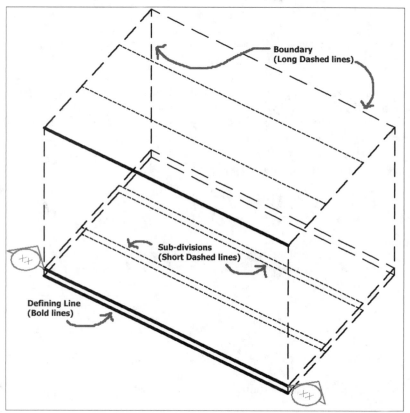

Figure 11–7 *The components of the Bldg Section Line and Bldg Elevation Line objects*

Although there are five Display Reps, which could potentially be configured differently, in most cases you can set the System Default of all five the same way. Assign both the Boundary and the Sub-divisions to one of the Primary Utility colors and the Defpoints layer. (You could also use a No Plot layer such as A-Anno-Nplt.) For example, the Boundary can be assigned to color Blue and the Subdivision Lines to color Magenta (see Figure 11–8). You can also change the linetype if you wish. The lineweight and plot style are unimportant since the Defpoints (or A-Anno-Nplt) layer assignment will keep them from plotting. This type of configuration will allow you to see all three components on screen at all times, but only plot the Section Defining Line itself.

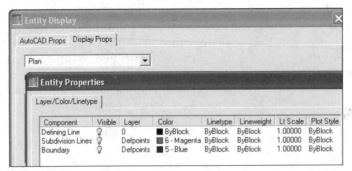

Figure 11–8 *Flag the Boundary and Subdivision Lines components with primary colors and set them to no plot in the default Display Props*

As an alternative, you can configure the Plan, Model and Reflected Display Reps this way and turn off the Boundary and Subdivision Lines components in the Plan 1-100 and Plan 1-50 displays, which are used in the "Plot" Display Sets (see Figure 11–9). The same net effect is achieved either way; the former method shows all components on screen at all times so that the Section Line is always visible and easy to "tweak," while the latter method gives a more accurate on-screen representation of what the final plot will look like.

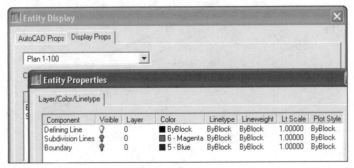

Figure 11–9 *Set Plan 1-100 and Plan 1-50 to have the Boundary and Sub-divisions turned off*

Whichever you decide, these are certainly settings for your template file.

SECTION LINE TAGS

A variety of pre-built tags can be used to identify Bldg Section Line and Bldg Elevation Line objects. If you use the **Documentation>Sections>Add Section Line** command, you will automatically get the **StandardSectionBubble** tag. If you use the **Documentation>Elevations>Add Elevation Line** command, you will automatically get the **StandardElevationBubble** tag. These tags are automatically generated by ADT on the fly. They are not stored in a template or library file. However, if you wish to customize them, simply add your own version to your office template using the *same* name. When the **Add Section Line** or **Add Elevation Line** command is executed, it will use your customized version instead, if a version with the same name is already resident in the file. There are several limitations and/or issues with these default tags:

- They only have View Blocks in the General Display Rep. This means that the tag will not appear in Model or Reflected view by default.

- The attributes within the View Blocks must be edited manually. The process requires editing the Multi-View Block Properties and manually inputting the desired value on the Attributes tab. This is very counter-intuitive to new and seasoned AutoCAD/ADT users alike.

- The **Add Section Line** command creates two bubbles, one at each end. Each one must be edited separately, because they are not linked together.

- If you would prefer that these tags were scale-dependent, you will need to edit them to be such. They use Annotation scaling by default. (Refer to Chapter 7 for information on both of these issues.)

On the positive side, these tags are anchored to the Section or Elevation line so that if the Section/Elevation Line object should move or rotate, the tags move and rotate with them. This is the only advantage to these default tags.

Several alternative tags are available in the Documentation Content library (**Documentation>Documentation Content>Elevation Marks** and **Section Marks**). These are Custom Command Content items. Each of these will give you the option of adding an AEC Section or Elevation Line at the time of insertion. In addition, they use standard AutoCAD Blocks; therefore, editing the attributes within them is as easy as double-clicking the tag. Unfortunately, these tags cannot be anchored to the Section/Elevation lines, which means that if you move the Section Line, you will need to move the tag manually as well. For more information on the Documentation Content and DesignCenter items, refer to Chapters 7 and 8.

Linking Tags to the Section Line with Property Set Data

There is another important issue not addressed by either the default Section/Elevation Line tags or any of the DesignCenter offerings. There is no way to easily coordinate a specific Section or Elevation Line in the drawing with the actual Section/Elevation object to which it is referenced. Editing the attributes in the tags does nothing to assist in this critical coordination.

Using Property Set Data, we can link the data displayed in the bubbles with the data displayed in the title bar beneath a particular Section/Elevation drawing and make it easy to edit. To make this work, both the bubble and the title bar must become "Schedule Tags" to the Bldg Section Line or Bldg Elevation Line object. This requires a Property Set Definition with properties for each of the pieces of required information, a Multi-View Block for the Section/Elevation bubble and another Multi-View Block for the title bar. All of these components could be added and linked manually, or a Custom Command Content item in the DesignCenter can be devised. Two such items have been provided in a folder named *SectElev Tags* in the *Chapter11* folder. If you wish to try these items, please see the instructions in the *ReadMe.txt* file in the same folder for installation of these items. Once you have installed them, they will operate like any other Content item; simply drag them from the DesignCenter to use.

Use the AIG Elev Mark and AIG Sect Mark

These steps assume that you have installed the AIG Elev Mark and AIG Sect Mark as instructed in the *ReadMe.txt* document in the *Chapter11\SectElev Tags* folder.

1. Begin a **New** drawing using your template file.
2. Click your custom **Office Standards** icon to go to the **Office Standards** folder of the DesignCenter.
3. Drag the **AIG Elev Mark** into the current drawing.
4. Follow the command line prompts.

The AECANNOSCHEDULETAGADD command is executed twice by this routine. This is because there are actually two separate "Tags" being added and anchored to the Elevation Line object. Therefore, be sure to click the Elevation Line object when prompted to "Select an Object to Tag" and then press ENTER. At this point, the AECANNOSCHEDULETAGADD command repeats, and a second Tag, (the Title Mark), is added. When you are finished, you should have something like Figure 11–10 on your screen.

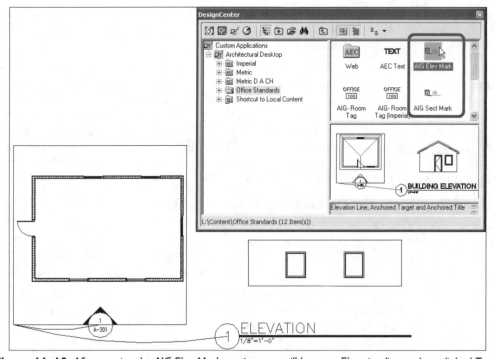

Figure 11–10 *After running the AIG Elev Mark routine, you will have an Elevation line and two linked Tags*

To edit the values in both the Elevation Tag and the title bar, right-click the Elevation Line object and choose **Edit Schedule Data**. (It will also work if you right-click the Tags.)

5. **Edit** the Schedule Data and then click **OK**.

When you click **OK**, the values will appear in both the Tag and the title bar.

THE 2D SECTION/ELEVATION CREATION PROCEDURE

Creating a 2D Section/Elevation object is metaphorically similar to a photo shoot. In a photo shoot, we need a photographer and a model. The model must be properly attired, lighting and other effects are applied, and finally a photograph is generated. If necessary or desirable, the photograph can be retouched to perfect the image.

Analogies can be drawn to each of these basic components of the photo shoot when generating a Section or Elevation object in ADT. Think of the Bldg Section Line and Bldg Elevation Line objects as the "photographer." Just as the photographer adjusts the camera lens to determine what will be included in the final image, so can we can adjust the Section/Elevation box, as we have already seen. The model is naturally your ADT Building Model (or the portions of it that you select). Models have vast wardrobes from which to choose for each photo shoot. Your ADT Model has a wardrobe of its own: Display Representation Sets.

When we generate a Section or Elevation in ADT, the Display Set can have a dramatic impact on results of the final output. Lighting and filters can make the difference between an outstanding photograph and a mediocre one. Section/Elevation Styles serve this crucial role in 2D Section/Elevation objects. Finally, the **Edit Linework** and **Merge Linework** commands allow us to "retouch" our Sections and Elevations once they are created. Keeping this analogy in mind as we continue through this chapter should help you understand the relationship of all the pieces clearly.

MODEL ISSUES (OBJECT SELECTION)

We have already discussed the Section/Elevation Line (photographer), now let's look more closely at the Model. When we generate 2D Section/Elevation objects from large models, it can take a several minutes to calculate the result. Two important issues relative to the Model can help improve performance: object selection and optimizing the Model itself. Below, in the "Optimizing the Model" section, we look at the optimization issues; here we will explore the issues of object selection.

The Bldg Section Line and Bldg Elevation Line objects determine the overall extent of the Model that will be included in the 2D Section/Elevation object, but equally important to Section output is the Selection Set within the Section box (see Figure 11–11).

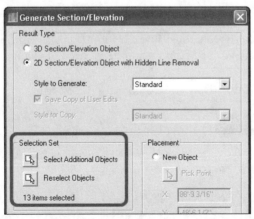

Figure 11–11 *The Selection Set area of the Generate Section/Elevation dialog box*

Naturally, performance will improve if fewer objects are selected. However, this is not always easily accomplished, considering that many 2D Section/Elevation objects are generated from XREFs. Try to avoid nested XREFs if possible, because it will give you more control over what to select in the Section file.

The Effects of Nested XREFs

1. In the *Chapter11\2D SectElev* folder, **Open** the file named *A-CM01-C.dwg*.
2. Right-click the Front Elevation and **Update** it.

Keep a timer handy and make note of how long it takes to update the Elevation without changing any of the settings.

3. Open the XREF Manager and click the Tree View icon (see Figure 11–12).

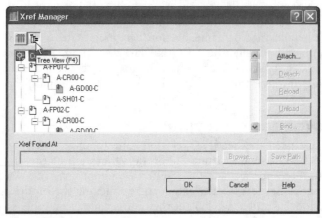

Figure 11–12 *View the XREFs in Tree View*

Notice that the XREF *A-SH01-C.dwg* is a nested XREF that occurs four separate times in the hierarchy. This is because it is attached to each of the floor plans, which are in turn attached to this file. The *A-SH01-C.dwg* file is the outer shell of the building. It must be included to generate a proper elevation. However, in the current XREF structure, the Shell is being called into the Composite Model by the individual Floor Plans. Therefore, we are forcing ADT to calculate all of the geometry contained in each Floor Plan, even though most of it will not appear in the Elevation. To remedy this situation, we need to change the Attached XREFs in the floor plan files to Overlays.

4. **Close** the *A-CM01-C.dwg* file. Do not save the changes.

Remove nested XREFs

 Note: If you would prefer to skip the XREF re-pathing exercise, open the Composite Model file in the *Nesting Removed* folder and skip to the "Update the Elevation" step sequence below.

1. **Open** the First Floor Plan (*A-FP01-C.dwg*).
2. **Open** the XREF Manager and double-click **Attach** next to the **A-CR00-C** and **A-SH01-C** XREFs (see Figure 11–13).

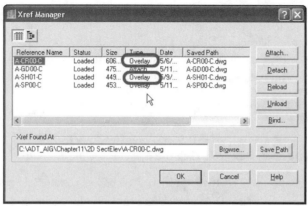

Figure 11–13 *Double-click Attach to change the Type to Overlay*

This will switch them to Overlays. You cannot do this to the A-GD01-C XREF since it is nested in the Core XREF (A-CR00-C). However, this will not be a significant detriment to performance in the Elevation file, and the benefits of leaving it nested to the core outweigh the minor performance gain.

3. Repeat the above steps on each of the other Floor Plan files: *A-FP02-C.dwg, A-FP03-C.dwg* and *A-FP04-C.dwg*.

4. **Save** and **Close** all Floor Plan files.

Now that we have changed these files to Overlays, they will no longer appear in the Composite Model. We will need to XREF them anew as Overlays.

5. **Open** the *A-CM01-C.dwg* file.

6. Verify that the current Layer is **A-Xref** and **Overlay** both **A-CR00-C** (Stair Core file) and **A-SH01-C** (Shell Wall and Windows) to 0,0,0 (see Figure 11–14).

Be sure to clear the **Retain Path** check box for both references.

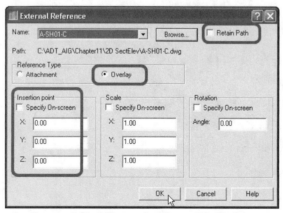

Figure 11–14 *Overlay the Core and Shell files to the Composite Model*

7. Using the **3D Array** command (**Modify>3D Operation>3D Array**), duplicate both XREFs for the upper floors. (Set the distance between Levels to 12'-0".)

You will be using the Levels option of the **3D Array** command with four Levels (creating three copies). The result will look identical to the original file. The difference is that this file is more efficient in the way it loads the XREFs. It also is more flexible because we can now select the exterior shell without the interior build out. The next step may be easier from a 3D viewpoint.

8. From the **View** menu, choose **3D Views>SW Isometric**.

Update the Elevation

1. Right-click the Front Elevation object and choose **Update**.

2. In the **Selection Set** area, click the **Reselect Objects** button.

Note: If **Reselect Objects** is not an available option, choose **Select Objects** instead.

This choice allows us to completely replace the existing selection set with a new one.

3. Select the objects shown in Figure 11–15 and then press ENTER.

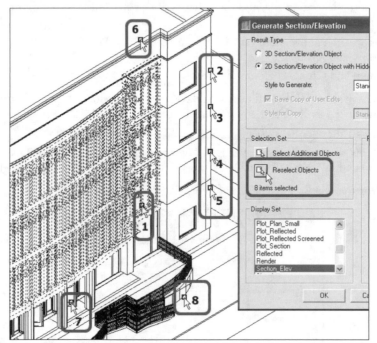

Figure 11–15 *Reselect objects for Elevation*

1 – Multi-floor Curtain Wall

2 through 5 – Outer Shell files

6 – Roof

7 – First Floor Curtain Wall

8 – Site Plan conditions (Entrance ramps and stairs)

4. Click **OK** to update the Elevation.

If you use your timer again, it should complete in about 50 percent of the time it took before. Refer back to Chapter 2 and the associated tables in Appendix B for more recommendations on file hierarchy and XREF structure.

Freezing Layers to Increase Performance

Once you have organized your XREF hierarchy in the optimal fashion, the only other way to limit what is included in the 2D Section/Elevation object is to freeze layers within the XREFs. In order to reap a benefit from this, you must be demand loading your XREFs, and Layer indexing should be turned on. (Both of these topics are covered in Chapter 2.) You will also want to have VISRETAIN turned on to ensure that the frozen layers remain frozen even after the file is saved and closed.

UNDERSTANDING THE RELATIONSHIP TO SETS

In the photographer analogy above, we equated the Display Set with the model's clothing. You will get a very different photograph in cutoff jeans than you will in your Sunday best. The same is true for Display Sets. By choosing a different Display Set in the Generate Section/Elevation dialog box, you can end up with vastly different results. However, in most cases only a select few Display Sets that will generate an acceptable Section or Elevation.

Experiment with Alternate Sets

1. In the *Chapter11\Nested Removed* folder, **Open** the file named: *A-CM02-C.dwg*.
2. Select the Building Section, right-click and choose **Update**.

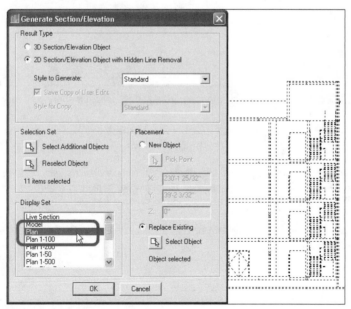

Figure 11–16 *Update the Section*

3. From the **Display Set** list, choose **Plan** and then click **OK** (see Figure 11–16).

Notice that only the Slabs, Elevators and Furniture on the third floor seem to remain. In reality, all of the other objects were rendered "flat" since we chose a 2D Plan Set. Most ADT objects rely on their Model Display Rep for generating Sections and Elevations. You may recall however, that the Slab object is a 3D object and, even though it does have a Plan Display Rep, it uses 3D graphics to display plans. This is why they are still showing (see Figure 11–17).

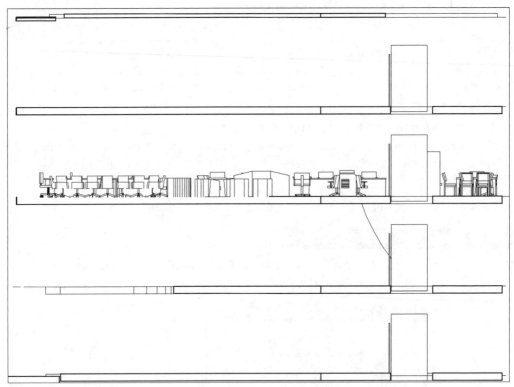

Figure 11–17 *A Section generated with the Plan Display Rep does not show very much*

The furniture and elevators appear because they are Multi-View Blocks that have View Blocks loaded for the vantage point from which the Section was cut (Right in this particular case).

 Note: The furniture plan was included in this Section object's Selection Set to help illustrate the point about Display Sets. In reality, to improve performance, we would want to exclude that XREF from the Selection Set as we did in the earlier exercise.

4. Update the Section again, but choose the **Render** Display Set this time.

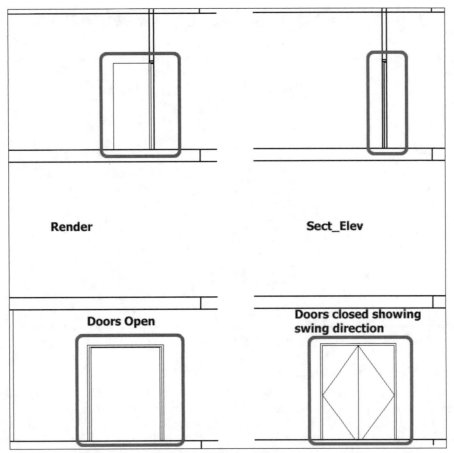

Figure 11–18 *Comparing the Render Set with Section_Elev*

The Section generated using Render is very similar to the original one generated with Section_Elev (see Figure 11–18). The only difference lies with Doors, Windows and Structural Members. Each of these objects has an Elevation Display Rep that varies from the Model Display Rep used in the Render Display Set.

5. Open the Display Manager and Select the **Section_Elev** Set (see Figure 11–19).

Nearly all of the Architectural objects are active in the Model Display Rep. Doors, Windows and Structural Members use Elevation as was already mentioned above, and a variety of Helper objects are also visible using the Display Rep General. You may want to turn off some of these Helpers. For instance, how likely is it that you will need to include Schedule Tables in the Selection Set of a 2D Section/Elevation object?

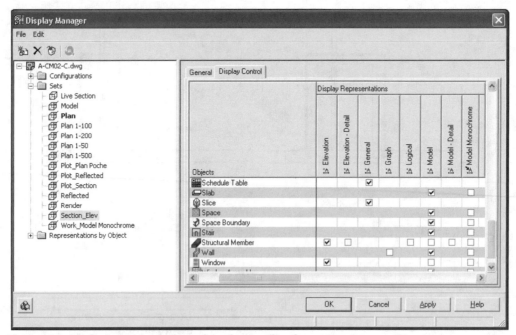

Figure 11-19 *Exploring the settings of the Section_Elev Set*

Template Display Set Settings for Section_Elev

In Chapter 5, we explored the default templates and consolidated many of their settings into the one we have been using throughout the book. As a result, we are using the Model Display Rep for most objects in Sections and Elevations. As we have already noted, the Render Display Set also uses the Model Display Rep for nearly all objects. The potential does exist for conflict between the needs of 3D renderings and 2D Section/Elevation objects. However, the likelihood is rare, considering how differently each of these drawing types is generated. Nonetheless, you might find it useful to compare the settings for the Section_Elev Display Set as it is configured in the Imperial (**I**), D A CH (**D**) and Advanced Implementation Guide (**G**). They are shown in side by side in Table 11-1.

Table 11-1 *Comparison of Display Sets*

Display Rep	Elevation	General	Model	Model ByBlock	Sub-divisions
2D Section/Elevation	I				
AEC Dimension					
AEC Polygon		I, D, G			
Anchor Bubble to Column Grid		I, G			
Anchor Free					
Anchor Lead Entity To Node	I, D, G				

Table 11–1 *Comparison of Display Sets (continued)*

Display Rep	Elevation	General	Model	Model ByBlock	Sub-divisions
Anchor Tag To Entity					
Area					
Area Group					
Bldg Elevation Line			I, D, G		
Bldg Section					I, D, G
Bldg Section Line			I, D, G		
Camera	I				
Ceiling Grid			I, D, G		
Clip Volume			I, D, G		
Clip Volume Result					I, D, G
Column Grid			I, D, G		
Curtain Wall			D, G	I	
Curtain Wall Unit			D, G	I	
Door	I, D, G				
Entity Reference		I, D, G			
Layout Curve		I, D, G			
Layout Grid 2D		I, D, G			
Layout Grid 3D		I, D, G			
Mask Block Reference		I, D, G			
Mass Element			D, G	I	
Mass Group			D, G	I	
Multi-View Block Reference		I, D			
Opening			I, D, G		
Railing			D, G	I	
Roof			D, G	I	
Roof Slab			D, G	I	
Schedule Table		I, G			
Slab			D, G	I	
Slice		I, G			
Space			D, G	I	

Table 11–1 *Comparison of Display Sets (continued)*

Display Rep	Elevation	General	Model	Model ByBlock	Sub-divisions
Space Boundary			D, G	I	
Stair			D, G	I	
Structural Member	I, G		D		
Wall			D, G	I	
Window	I, D, G				
Window Assembly			D, G	I	

As you can see, the major difference lies in the use of the Model ByBlock Display Rep by the Imperial template. This is used primarily to keep the settings for rendering and Sections separate from one another. However, this is not necessary, and it needlessly creates additional Display Reps to manage. It is recommended that you use the Model Display Rep for Sections instead, like the settings shown for D A CH and AIG in Table 11–1. Most of the other settings are substantially similar in all three files.

Remember, you will often be looking at the plan view of your building while generating Sections and Elevations that are "looking" at the Model. For this reason, it is important to become familiar with whatever set you ultimately use. Knowledge of which Display Rep is being used to generate the graphics in Section or Elevation of a particular object makes it much easier to make changes to those objects. This raises another issue related to the default templates: None of them has a configuration that *shows* the Section_Elev Set on screen. This is undesirable, since it will be impossible to preview the Model in the same way the Section will *see* it. It is recommended that you edit your Work Display Configuration to show Section_Elev in the Front, Left, Right and Back view directions.

Setting Work to Show Section_Elev

The following steps are best performed directly in your *Plan-Sect-Elev.dwt* template file. If you have built a different template, you can substitute it here. You will also need to perform these steps in any existing files where you want them to apply as well.

1. **Open** the *Plan-Sect-Elev.dwt* template file (see Figure 11–20).

Tip: To open a template file, remember to change the **Files of Type** to **Drawing Template File**.

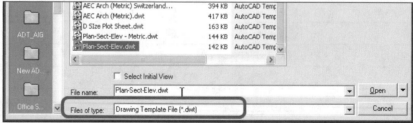

Figure 11–20 *Open the Plan-Sect-Elev.dwt template file*

2. Open the Display Manager and expand **Configurations**.
3. Select **Work** and then click the Configuration tab on the right.
4. Using SHIFT, select **Left, Right, Front** and **Back**.
5. Click next to any of the four highlighted view directions and choose **Section_Elev** from the list (see Figure 11–21).

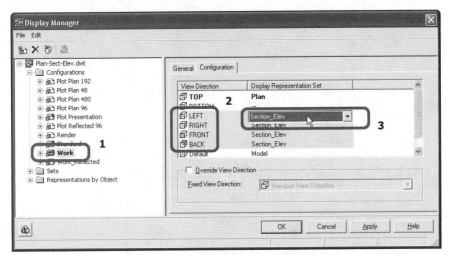

Figure 11–21 *Select all of the elevation views and assign them to the Section _Elev Set*

You must be careful when updating your template file. Any errors can be transferred to numerous drawings before the error is detected. Always double-check your work before you save the template, and keep a backup copy of the original just in case.

Recall from Chapter 4, when we discussed Layer Standards, that it is best to purge any Layer Styles from the template file before saving it. This will ensure that new drawings always get the latest version of the Layer Standards and Layer Keys.

6. At the command line, type **aecpurgelayerstyles** and then press ENTER (see Figure 11–22).

```
AECPURGELAYERSTYLES
This will remove all Layer Key Styles and Layer Standards from the drawing.
Do you wish to continue? [Yes/No] <Y>:
99'-9 11/32", -12'-0 3/16", 0'-0"          SNAP GRID ORTHO POLAR OSNAP OTRACK LWT MODEL
```

Figure 11–22 *Purge Layer Styles prior to saving the template*

7. Type **y** and press ENTER.
8. **Save** and **Close** your template.

Repeat the same steps in the *A-CM01-C.dwg* file that we worked in earlier. You will then be able to switch to one of those views and preview any changes you make to the Display Reps used in Section_Elev Display Set. This will be very helpful in the next sequence.

OPTIMIZING THE MODEL

Consider some of the following options to help optimize your models and make your 2D Section/Elevation objects perform better:

Optimize the Profile Shape of Railings

Return to the *A-CM01-C.dwg* file in the *Nesting Removed* folder. If you zoom in on entrance portion of the Elevation that was created from this composite model, you can see that there are quite a few extraneous lines in the railings (see Figure 11–23).

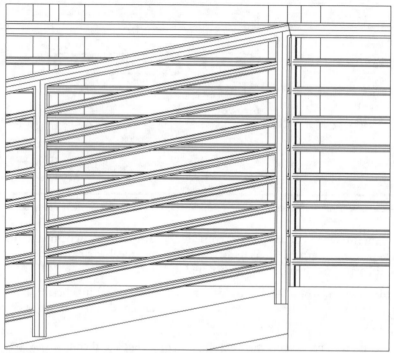

Figure 11–23 *Excessive tessellation in the railings makes this Elevation much too "muddy"*

This tessellation is caused from using a round profile shape for the pipe railings. If you think about this situation critically, you can ask yourself "when will this level of detail ever be required of these railings?" For plans and elevations, we will usually require only two lines to represent the thickness of the railing. Therefore, we should open the Site plan file where these Railing objects reside and change the shape of all components to Rectangular. This is purely for the purpose of the graphical presentation. Details of the railings will either be drawn separately or will be much easier to edit back to round than to remove hundreds of tessellation lines from the Elevation drawing. To edit this, do the following:

1. **Open** the file containing the Railings (A-SP00-C.dwg).

2. Edit the Railing style (right-click any Railing and choose **Edit Railing Style**) and on the Components tab, change the **Profile Name** of each Component to ***Rectangular*** (see Figure 11–24).

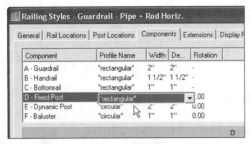

Figure 11–24 *Change the profile name to Rectangular*

3. **Close** and **Save** the file and reload the XREFs back in the Composite Model.
4. Update the Elevation (see Figure 11–25).

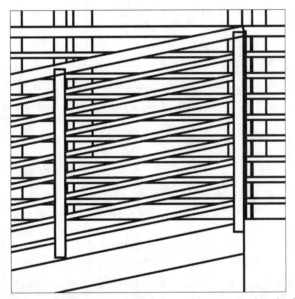

Figure 11–25 *There are now fewer lines and much less editing required in the Elevation*

Certain spots of the Railing will still require manual cleanup. This is done either with the **Edit Linework** and **Merge Linework** commands or by exploding the Elevation. These topics are covered below.

Note: A bonus of this technique is that the Elevation updates more quickly.

Curved Curtain Wall Infills

A similar situation occurs with the Infill Panels along the curved Curtain Wall (see Figure 11–26). When you use the Simple Panel Infill component, it will follow the same curve as the Curtain Wall itself. Nested Style Infill components do not do this.

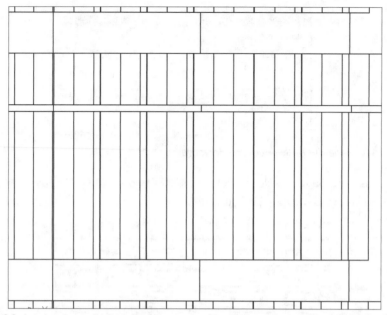

Figure 11–26 *It is hard to tell which lines are vertical mullions and which are tessellation lines*

Therefore, if we were to edit the Curtain Wall and reference a Window, Curtain Wall Unit or AEC Polygon Style for the glazing instead, it would no longer curve, and the quantity of unnecessary tessellation lines is again reduced. To do this, perform the following steps:

1. **Ope**n the Curtain Wall Shell file (*A-SH02-C.dwg*).

2. Select the curved Curtain Wall, right-click and **Edit** the Style. Click on the Design Rules tab.

3. Highlight **Infills** on the left, and then **Default Infill** on the top right. Change the **Infill Type** to **Style** and choose the **Standard** AEC Polygon Style (see Figure 11–27).

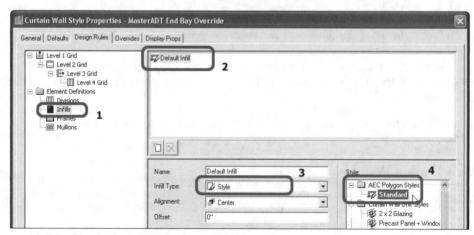

Figure 11–27 *Change the Infill Type to an AEC Polygon*

4. Click **OK** to close all dialog boxes, save and close the file. Reload the XREFs and update the Elevation (see Figure 11–28).

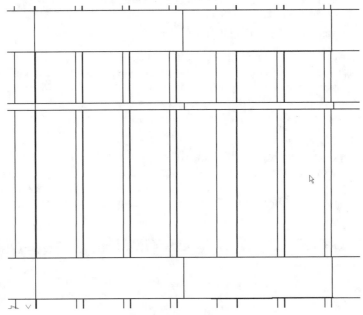

Figure 11–28 *The excess tessellation lines have been removed*

A few additional bonuses from this approach are as follows:

- Rendering times in VIZ will be reduced since the Infill Panel is now a single face rather than a six-sided box.
- AEC Polygons can display a color fill using the full "true color" spectrum. This works in both model and paper space.

The drawback to this approach is that AEC Polygon Infills will display as a single line in plan view. This is okay for small scales, but you may need to use a custom Display Block for the larger scales.

UNDERSTANDING STYLES AND DESIGN RULES

Even with the optimization tricks, there is still plenty of work to be done on our Elevation. We need to improve the overall "read" of the drawing. Currently the entire 2D Section/Elevation object is displaying on a single layer, color, linetype and lineweight. The first thing we can do to begin introducing lineweights (and therefore contrast) to the Elevation is to add sub-divisions. This will require some global System Default level display settings first.

Set Up 2D Section/Elevation System Default Display Props

1. Open the Display Manager, click on **Representations by Object** and then select **2D Section/Elevation**.

2. Double-click the **General** Display Rep on the right to edit it.

There are many ways that we could configure the Display Properties. Recalling many of the factors discussed in Chapters 2 and 3, set up the Layer, Color, Linetypes, Lineweight and Plot Styles of each component. Even though there are ten sub-divisions, in most cases you will not need more than three or four. At this stage, we are setting up the System Default settings. These will be the basis for any 2D Section/Elevation Style overrides that we create.

Regardless of whether you assign Lineweights by Layer, Color or Lineweight property, be sure to turn *on* all components, including the "Erased" component. Let's also use our alert colors for these components as well. Make the Erased component Red and the Unknown Component Magenta (see Figure 11–29). Be sure to use colors not in use anywhere else in your standards for these settings. Refer to Figures 3–27 and 3–28 in Chapter 3 for more information on the colors that are in use.

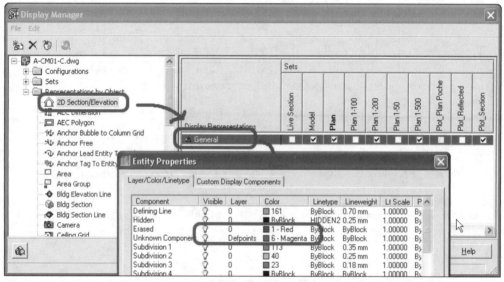

Figure 11–29 *Edit the Display props for 2D Section/Elevation objects*

3. Click **OK** when finished configuring the settings.

You can use many of the same settings for the 3D Section/Elevation object. You will also need to add these settings to your template file. Follow steps similar to the process used above to update the Work Display Configuration. Don't forget the AECPURGELAYERSTYLES command before saving.

Add Sub-divisions

Once you have your Display Props configured, you are ready to add sub-divisions. To do so, perform the following steps:

1. Back in the drawing, right-click the Elevation Line and choose **Elevation Line Properties**.

2. On the Dimensions tab, in the **Graphic Divisions from Cut Plane** area, **Add** two sub-divisions (see Figure 11–30). This will give you three zones, which is usually sufficient. Do not worry about the exact dimensional locations; you can use grips to adjust them.

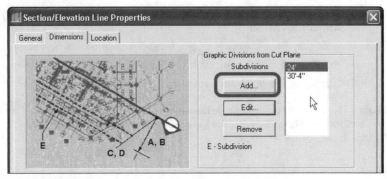

Figure 11–30 *Add Sub-divisions*

3. Tweak the locations of each sub-division line in the drawing with grips (see Figure 11–31).

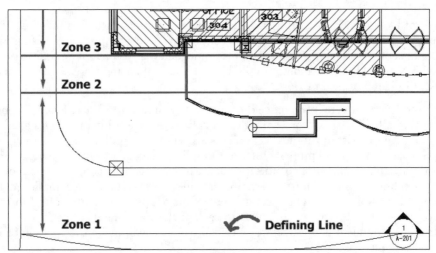

Figure 11–31 *Fine-tune the location of each sub-division*

4. Update the Elevation to see the effect.

Ordinarily, the zone closest to the Defining line would have the heaviest lineweight while the zone furthest away would have the lightest. In this case, it might actually be nice to have the Railings use a lighter lineweight. This will be easy to do, now that we have set up the sub-divisions and effectively isolated the railings to the first sub-division. What we will need to complete the task is to create a 2D Section/Elevation Style that has the first sub-division set to a lighter pen.

Build a 2D Section/Elevation Style

Create and Edit 2D Section/Elevation Styles in the Style Manager in the same way as you create other Styles.

1. Open the Style manager to 2D Section/Elevation Styles, create a New Style and name it **South Elevation**.
2. Begin with Display Props. Attach a Style level override to this Style and edit it.
3. Turn off the **Erased** component, and set Sub-division 1 to use a lighter pen.
4. Click **OK** back the drawing and update the Elevation.

 Note: This Style has a specific name because of the need to override the pen properties of the first sub-division. Normally you could use a simple name like "Construction Documents" or "Design Development."

Design Rules

In the photographer analogy above, the Section Style was equated with a photographer's filter, lighting and other "special" effects. Design Rules are very similar to camera filters. They have the ability to dramatically alter the mood and look of the photo but require a great deal of patience and knowledge to use properly.

The topic of Design Rules will bring our ADT Implementation discussion full circle. Early in the second and third chapters, lineweight and color were raised as important issues for which to develop global office standards. It was also recommended that you adopt an industry standard such as the National CAD Standard (in the United States) if an appropriate one exists in your country, jurisdiction or field of work.

Sections and Elevations are interpretive drawings. They show an abstracted view of a building (or portion of a building) for conveying design and construction intent. A Design Rule is a simple tool used to assist in this abstraction process. With Design Rules, we can help the 2D Section/Elevation command properly interpret Building Model data. This is done by applying colors to individual components of the Building Model and then mapping those colors to a particular rendition in the Section or Elevation drawing. This is typically accomplished through 2D Section/Elevation Styles. There can be quite a bit of configuration involved in this. For this reason, you should try to assign colors to each of the components in your standard template. If each component is pre-assigned a default color, it will make the creation of Design Rules much simpler.

Refer again to Table 11–1. A much more detailed version is included in the *Chapter11* folder in an Excel file named *Component Colors.xls* (see Figure 11–32).

	A	B	C	Q	R	S	T	U	V	W	X	Y	Z	AA	AB	AC	AD	AE	AF
2	**Architectural Objects**	Custom Display Blocks	Display Rep	Default Mullion	Door Panel	Frame	Stop	Swing	Glass	Entity	Bounding Box	Marker	Opening	Baluster	Post	Guardrail	Handrail	Roof	Eave
13	Door	**X**	Elevation*		Blk	Blk	Blk	Blk	Blk										
15	Door	**X**	Model**		41	251	251	Blk	141										
17	Mass Element		Model							254	Blk								
19	Mass Group		Model							53		Blk							
21	Opening	**X**	Model										Blk						
23	Railing	**X**	Model											143	143	143	143		

AIG Model Rep Color Matrix / D A CH Model Rep Color M

Figure 11–32 *The Excel file Component Colors.xls shows the color assigned to every Architectural Object in the Model Rep*

This file shows each of the Architectural objects (and a few others) and their respective Sub-components for each Display Rep listed in Table 11–1. There are three versions of the table, on three separate tabs:

- Table XL11–1 – AIG Model Rep Color Matrix
- Table XL11–2 – D A CH Model Rep Color Matrix
- Table XL11–3 – Imperial Model Rep Color Matrix

The tables each detail the sub-components included for each object as well as their respective colors and on/off state. A count of all components reveals approximately 70 sub-components for 21 ADT object types. This count groups all of the Wall and Structural Member sub-components together, for instance, Boundary 1-20 for Walls rather than 20 separate entries. ADT has a palette of 255 colors. At this point in the Implementation Guide, we have used just shy of 100 of these. That leaves well over 155 colors to work with for our Model Display Rep Assignments.

Remember that the Model Display Rep is being used to generate the 2D Section/Elevation objects. It is not being viewed and printed directly. It is also typically used for rendering purposes. The Imperial template color assignments in the Model Display Rep were devised for this purpose. (Refer to Table B5–4, "Model Display Rep Colors" in Appendix B included on the CD ROM in PDF format.) When an ADT Model is imported into Autodesk VIZ, the colors of the Model Display Rep automatically map to materials. In both the Section and Rendering scenarios, the traditional ByLayer CAD Management strategy would not prove beneficial. Therefore, our goal here is devise a list of color assignments that work well for generating Sections with Design Rules and for renderings. It turns out that this is not that difficult to achieve. Let's look at a few possible Design Rules.

Add a Design Rule

1. In the *Chapter11\Design Rules* folder, **Open** the file named *A-CM01-R.dwg*.

The drawing opens to a paper space layout named Plot Preview, showing a Building Elevation at 1/4"=1'-0". If you **Zoom** around and examine the Elevation, you will notice several areas where the lineweights get muddy. (Be sure the LWT toggle button in the AutoCAD Status bar is depressed to preview the lineweights.) This is particularly noticeable at the Windows Muntins on the second floor, the Bow Window (see Figure 11–33) and the Railing. If you click the Model tab and examine the Bldg Elevation Line used to generate this Elevation, you will notice that there is a single sub-division. This accounts for the chimney's being bolder than the rest of the Elevation. We can use Design Rules to lighten the Muntins and frames of the Windows and the balusters of the Railings.

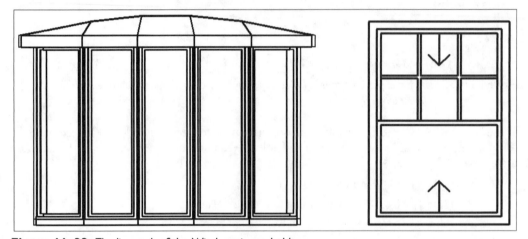

Figure 11–33 *The linework of the Windows is too bold*

2. Select the Elevation, right-click and choose **Edit 2D Section/Elevation Style**.
3. Click the Design Rules tab.
4. Click the **Add** button and then in the **Color** column, click on **1-Red**.
5. In the Select Color dialog box, type **92** and then click **OK** (see Figure 11–34).

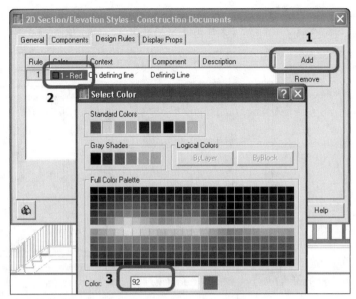

Figure 11–34 *Choose a Color for Rule 1*

Color 92 is the color of the Window Layer. If you consult Table XL11–3 (upon which the settings for this file are based), you will see that Windows have their colors set to ByBlock for all sub-components in the Elevation Display Rep. Therefore, assigning a Design Rule to Color 92 will allow us to effect a change to all of the Windows in the file.

    6. In the **Context** column, choose **Any Visible**.

This will apply this rule to all Windows regardless of the sub-division into which they fall, as long as they are "visible" from this vantage point. The **Component** column establishes where the lines meeting the criteria (Color 92 and Visible) are moved.

    7. From the **Component** column, choose **Subdivision 3** and type a description (see Figure 11–35).

A description such as "Lighten Window details" will help us remember the purpose of this rule in the future.

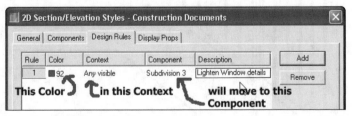

Figure 11–35 *Complete the criteria for the rule*

    8. Click **OK** to return to the drawing and **Update** the Elevation to see the effect.

You will be able to "see" the effect much better from the Layout tab. Be sure to click the Plot Preview layout tab and examine the effect on the Windows (see Figure 11–36).

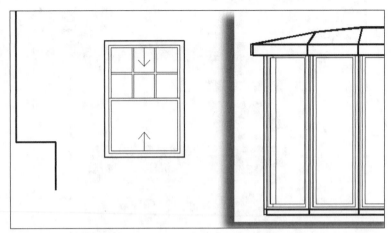

Figure 11–36 *The internal details and Muntins display in a lighter pen weight*

Design Rules and Template Colors

The Railing is a bit more complicated. With the Railing, only the Balusters are too heavy (see Figure 11–37). Therefore, we need a Design Rule that applies to them alone. The only way we can do that is if the Balusters are assigned their own unique color.

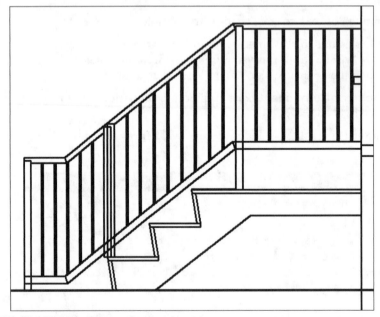

Figure 11–37 *The Railing's Balusters are too heavy*

If we examine the information contained in Tables XL11–1, XL11–2 and XL11–3, we will notice that Railings show their Model Display Rep in the Section_Elev Set. Furthermore, we see that all four sub-components of the Railing's Model Display Rep are assigned to the same color. This has presumably been done under the assumption that all four components would typically be made of the same material. By assigning them all to the same color, it would be very simple to assign a material in VIZ. However, as we can see from the Design Rule that we just added, we lose the ability to control the individual components separately in the Sections and Elevations if they are all the same color.

This leaves us with one question: Must all four sub-components be assigned the same color in order to map them to a single material in VIZ? The answer is most definitely "no." With little or no effort, we could map a single material to all four subcomponents of a Railing (or any object) in VIZ, even if they were four separate colors. The conclusion that we can draw from this is the following:

 GUIDING PRINCIPLE: We will gain the most amount of flexibility from our Models if we assign a unique color to each (Architectural) component of the Model Display Rep.

This approach has its drawbacks. In the instances where it is desirable to perform the same Design Rule on all components (as in the Window above) or assign the same material in VIZ, there is a little more setup work required. However, the majority of this setup can be saved in templates and reused seamlessly from one project to the next. The advantage is that users who may already be uneasy at the very mention of a "Design Rule" will rarely if ever need to add or edit them, because they will already be present in the template and Styles. We will discuss more details on the VIZ front in Chapter 13, available in PDF format on the Autodesk Architectural Desktop: Advanced Implementation Guide CD ROM.

Add a Sub-component Design Rule

1. In the *Chapter11\Design Rules* folder, **Open** the file named *A-FP01-R.dwg*.

This First Floor file contains the Railing objects. Since the template that this file used did not have separate colors assigned to each Railing sub-component, we must edit this first, before we can add a new Design Rule.

2. Open the Display Manager, expand **Representations by Object** and select **Railing** from the list of **Objects**.
3. On the right, double-click **Model** to edit it.

All four components are assigned Color 143. Refer back to Figure 3–27 in Chapter 3 and notice that the colors immediately adjacent to 143, Colors 123, 133 and 153, are all unused. You can choose any unused color you like, but selecting ones in the same color range will keep things organized and easy to understand.

4. Moving down the list from top to bottom, assign each component a unique color (see Figure 11–38).

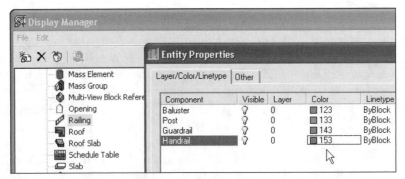

Figure 11–38 *Assign a set of unique colors to each component*

5. **Close** the Display Manager, **Save** and **Close** the file.

 Note: If you did not use Color 123 for your Balusters, be sure to write down the color you did use.

We now need to add a new Design Rule to the Composite file.

6. **Reload** the First Floor XREF.

7. **Edit** the Elevation Style again, and **Add** another Design Rule.

The Rule should look for **Any Visible** Color 123, and move it to the **Subdivision 3** Component (see Figure 11–39).

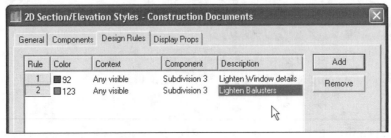

Figure 11–39 *Add the Balusters Design Rule*

8. Click **OK**, and **Update** the Elevation (see Figure 11–40).

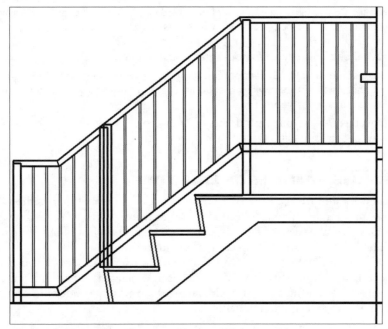

Figure 11–40 *The two lines of the Balusters no longer "bleed" together*

 Note: The Railings in this file use a square profile as recommended in the section above. With a round profile, this Rule would provide little help.

This discussion brings us back to the color assignments in our templates. You can now begin to appreciate how wide your choices regarding Color will become. Do not be afraid to push the limits of your 255-color spectrum. The flexibility you will gain will be well worth the effort.

You are highly encouraged to assign every Model Display Rep sub-component its own unique color. (It will require nearly 70 separate colors.) Although you are unlikely to need 70 separate Design Rules in any given drawing, if you or anyone else develops a need for a new rule, imagine how much more likely they will be to consider it if all that must be done is to add the rule to the current 2D Section/Elevation Style. If they must open several drawings, edit colors of Display Reps that they do not even see in a viewport and then add a rule back in the original file, who could blame them for saying that "exploding is easier."

For Walls, the System Default will likely use a default color. However, if you are using multi-component Wall Styles (see Chapter 9), you will want to add a Color column to your Wall Materials list. This list was mentioned in Chapter 9 as containing every component and its cleanup priority. Adding the colors to that list will allow anyone building Wall Styles to choose the correct color for each material they use.

To prepare your template for optimal use of 2D Section/Elevation objects, do the following:

1. Open the Template and assign a unique color to each "Architectural" sub-component of the Model Display Rep. Use Tables XL11–1, XL11–2 and XL11–3 as a guide.

Note: Components like Defect Warning, Node, Cell, and Marker are not Architectural components and should be placed on the Defpoints layer with an alert color assigned.

2. Save the template file. Do not forget to purge the Layer Key Styles.

3. Build some sample 2D Section/Elevation Styles with some typical Design Rules. For instance, make Door Swings dashed and lighten Muntin and Baluster components. Erase extraneous geometry.

4. Document what you created, especially the colors you chose for everything. This way, people can edit a Design Rule if they wish.

WHEN TO USE EDIT/MERGE LINEWORK

The **Edit Linework** and **Merge Linework** commands do the same thing as the Design Rules do, except that they require manual selection. In other words, when a 2D Section/Elevation is generated, a Design Rule "selects" a group of lines in the Section based on their parent object's original color and location within the Section box, and "moves" those lines to a different display component of the Section. With the **Edit Linework** and **Merge Linework** commands, we are *manually* selecting lines in the Section after it has been generated and assigning them to a new display component within the Section. **Edit Linework** and **Merge Linework** are analogous to photographic post processing and retouching.

Edit Linework and **Merge Linework** can be effective for simple edits. The major drawback to these tools is that they only allow you select one object at a time. In addition, you are not really "editing" linework, but rather "re-indexing" it. For these reasons, many users end up exploding their 2D Section/Elevation objects. In some cases, this is the appropriate course of action, but you should try to exhaust other possibilities first.

Perhaps the most common reason for exploding 2D Section/Elevation objects is the perception that one cannot preserve the edits made with the **Edit Linework** and **Merge Linework** commands. This is simply not true. All edits can be saved and, in most cases, reused.

Understanding Edit Linework and Merge Linework

1. In the *Chapter11* folder, **Open** the file named *EditMerge.dwg*.

2. Select the Elevation, right-click and choose **Edit Linework**.

3. Select one or more lines from one of the Windows and press ENTER.

4. When prompted to "Enter Component Index," type **3** and then press ENTER (see Figure 11–41).

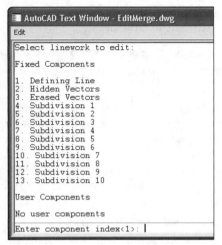

Figure 11–41 *Erase lines, by indexing them to the "Erased" component*

Component Index 3 corresponds to the Erased component. Earlier in the "Understanding Styles and Design Rules" section, we turned this component on and assigned it color Red in the System Default. That was done to facilitate the Edit/Merge Linework procedure. The Elevation in this file uses a 2D Section/Elevation Style named Construction Documents. This Style has the Erased component turned *Off* as you might expect. When we choose the Erased component in the **Edit Linework** command, it "looks" erased, because the component is typically turned off.

5. Right-click the Elevation and choose **Update**.

Notice the **Save Copy of User Edits** check box at the top of the dialog box (see Figure 11–42). This feature is not widely understood. Notice that the Style chosen for this feature by default is Standard. Standard in this drawing does not have any "Style level" overrides; therefore, all components are turned on, including Erased. Furthermore, we made the Erased Component Red, so that it would stand out. The reason that we did all of this is so that when we update, we can save these User Edits (erasures 90 percent of the time) to another Section object in the Standard Style.

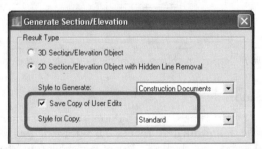

Figure 11–42 *Save a Copy of User Edits to the Standard Style that has the Erased Component set to Red and visible*

6. Click **OK** to update the Elevation.

It will *appear* as though the User Edits have been lost. In fact, they are preserved in a second Elevation object that is directly on top of the original. The lines you erased should be Red.

7. Select the Elevation, right-click and choose **Merge Linework**.

8. When prompted, choose the copied Elevation (the one with Red lines) and press ENTER.

The two Elevations merge to become one, and the lines "re-erase." This will work the majority of the time. Sometimes, rounding errors can occur in the Elevation that prevent the merge from working at 100 percent. However, it should still be better than if you had to perform all edits over again. If you move an object that was previously edited, the merge feature will not work, because it will "lose track" of the objects that it edited. However, you can often manually move the edited copy to the new location before merging and still achieve satisfactory results. To try this out, do the following:

1. Erase some portion of one of the Windows.

2. Move the original Window in the Model and then update the Elevation. Be sure to save a copy of user edits to Standard.

The Edit Copy will not be synchronized with the original.

3. Move the Edit Copy, snapping it to the new location.

4. Merge Linework (see Figure 11–43).

You may need to redo a few of the lines, but most of the editing should be reapplied, even in the new location.

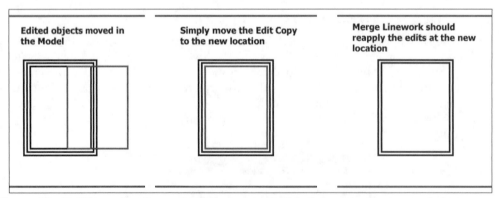

Figure 11–43 *When editing "moves" an object, move the Edit Copy to the new location before merging*

The Lines between Floors

Many users are bothered by the "lines" between floors that occur when an Elevation is generated from an XREF model. Unfortunately, there is no automatic way to eliminate them. However, there are a few ways to deal with them. The obvious solution is to "erase" them with the **Edit Linework** command. Now that we have seen that those edits can be preserved, this is not as undesirable a technique as it may at first seem. However, erasing the lines may not always be the best solution. In many Elevations, you will later add a line to demarcate the Finished Floor line. Consider simply changing this line to a dashed line or other appropriate linetype for this purpose (see Figure 11–44). It will utilize the same process as erasing it would, but it saves the step of adding the line back manually later.

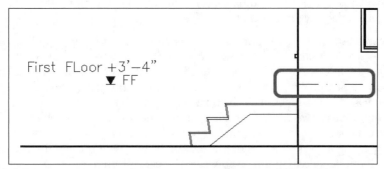

Figure 11–44 *Edit the line between floors to demarcate the Finished Floor line*

Another technique that you can try is to add the XREFs to a Mass Group object (see Figure 11–45). If the XREFs meet flush, the Group will "smooth" the line between the floors away (most of the time).

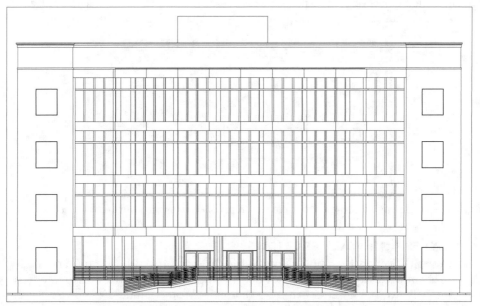

Figure 11–45 *A Mass Group can help to eliminate the "lines between floors"*

TRACING AND EXPLODING

Despite all of the techniques covered in this chapter, there comes a time while working on a Section or Elevation when you simply need to work directly on the Elevation itself rather than in the Model. There are two ways to achieve this "hands-on" approach. You can trace over the

2D Section/Elevation object, leaving it intact underneath, or you can explode it. It is recommended that, even if you choose to explode, you always keep a copy of the linked 2D Section/Elevation object for coordination purposes. You can simply move this copy to a frozen layer and recall it when you wish to coordinate your manually drawn copy to any changes that have occurred in the model since it was exploded.

Before you explode, however, consider tracing. By simply drawing on top of the live 2D Section/Elevation, you can add a bold line around an Elevation and add details that do not, or should not, appear in the model. You can also add note, dimensions, annotation and hatching. The most important thing to remember about the 2D Section/Elevation tool is that even on very large models, ADT can generate the base geometry required for any Section or Elevation much more quickly than we can with traditional techniques. Therefore, even if you choose to explode it, you are still ahead of the game in production time. Just be careful not to squander these gains by trying to make the live object too "perfect." There is a certain level of detail that this object will not be able to attain, even with the best collection of styles and Design Rules. Every professional team knows when to "punt." If you have a good kicker, you can even turn this into a scoring opportunity.

HATCHING

Never allow hatching that you create on top of a 2D Section/Elevation object to be associated with the Section object directly. This can cause performance problems. Use the Advanced tab of the **Bhatch** command to **Retain Boundaries** when generating hatching (see Figure 11–46).

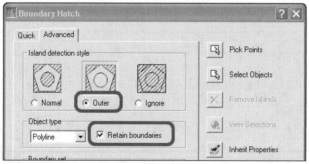

Figure 11–46 *Setting hatching to associate with a polyline rather than the Section object*

WALL TYPES AND WALL SECTIONS

You can use 2D Section/Elevation objects as the "starter" for Wall Sections and Wall Type details. It is not necessary to use complex multi-component Wall Styles. If you do, however, the section will give you an even better starting point. Some sample Wall Type details have been included in the *Chapter11\Wall Types* folder.

BATCH UPDATE

If you are an Autodesk subscription member, the Productivity Enhancement Extension includes a "Batch Update" routine for Sections and Elevations. This tool can be a tremendous productivity boost when you have a large project with many Sections and Elevations. If you do not have the Extension, you should contact your reseller about getting a copy.

USING LIVE SECTIONS

Live Sections are a terrific design tool, with one major drawback: they don't work through XREFs in this release. Unfortunate as this is, if you are working in a complex file that would benefit from cross-sectional study, consider setting up a Live Section. Live Sections use the same Bldg Section Line as the 2D Section/Elevation and 3D Section/Elevation objects. Simply right-click an existing Section Line to create a Live Section Configuration. This is the only way to create a Live Section—you cannot set this up in the Display Manager alone. However, once you have created a Live Section Configuration, you can edit it in the Display Manager as required. To view your Live Section, simply make the Live Section Configuration active in the same way you would any other Display Configuration. Most Architectural objects have a Section Display Rep that is used by the Live Section Configuration. You can edit the Display Properties of these Display Reps to add hatching to the cut line. The best part of a Live Section is that it is "*live.*" What this means is that unlike the other two Section/Elevation objects, which generate a graphical report of the building model in a separate object, the Live Section is a display mode of the actual Model data. Editing this data affects all views of the Model.

 CHECKLIST

I 3D Section/Elevation Objects

1. If you intend to use them for Spanning Curtain Walls, documentation of the procedure is the only implementation strategy that you can reasonably deploy.

II 2D Section/Elevation Objects

1. Configure the Bldg Section Line object with the Boundary and Sub-Divisions components set to a "No Plot" layer, with "Alert" colors assigned.

   a. Alternatively, turn these components off in the Plotting Display Reps of Plan 1:100 and Plan 1:50.

2. Configure the Bldg Elevation Line object with *all* components set to a "No Plot" layer, with "Alert" colors assigned.

   a. Alternatively, turn these components off in the Plotting Display Reps of Plan 1:100 and Plan 1:50.

3. Save all of these settings to your template file.

4. Develop a procedure for determining which XREFs to include in Composite Models used for Sections and Elevations.

5. Use the Model Display Rep in your template file for both Section/Elevation objects and renderings.

6. Modify the Work configuration in the template to show the Sect_Elev Set in all Elevation View Directions.

7. In your template file, assign a unique color to each sub-component of each Architectural object in the Model Display Rep.

8. Remember to **Purge** Layer Standards and Key Styles from the template before saving.

SUMMARY

Even though it is considered obsolete by most, the 3D Section/Elevation object can offer an alternative to the Display Rep customization technique for Spanning Curtain Walls.

2D Section/Elevation objects offer a true 2D linked object that is Style-based and flexible.

Simplify your XREF structure to speed up 2D Section/Elevation object update.

Freeze extraneous layers and simplify Model geometry to speed up 2D Section/Elevation update.

Design Rules link Model colors to 2D Section/Elevation display conditions.

One set of colors can be used to satisfy the needs of both 2D Section/Elevation objects and VIZ renderings.

Explode only when necessary. It is better to work on top, and keep the live section to check coordination.

Interoperability

This section is devoted to the subject of interoperability. It is divided into two chapters, the first of which discusses the many issues that can arise when drawings are sent to outside firms that do not use ADT. The second chapter, available only on the Autodesk Architectural Desktop: Advanced Implementation Guide CD ROM, covers the use of Autodesk Architectural Desktop models inside Autodesk VIZ. The topic of standardization of the "color to material" mapping procedure is central to the chapter.

Section V is organized as follows:

Chapter 12	Sharing Files with Consultants
Chapter 13	Using ADT Models in Autodesk VIZ*

* Available on the Autodesk Architectural Desktop: Advanced Implementation Guide CD ROM

WHO SHOULD READ THIS SECTION?

This section is directed toward the **CAD Software User**. Most topics are of equal interest to the **CAD Manager** and the **Project Data Coordinator**.

Sharing Files with Consultants

INTRODUCTION

In nearly every project, you will be faced at some point with the need to share digital design data with other individuals or software packages. For instance, preparing drawings for distribution to MEP and other consultants offers some challenges in ADT. The objects that make ADT so powerful do not exist in the base version of AutoCAD or in other CAD software. We must therefore translate ADT drawings to an AutoCAD-friendly format beforehand. This chapter is focused on the various issues related to distributing ADT drawings to consultants who may be using a variety of alternative ADT and AutoCAD versions.

OBJECTIVES

In this chapter, we will discover what is required to translate drawings into formats readable by consultants not using ADT. Due to the potentially destructive nature of the commands being covered, be sure to practice on files unrelated to any "real" projects until you are familiar all the issues covered in this chapter. The following summarizes topics covered in this chapter:

- Understand file formats
- Understand Object Enablers
- Learn about **Explode AEC Objects** and when to use it
- Learn to script —AECOBJECTEXPLODE
- Use **eTransmit** to send files

UNDERSTANDING OBJECT ENABLERS

Because AutoCAD, previous versions of ADT, and AutoCAD based "Desktop" products are not capable of properly displaying ADT 3.3 objects, sharing drawings with consultants who use these other products requires an additional piece of software called an Object Enabler. An Object Enabler is a free plug-in provided by Autodesk that "enables" otherwise unknown ADT objects directly in the vanilla version of AutoCAD (or other "Desktop" products). Users of an Object Enabler will be able to correctly view and print ADT objects, as well as perform limited grip editing functions upon ADT objects contained within the file. They will not be able to add new ADT objects to the drawing. Without an Object Enabler, sharing drawings with consultants using other versions of AutoCAD will result in the use of Proxy Graphics (simplified geometric approximations of the actual ADT object) or the need to Explode AEC Objects.

To determine which options are available for your particular situation, consult the following tables: (Also be sure to visit: http://pointa.autodesk.com/local/enu/portal/prodcent/obj_enabler.htm to see the latest Object Enabler Comparison Charts.)

When the Consultant Is Using Autodesk Architectural Desktop

Use the following chart to determine the relevant issues.

Consultant Software and Version	Translation Required
Architectural Desktop R3 or R3.3	No translation necessary
Architectural Desktop R2i	See steps for 2000i below
Architectural Desktop R2	Not compatible (use proxy graphics or Explode AEC Objects)
Architectural Desktop R1	proxy graphics or Explode AEC Objects

 Note: An additional option, if the consultants are using ADT R2i, is to install the Object Enabler V2.53i on ADT R3. This will allow ADT R3 to open 2i files to without updating them to R3.

When the Consultant Is Using Standard AutoCAD

There are three ways to share ADT R3 data with them: the consultant can install an Object Enabler, you can save proxy graphics into the ADT R3 file, or you can use the Explode AEC Objects routine prior to sending the consultant the file.

Consultant Software and Version	Translation Required
AutoCAD 2000i	Object Enabler V3.02 (log on to *http://pointa.autodesk.com/*) or proxy graphics or Explode AEC Objects
AutoCAD 2000	Object Enabler V3.02 (log on to *http://pointa.autodesk.com/*) or proxy graphics or Explode AEC Objects
AutoCAD R14	proxy graphics or Explode AEC Objects

When the Consultant Is Using a Non-AutoCAD CAD Package

The only option is to use Explode AEC Objects and save the drawing to either DWG or DXF format.

Consultant Software and Version	Translation Required
Non-AutoCAD CAD Software that reads DWG format	Explode AEC Objects, save as R14 DWG
Non-AutoCAD CAD Software that does not read DWG format	Explode AEC Objects, save as R14 DXF

UNDERSTANDING PROXY GRAPHICS

ADT objects generate the graphics used to represent themselves in different views automatically and based on the parameters of the specific object. The generic AutoCAD package (2002, 2000i, 2000 and R14) is incapable of generating these graphics and would simply ignore the "unknown" objects under typical circumstances. With proxy graphics turned on, ADT creates two sets of graphics for the ADT object when the file is saved, the one used by ADT and the one saved with the file for generic AutoCAD. As the name *proxy* implies, proxy graphics are stand-in objects that appear on screen only when the host application that created them is not present on the machine. In this case, that host application is Autodesk Architectural Desktop R3. A proxy graphic is not editable in any way. Therefore, if the recipient of the file with proxies needs to edit the data, this would not be a good choice of formats.

For persons needing only to view the data in the file, for review purposes for instance, proxies can provide a good solution. This is because no special requirements are placed on recipients. They simply open files as they normally would. One small caveat: a message will typically appear alerting the recipient of the presence of proxy graphics and requesting their input on how to deal with them. The choices are to display the proxies, show the bounding box only, or ignore them. Please instruct your recipients to choose to display proxy graphics. If they choose not to display them, they will see nothing and will most likely call you to ask why you sent an empty file. If they choose bounding box, each object in the file will appear as a box that matches the overall size of the total object. This is good if the file is large and takes a long time to open, but it will not reveal very detailed information.

Avoid using proxy graphics if you can. Again, these are stand-in graphics, not the real thing. They cannot be edited and they provide only limited visual information. Finally, proxies nearly double the size of ADT files. Depending on the file size without proxies, this can be significant.

TURNING PROXY GRAPHICS OFF

1. Type **proxygraphics** at the command line and then press ENTER.
2. Type **0** and press ENTER again.
3. To make this setting permanent, change the setting in your template file.

EXPLODE AEC OBJECTS

This routine completely eliminates the AEC objects from a drawing file and replaces them with lines, arcs and circles. Use this only if the file will not be returned to you for editing. The first two exercises in this lesson cover the usage of **Explode AEC Objects**.

USING EXPLODE AEC OBJECTS FROM THE MENU

To translate an ADT file into a generic AutoCAD R14 drawing requires a special tool called **Explode AEC Objects**. Using the generic AutoCAD **Explode** command is not sufficient. Please use the steps that follow instead. If a DXF is needed, follow these steps and then **Save As** DXF.

Install the CD Files and Open a Sample File

1. Install the files for Chapter 12 located on the Autodesk Architectural Desktop: Advanced Implementation Guide CD ROM.

Refer to the "How to Use the CD" section in the Preface for instructions on installing the sample files included on the CD.

A folder named *Chapter12* was created containing a collection of sample files that we will use throughout this chapter.

2. Launch Autodesk Architectural Desktop R3.

3. Open the file named *AECExplode01.dwg*.

Explore the AutoCAD Properties Palette

1. From the **Tools** menu, choose **Properties** (see Figure 12–1).

Of the great features of the AutoCAD Properties palette (refer to any good AutoCAD reference or the online help), its ability to be used as an instant **List** command stands out.

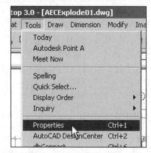

Figure 12–1 *Open the Properties palette*

2. Select any wall in the drawing (see Figure 12–2).

Notice that **Wall** appears at the top of the Properties palette. You instantly know that the type of object selected is a wall, which is faster than the traditional AutoCAD **List** command.

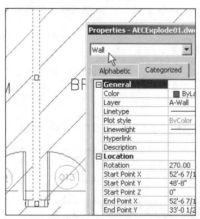

Figure 12–2 *Use Properties in place of List*

3. Select several objects (see Figure 12–3).

Notice that something like "All (10)" will appear.

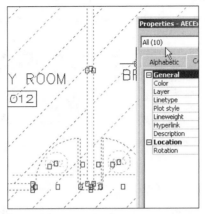

Figure 12–3 *Listing several objects at once*

    4. Click on the list headed by **All (10)** (see Figure 12–4).

Notice that each type of object in the selection is listed separately, with a quantity next to each one.

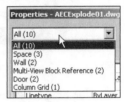

Figure 12–4 *View separate object types within the selection*

Again, comparing this functionality to that of the AutoCAD **List** command, this is quicker and more immediate. There is one additional benefit: Unlike with **List**, an object's properties can be edited directly in the Properties palette.

    5. Test it on the selection you have active.

    6. Deselect all objects by right-clicking and choosing **Deselect All**, or simply press
        ESC (see Figure 12–5).

Leave the Properties palette open on screen throughout this exercise.

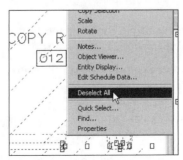

Figure 12–5 *Deselect with the right-click menu*

Explode AEC Objects

1. From the **Desktop** menu, choose **Utilities>Explode AEC Objects** (see Figure 12–6).

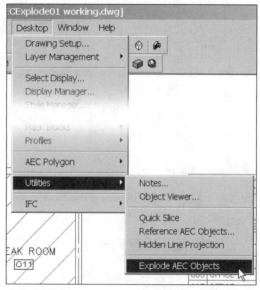

Figure 12–6 *Choose Explode AEC Objects*

2. In the Explode AEC Objects dialog box, make sure there is a check mark in the **Explode AEC Objects To Primitive Graphics** check box and that the **Current View** option is selected (see Figure 12–7).

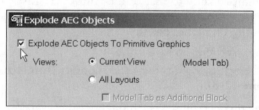

Figure 12–7 *Explode objects to primitive graphics*

Choosing **All Layouts** generates multiple copies of the exploded graphics, one on top of the other, for each layout view in the drawing. This is not recommended.

3. Put a check mark in the **Explode To Anonymous Blocks** check box (see Figure 12–8).

An anonymous block will be created for each AEC object in the file. Use this setting when you wish to preserve the quantity and composition of the original objects rather than their sub-components.

4. Do not check **Maintain Resolved Layer, Color, Linetype**.

Figure 12–8 *Explode to anonymous blocks*

> 5. Verify that there is a check mark in each of the remaining settings and then click **OK** (see Figure 12–9).

Erase AEC Objects removes all AEC objects from the drawing file. This will guarantee that the file will contain only lines, arcs, and circles. Use **Shorten Symbol Names for R14 Compatibility** if you are sending the file to someone using AutoCAD R14. Leave the **Length** set to 31; this is the maximum symbol name length for R14 and is the recommended setting.

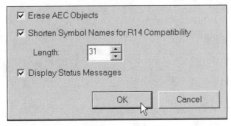

Figure 12–9 *Click OK to complete the command*

If **Display Status Messages** is turned on, a stream of messages will appear at the command line (see Figure 12–10). When the command prompt returns, you will know that the process is complete. This is recommended because the process can take a while to complete and this will serve as a status message.

```
Exploding graphics for object     <AecDbSpace        1EAD>
Exploding graphics for object     <AecDbSpace        1EB3>
```

Figure 12–10 *Command line feedback*

Save As to Preserve Original

> 1. From the **File** menu, choose **Save As** (see Figure 12–11).

   Caution: It is critical that you do a **Save As**!

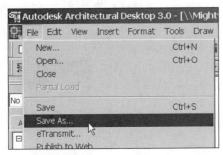

Figure 12–11 *Don't Forget to choose Save As*

Explode AEC Objects has completely eliminated all remnants of the original intelligent AEC objects. Doing **Save As** will guarantee that you preserve your original file.

2. Change the **Files of type** to **AutoCAD R14/LT98/LT97 Drawing (*.dwg)** (see Figure 12–12).

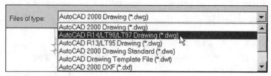

Figure 12–12 *Choose R14 Format if required*

3. Type a new file name and then click **Save** (see Figure 12–13).

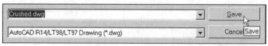

Figure 12–13 *Type a file name*

4. When the AutoCAD warning appears, simply click **OK** (see Figure 12–14).

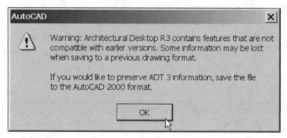

Figure 12–14 *Saving in R14 format will lose ADT object data*

ADT objects are not preserved when you save to R14 format. However, because we used Explode AEC Objects to remove all of the ADT objects, the warning is irrelevant in this case.

Explore the Results

1. Select one of the walls in the file.

   Notice that the object type is now a Block Reference rather than a Wall (see Figure 12–15).

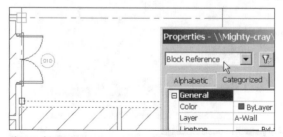

Figure 12–15 *The resulting objects are anonymous blocks*

This is because we chose to **Explode To Anonymous Blocks** in the steps above. Anonymous blocks preserve the original object structure and the layering of those original objects.

> 2. Explode the wall. Then click on one of the resulting lines (see Figure 12–16).

Notice that the anonymous block has been reduced to lines. Notice however, that the lines are on Layer 0 and Color is ByBlock.

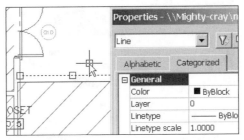

Figure 12–16 *Explode the anonymous block*

Repeat Explode AEC Objects

> 1. Open the original *AECExplode01.dwg* file again.
> 2. From the **Desktop** menu, choose **Utilities>Explode AEC Objects** again to change the settings.
> 3. Clear the **Explode To Anonymous Blocks** check box.
> 4. Check **Maintain Resolved Layer, Color, Linetype** (see Figure 12–17).

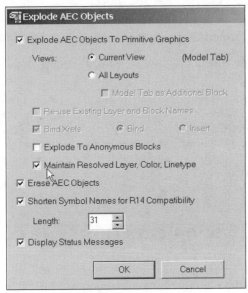

Figure 12–17 *Repeat the command with slightly different settings*

5. Click **OK** to run the command.

6. Save the drawing as a new name before proceeding.

7. Select a wall again (see Figure 12–18).

Notice that this time objects were exploded to individual lines. Notice also that these lines have now acquired the layer, color, and linetype of the original object.

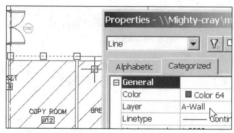

Figure 12–18 *This time the objects are lines*

8. Reopen the file and try it once more. Check both **Explode To Anonymous Blocks** and **Maintain Resolved Layer, Color, Linetype**.

When choosing both, object structure will be maintained initially, but if one of the anonymous blocks is exploded, the internal objects will be on the correct layers.

9. Remember to **SaveAs**!

ADDITIONAL THOUGHTS ON EXPLODE AEC OBJECTS

Regardless of the exact combination of settings chosen, **Explode AEC Objects** should be used only if files are being sent to consulting firms outside your organization and not returned. If the consultants you work with need to edit the files and return them to you, then ideally, they should upgrade to Architectural Desktop.

Caution: Once objects have been exploded, there is no way to "re-objectify."

CAD collaboration with vendors, clients, joint venture partners, and contractors is a big area of concern in the AEC industry. Before a new project is begun, schedule a Project CAD Standards Kick-Off meeting to discuss standards and setup. In this way, a clear guideline of the rules, proper workflow, and procedure can be established with every team member, inside and outside your office. If you find yourself involved with a situation where mass conversions are necessary, use the procedure outlined below to automate the process.

USING EXPLODE AEC OBJECTS FROM THE COMMAND LINE

Although **Explode AEC Objects** is very powerful, it can become a bit tedious to run this command across the multiple drawings that might be required for a submission to consultants. In this exercise, you will learn to automate the conversion process.

Run Explode AEC Objects from the Command line

Most ADT commands are executed in dialog boxes. Although user friendly, dialog boxes do not lend themselves to automation in ADT. ADT supports script files, which allow a series of commands to be run at the command line automatically. In order to be used in a script file, the command to be scripted must be able to have its dialog boxes suppressed and to be run at the command line. There are two ways to accomplish this depending on the situation:

- Turn off dialog box system variables (such as FILEDIA, CMDDIA, and ATTDIA).
- Use a hyphen (-) in front of a command name and type it in. At the command line type: **-aecobjexplode** and then press ENTER (see Figure 12–19).

Figure 12–19 *Type -aecobjexplode and press* ENTER

Rather than a dialog box as would appear when the command is selected from the **Desktop>Utilities** menu, the options for the command will appear as questions at the command line. This is a dry run to understand the questions that will be asked and to record appropriate responses for the task of developing a script.

Write down the answers to each of the questions as they occur—you will need this information when writing the script.

1. At the Explode AEC objects to primitive graphics question, type **y** and then press ENTER (see Figure 12–20).

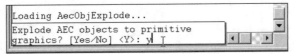

Figure 12–20 *Write down the answer to this command prompt*

2. At the Views to explode question, type **c** and then press ENTER (see Figure 12–21).

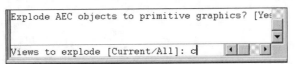

Figure 12–21 *Choose Current and write down the answer to this command prompt*

3. At the Explode to anonymous blocks question, type **y** and then press ENTER (see Figure 12–22).

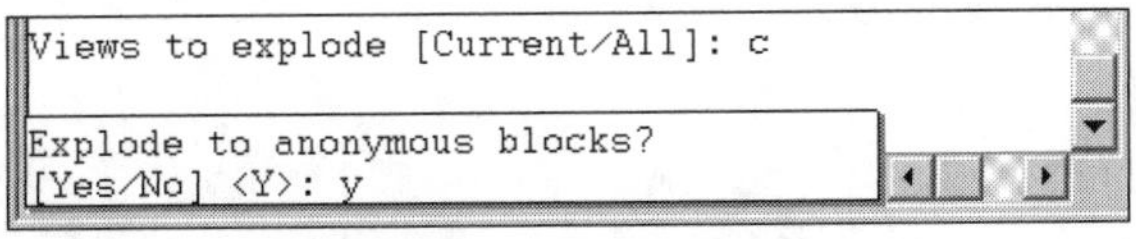

Figure 12–22 *Write down the answer to this command prompt*

4. At the Maintain Resolved Layer, Color, and Linetype question, type **y** and then press ENTER (see Figure 12–23).

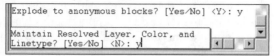

Figure 12–23 *Write down the answer to this command prompt*

5. At the Erase AEC objects question, type **y** and then press ENTER (see Figure 12–24).

Figure 12–24 *Write down the answer to this command prompt*

6. At the Use short symbol names for R14 compatibility question, type **y** and then press ENTER (see Figure 12–25).

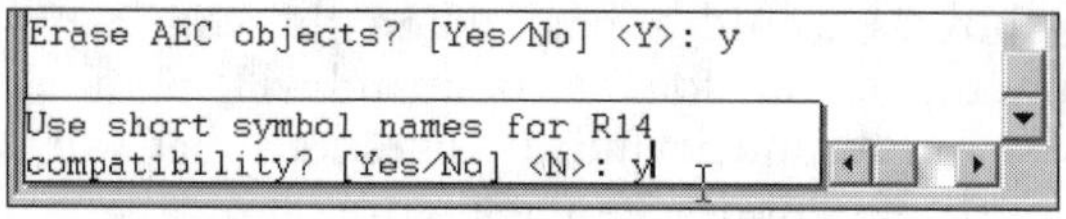

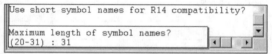

Figure 12–25 *Write down the answer to this command prompt*

7. At the Maximum length of symbol names question, type 31 and then press ENTER (see Figure 12–26).

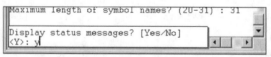

Figure 12–26 *Choose the maximum allowed and write down the answer to this command prompt*

8. At the Display status messages question, type **y** and then press ENTER (see Figure 12–27).

Figure 12–27 *Write down the answer to this command prompt*

The responses to the questions are as follows:

    y

    c

    y

    y

    y

y

31

y

Create a File List

In order for the script to be run across many files, a list of file names with their locations will be needed. The Windows command line prompt can be used for this.

1. From the Windows Start menu, choose **Run** (see Figure 12–28).

Figure 12–28 *Choose Run from the Start menu*

2. In the Run dialog box, type **CMD** in the **Open** field and then click **OK** (see Figure 12–29).

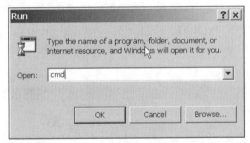

Figure 12–29 *The Run dialog box*

You will need to know the location of the files you wish to convert. For this example, they are in the *C:\ADT_AIG\Chapter12* folder. Substitute the appropriate drive letter and folder path in your own example.

3. Type **cd c:\ADT_AIG\Chapter12\Explode** and then press ENTER.

4. Type **dir /b /s >AECExplode.SCR** and then press ENTER (see Figure 12–30).

Figure 12–30 *Enter the DIR (directory) command with switches*

5. Close the Command Window.

For complete information on the CD and DIR commands, consult a DOS command reference.

Build the Script

The previous step created a file with a list of drawings in it. We need to open that file and add the lines of the script to it.

1. From the Windows Start menu, choose **Programs>Accessories>Notepad** (see Figure 12–31).

 Note: Any text editor such as Microsoft Word can be used, as long as the file created is saved in plain ASCII text format (DOS Text in Word).

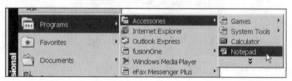

Figure 12–31 *Open Notepad*

2. In Notepad, or whichever text editor you are using, choose **Open** from the **File** menu. Locate the file named *AECExplode.scr* created in the last sequence (see Figure 12–32).

 Tip: You might need to change the **Files of Type** list to **All Files** in order to see the file.

Figure 12–32 *Locate the text file dumped by the DIR (directory) command*

3. Delete any entry in the list that is not a drawing file (does not end with ".dwg") (see Figure 12–33).

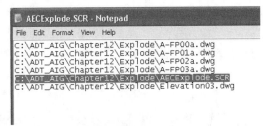

Figure 12–33 *Eliminate extraneous lines in the text file*

The lines of the script file need to be inserted in the text file for each drawing file in the list.

4. Place the cursor at the start of the file, type **Filedia 0** and then press ENTER. (That is **Filedia** SPACE **0** ENTER.)

5. Type **Open** and then press ENTER.

The file should look like the Figure 12–34.

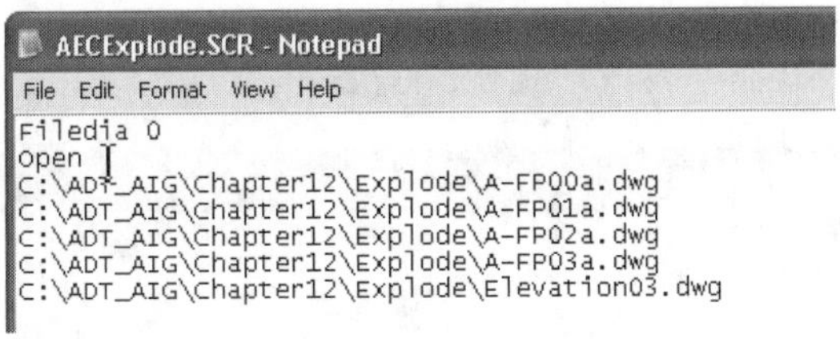

Figure 12–34 *Add the first few lines to the script*

FILEDIA set to 0 turns off dialog boxes for **Open** and **Save** commands. **Open** will open the first file in the list. Typing a SPACE or pressing ENTER after each item is important because these will be read as ENTER by AutoCAD. SPACE or ENTER can be used; it is a matter of personal preference. If an item needs to include a SPACE, such as folder names with spaces, enclose the entire item in quotes, like this: "*C:\Program Files\Autodesk Architectural Desktop 3*"

6. Place the cursor after the "dwg" of the first file in the list and press ENTER.

7. Type **-aecobjexplode** followed by a SPACE.

8. Type each answer written down in the AECOBJEXPLODE sequence above followed by a SPACE. **y c y y y y 31 y** and then press ENTER.

The screen should look like Figure 12–35.

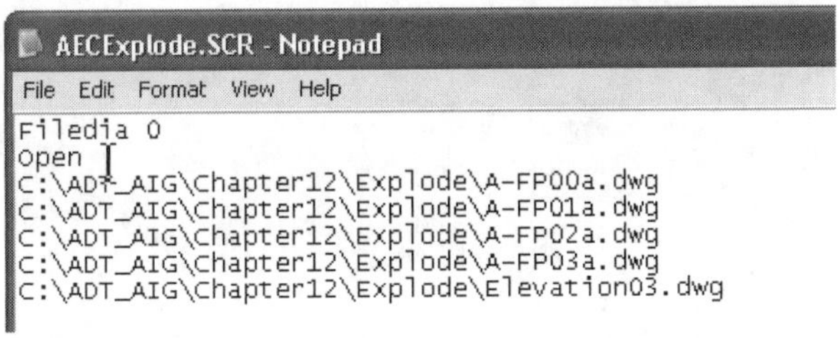

Figure 12–35 *Add several more lines to the script*

The cursor should be positioned on the blank line. (If there is no blank line between the -aecobjexplode line and the next file, add one by pressing ENTER to add a new blank line.

9. Type **Saveas R14 y** and then press ENTER (see Figure 12–36).

 Important: There must be two spaces following R14: Saveas SPACE R14 SPACE SPACE y ENTER

```
AECExplode.SCR - Notepad
File  Edit  Format  View  Help
Filedia 0
Open
C:\ADT_AIG\Chapter12\Explode\A-FP00a.dwg
-aecobjexplode y c y y y y 31 y
saveas r14   y

C:\ADT_AIG\Chapter12\Explode\A-FP01a.dwg
C:\ADT_AIG\Chapter12\Explode\A-FP02a.dwg
C:\ADT_AIG\Chapter12\Explode\A-FP03a.dwg
C:\ADT_AIG\Chapter12\Explode\Elevation03.dwg
```

Figure 12–36 *Continue*

10. After the Saveas line, type Open, but do not press ENTER this time (see Figure 12–37).

```
AECExplode.SCR - Notepad
File  Edit  Format  View  Help
Filedia 0
Open
C:\ADT_AIG\Chapter12\Explode\A-FP00a.dwg
-aecobjexplode y c y y y y 31 y
saveas r14   y
Open
C:\ADT_AIG\Chapter12\Explode\A-FP01a.dwg
C:\ADT_AIG\Chapter12\Explode\A-FP02a.dwg
C:\ADT_AIG\Chapter12\Explode\A-FP03a.dwg
C:\ADT_AIG\Chapter12\Explode\Elevation03.dwg
```

Figure 12–37 *One complete sequence*

11. Copy everything between –aecobjexplode and Open. Paste it after the next filename.

The screen should look like Figure 12–38.

```
AECExplode.SCR - Notepad
File  Edit  Format  View  Help
Filedia 0
Open
C:\ADT_AIG\Chapter12\Explode\A-FP00a.dwg
-aecobjexplode y c y y y y 31 y
saveas r14   y
Open
C:\ADT_AIG\Chapter12\Explode\A-FP01a.dwg
-aecobjexplode y c y y y y 31 y
saveas r14   y
Open
C:\ADT_AIG\Chapter12\Explode\A-FP02a.dwg
C:\ADT_AIG\Chapter12\Explode\A-FP03a.dwg
C:\ADT_AIG\Chapter12\Explode\Elevation03.dwg
```

Figure 12–38 *Copy and paste the lines to the next file*

12. Paste the contents of the clipboard after each of the remaining drawing file listings including the last one (see Figure 12–39).

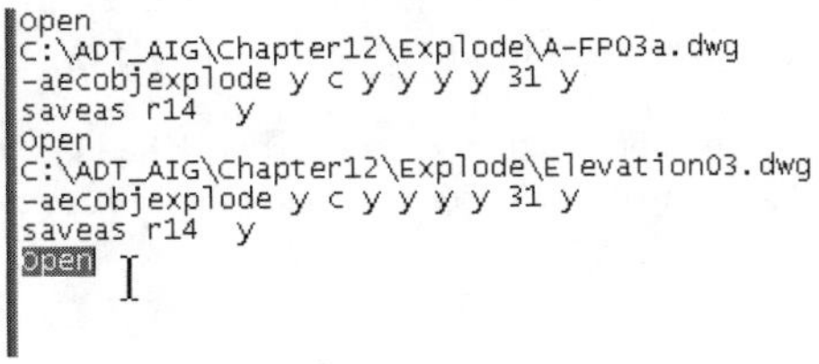

```
open
C:\ADT_AIG\Chapter12\Explode\A-FP03a.dwg
-aecobjexplode y c y y y y 31 y
saveas r14  y
open
C:\ADT_AIG\Chapter12\Explode\Elevation03.dwg
-aecobjexplode y c y y y y 31 y
saveas r14  y
open
```

Figure 12–39 *Finish the script code*

13. Delete the word "Open" at the very end.
14. Type **Filedia 1** (**Filedia** SPACE **1**) and then press ENTER.
15. Type **Closeall** and then press ENTER.
16. **Save** the script.

The final script appears as in Figure 12–40.

```
AECExplode.SCR - Notepad
File  Edit  Format  View  Help
Filedia 0
open
C:\ADT_AIG\Chapter12\Explode\A-FP00a.dwg
-aecobjexplode y c y y y y 31 y
saveas r14  y
open
C:\ADT_AIG\Chapter12\Explode\A-FP01a.dwg
-aecobjexplode y c y y y y 31 y
saveas r14  y
open
C:\ADT_AIG\Chapter12\Explode\A-FP02a.dwg
-aecobjexplode y c y y y y 31 y
saveas r14  y
open
C:\ADT_AIG\Chapter12\Explode\A-FP03a.dwg
-aecobjexplode y c y y y y 31 y
saveas r14  y
open
C:\ADT_AIG\Chapter12\Explode\Elevation03.dwg
-aecobjexplode y c y y y y 31 y
saveas r14  y
Filedia 1
Closeall
```

Figure 12–40 The complete text of the script

A complete version of the script is included with the sample files of this lesson. If the files you are using use XREFs, either bind them into the files first or adjust the script to deal with the XREFs as it runs. Repeat the dry run portion of the lesson on XREF files to see where to make the edits.

17. Save whatever files are on screen.
18. From the **Tools** menu, choose **Run Script**.
19. Locate the script file and open it. Sit back and watch it go!

If it appears to "hang" at the end, there might be an unsaved file open. Press ESC once, and the Save Drawing dialog box will come forward.

FINAL WORD

Script files offer a very powerful way of performing repetitive tasks automatically. When they work as expected, the time saving and productivity gained are amazing. When they don't, finding the error can be maddening. Stick with it. The payoff when they work is worth it!

Tip: Before using this command in production, copy your work files into a separate translation directory. This will ensure that your script file does not accidentally copy over the original work.

USING ETRANSMIT

eTransmit is a useful tool for gathering all of the file dependencies of the current file, packaging them together, and sending them by email, all in one routine. Use this routine to send documents to consultants through email.

Send Files with eTransmit

1. Open the *etransmit01.dwg*.
2. From the **File** menu, choose **eTransmit** (see Figure 12–41).

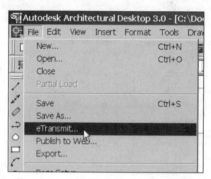

Figure 12–41 *Launch eTransmit*

3. In the Create Transmittal dialog box, click the General tab and type a message in the **Notes** field (see Figure 12–42).

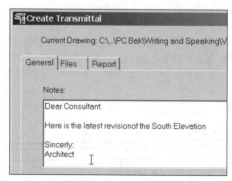

Figure 12–42 *Type a message to go with the transmittal*

4. From the **Type** list, choose the type of file you wish to create.
   - A Folder will copy all files to a new folder.
   - A Self-extracting executable will make a single "double-clickable" file containing all the files.
   - A Zip file will create a single file containing all of the files, but it requires your recipient to have un-zipping software such as WinZip to open.

5. Choose Self-extracting executable (*.exe) (see Figure 12–43).

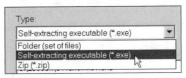

Figure 12–43 *Choose a self extracting Zip file*

 Note: Because of the proliferation of viruses, more and more companies are blocking self-executing file types (EXE files). If this is the case with your firm or your recipient's firm, use the ZIP format instead.

If you choose, you can lock the (*.exe) file with a password by clicking the **Password** button (see Figure 12–44).

Figure 12–44 *Add a password if you like*

6. Click the **Browse** button, locate the folder where you wish to save the (*.exe) file, type a name for the archive, and click the **Save** button (see Figure 12–45).

The name and path you choose will show up in the **Location** box of the Create Transmittal dialog box.

Figure 12–45 *The name you typed in the Location box*

If desired, put a check mark in the **Convert drawings to** check box and choose R14 format (see Figure 12–46). If you pick this option, bear in mind the issues raised in the previous two exercises and convert drawings accordingly.

 Note: Do not choose this option for this example.

Figure 12–46 *Choose this option only if you have properly exploded the file first*

7. Put a check mark in the **Send e-mail with transmittal** check box (see Figure 12–47).

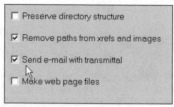

Figure 12–47 *Choose to send email with the transmittal*

8. Click the Files tab (see Figure 12–48).

Notice the list of files that will be included in this transmittal. You can remove files from the archive by deselecting them.

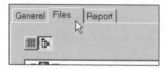

Figure 12–48 *Click on Files*

9. Deselect the *txt.shx* file. All AutoCAD users have this font, and it is unnecessary to send it (see Figure 12–49).

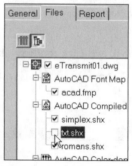

Figure 12–49 *Remove files that are not necessary to send*

If you wish, clicking the **Add File** button would allow you to add additional files to the archive (see Figure 12–50). Do not do this for this example.

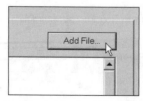

Figure 12–50 *Click the Add File button to optionally add additional files to the archive*

10. Click the Report tab.

Review the report. This will be included in a text file with the transmittal (see Figure 12–51).

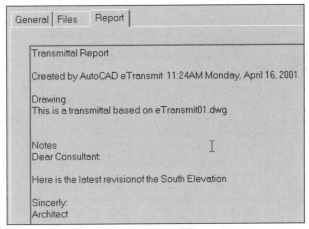

Figure 12–51 *A report will be included with the transmittal*

11. Click the **OK** button to launch your email program and send the transmittal (see Figure 12–52).

Figure 12–52 *Click OK to send it*

12. When your email program launches, type your own email address in the **To** field and change the subject if you wish (see Figure 12–53).

13. Click **Send**.

14. Launch your email program and check your mail. You should receive the message shortly.

15. To access the archive, download it from your email to a folder and double-click the (*.exe) file.

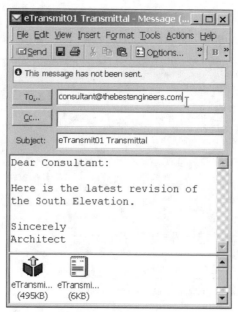

Figure 12–53 *Send the email to yourself to test it out*

 SHARING FILES WITH CONSULTANTS CHECKLIST

I. Object Enablers

1. When the consultant is using *Autodesk Architectural Desktop*

   a. **Architectural Desktop R3 or R3.3** – No translation necessary

   b. **Architectural Desktop R2i** – See steps for 2000i below

   c. **Architectural Desktop R2** – Not Compatible (Use Proxy graphics or AEC Object Explode.)

   d. **Architectural Desktop R1** – See steps for r14 below

2. When the consultant is using *Standard AutoCAD*

   a. **AutoCAD 2000i** – Object Enabler V3.02 Log on to http://pointa.autodesk.com/ Or Proxy Graphics or AEC Object Explode.

   b. **AutoCAD 2000** – Object Enabler V3.02 Log on to http://pointa.autodesk.com/ Or Proxy Graphics or AEC Object Explode.

   c. **AutoCAD R14** – Proxy Graphics or AEC Object Explode.

3. When the consultant is using a *non-AutoCAD CAD Package*

   a. **Non-AutoCAD CAD Software that reads DWG format** – AEC Object Explode, saveas R14 **DWG**.

   b. **Non-AutoCAD CAD Software that does not read DWG format** – AEC Object Explode, saveas R14 **DXF**.

II. Proxy Graphics

1. If consultants are using Object Enabler, or you plan to use AEC Object Explode, then turn Proxy Graphics off in all files and Templates.

   a. To do this, type PROXYGRAPHICS at the command line and set it to **0**.

III. AEC Object Explode

1. Run Explode AEC Objects from the Desktop Menu.
2. Choose the level to explode to.

   a. "Explode AEC Objects To Primitive Graphics" or

   b. "Explode To Anonymous Blocks"

3. Choose "Current View" (Never choose "All Views" as it will create several superimposed copies of the exploded file.)
4. Choose whether to maintain "Maintain Resolved Layer, Color, Linetype."
5. Choose to Erase AEC Objects
6. Shorten Symbol Names for R14 Compatibility if planning to save as R14.
7. Do choose to Display Status Messages
8. Remember to Save As to a new file! If you don't you will kill your original!
9. If processing several drawings, consider using a Script File.

IV. Using eTransmit

1. Run eTransmit
2. Type a message in the Notes field
3. Choose the type of file you wish to create

   a. Self-extracting executable (*.exe)

   b. Zip file (*.zip)

   c. Folder full of files

4. If desired, choose Convert drawings and choose R14 format.
5. If desired choose Send e-mail with transmittal checkbox.
6. Choose which files to include with the transmittal. Eliminate unnecessary Font files before sending.
7. Click Send

SUMMARY

The best collaboration scenario is when all parties use the same version of the software.

Several Object Enablers for generic AutoCAD are available free of charge on the Autodesk Point A Web site.

Proxy graphics can be used to allow consultants to view but not edit AEC objects.

Explode AEC Objects reduces drawings to editable primitive AutoCAD entities, which can be saved as R14 format.

Do not send files that have been processed with **Explode AEC Objects** if they need to be returned to you.

eTransmit offers a one-step process to gather and send by email all files required to correctly open a drawing.

SECTION VI

Conclusion

This section is contains the wrap-up for *Autodesk Architectural Desktop: Advanced Implementation Guide*. This section includes a single conclusion chapter and two appendices, which are both available on the Autodesk Architectural Desktop: Advanced Implementation Guide CD ROM.

Section VI is organized as follows:

Conclusion

Appendix A   Online Resources*

Appendix B   Tables*

* Available on the Autodesk Architectural Desktop: Advanced Implementation Guide CD ROM

WHO SHOULD READ THIS SECTION?

This section is intended for everyone.

CONCLUSION

PULLING IT ALL TOGETHER

We are now approaching the end of our journey through the process of ADT implementation. We have covered many procedures and techniques throughout the course of this manual. In conclusion, we will summarize the high points and give you an overall summary and wrap-up of the entire process.

CREATE A BIG PLAN, TAKE "BABY STEPS" TO ACHIEVE IT

In the Introduction to this manual, we began our journey with a brief listing of many of the "high level" benefits of a comprehensive Building Model strategy; among them were better communication, understanding and coordination of the design and construction processes. We followed this with a short hands-on tour of the specific results that these could bring to the ADT process today. In the remaining chapters, we continued to progressively drill down into the specifics. Here at the end of this book, you now have the knowledge required to duplicate the results shown throughout and take them much further still! It is hoped that the completion of this text marks the beginning your firm's realization of a more efficient and productive Architectural Desktop implementation, with an eye toward the future promise of Model-based design.

To build an implementation plan for any process, it is important to build a road map detailing what you hope to accomplish. With it, you can stay focused on the tasks required to succeed and chart your progress along the way. There is no single correct way to implement ADT (or any other process) in an organization. However, as we have seen throughout this book, there are many common threads, that when woven together, will yield a coherent, finely worked tapestry.

The scope of your particular implementation strategy in the long term will involve a variety of factors. Your firm's size, composition, culture and body of work will all be important considerations. Many of these have been discussed in detail in Chapter 2. In the short term, the firm must remain profitable during the transition. Upon consideration of these factors, it is very common and most often advisable to develop a "phased approach" to ADT implementation.

Regardless of your ultimate goals regarding 3D Model-based design, many firms find initial success (and upper management buy-in) from implementing the tools most likely to yield the highest dividend in 2D production. For this reason, most firms will begin with Walls, Doors, Windows and little else. Others may see benefits to *today's* processes in immediately adopting 3D Building Model design strategies. Regardless of your specific approach, keep the following points in mind:

- Have a "big" goal in mind and many smaller goals designed to help you achieve the big one.

- Start with strategies that are likely to be received in the most positive light by your colleagues. (Walls and Doors provide immediate productivity gains with a relatively small learning curve.)
- Try to avoid rolling out new tools until they are ready. (If a tool requires customization to behave the way your team needs or expects it to, rolling it out sooner will only sour their opinion of it.)

All of these points can be summed thus:

GUIDING PRINCIPLE: Carefully plan your implementation and then patiently execute each phase.

Much of what we have discussed can be distilled into the following points:

PLANNING

Planning occurs at a few levels when ADT is deployed. We must decide which projects will use ADT, which members of our staff will comprise those project teams, and which items and specific features of the software we plan to implement and when. For these reasons, we began our discussions with a high-level look at the potential of the Building Model process, installation and configuration of the software, Project and Building Model Structure file structure, ADT object composition and CAD Standards Assessment. Each of these topics has something to offer us as we plan our implementation strategy. Remember, the planning phase is about developing the "big" picture and detailing the small steps needed to achieve it.

TEMPLATES AND LIBRARIES

The configuration of layers and the Display System will have impact not just on the production phase of the project but also throughout the life the Building Model. These are the tools that we use to organize and filter the data within our files. Careful consideration of all of the possibilities that this implies will help us to build robust template and library files. The meticulous creation of these resources will prove one of the most valuable activities in the entire implementation process. The value of a well-conceived template cannot be over stated.

Let's also remember the importance of Property Sets and Object Styles to our Building Model design and production procedures. Develop well-conceived strategies for these items prior to embarking on the development of an extensive Content Library of standard resources. So much of the power and ability of ADT lies in these resources.

INTEROPERABILITY

No implementation plan can be considered complete without consideration of the way that data flows into or out of the process. Whether we consider output to paper or sharing of digital files, interoperability among the various parties that service a project is a key concern. In an ideal scenario, all players would be using the same or "seamlessly" compatible solution to generate and read data. In reality, this is rarely the case. Pre-planning and coordination are often the solution to this dilemma.

PUBLISHING YOUR STANDARDS

Your work is not complete once the templates and library files have been posted to the server. To be certain that the new procedures succeed, you must document them. CAD Manuals do not have to be several hundred pages nor is it necessary to list every single Layer and Display Control setting used. To be useful as an ongoing resource to you and the other members of your firm, a CAD Manual ought to contain clear and concise summaries of the most critical bits of information. Diagrams, bold type, short summaries and concise tables will be much better received than pages of prose or endless lists. Remember, the goal is to assist people in using the procedures and templates you have set up, not to clutter their desk with another heavy binder.

Finally, no procedure is ever completely "finished," particularly in a field like building design, where solutions are rarely repeated without "tweaking." Just make sure that you spend your efforts tweaking things that will have benefit in the long haul. No one ever walks into a building and exclaims, "Wow, what a great set of CAD Standards!"

TEMPLATE CHECKLIST

I Plot Style Choice

1. Choose the appropriate Plot Style type.
2. Choose your standard lineweights.
3. Decide how to assign lineweights (ByLayer recommended).

II Format Items

1. Add or import all Text Styles.
2. Add or import all Dimension Styles.
3. Set DIMASSOC to **2**.
4. Import common Linetypes

III Layouts and Named Page Setups

1. Set up "Working" layouts.
2. Set up Plotting layouts.
3. Build sheet size "Live Areas."
4. Build common Page Setups.

IV Layers

1. hard-coded layers that do not use Layer Keys.

V Display Control

1. Set all System Default Display Control settings to all Display Reps and objects.
2. Reserve primary colors for "Alert" conditions.
3. Build or import all Configurations and Sets.

VI Drawing Setup

1. Set units to **Inches [Millimeters]**.
2. Precision to **1/32 [0.0000]**.
3. Set preferred choices on the remainder of the Units tab.
4. Choose **Save as Default**.

VII Options

1. Set your AEC DwgDefaults.
2. Set your AEC Stair Defaults.
3. Set your AEC Dimension Defaults.

VIII Misc

1. Create a "Standard" Schedule Table Style containing typical Text and Formatting settings.
2. Create or import most common office standard AEC Object Styles. (Include a 2D Section/Elevation Style with built-in Design Rules.)
3. Set VISRETAIN and XEDIT to preferred values (both ON is recommended).
4. Purge AEC Layer Styles (AECPURGELAYERSTYLES) before saving.

SUMMARY

The following summarizes many of the recommendations made throughout this book:

Perform a Clean Install of Architectural Desktop and build all user machines the same way.

Place all common resources on the network accessible to all users.

Use the Lineweight property, assigned ByLayer and within Display Control to establish plotted lineweights.

There are 255 colors, use them!

Try to avoid using the same color for more than one purpose.

Use Named Plot Styles and limit their influence to ink color, line end and line join style.

If using Color-dependent Plot Styles, assign your collection of lineweights across the entire spectrum of colors. (In other words, use all the colors, regardless of the number of lineweights.)

Use the primary colors (1 through 6) and the "Normal" Plot Style for "non-plotting," "alert" and "diagnostic" layers and display components.

Set up Layer Key Styles with pre-assigned Lineweight and Plot Style assignments. Each Layer key should use a unique color regardless of Plot Style type.

Add as few template and "hard-coded" Content layers to your templates as possible.

Simplify the Display System by eliminating "excess" custom Display Reps, Sets and Configurations.

Consider using Scale-dependent Configurations like the ones in the D A CH template.

Use Display Control to adjust lineweights for Scale-dependent displays.

In the Plan Display Rep, assign layers to sub-components and make their colors ByLayer.

In the Model Display Rep, use explicit color mapping for sub-components (by color, not ByLayer).

Consolidate the needs of VIZ and 2D Section/Elevation objects in the Model Display Rep and assign a unique color to every sub-component.

Use Style-based Property Sets wherever possible.

Limit each Schedule Table Style to the use of one Style-based and one Object-based Property Set.

Add Style-based Property Sets to all appropriate Content items in the library.

Use Scale-dependent tags for appropriate symbols such as Room Tags.

Try to store resource items like Text, Dimension Styles, and other "dependent" items in as few default locations as possible.

Analyze the contents of the Content library carefully and add, edit and purge as required.

Develop a procedure for ongoing update and maintenance of the Content Library.

Don't forget to include any Property Sets and Custom Display Reps required by custom AEC Content.

Keep styles as simple as possible and as complex as necessary.

Use simple "single-line" Wall Styles wherever possible. (Multiple Walls of the same component composition with different names can be used for type and tagging purposes as required.)

Name styles descriptively in the library and in projects with the appropriate project designation, such as Type "A," Type "B1," etc.

Reference Style-based Property Sets for "Types" within Tags to make tagging easier (for instance Wall Type, Door Type, etc.).

Build a Content Library housing Area and Area Group object styles and resources for Area calculations and takeoffs.

Perform Area object calculations in a separate set of drawing files through XREFs.

Use Schedule Tables exported to Excel for reporting rather than Area Evaluations.

Assign every sub-component of every object in the Model Display Rep a unique color to facilitate Design Rules for Sections and material mapping in VIZ.

Build Styles that reference these colors in a set of standard Design Rules.

Build a set of Standard VIZ Material Libraries referencing these colors.

Coordinate with consultants on the platform/format issues early in the project to develop a procedure for successful document sharing.

Document all procedures and resources in a concisely written, clearly illustrated CAD Procedures Manual. (Keep it under 100 pages.)

INDEX

Note: Entries in SMALL CAPS indicate keyed-in commands; page numbers in **bold** indicate material in Tables; page numbers with "CD" indicate material from Chapter 13 found on the CD-ROM.

LICENSE AGREEMENT FOR AUTODESK PRESS
A Thomson Learning Company

Educational Software/Data

You the customer, and Autodesk Press incur certain benefits, rights, and obligations to each other when you open this package and use the software/data it contains. BE SURE YOU READ THE LICENSE AGREE-MENT CAREFULLY, SINCE BY USING THE SOFTWARE/DATA YOU INDICATE YOU HAVE READ, UNDERSTOOD, AND ACCEPTED THE TERMS OF THIS AGREEMENT.

Your rights:

1. You enjoy a non-exclusive license to use the enclosed software/data on a single microcomputer that is not part of a network or multi-machine system in consideration for payment of the required license fee, (which may be included in the purchase price of an accompanying print component), or receipt of this software/data, and your acceptance of the terms and conditions of this agreement.

2. You own the media on which the software/data is recorded, but you acknowledge that you do not own the software/data recorded on them. You also acknowledge that the software/data is furnished "as is," and contains copyrighted and/or proprietary and confidential information of Autodesk Press or its licensors.

3. If you do not accept the terms of this license agreement you may return the media within 30 days. However, you may not use the software during this period.

There are limitations on your rights:

1. You may not copy or print the software/data for any reason whatsoever, except to install it on a hard drive on a single microcomputer and to make one archival copy, unless copying or printing is expressly permitted in writing or statements recorded on the diskette(s).

2. You may not revise, translate, convert, disassemble or otherwise reverse engineer the software/data except that you may add to or rearrange any data recorded on the media as part of the normal use of the software/data.

3. You may not sell, license, lease, rent, loan, or otherwise distribute or network the software/data except that you may give the software/data to a student or and instructor for use at school or, temporarily at home.

Should you fail to abide by the Copyright Law of the United States as it applies to this software/data your license to use it will become invalid. You agree to erase or otherwise destroy the software/data immediately after receiving note of Autodesk Press' termination of this agreement for violation of its provisions.

Autodesk Press gives you a LIMITED WARRANTY covering the enclosed software/data. The LIMITED WARRANTY can be found in this product and/or the instructor's manual that accompanies it.

This license is the entire agreement between you and Autodesk Press interpreted and enforced under New York law.

Limited Warranty

Autodesk Press warrants to the original licensee/ purchaser of this copy of microcomputer software/ data and the media on which it is recorded that the media will be free from defects in material and workmanship for ninety (90) days from the date of original purchase. All implied warranties are limited in duration to this ninety (90) day period. THEREAFTER, ANY IMPLIED WARRANTIES, INCLUDING IMPLIED WARRANTIES OF MERCHANTABILITY AND FITNESS FOR A PARTICULAR PURPOSE ARE EXCLUDED. THIS WARRANTY IS IN LIEU OF ALL OTHER WARRANTIES, WHETHER ORAL OR WRITTEN, EXPRESSED OR IMPLIED.

If you believe the media is defective, please return it during the ninety day period to the address shown below. A defective diskette will be replaced without charge provided that it has not been subjected to misuse or damage.

This warranty does not extend to the software or information recorded on the media. The software and information are provided "AS IS." Any statements made about the utility of the software or information are not to be considered as express or implied warranties. Delmar will not be liable for incidental or consequential damages of any kind incurred by you, the consumer, or any other user.

Some states do not allow the exclusion or limitation of incidental or consequential damages, or limitations on the duration of implied warranties, so the above limitation or exclusion may not apply to you. This warranty gives you specific legal rights, and you may also have other rights which vary from state to state. Address all correspondence to:

AutodeskPress
3 Columbia Circle
P. O. Box 15015
Albany, NY 12212-5015